Lecture Notes in Computer Science $\quad$ 11555

Commenced Publication in 1973
Founding and Former Series Editors:
Gerhard Goos, Juris Hartmanis, and Jan van Leeuwen

Editorial Board Members

More information about this series at http://www.springer.com/series/7407

Huchuan Lu · Huajin Tang ·
Zhanshan Wang (Eds.)

Advances in Neural Networks – ISNN 2019

16th International Symposium on Neural Networks, ISNN 2019
Moscow, Russia, July 10–12, 2019
Proceedings, Part II

 Springer

Editors
Huchuan Lu
Dalian University of Technology
Dalian, China

Huajin Tang
Sichuan University
Chengdu, China

Zhanshan Wang
Northeastern University
Shenyang, China

ISSN 0302-9743   ISSN 1611-3349 (electronic)
Lecture Notes in Computer Science
ISBN 978-3-030-22807-1   ISBN 978-3-030-22808-8 (eBook)
https://doi.org/10.1007/978-3-030-22808-8

LNCS Sublibrary: SL1 – Theoretical Computer Science and General Issues

This Springer imprint is published by the registered company Springer Nature Switzerland AG
The registered company address is: Gewerbestrasse 11, 6330 Cham, Switzerland

Preface

This volume contains the papers presented at ISNN 2019: the 16th International Symposium on Neural Networks held during July 10–12, 2019, in Moscow. Located by the Moskva River, Moscow is the capital and largest city in Russia with a population of over 13 million. Thanks to the success of the previous events, ISNN has become a well-established series of popular and high-quality conferences on the theory and methodology of neural networks and their applications. ISNN 2019 aimed to provide a high-level international forum for scientists, engineers, and educators to present the state of the art of neural network research and applications in related fields. The symposium also featured plenary speeches given by world renowned scholars, regular sessions with a broad coverage, and special sessions focusing on popular topics.

This year, the symposium received more submissions than previous years. Each submission was reviewed by at least two, and on average, 4.5 Program Committee members. After the rigorous peer reviews, the committee decided to accept 111 papers for publication in the *Lecture Notes in Computer Science* (LNCS) proceedings. These papers cover many topics of neural network-related research including learning system, graph model, adversarial learning, time series analysis, dynamic prediction, uncertain estimation, model optimization, clustering, game theory, stability analysis, control method, industrial application, image recognition, scene understanding, biomedical engineering, hardware. In addition to the contributed papers, the ISNN 2019 technical program included three keynotes and plenary speeches by renowned scholars: Prof. Andrzej Cichocki (IEEE Fellow, Skolkovo Institute of Science and Technology, Moscow, Russia), Prof. Yaochu Jin (IEEE Fellow, University of Surrey, Guildford, UK), and Prof. Nikhil R. Pal (IEEE Fellow, Indian Statistical Institute, Calcutta, India).

Many organizations and volunteers made great contributions toward the success of this symposium. We would like to express our sincere gratitude to Skolkovo Institute of Science and Technology and City University of Hong Kong for their sponsorship, the International Neural Network Society, Asian Pacific Neural Network Society, Polish Neural Network Society, and Russian Neural Network Society for their technical co-sponsorship. We would also like to sincerely thank all the committee members for their great efforts in organizing the symposium. Special thanks to the Program Committee members and reviewers whose insightful reviews and timely feedback ensured the high quality of the accepted papers and the smooth flow of the symposium. We would also like to thank Springer for their cooperation in publishing the proceedings in the prestigious LNCS series. Finally, we would like to thank all the speakers, authors, and participants for their support.

June 2019

Huchuan Lu
Huajin Tang
Zhanshan Wang

Organization

Program Committee

Ping An	Shanghai University, China
Sabri Arik	Istanbul University, Turkey
Xiaochun Cao	Institute of Information Engineering, Chinese Academy of Sciences, China
Wenliang Chen	Soochow University, China
Long Cheng	Institute of Automation, Chinese Academy of Sciences, China
Zheru Chi	The Hong Kong Polytechnic University, China
Yang Cong	Chinese Academy of Science, China
Jose Alfredo Ferreira Costa	Federal University of Rio Grande do Norte, Brazil
Ruxandra Liana Costea	Polytechnic University of Bucharest, Romania
Chaoran Cui	Shandong University of Finance and Economics, China
Cheng Deng	Xidian University, China
Lei Deng	University of California, Santa Barbara, USA
Jing Dong	National Laboratory of Pattern Recognition, China
Minghui Dong	Institute for Infocomm Research, Singapore
Jianchao Fan	National Marine Environmental Monitoring Center, China
Xin Fan	Dalian University of Technology, China
Liang Feng	Chongqing University, China
Wai-Keung Fung	Robert Gordon University, UK
Shaobing Gao	Sichuan University, China
Shenghua Gao	ShanghaiTech University, China
Shiming Ge	Chinese Academy of Sciences, China
Tianyu Geng	Sichuan University, China
Xin Geng	Southeast University, China
Xiaofeng Gong	Dalian University of Technology, China
Jie Gui	Chinese Academy of Sciences, China
Chengan Guo	Dalian University of Technology, China
Ping Guo	Beijing Normal University, China
Zhenhua Guo	Tsinghua University, China
Zhenyuan Guo	Hunan University, China
Zhishan Guo	University of Central Florida, USA
Huiguang He	Institute of Automation, Chinese Academy of Sciences, China
Ran He	National Laboratory of Pattern Recognition, China
Jin Hu	Chongqing Jiaotong University, China
Jinglu Hu	Waseda University, Japan

Xiaolin Hu	Tsinghua University, China
Bonan Huang	Northeastern University, China
Wei Jia	Chinese Academy of Sciences, China
Muwei Jian	Ocean University of China, China
Danchi Jiang	University of Tasmania, Australia
Min Jiang	Xiamen University, China
Shunshoku Kanae	Junshin Gakuen University, Japan
Sungshin Kim	Pusan National University, South Korea
Wanzeng Kong	Hangzhou Dianzi University, China
Chiman Kwan	Signal Processing, Inc., USA
Man Lan	East China Normal University, China
Xinyi Le	Shanghai Jiao Tong University, China
Chuandong Li	Southwest University, China
Guoqi Li	Tsinghua University, China
Michael Li	Central Queensland University, Australia
Tianrui Li	Southwest Jiaotong University, China
Qiuhua Lin	Dalian University of Technology, China
Xiangbo Lin	Dalian University of Technology, China
Jian Liu	University of Leicester, UK
Ju Liu	Shandong University, China
Lei Liu	Northeastern University, China
Qingshan Liu	Southeast University, China
Zhenwei Liu	Northeastern University, China
Huchuan Lu	Dalian University of Technology, China
Wenlian Lu	Fudan University, China
Yanhong Luo	Northeastern University, China
Garrick Orchard	Intel, USA
Yunpeng Pan	Georgia Institute of Technology, USA
Yaxin Peng	Shanghai University, China
Zhouhua Peng	Dalian Maritime University, China
Yu Qi	Zhejiang University, China
Sitian Qin	Harbin Institute of Technology at Weihai, China
Qinru Qiu	Syracuse University, USA
Hong Qu	University of Electronic Science and Technology of China, China
Pengju Ren	Xi'an Jiaotong University, China
Sergey Sukhov	Russian Academy of Sciences, Russia
Norikazu Takahashi	Okayama University, Japan
Huajin Tang	Sichuan University, China
Qing Tao	Institute of Automation, Chinese Academy of Sciences, China
Feng Wan	University of Macau, SAR China
Jun Wang	City University of Hong Kong, SAR China
Lidan Wang	Southwest University, China
Zhanshan Wang	Northeastern University, China

Qinglai Wei	Institute of Automation, Chinese Academy of Sciences, China
Huaqiang Wu	Tsinghua University, China
Yong Xia	Northwestern Polytechnical University, China
Xiurui Xie	Institute for Infocomm Research, Singapore
Shaofu Yang	Southeast University, China
Mao Ye	University of Electronic Science and Technology of China, China
Qiang Yu	Tianjin University, China
Shan Yu	Institute of Automation, Chinese Academy of Sciences, China
Wen Yu	Cinvestav, Mexico
Zhaofei Yu	Peking University, China
Zhi-Hui Zhan	South China University of Technology, China
Jie Zhang	Newcastle University, UK
Malu Zhang	National University of Singapore, Singapore
Nian Zhang	University of the District of Columbia, USA
Tielin Zhang	Institute of Automation, Chinese Academy of Sciences, China
Xiaoting Zhang	Boston University, USA
Xuetao Zhang	Xi'an Jiaotong University, China
Yongqing Zhang	Chengdu University of Information Technology, China
Zhaoxiang Zhang	Institute of Automation, Chinese Academy of Sciences, China
Wenda Zhao	Dalian University of Technology, China
Chengqing Zong	Institute of Automation, Chinese Academy of Sciences, China

Additional Reviewers

Bai, Ao	Dang, Suogui	Huang, Haoyu
Bao, Gang	Deng, Xiaodan	Huang, Jiangshuai
Bhuiyan, Ashik Ahmed	Dong, Jiahua	Huang, Li
Bian, Jiang	Dong, Junfei	Huang, Yaoting
Cai, Qing	Fan, Huijie	Huang, Zitian
Che, Hangjun	Fan, Xiaofei	Huimin, Xu
Chen, Boyu	Fang, Xiaomeng	Jia, Haozhe
Chen, Haoran	Gu, Peng	Jiang, Peilin
Chen, Junya	Gu, Pengjie	Jiang, Xinyu
Chen, Xiaofeng	Guo, Zhenyuan	Ju, Xiping
Chen, Yi	Han, Kezhen	Kong, Wanzeng
Chen, Yuanyuan	Harsha Vardhan Reddy	Li, Cong
Chen, Zhaodong	Velmula, Harsha	Li, Dengju
Cui, Hengfei	Hu, Jin	Li, Jiaxin
Cui, Lili	Hu, Zhanhao	Li, Lixiang

Li, Lukai
Li, Zhe
Liang, Hongjing
Liang, Ling
Liang, Yudong
Lin, Jilan
Liu, Faqiang
Liu, Jiayuan
Liu, Kang
Liu, Lining
Liu, Liu
Liu, Lu
Liu, Mingkai
Liu, Sichao
Liu, Yingying
Liu, Yongsheng
Lu, Bao-Liang
Lu, Junwei
Luo, Xiaoling
Luo, Xuefang
Ma, Xin
Ma, Yuhao
Mao, Xin
Mikheev, Aleksandr
Mohammed, Mahmoud
Niu, Dong
Niu, Weina
Pan, Zihan
Pang, Dong
Peng, Jianxin
Qu, Zheng
Ren, He
Ren, Yanhao
Rong, Nannan

Shan, Qihe
Shi, Zhan
Song, Shiming
Sun, Pengfei
Sun, Xiaoxuan
Tan, Guoqiang
Tang, Chufeng
Tang, Tang
Tang, Tianqi
Tian, Yufeng
Tian, Zhiqiang
Vaidhun, Sudharsan
Wan, Fukang
Wang, Bangyan
Wang, Dingheng
Wang, Hongfeng
Wang, Jiru
Wang, Jue
Wang, Lin
Wang, Songsheng
Wang, Xiaoping
Wang, Yaoyuan
Wang, Yixuan
Wang, Yuan
Wang, Yulong
Wang, Zengyun
Wei, Wang
Wen, Shiping
Wen, Xianglan
Wen, Yinlei
Wu, Jian
Wu, Jibin
Wu, Yanming
Wu, Yicheng

Wu, Yining
Wu, Yujie
Xie, Xinfeng
Xie, Xiurui
Xie, Yutong
Xing, Yanli
Xu, Qi
Xu, Yang
Xujun, Yang
Yan, Mingyu
Yang, Yukuan
Yang, Zeheng
Yang, Zhaoxu
Yang, Zheyu
Yao, Xianshuang
Yao, Yanli
Yu, Xue
Yu, Yikuan
Yuan, Mengwen
Zeng, Zhigang
Zhan, Qiugang
Zhang, Haodong
Zhang, Houzhan
Zhang, Jianpeng
Zhang, Malu
Zhang, Shizhou
Zhang, Xin
Zhang, Yun
Zhao, Bo
Zheng, Hao
Zheng, Wei-Long
Zhou, Bo
Zhou, Yue
Zhu, Mingchao

Contents – Part II

Signal Processing, Industrial Application, and Data Generation

Image Recognition, Scene Understanding, and Video Analysis

Bio-signal, Biomedical Engineering, and Hardware

Contents – Part I

Time Series Analysis, Dynamic Prediction, and Uncertain Estimation

Model Optimization, Bayesian Learning, and Clustering

Game Theory, Stability Analysis, and Control Method

Decentralized Robust Optimal Control for Modular Robot Manipulators Based on Zero-Sum Game with ADP

Bo Dong[1,2(✉)], Tianjiao An[1], Fan Zhou[1], Shenquan Wang[1], Yulian Jiang[1], Keping Liu[1], Fu Liu[3], Huiqiu Lu[3], and Yuanchun Li[1]

[1] Department of Control Science and Engineering,
Changchun University of Technology, Changchun, China
dongbo@ccut.edu.cn
[2] State Key Laboratory of Management and Control for Complex Systems,
Institute of Automation, Chinese Academy of Sciences, Beijing, China
[3] Department of Control Science and Engineering,
Jilin University, Changchun, China

Abstract. In this paper, a decentralized robust zero-sum optimal control method is proposed for modular robot manipulators (MRMs) based on the adaptive dynamic programming (ADP) approach. The dynamic model of MRMs is formulated via joint torque feedback (JTF) technique that is deployed for each joint module, in which the local dynamic information is used to design the model compensation controller. An uncertainty decomposition-based robust control is developed to compensate the model uncertainties, and then the robust optimal control problem of MRMs with uncertain environments can be transformed into a two-player zero-sum optimal control one. According to the ADP algorithm, the Hamilton-Jacobi-Isaacs (HJI) equation is solved by constructing action-critic neural networks (NNs) and then the approximate optimal control policy derivation is possible. Experiments are conducted to verify the effectiveness of the proposed method.

Keywords: Modular robot manipulators ·
Adaptive dynamic programming · Optimal control · Zero-sum game

1 Introduction

Modular robot manipulators have attracted wildly attentions in robotics community since they are possessed of better structural adaptability and flexibility than conventional robot manipulators. An MRM is composed by standardized robotic modules, which consist of actuators, speed reducers, sensors and communication units. These modules can be assembled to desired configurations via standard mechanical interfaces according to the requirements of various tasks. Benefitting from the advantages above, MRMs are often utilized in dangerous and complex environments. Correspondingly, appropriate control systems are

© Springer Nature Switzerland AG 2019
H. Lu et al. (Eds.): ISNN 2019, LNCS 11555, pp. 3–14, 2019.
https://doi.org/10.1007/978-3-030-22808-8_1

required to guarantee the robustness and precision of MRMs with uncertain environments.

In order to seek out appropriate control for MRMs, joint torque feedback techniques have attached attentions of both robotic researchers and industrial manufacturers. JTF technique can not only eliminate the necessity of contact model in motion control of robot-environment interactive tasks, it also facilitates in suppressing the effect of uncertain contact force created between the end-effector and payloads [1–3]. However, the existed JTF-based motion/force control methods have not considered the issue of optimal implementation between the error performance of joint motion and the output torque of actuator. Indeed, an ideal controllers for MRMs should be possessed of the properties that ensure the robustness of the robotic systems, and simultaneously take into account the optimal realization of the composite of error performance and output power.

Adaptive dynamic programming methodology is considered as one of the key directions to address the optimal control problems in complex systems. Some investigations reported the latest research progress of ADP-based optimal control methods for robot manipulators systems [4–6]. These strategies both follow the premise that the dynamic models of the robotic systems are completely unknown, thus the application of these methods are limited to address the optimal control problems of specific classes of robotic systems without implementing dynamic compensation. Dong et al. address the optimal control problems of MRMs by combining the model-based compensation control with ADP-based learning control [7], and their researches are further expanded to deal with the optimal tracking control issues of MRMs with uncertain environments [8]. However, the existing methods consider the disturbance torques, which are introduced by environmental contacts, as a class of dynamic uncertainties, and ignoring the intractability of decomposing the effects of model uncertainties and environmental contact disturbance targetedly. Indeed, there is less discussion on ADP-based robust optimal control for robot manipulators with both dynamic decomposition and optimal compensation, especially, for MRMs with uncertain environmental contact.

In this paper, a novel decentralized robust zero-sum neuro-optimal control method is presented for MRMs with uncertain environments. First, the dynamic model of the MRM systems is formulated via JTF technique, and a model-based compensation controller is designed by effectively utilizing the local dynamic information of each joint module. Second, a decentralized robust controller is proposed to deal with the compensation control issue of model uncertainties. Then, the robust optimal control problem of uncertain environment-contacted MRMs is transformed into a two-player zero-sum optimal control one. The ADP algorithm is employed to solve the HJI equation, in which the cost function, optimal control policy and worst disturbance can be approximated by constructing one critic NN and two action NNs, and then the decentralized robust zero-sum neuro-optimal control is developed. Finally, experiments are performed to verify the effectiveness of the proposed method.

2   Dynamic Model Formulation

We consider the MRMs comprise n joint modules installed in series, and each integrated with a DC motor, a speed reducer, two encoders and a joint torque sensor. By referencing the modelling methods of JTF-based robot manipulators with multi-degree of freedom (DOF) [9,10], we formulate the dynamic model of the MRM system as a synthesis of interconnected robotic joint subsystems, in which the ith joint subsystem dynamic model is represented as:

$$I_{mi}\gamma_i \ddot{q}_i + f_i(q_i, \dot{q}_i) + z_i(q, \dot{q}, \ddot{q}) + \frac{\tau_{si}}{\gamma_i} + d_i(q_i) = \tau_i, \qquad (1)$$

where the subscript "i" indicates the ith subsystem; I_{mi} is the moment of inertia of actuator rotor about the rotation axis; γ_i denotes the gear ratio; $q_i, \dot{q}_i$ and $\ddot{q}_i$ are the joint angular position, velocity and acceleration respectively; $f_i(q_i, \dot{q}_i)$ represents the joint lumped frictional torque; $z_i(q_i, \dot{q}_i, \ddot{q}_i)$ denotes the interconnected dynamic coupling (IDC) among the joint subsystems; τ_{si} represents the coupling torque at the torque sensor location; $d_i(q_i)$ is the disturbance torque and τ_i indicates the control torque.

(1) Joint friction

The friction term can be defined as a function with respect to joint angular position and velocity, which is given as:

$$f_i(q_i, \dot{q}_i) \approx \hat{f}_{bi}\dot{q}_i + \left(\hat{f}_{si}e^{(-\hat{f}_{\tau i}\dot{q}_i^2)} + \hat{f}_{ci}\right)\mathrm{sgn}(\dot{q}_i) + f_{pi}(q_i, \dot{q}_i) + Y(\dot{q}_i)\tilde{F}_i, \qquad (2)$$

where $\tilde{F}_i = \left[f_{bi} - \hat{f}_{bi} \ \ f_{ci} - \hat{f}_{ci} \ \ f_{si} - \hat{f}_{si} \ \ f_{\tau i} - \hat{f}_{\tau i}\right]^T$ indicates the parametric uncertainty of the friction; f_{bi}, f_{si}, $f_{\tau i}$ and f_{ci} represent the ideal parameters of the friction model; $f_{pi}(q_i, \dot{q}_i)$ denotes the position dependency of friction and the other friction modeling errors; $\mathrm{sgn}(\dot{q})$ is a classical sign function; $\hat{f}_{bi}$, $\hat{f}_{si}$, $\hat{f}_{\tau i}$ and $\hat{f}_{ci}$ are estimated parameters of the friction model and $Y(\dot{q}_i)$ is defined as:

$$Y(\dot{q}_i) = \left[\dot{q}_i, \ \mathrm{sgn}(\dot{q}_i), \ e^{(-\hat{f}_{\tau i}\dot{q}_i^2)}\mathrm{sgn}(\dot{q}_i), \ -\hat{f}_{si}\dot{q}_i^2 e^{(-\hat{f}_{\tau i}\dot{q}_i^2)}\mathrm{sgn}(\dot{q}_i)\right]. \qquad (3)$$

(2) Interconnected dynamic coupling

The IDC term $z_i(q, \dot{q}, \ddot{q})$, which is considered a complex nonlinear function, can be defined with respect to the positions, velocities and accelerations of the lower joint modules (from joint 1 to $i - 1$), that is given as:

$$z_i(q, \dot{q}, \ddot{q}) = I_{mi}\sum_{j=1}^{i-1} D_j^i \ddot{q}_j + I_{mi}\sum_{j=2}^{i-1}\sum_{k=1}^{j-1}\Theta_{kj}^i \dot{q}_k \dot{q}_j = U_{zi} + V_{zi},$$

$$U_{zi} = \sum_{j=1}^{i-1}\left[I_{mi}\hat{D}_j^i \ I_{mi}\right]\left[\ddot{q}_j \ \ \tilde{D}_j^i \ddot{q}_j\right]^T, V_{zi} = \sum_{j=2}^{i-1}\sum_{k=1}^{j-1}\left[I_{mi}\hat{\Theta}_{kj}^i \ I_{mi}\right]\left[\dot{q}_k \dot{q}_j \ \ \tilde{\Theta}_{kj}^i \dot{q}_k \dot{q}_j\right]^T,$$

$$\qquad (4)$$

where D_j^i and Θ_{kj}^i are given as $D_j^i = z_{mi}^T z_{lj}$ and $\Theta_{kj}^i = z_{mi}^T(z_{lk} \times z_{lj})$ respectively, z_{mi}, z_{lj} and z_{lk} are the unity vectors along the axis of rotation of the

ith robot, jth joint and kth joint respectively. Moreover, we have the relations $\hat{D}_j^i = D_j^i - \tilde{D}_j^i$ and $\hat{\Theta}_{kj}^i = \Theta_{kj}^i - \tilde{\Theta}_{kj}^i$, where $\tilde{D}_j^i$ and $\tilde{\Theta}_{kj}^i$ are the alignment errors.
(3) Joint torque and disturbance torque

The coupled joint torques τ_{si}, which are measured by using the embedded torque sensors, can be represented as the following form:

$$\tau_{si} = \tau_{sri} + \tau_{sci}, \tag{5}$$

where τ_{sri} indicates the joint torque that is obtained in free space and τ_{sci} denotes the torque produced by environmental contact in constrained space. Besides, the disturbance torque term $d_i(q_i)$ can be defined as:

$$d_i(q_i) = d_{is}(q_i) + d_{ic}(q_i), \tag{6}$$

where $d_{is}(q_i)$ represents the disturbance of torque sensor and $d_{ic}(q_i)$ indicates unmeasured environmental contact torque in joint space.

Assumption 1. The uncertainty term $\tilde{F}_i$ is bounded by $\left|\tilde{F}_i\right| \leq \rho_{Fin}$, the friction modeling error is bounded by $\left|Y(\dot{q}_i)\tilde{F}_i\right| \leq Y(\dot{q}_i)\rho_{Fin}$ and $f_{pi}(q_i, \dot{q}_i)$ is bounded by $|f_{pi}(q_i, \dot{q}_i)| \leq \rho_{fpi}$.

Assumption 2. The terms U_{zi} and V_{zi} are bounded by $|U_{zi}| \leq \rho_{Ui}, |V_{zi}| \leq \rho_{Vi}$.

Assumption 3. The disturbance $d_{is}(q_i)$ is bounded by $|d_{is}(q_i)| \leq \rho_{dsi}$ and the unmeasured environmental contact torque $d_{ic}(q_i)$ is bounded by $|d_{ic}(q_i)| \leq \rho_{dci}$.

Then, define $B_i = (I_{mi}\gamma_i)^{-1} \in R^+$ and the state vector $x_i = \begin{bmatrix} x_{i1} & x_{i2} \end{bmatrix}^T = \begin{bmatrix} q_i & \dot{q}_i \end{bmatrix}^T \in R^{2\times 1}$ and the control input $u_i = \tau_i \in R^{1\times 1}, i = 1, 2, \ldots n$, the state space of ith subsystem can be formulated as

$$S_i = \begin{cases} \dot{x}_{i1} = x_{i2} \\ \dot{x}_{i2} = \phi_i(x_i) + \upsilon_i(x) + p_i(x_i) + B_i u_i , \\ y = x_{i1} \end{cases} \tag{7}$$

where $\phi_i(x_i)$, $\upsilon_i(x)$ and $p_i(x_i)$ are represented as follows:

$$\begin{aligned} \phi_i(x_i) &= B_i \begin{pmatrix} -\left(\hat{f}_{si}e^{(-\hat{f}_{ri}\dot{q}_i^2)} + \hat{f}_{ci}\right)\mathrm{sgn}(\dot{q}_i) \\ -\hat{f}_{bi}\dot{q}_i - \dfrac{\tau_{si}}{\gamma_i} \end{pmatrix} \\ \upsilon_i(x) &= B_i\left(-U_{zi} - V_{zi} - f_{pi}(q_i, \dot{q}_i) - Y(\dot{q}_i)\tilde{F}_i\right) \\ p_i(x_i) &= B_i\left(-d_{is}(q_i) - d_{ic}(q_i)\right) \end{aligned} \tag{8}$$

Next, a decentralized robust zero-sum optimal control method is presented for MRMs to ensure that both position and velocity of the closed-loop robotic systems are asymptotically stable.

3 Decentralized Robust Zero-Sum Neuro-Optimal Control Based on ADP

Consider the MRM systems with the dynamic model formulation (1) and the state space description (7), if the contact disturbance $p_i(x_i)$ is viewed as a system control input, the robust optimal control problem can be transformed into a two-player zero-sum optimal control issue, in which the continuously differentiable infinite horizon local cost function can be defined as follows:

$$
J_i(s_{i0}(e_i)) = \int_0^\infty \left\{ \begin{array}{l} s_i(e_i(\tau))^T Q_i s_i(e_i(\tau)) + u_i(\tau)^T R_i u_i(\tau) \\ - \gamma_{pi}^2 p_i(x_i(\tau))^T p_i(x_i(\tau)) \end{array} \right\} d\tau, \tag{9}
$$

where

$$
s_i(e_i) = \dot{e}_i + \alpha_{ei} e_i \tag{10}
$$

is a filtered error function with the initial filtered error $s_{i0}(e_i) = s_i(0)$; $e_i = x_{i1} - x_{id}$ and $\dot{e}_i = x_{i2} - \dot{x}_{id}$ indicate the position and velocity tracking error of the ith joint module respectively; x_{id}, $\dot{x}_{id}$ and $\ddot{x}_{id}$ represent the desired angular position, velocity and acceleration that are known and bounded reference states. $Q_i^T = Q_i$, $R_i^T = R_i$ denote the positive constant matrixes; α_{ei} and γ_{pi} are determined positive constants.

According to (9), one can define the local Hamiltonian as:

$$
\begin{aligned}
H_i(s_i, u_i, p_i, \nabla J_i) &= s_i^T Q_i s_i + u_i^T R_i u_i \\
&\quad + \nabla J_i^T(s_i)(\phi_i + v_i + p_i + \alpha_{ei}\dot{e}_i + B_i u_i - \ddot{x}_{id}) - \gamma_{pi}^2 p_i^T p_i,
\end{aligned} \tag{11}
$$

where $\nabla J_i(s_i) = (\partial J_i(s_i)/\partial s_i)$ denotes the partial derivative of $J_i(s_i)$ with regard of s_i. Define the utility function as

$$
\Omega_i(s_i, u_i, p_i) = s_i^T Q_i s_i + u_i^T R_i u_i - \gamma_{pi}^2 p_i^T p_i. \tag{12}
$$

Then, one obtain that the optimal local cost function J_i^* satisfies:

$$
J_i^*(s_i) = \min_{u_i} \max_{p_i} \int_0^\infty \Omega_i(s_i, u_i, p_i)\, d\tau = \max_{p_i} \min_{u_i} \int_0^\infty \Omega_i(s_i, u_i, p_i)\, d\tau. \tag{13}
$$

According to (11) and (13), the local optimal control pair (u_i^*, p_i^*) satisfies the following HJI equation:

$$
0 = \nabla J_i^{*T}(s_i)(\phi_i + v_i + p_i^* + \alpha_{ei}\dot{e}_i + B_i u_i^* - \ddot{x}_{id}) + \Omega_i(s_i, u_i^*, p_i^*). \tag{14}
$$

On the basis of the principle of optimality, the local optimal control pair (u_i^*, p_i^*) can be represented as follows:

$$
u_i^* = -\frac{1}{2} R_i^{-1} B_i^T \nabla J_i^*, \quad p_i^* = \frac{1}{2\gamma_{pi}^2} \nabla J_i^*. \tag{15}
$$

Rewriting the optimal control law u_i^* of the ith subsystem as:

$$
u_i^* = u_{i1} + u_{i2} + u_{i3}^* \tag{16}
$$

to deal with modeled dynamic term ϕ_i, model uncertainty term υ_i and contact disturbance term p_i respectively, then, the HJI equation can be modified as:

$$0 = \nabla J_i^* \left(\begin{array}{c} \phi_i + \upsilon_i + p_i^* + \alpha_{ei}\dot{e}_i - \ddot{x}_{id} \\ + B_i u_{i1} + B_i u_{i2} + B_i u_{i3}^* \end{array} \right) + \Omega_i \left(s_i, u_i^*, p_i^* \right). \tag{17}$$

Therefore, the zero-sum optimal control problem has been transformed into the one of obtaining the decentralized control laws u_{i1}, u_{i2} and u_{i3}^* respectively, and therefore, to realize the optimal compensation of model uncertainty and environmental contact disturbance of the MRM system.

First, we can design control law u_{i1}, that is given as:

$$u_{i1} = - \left(- \left(\hat{f}_{si} e^{\left(-\hat{f}_{\tau i} x_{i2}^2 \right)} + \hat{f}_{ci} \right) \text{sgn}\left(x_{i2} \right) - \hat{f}_{bi} x_{i2} - B_i^{-1} \ddot{x}_{id} - \frac{\tau_{si}}{\gamma_i} + B_i^{-1} \alpha_{ei}\dot{e}_i \right). \tag{18}$$

Second, we consider that the MRM systems, which work in free space, is possessed of accurate joint torque sensing. Thus, substituting (18) into (17), one obtains:

$$0 = \Omega_i \left(s_i, u_i^*, p_i^* \right) + \nabla J_i^* \left(\upsilon_i + p_i^* + B_i u_{i2} + B_i u_{i3}^* \right). \tag{19}$$

Note that we have $p_i^* = u_{i3}^* = 0$ when the MRM system works in free space, therefore, combining the time derivative of (10) and the model-based compensation control (18), one obtains

$$\dot{s}_i = B_i \left(- f_{pi}\left(x_{i1}, x_{i2} \right) - U_{zi} - Y\left(x_{i2} \right) \tilde{F}_i - V_{zi} \right) + B_i u_{i2}. \tag{20}$$

Here, we introduce the decentralized robust control u_{i2} to compensate the effects of model uncertainties. For the unmodeled friction in (2), we decomposed $\tilde{F}_i$ as the summation of constant part and variable part, which is given as:

$$\tilde{F}_i = \tilde{F}_{ic} + \tilde{F}_{iv}, \tag{21}$$

where $\tilde{F}_{ic}$ indicates an unknown constant vector and $\tilde{F}_{iv}$ denotes an unknown variable vector that is bounded as:

$$\left| \tilde{F}_{iv} \right| \leq \rho_{Fivn}, n = 1, 2, 3, 4, \tag{22}$$

in which $\rho_{Fivn} = \left[\rho_{Fiv1} \ \rho_{Fiv2} \ \rho_{Fiv3} \ \rho_{Fiv4} \right]^T$ is a determined constant vector. Then, one can design the robust control law u_{i2f}, which is given as:

$$u_{i2f} = u_{i2fu} + Y\left(x_{i2} \right)\left(u_{i2fc} + u_{i2fv} \right), \tag{23}$$

where u_{i2fu} is designed to compensate friction term $f_{pi}\left(x_{i1}, x_{i2} \right)$; u_{i2fc}, u_{i2fv} are used to compensate the effect of uncertainties $\tilde{F}_{ic}$ and $\tilde{F}_{iv}$ respectively:

$$u_{i2fu} = \begin{cases} -\rho_{fpi} \frac{s_i}{|s_i|} & if \ |s_i| > \varepsilon_{ifu} \\ -\rho_{fpi} \frac{s_i}{\varepsilon_{ifu}} & if \ |s_i| \leq \varepsilon_{ifu} \end{cases}, \tag{24}$$

$$u_{i2fc} = -\varepsilon_{ifc}\left(\int_0^t Y(x_{i2})s_i d\tau + s_i\right), \tag{25}$$

$$u_{i2fv} = \begin{cases} -\rho_{Fivn}\dfrac{Y(x_{i2})s_i}{|Y(x_{i2})s_i|} & if \ |Y(x_{i2})s_i| > \varepsilon_{ifvn} \\ -\rho_{Fivn}\dfrac{Y(x_{i2})s_i}{\varepsilon_{ifvn}} & if \ |Y(x_{i2})s_i| \le \varepsilon_{ifvn} \end{cases}, n = 1,2,3,4 \tag{26}$$

where ε_{ifu}, ε_{ifc} and ε_{ifvn} are determined positive parameters. Moreover, we decomposed the terms of U_{zi} and V_{zi} as:

$$U_{zi} = U_{zic} + U_{ziv}, \quad V_{zi} = V_{zic} + V_{ziv}, \tag{27}$$

where U_{zic} and V_{zic} are with known up-bounds; U_{ziv} and V_{ziv} are bounded as:

$$|U_{ziv}| \le \rho_{Uiv}, \quad |V_{ziv}| \le \rho_{Viv}, \tag{28}$$

where ρ_{Uiv} and ρ_{Viv} are constant up-bounds to be determined.

Then, by adopting the decomposition-based analysis of the uncertainties, we can design the robust control law u_{i2h}, which is formulated as:

$$u_{i2h} = u_{i2hU} + u_{i2hV}, \tag{29}$$

where u_{i2hU} is used to compensate the effect of the term U_{zi}, that is given as:

$$u_{i2hU} = u_{i2hUc} + u_{i2hUv},$$

$$u_{i2hUc} = -\varepsilon_{iUs}\left(\int_0^t \varepsilon_{iUs}s_i d\tau + s_i\right), u_{i2hUv} = \begin{cases} -\rho_{Uiv}\dfrac{\varepsilon_{iUs}s_i}{|\varepsilon_{iUs}s_i|} & if \ |\varepsilon_{iUs}s_i| > \varepsilon_{iUB}, \\ -\rho_{Uiv}\dfrac{\varepsilon_{iUs}s_i}{\varepsilon_{iUB}} & if \ |\varepsilon_{iUs}s_i| \le \varepsilon_{iUB} \end{cases} \tag{30}$$

where ε_{iUs} and ε_{iUB} are positive parameters to be determined. Moreover, one can design the control law u_{i2hV} to compensate the effect of the term V_{zi}:

$$u_{i2hV} = u_{i2hVc} + u_{i2hVv}$$

$$u_{i2hVc} = -\varepsilon_{iVs}\left(\int_0^t \varepsilon_{iVs}s_i d\tau + s_i\right), u_{i2hVv} = \begin{cases} -\rho_{Viv}\dfrac{\varepsilon_{iVs}s_i}{|\varepsilon_{iVs}s_i|} & if \ |\varepsilon_{iVs}s_i| > \varepsilon_{iVB}, \\ -\rho_{Viv}\dfrac{\varepsilon_{iVs}s_i}{\varepsilon_{iVB}} & if \ |\varepsilon_{iVs}s_i| \le \varepsilon_{iVB} \end{cases} \tag{31}$$

where ε_{iVs} and ε_{iVB} are positive parameters to be determined. Then, by combining (23) with (29), one obtain the decentralized control law u_{i2}, which is given as follows:

$$u_{i2} = u_{i2f} + u_{i2h}. \tag{32}$$

Then, we introduce the critic NN, u-action NN and p-action NN to approximate the performance index function $J_i(s_i)$, the optimal control law u_{i3}^* and the worst contact disturbance p_i^* respectively.

By employing RBF NNs, the ideal critic NN is represented as:

$$J_i(s_i) = w_{ci}^T \sigma_{ci}(s_i) + \varepsilon_{ci}, \tag{33}$$

where w_{ci} is the unknown ideal NN weight, ε_{ci} represent the finite approximation error of the critic NN, $\sigma_{ci}(s_i)$ indicates a standard activation function. Besides, the gradient of the approximated cost function is given as:

$$\nabla J_i(s_i) = \nabla \sigma_{ci}(s_i)^T w_{ci} + \nabla \varepsilon_{ci}^T, \tag{34}$$

where $\nabla \sigma_{ci}(s_i) = \partial \sigma_{ci}(s_i)/\partial s_i$ indicates the gradients of the activation function and $\nabla \varepsilon_{ci}$ is the gradient error. Then, one can rewrite the Hamiltonian and obtain the following relation:

$$H_i(s_i, u_i, p_i, w_{ci}) = \Omega_i(s_i, u_i, p_i) + \left(w_{ci}^T \nabla \sigma_{ci}(s_i) \right) \dot{s}_i - e_{cHi} = 0, \tag{35}$$

where e_{cHi} denotes the approximation error of the critic NN:

$$e_{cHi} = \Omega_i + \left(w_{ci}^T \nabla \sigma_{ci} \right) \dot{s}_i. \tag{36}$$

Let $\hat{w}_{ci}$ be the estimated weight vector of w_{ci}, therefore, one obtains:

$$\hat{J}_i(s_i) = \hat{w}_{ci}^T \sigma_{ci}(s_i). \tag{37}$$

Then, one can define the approximate Hamilton function as:

$$\hat{H}_i(s_i, u_i, p_i, \hat{w}_{ci}) = \Omega_i(s_i, u_i, p_i) + \left(\hat{w}_{ci}^T \nabla \sigma_{ci}(s_i) \right) \dot{s}_i. \tag{38}$$

Define the weight approximation error $\tilde{w}_{ci} = w_{ci} - \hat{w}_{ci}$, then, one obtains:

$$e_{ci} = e_{cHi} - \tilde{w}_{ci}^T \nabla \sigma_{ci}(s_i) \dot{s}_i. \tag{39}$$

Define the residual error function $E_{ci} = \frac{1}{2} e_{ci}^2$, which is minimized to adjust the weight vector of the critic NN, that is updated by

$$\dot{\hat{w}}_{ci} = -\alpha_{ci} \left(\frac{\partial E_{ci}}{\partial \hat{w}_{ci}} \right) = -\alpha_{ci} \frac{p_{ci} \left(p_{ci}^T \hat{w}_{ci} + \Omega_i \right)}{\left(p_{ci}^T p_{ci} + 1 \right)^2}, \tag{40}$$

where $\alpha_{ci} > 0$ is the learning rate of the critic NN, and $p_{ci} = \nabla \sigma_{ci}(s_i) \dot{s}_i$.

Then, we employ the action NNs to approximate u_{i3}^* and p_i^* respectively. The ideal u- and p-action NNs are given as follows:

$$u_{i3}^* = w_{ai}^T \sigma_{ai}(s_i) + \varepsilon_{ai}, \quad p_i^* = w_{pi}^T \sigma_{pi}(s_i) + \varepsilon_{pi}, \tag{41}$$

where w_{ai}, w_{pi} represent the ideal weight vectors, σ_{ai}, σ_{pi} denote the activation functions and ε_{ai}, ε_{pi} are the finite approximation errors of u- and p-action respectively. Let $\hat{w}_{ai}$ and $\hat{w}_{pi}$ be the estimation weight vectors of w_{ai}, w_{pi}, then we have:

$$\hat{u}_{i3} = \hat{w}_{ai}^T \sigma_{ai}(s_i), \quad \hat{p}_i = \hat{w}_{pi}^T \sigma_{pi}(s_i). \tag{42}$$

Since $\left(\partial \hat{H}_i(s_i, \hat{u}_i, \hat{p}_i, \hat{w}_{ci})/\partial \hat{u}_i \right) = 0$ and $\left(\partial \hat{H}_i(s_i, \hat{u}_i, \hat{p}_i, \hat{w}_{ci})/\partial \hat{p}_i \right) = 0$, so that $\hat{u}_{i3}$ and $\hat{p}_i$ can be further described as follows:

$$\hat{u}_{i3} = -\frac{1}{2} R_i^{-1} B_i^T \nabla \sigma_{ci}^T(s_i) \hat{w}_{ci}, \quad \hat{p}_i = \frac{1}{2\gamma_{pi}^2} \nabla \sigma_{ci}^T(s_i) \hat{w}_{ci}, \tag{43}$$

where the update laws for u- and p-action NN weights are given as:

$$\dot{\hat{w}}_{ai} = -\alpha_{ai}\sigma_{ai}(s_i) \cdot \left(\hat{w}_{ai}^T \sigma_{ai}(s_i) + \frac{1}{2}R_i^{-1}B_i^T \nabla \sigma_{ci}^T(s_i)\hat{w}_{ci} \right)^T, \qquad (44)$$

$$\dot{\hat{w}}_{pi} = -\alpha_{pi}\sigma_{pi}(s_i) \cdot \left(\hat{w}_{pi}^T \sigma_{pi}(s_i) - \frac{1}{2\gamma_{pi}^2}\nabla \sigma_{ci}^T(s_i)\hat{w}_{ci} \right)^T, \qquad (45)$$

where α_{ai} and α_{pi} are the positive learning rates of u- and p-action NNs.

Then, by combining (18), (32) with (43), the proposed decentralized robust zero-sum neuro-optimal control law u_i^* is given as:

$$u_i^* = - \begin{pmatrix} -\hat{f}_{bi}x_{i2} - B_i^{-1}\ddot{x}_{id} - \dfrac{\tau_{si}}{\gamma_i} + B_i^{-1}\alpha_{ei}\dot{e}_i \\ -\left(\hat{f}_{sie}^{(-\hat{f}_{\tau i}x_{i2}^2)} + \hat{f}_{ci}\right)\operatorname{sgn}(x_{i2}) \end{pmatrix} - (-u_{i2hU} - u_{i2hV})$$
$$- (-u_{i2fu} - Y(x_{i2})(u_{i2fc} + u_{i2fv})) - \frac{1}{2}R_i^{-1}B_i^T \nabla \sigma_{ci}^T \hat{w}_{ci} \qquad (46)$$

Theorem 1. *Consider an n-DOF modular robot manipulator, with the dynamic model of the ith joint subsystem formulated in (1); the model uncertainties existed in (2), (4) and the environmental contact disturbance given in (6). The closed-loop robotic system is asymptotically stable under the decentralized robust zero-sum neuro-optimal control law proposed by (46).*

Proof. A Lyapunov function candidate is selected as:

$$V_g(t) = \sum_{i=1}^{n} V_{gi}(t) = \sum_{i=1}^{n} \begin{pmatrix} \delta_{si}s_i^T s_i + \delta_{\Omega i}J_i(s_i) \\ + \delta_{pi}\int_t^\infty \gamma_{pi}^2 p_i^T(\tau)p_i(\tau)d\tau \end{pmatrix}, \qquad (47)$$

where $\delta_{si} > 0$ and $\delta_{pi} \geq \delta_{\Omega i} > 0$ are determined positive definite constants. Then, one obtains the time derivative of (47) that is given as:

$$\dot{V}_g(t) \leq \sum_{i=1}^{n} \left(2\delta_{si} \left(\|s_i\|^2 + \frac{1}{2}\|B_i\|^2\|u_{i3}^*\|^2 \right) \right) - \sum_{i=1}^{n}(-\delta_{si} - (\delta_{\Omega i} - \delta_{pi})\gamma_{pi}^2)\|p_i\|^2$$
$$- \sum_{i=1}^{n} \delta_{\Omega i} \begin{pmatrix} \lambda_{\min}(Q_i)\|s_i\|^2 + \lambda_{\min}(R_i)\|u_{i1}\|^2 \\ + \lambda_{\min}(R_i)\|u_{i2}\|^2 + \lambda_{\min}(R_i)\|u_{i3}^*\|^2 \end{pmatrix}$$
$$= -\sum_{i=1}^{n}(\delta_{\Omega i}\lambda_{\min}(Q_i) - 2\delta_{si})\|s_i\|^2 - \sum_{i=1}^{n}(\delta_{\Omega i}\lambda_{\min}(R_i))\|u_{i1}\|^2$$
$$- \sum_{i=1}^{n}(\delta_{\Omega i}\lambda_{\min}(R_i))\|u_{i2}\|^2 - \sum_{i=1}^{n}\left(\delta_{\Omega i}\lambda_{\min}(R_i) - \delta_{si}\|B_i\|^2\right)\|u_{i3}^*\|^2$$
$$- \sum_{i=1}^{n}(-\delta_{si} - (\delta_{\Omega i} - \delta_{pi})\gamma_{pi}^2)\|p_i\|^2$$

$$(48)$$

Therefore, we know that $\dot{V}_g(t) \leq 0$ when the following condition holds:

$$\lambda_{\min}(Q_i) \geq \frac{2\delta_{si}}{\delta_{\Omega i}}, \quad \lambda_{\min}(R_i) \geq \frac{\delta_{si}}{\delta_{\Omega i} I_{mi}^2 \gamma_i^2}, \quad \delta_{pi} \geq \delta_{\Omega i} + \frac{\delta_{si}}{\gamma_{pi}^2}. \qquad (49)$$

Besides, from (48), one obtains that if (49) is satisfied, then $\dot{V}_g(t) < 0$ for any $s_i \neq 0$. Therefore, by the Lyapunov theory, the closed-loop robotic system is asymptotically stable under the proposed decentralized robust zero-sum neuro-optimal control (46). This completes the proof of the Theorem 1.

4   Experimental Results

In this section, the experimental results are presented based on the experimental platform as illustrated in [8], and the experimental results are collected to verify the effectiveness of the proposed control method. We consider the environmental contact as a kind of instantaneous collision between the MRM link and the collision object that may create instantaneous contact forces.

Table 1. Parameters setting

Name	Value	Name	Value	Name	Value
$\hat{f}_{bi}$	12 mNms/rad	$\hat{f}_{ci}$	30 mNm	$\hat{f}_{si}$	40 mNm
$\hat{f}_{\tau i}$	20 s²/rad²	γ_i	100	I_{mi}	120 gcm²
ρ_{Fiv1}	30 mNms/rad	ρ_{Fiv2}	60 mNm	ρ_{Fiv3}	80 mNm
ρ_{Fiv4}	50 s²/rad²	ρ_{dsi}	10 mNm	ρ_{dci}	50 mNm
ρ_{Viv}	2.3	ρ_{fpi}	0.5	ρ_{Uiv}	2.6
α_{ei}	0.5	ε_{ifu}	0.1	ε_{ifc}	1
$\alpha_{ci}, \alpha_{pi}, \alpha_{ai}$	0.8	$\varepsilon_{ifv1}, \varepsilon_{ifv2}$	0.01	$\varepsilon_{ifv3}, \varepsilon_{ifv4}$	0.01
γ_{pi}	1.87	$\varepsilon_{iUs}, \varepsilon_{iVs}$	0.02	$\varepsilon_{iVB}, \varepsilon_{iUB}$	0.02

The parameter definition is represented as follows: The critic NN is chosen as 2-5-1 with 2 input neurons, 5 hidden neurons and 1 output neuron, and the weight vectors are selected as $\hat{w}_{c1} = [\hat{w}_{c11} \ \hat{w}_{c12} \ \hat{w}_{c13} \ \hat{w}_{c14} \ \hat{w}_{c15}]^T$ for joint 1 and $\hat{w}_{c2} = [\hat{w}_{c21} \ \hat{w}_{c22} \ \hat{w}_{c23} \ \hat{w}_{c24} \ \hat{w}_{c25}]^T$ for joint 2, with the initial values $\hat{w}_{c1} = \hat{w}_{c2} = [0]$. The NN structures and initial weight values of the and action NNs are selected as the same of the critic NNs. The other model parameters, uncertainty up-bound parameters and control parameters are given in Table 1.

Figure 1 illustrate the position and velocity error curves under the situation of instantaneous collision contact. In this figure, obvious position and velocity deviations are captured, which are due to the effect of collision force. Moreover, we also observe that the error values decreased to normal ranges with very short time periods, which may attribute to the performance of the proposed zero-sum

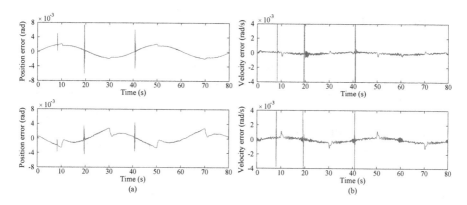

Fig. 1. Error curves under the collision contact, (a) position errors, (b) velocity errors.

neuro-optimal controller that realized the optimal compensation of unknown contact disturbance.

Figure 2(a) illustrate the control torque curves under the situation of instantaneous collision contact. From this figure, the reliable control torque outputs are obtained, in which the instant increase of control torques keep in safe limits. This is attribute the contribution of the proposed optimal control that realizes the optimization of tracking errors and output torques.

Figure 2(b) illustrate the variations of the estimated critic NN weights under the proposed method. From this figure, one obtain that the critic NN weights may converge in certain ranges for each isolated joint subsystem. With the critic NN weight estimations, the HJI equation and the optimal control law can be solved and derived respectively.

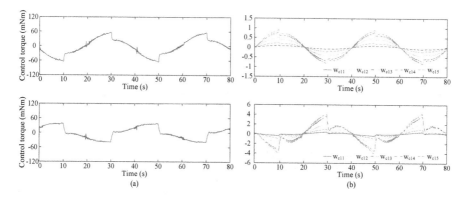

Fig. 2. Control torque and weight adjustment curves, (a) control torque, (b) weight adjustment.

5 Conclusion

In this paper, we develop a decentralized robust zero-sum neuro-optimal control scheme for MRMs. Based on local dynamic information of each joint module and the JTF technique, a model-based compensation controller and an uncertainty decomposition-based robust controller are designed for controlling MRM systems. Based on ADP algorithm, the HJI equation is solved by constructing the action-critic NNs, and then the decentralized robust zero-sum neuro-optimal control is developed. Finally, experimental results are illustrated to verify the effectiveness and advantages of the proposed method.

Acknowledgements. This work is supported by the National Natural Science Foundation of China (Grant nos. 61374051, 61773075 and 61703055) and the Scientific Technological Development Plan Project in Jilin Province of China (Grant nos. 20170204067GX, 20160520013JH and 20160414033GH).

References

1. Albu-Schaffer, A., Ott, C., Hirzinger, G.: A unified passivity-based control framework for position, torque, and impedance control of flexible joint robots. Int. J. Robot. Res. **26**(1), 23–39 (2007)
2. Liu, G., Liu, Y., Goldenberg, A.A.: Design, analysis, and control of a spring-assisted modular and reconfigurable robot. IEEE/ASME Trans. Mech. **16**(4), 695–706 (2011)
3. Lee, H., Kim, S., Chang, H., Kim, J.: Development of a compact optical torque sensor with decoupling axial-interference effects for pHRI. Mechatronics **52**, 90–101 (2018)
4. He, W., Dong, Y.: Adaptive fuzzy neural network control for a constrained robot using impedance learning. IEEE Trans. Neural Netw. Learn. Syst. **29**(4), 1174–1186 (2018)
5. Roveda, L., Pallucca, G., Pedrocchi, N., Braghin, F., Tosatti, L.M.: Iterative learning procedure with reinforcement for high-accuracy force tracking in robotized tasks. IEEE Trans. Ind. Electron. **14**(4), 1753–1763 (2018)
6. Leottau, D., Ruiz-del-Solar, J., Babuska, R.: Decentralized reinforcement learning of robot behaviors. Artif. Intell. **256**, 130–159 (2018)
7. Dong, B., Zhou, F., Liu, K., Li, Y.: Decentralized robust optimal control for modular robot manipulators via criticidentifier structure-based adaptive dynamic programming. Neural Comput. Appl. (2018). https://doi.org/10.1007/s00521-018-3714-8
8. Dong, B., Zhou, F., Liu, K., Li, Y.: Torque sensorless decentralized neuro-optimal control for modular and reconfigurable robots with uncertain environments. Neurocomputing **282**, 60–73 (2018)
9. Imura, J., Yokokohji, Y., Yoshikawa, T., Sugie, T.: Robust control of robot manipulators based on joint torque sensor information. Int. J. Robot. Res. **13**(5), 434–442 (1994)
10. Dong, B., Liu, K., Li, Y.: Decentralized control of harmonic drive based modular robot manipulator using only position measurements: theory and experimental verification. J. Intell. Robot. Syst. **88**, 3–18 (2017)

Leader-Following Consensus of Nonlinear Multi-agent System via a Distributed ET Impulsive Control Strategy

Yaqi Wang, Jianquan Lu$^{(\boxtimes)}$, Jinling Liang, and Jinde Cao

Jiangsu Provincial Key Laboratory of Networked Collective Intelligence, and School of Mathematics, Southeast University, Nanjing 210096, China
`yqwtengzhou@126.com`, {`jqluma,jinlliang,jdcao`}`@seu.edu.cn`

Abstract. This paper investigates the leader-following consensus problem of nonlinear multi-agent systems with directed topology by a novel delayed impulsive controller. The impulsive moments are determined by an event-triggered condition. Meanwhile, we also consider the delays in impulsive term on account of the network's limited communication. Some sufficient conditions are derived to achieve the leader-following consensus, and the Zeno behavior dose not exhibit. A numeral example is derived to show the validity of our results.

Keywords: Multi-agent systems · Event-triggered · Impulsive control · Delayed impulses

1 Introduction

Many natural phenomenon and man-made networks can be modeled by leader-following multi-agent systems (MASs), such as [1–4]. The leader-following consensus (LFC) problems of MASs have received much attention during the past decades due to the widely applications in signal process, coordination of unmanned vehicles, wireless senor networks, and so on [5–7]. Recently, many literatures adopted the existing distributed control techniques to study the LFC problems. However, common control protocols need continuous monitoring and communication between agents, which may caused the redundant cost and packet dropout in the large-scale MASs. Hence, designing an effective control protocol is a great challenge for the LFC problems.

Impulsive control method, which describes the abrupt change at certain instants, has been widely used to deal some practical problems. There existed many works concerning the impulsive consensus [8–10]. Besides, some researchers focused on the influence of impulsive time sequence on the consensus. Lu *et al.* presented a novel concept named average impulsive interval to describe the frequency of impulses in [11]. Then Wang *et al.* [12] extended the concept to a form of limited and given a novel notion of average impulsive gain. Furthermore, the

© Springer Nature Switzerland AG 2019
H. Lu et al. (Eds.): ISNN 2019, LNCS 11555, pp. 15–24, 2019.
https://doi.org/10.1007/978-3-030-22808-8_2

concept of impulsive time window was used to determine impulsive moments previously [13]. The impulsive instants in the above studies are restricted into some regions, but the accurate instants when the impulses occur are not determined.

On the other hand, event-triggered (ET) control method, that the controller updates only at ET instants which determined by some predefined conditions, is a very useful in some practise industrial field, due to this control mechanism can greatly save the communication cost [14]. In [15], the LFC problems of second-order MASs with or without switching topology were addressed using distributed ET sampling strategy. Although the ET control protocol is a valid approach to avoid continuous communication, it is not easy to implement this control method due to the limited network bandwidth and limited information processing capability.

From the above discussion, we can see that ET method can accurately determined the time when the agent broadcasts its state or exchange information with its neighbors. Considering the advantage and disadvantage of impulsive controller, using appropriate ET conditions to decide the impulsive instants is a very available method. Many existing results discussed the ET impulsive control protocol. The $\mathcal{L}_\infty$ leader-following consensus of continuous-time MASs via a ET observer-type protocol under an impulsive framework was concerned in [16]. An ET impulsive control protocol was designed to enhance the property of differential evolution algorithms in [17]. Subsequently, Zhu $et\ al.$ [18] used the event-based impulsive control method to study the exponential stabilization and applied the results to memristive neural networks. Recently, by the ET impulsive control method, the LFC for MASs was derive in [19]. Although these work have discussed the event-based impulsive control schemes, there are few literatures considering the delay naturally caused by network communication between sensors, actuators and controllers.

Motivated by the above discusses, this paper studies the LFC problem for nonlinear MASs with delay impulses via designing a ET impulsive controller. The rest of the paper is organized as following: In Sect. 2, several definitions and important lemmas are given. Main results are formulated in Sect. 3. Then, we provide a numerical simulation to show the valid of our results in Sect. 4. Section 5 concludes our results and discussed the further work.

2   Preliminaries and Problem Statement

The network's topology described by a directed graph $\mathcal{G} = \{\mathcal{V}, \mathcal{E}\}$, where $\mathcal{V} = \{v_1, v_2, \ldots, v_N\}$ is the set of nodes, and the edges set is $\mathcal{E} \subseteq \mathcal{V} \times \mathcal{V}$. If node j can receive information from node i, then $(i, j) \in \mathcal{E}$. The corresponding Laplacian matrix is defined as follows: $l_{ij} > 0$ if $(i, j) \in \mathcal{E}$, $l_{ij} = 0$ otherwise. We used the diagonal matrix $D = diag\{d_1, d_2, \ldots, d_N\}$ to describe the relationship between the the follower nodes and leader node, i.e. $d_i > 0$ means the agent i can take over information from the leader node, otherwise $d_i = 0$.

Consider the following nonlinear multi-agent network:

$$\dot{x}_i(t) = Bx_i(t) + C\tilde{f}(x_i(t)) + u_i(t), \tag{1}$$

where $i = 1, 2, \cdots, N$, B and $C \in \mathbb{R}^{n \times n}$, the nonlinear function $\tilde{f}(\cdot) = (f_1(\cdot), f_2(\cdot), \cdots, f_n(\cdot))^T : \mathbb{R}^n \to \mathbb{R}^n$, and $u_i(t)$ is control input which will be designed in the following.

The objective leader agent's dynamics is

$$\dot{x}_0(t) = Bx_0(t) + C\tilde{f}(x_0(t)). \tag{2}$$

Definition 1. *The leader-following exponential consensus of the MASs (1) with the leader node $x_0(t)$ is said to be achieved if* $\|e(t)\| \leq Me^{-\alpha(t-t_0)}$, $i = 1, 2, \cdots, N$, *where M and α are given positive constants, $e_i(t) = x_i(t) - x_0(t)$.*

In this paper, we will adopt the impulsive control method to achieve the leader-following exponential consensus, and the impulsive instants are decided by an ET condition. Next, we will give the ET condition. Let $s_i(t) = \sum_{j \in \mathcal{N}_i} a_{ij}(x_j(t) - x_i(t)) + d_i(x_0(t) - x_i(t))$ and $\varepsilon(t) = s_i(t_k^i) - s_i(t)$. The communication events of agent i are triggered at time sequence

$$t_k^i = \inf\{t : t > t_{k-1}^i, z_i(t) \geq 0\}, \tag{3}$$

where $z_i(t) = \|\varepsilon(t)\| - \beta \|s_i(t_k^i)\|$. Then, $u_i(t)$ is defined as

$$u_i(t_k^i) = cs_i(t_k^i - \tau_k^i). \tag{4}$$

and $u_i(t) = 0$, $t \neq t_k^i$. For $e_i(t) = x_i(t) - x_0(t)$, then the error systems can be rewritten as

$$\begin{cases} \dot{e}_i(t) = Be_i(t) + Cf(e_i(t)), t \neq t_k^i, \\ \Delta e_i(t_k^i) = e_i(t_k^+) - e_i(t_k^-) = -cs_i(t_k^i - \tau_k^i). \end{cases} \tag{5}$$

where $f(e_i(t)) - \tilde{f}(x_i(t)) - \tilde{f}(x_0(t))$. Then

$$\begin{cases} \dot{e}(t) = I_N \otimes Be(t) + I_N \otimes CF(e(t)), t \neq t_k, \\ \Delta e(t_k) = e(t_k^+) - e(t_k^-) = -c\delta(k)s(t_k - \tau_k), \end{cases} \tag{6}$$

where $\zeta = \{t_1, t_2, \cdots\}$ is rearrangement of all t_k^i according to chronological order, $F(e(t)) = (f^T(e_1(t)), f^T(e_2(t)), \cdots, f^T(e_N(t)))^T$, and $e(t) = (e_1^T(t), e_2^T(t), \cdots, e_n^T(t))^T$. $\delta_i(k) = 1$ if event occurs for agent i at t_k, $\delta_i(k) = 0$ otherwise. We use $\delta(k)$ to describe the triggered situation for every agent.

$$\delta(k) = \begin{pmatrix} \delta_1(k) \ \delta_2(k) \ \cdots \ \delta_N(k) \\ \delta_1(k) \ \delta_2(k) \ \cdots \ \delta_N(k) \\ \vdots \quad \vdots \quad \cdots \quad \vdots \\ \delta_1(k) \ \delta_2(k) \ \cdots \ \delta_N(k) \end{pmatrix}.$$

From the above matrices, it is easy to see that there are no more than 2^N possibilities for $\delta(k)$. We assume that $\delta(k) \in \Omega = \{\delta_1, \delta_2, \ldots, \delta_{2^N}, \}$.

Assumption 1. *For nonlinear function $\tilde{f}(\cdot)$, there exists a constant $L > 0$ such that for any $y_1, y_2 \in \mathbb{R}^n$ one has*

$$\|\tilde{f}(y_1) - \tilde{f}(y_2)\| \leq L\|y_1 - y_2\|. \tag{7}$$

3   Main Results

This section will discuss the LFC of systems (1) and (2) under the ET impulsive controller. Firstly, we will prove that there is no Zeno-behavior.

Lemma 1. *Consider the systems (1), (2) and impulsive controller (4) under the event condition (3). The Zeno behavior is excluded for the closed loop systems.*

Proof. Computing the right-upper Dini derivative of $\|e_i(t)\|$ on the interval $[t_k^i, t_{k+1}^i)$, one has

$$
\begin{aligned}
D^+ \varepsilon_i(t) &= \|\dot{\varepsilon}(t)\| \\
&= \sum_{j \in \mathcal{N}_i} a_{ij}(\dot{x}_j(t) - \dot{x}_i(t)) + d_i(\dot{x}_0(t) - \dot{x}_i(t)) \\
&\leq \|B + CL\|[\sum_{j \in \mathcal{N}_i} a_{ij}(x_j(t) - x_i(t)) \\
&\quad + d_i(x_0(t) - x_i(t))] \\
&\leq \tilde{a}\|s_i(t)\| \\
&\leq \tilde{a}\|\varepsilon_i(t)\| + \tilde{a}\|s_i(t_k^i)\|,
\end{aligned}
\tag{8}
$$

where $\tilde{a} = \|A + BL\|$. Hence, one has

$$
\|\varepsilon_i(t)\| \leq \|s_i(t_k^i)\|(e^{\tilde{a}(t - t_k^i)} - 1),
$$

Furthermore, we have

$$
\beta\|s_i(t_k^i)\| = \|\epsilon_i(t_{k+1}^i)\| \leq \|s_i(t_k^i)\|(e^{\tilde{a}(t_{k+1}^i - t_k^i)} - 1),
$$

and

$$
t_{k+1}^i - t_k^i \geq \frac{\ln(\beta + 1)}{\tilde{a}}.
$$

Hence, $\eta = \inf_{i=1,2,\cdots N, k \in \mathbb{N}^+}\{t_{k+1}^i - t_k^i\} > 0$, i.e. the Zeno-behavior is excluded.

Theorem 1. *Suppose that Assumption 1 and $t_k - t_{k-1} > \tau_k$ hold. Consider the MASs (1) and the leader node (2) with the ET impulsive controller (4). The corresponding error systems (5) is exponential stability if there exist constants α, $p > 0$, $\omega > 0$, $0 < \rho < 1$, positive-definition matrices $P > 0$ and $\Psi > 0$ such that the for any $\delta \in \Omega$, the following inequalities hold:*

$$
\begin{pmatrix} PB + B^T P + L^2 \Psi - \alpha P & PC \\ * & -\Psi \end{pmatrix} < 0,
\tag{9}
$$

$$
\begin{pmatrix} -(\varrho - \omega\beta_2^2)I_N & I_N - cH\delta & 0 \\ * & -I_N & cH\delta \\ * & * & -\omega I_N \end{pmatrix} < 0,
\tag{10}
$$

$$\frac{\ln \rho}{\eta} + \alpha < 0, \tag{11}$$

where $\beta_2 = (\|B\| + \|C\Psi\|)\sqrt{\frac{\lambda_N(P)}{\lambda_1(P)}}\tau.$

Proof. Let $V(t) = \sum\limits_{i=1}^{N} e_i^T(t)Pe_i(t)$. Considering the derivative along the error systems (6), it follows from Assumption 1 that

$$
\begin{aligned}
\dot{V}(e(t)) &= e^T(t)(I_n \otimes (PB + B^T P))e(t) \\
&\quad + e^T(t)(I_n \otimes PC)F(e(t)) \\
&\quad + F^T(e(t))(I_N \otimes C^T P)e(t) \\
&\leq e^T(t)(I_N \otimes (PB + B^T P))e(t) \\
&\quad + e^T(t)(I_N \otimes PC)\Psi^{-1}(I_N \otimes C^T P)e(t) \\
&\quad - F^T(e(t))\Psi F(e(t)) \\
&\leq \alpha V(e(t)).
\end{aligned}
$$

Hence, we have $\dot{V}(e(t)) \leq \alpha V(e(t))$.

Let $V_0 = \sup\limits_{t \in [t_0 - \tau, t_0)} \|V(t)\|$. Firstly, there exists a constant $\mu_1 > 0$ such that $V(t) \leq \mu_1 V_0 e^{-\epsilon(t - t_0 - \tau)}$ holds for $t \in [t_0 - \tau, t_0 + \tau)$. Let $\widetilde{V}(t) = V(t)e^{\epsilon(t - t_0 - \tau)}$ and $\mu_2 = \frac{\mu_1}{\varrho}$, where $0 < \varrho < 1$. For $t \in [t_0 + \tau, t_1)$, we will prove that $\widetilde{V}(t) \leq \mu_2 V_0$. If it is not true, i.e. there exist t such that $\widetilde{V}(t) > \mu_2 V_0$, then we have $\widetilde{V}(t^*) = \mu_2 V_0$ and $\widetilde{V}(t^{**}) = \varrho\mu_2 V_0$, where $t^* = \inf\{t \in [t_0 + \tau, t_1], \widetilde{V}(t) \geq \mu_2 V_0\}$ and $t^{**} = \sup\{t \in [t_0 + \tau, t_1], \widetilde{V}(t) \leq \varrho\mu_2 V_0\}$.

Then if ϵ is sufficient small, we can get

$$
\begin{aligned}
\dot{\widetilde{V}}(t) &= e^{\epsilon(t - t_0 - \tau)}(\epsilon V(t) + \dot{V}(e(t))) \\
&< \alpha e^{\epsilon(t - t_0 - \tau)}V(e(t)) \\
&= \alpha\widetilde{V}(e(t)).
\end{aligned}
$$

Hence, it follows from (11) that

$$
\begin{aligned}
\widetilde{V}(e(t^*)) &< \widetilde{V}(e(t^{**}))e^{\alpha(t^* - t^{**})} \\
&\leq \varrho\mu_2 V_0 e^{\alpha\eta} \\
&< \mu_2 V_0,
\end{aligned}
$$

which is contradict with $\widetilde{V}(t^*) = \mu_2 V_0$. Hence, for $t \in [t_0 + \tau, t_1)$ we have

$$\widetilde{V}(t) \leq \mu_2 V_0. \tag{12}$$

Suppose that (12) holds for $t \in [t_0 - \tau, t_m]$, next, we will prove that (12) holds for $t \in [t_m, t_{m+1})$. Actually, if we have

$$\widetilde{V}(t_m) \leq \varrho\mu_2 V_0, \tag{13}$$

then for any $t \in [t_m, t_{m+1})$

$$\begin{aligned}
\widetilde{V}(e(t)) &\leq \widetilde{V}(t_m) e^{\alpha(t-t_m)} \\
&\leq \varrho \mu_2 V_0 e^{\alpha \eta} \\
&\leq \mu_2 V_0.
\end{aligned} \tag{14}$$

Hence we only need to prove (13) holds.

Denote $E(t_m^-) = e(t_m^-) - e(t_m^- - \tau_m)$, then we have

$$\begin{aligned}
e(t_m^+) &= e(t_m^-) - cH\delta(m)e(t_m^- - \tau_m) \\
&= ((I_N - cH\delta(m)) \otimes I_n)e(t_m^-) \\
&\quad + cH\delta(m) \otimes I_n E(t_m^-).
\end{aligned}$$

It follows from

$$\|(I_n \otimes \sqrt{P})e(t)\|^2 \leq e^{-\epsilon(t-t_0-\tau)}\mu_2 V_0, \ t \in [t_0, t_m) \tag{15}$$

that

$$\|e(t)\|^2 \leq \|\frac{1}{\lambda_1(P)} e^{-\epsilon(t-t_0-\tau)}\mu_2 V_0. \tag{16}$$

For $t \in [t_{m-1}, t_m)$, one has

$$\begin{aligned}
&\|(I_n \otimes \sqrt{P})\dot{e}(t)\| \\
&\leq \sqrt{\lambda_N(P)}(\|B\| + \|C\Psi\|)\|e(t)\| \\
&\leq (\|B\| + \|C\Psi\|)\sqrt{\frac{\lambda_N(P)}{\lambda_1(P)}}\sqrt{e^{-\epsilon(t-t_0-\tau)}\mu_2 V_0},
\end{aligned}$$

and

$$\begin{aligned}
&\|(I_n \otimes \sqrt{P})E(t_m^-)\| \\
&\leq \int_{t_m-\tau_m}^{t_m} \|(I_n \otimes \sqrt{P})\dot{e}(s)\|ds \\
&\leq (\|A\| + \|B\Psi\|)\sqrt{\frac{\lambda_N(P)}{\lambda_1(P)}}\tau_m\sqrt{e^{-\epsilon(t_m-t_0-\tau)}\mu_2 V_0} \\
&\leq \beta_2\sqrt{e^{-\epsilon(t_m-t_0-\tau)}\mu_2 V_0}.
\end{aligned} \tag{17}$$

It follows from the definition of $\widetilde{V}(e(t))$ that (13) holds if and only if the following LMI holds

$$\begin{pmatrix} -\varrho\mu_2 V_0 e^{-\epsilon(t-t_0-\tau)} & e^T(t_m)(I_N \otimes P) \\ * & -I_N \otimes P \end{pmatrix} < 0, \tag{18}$$

which can be decomposed into the following LMI

$$\begin{pmatrix} -\varrho\mu_2 V_0 e^{-\epsilon(t-t_0-\tau)} & e^T(t_m^-)((I_N - cH\delta(m))^T \otimes P) \\ * & -I_N \otimes P \end{pmatrix}$$

$$+ \begin{pmatrix} E^T(t_m^-)(I_N \otimes \sqrt{P}) \\ 0 \end{pmatrix} \begin{pmatrix} 0 & c(H\delta(m))^T \otimes \sqrt{P} \end{pmatrix}$$

$$+ \begin{pmatrix} 0 \\ cH\delta(m) \otimes \sqrt{P} \end{pmatrix} \begin{pmatrix} (I_N \otimes \sqrt{P})E(t_m^-) & 0 \end{pmatrix} < 0.$$

Then for any constant $\omega > 0$, one has

$$\begin{pmatrix} E(t_m^-)(I_N \otimes \sqrt{P}) \\ 0 \end{pmatrix} \begin{pmatrix} 0 & c(H\delta(m))^T \otimes \sqrt{P} \end{pmatrix}$$

$$+ \begin{pmatrix} 0 \\ cH\delta(m) \otimes \sqrt{P} \end{pmatrix} \begin{pmatrix} (I_N \otimes \sqrt{P})E(t_m^-) & 0 \end{pmatrix}$$

$$\leq \omega \begin{pmatrix} E^T(t_m^-)(I_N \otimes \sqrt{P}) \\ 0 \end{pmatrix} \begin{pmatrix} (I_N \otimes \sqrt{P})E(t_m^-) & 0 \end{pmatrix}$$

$$+\omega^{-1} \begin{pmatrix} 0 \\ cH\delta(m) \otimes \sqrt{P} \end{pmatrix} \begin{pmatrix} 0 & c(H\delta(m))^T \otimes \sqrt{P} \end{pmatrix}.$$

Hence we only need to prove the following inequality holds

$$\begin{pmatrix} -\varrho\mu_2 V_0 e^{-\epsilon(t-t_0-\tau)} + \omega E^T(t_m^-)(I_N \otimes P)E(t_m^-) & e^T(t_m^-)(I_N - c(H\delta(m))^T \otimes P) & 0 \\ * & -I_N \otimes P & cH\delta(m) \otimes \sqrt{P} \\ * & * & -\omega I_N \end{pmatrix} < 0. \tag{19}$$

Actually,

$$\begin{pmatrix} -(\varrho - \omega\beta_2^2)\mu_2 V_0 e^{-\epsilon(t-t_0-\tau)} & e^T(t_m^-)(I_N - c(H\delta(m))^T \otimes P) & 0 \\ * & -I_N \otimes P & cH\delta(m) \otimes \sqrt{P} \\ * & * & -\omega I_N \end{pmatrix} < 0. \tag{20}$$

For (17), if ω is small enough, we can from the LMI (20) to obtain (19).

Actually, based on (10), we can conclude that the following inequality holds, where $\Upsilon = diag\{e^T(t_m^-)(I_N \otimes \sqrt{P}), I_N \otimes \sqrt{P}, I_N \otimes I_n\}$.

$$\Upsilon^T \begin{pmatrix} -(\varrho - \omega\beta_2^2)I_N & I_N - c(H\delta(m))^T & 0 \\ * & -I_N & cH\delta(m) \\ * & * & -\omega I_N \end{pmatrix} \Upsilon < 0. \tag{21}$$

It follows from (14) that (19) holds for some small enough ε. Hence, according to the above analysis, we can obtain (13). Furthermore, we have proved that for $t \in [t_0 - \tau, +\infty)$ one has

$$V(e(t)) \leq \mu_2 V_0 e^{-\varepsilon(t-t_0-\tau)}. \tag{22}$$

Therefore

$$e(t) \le \sqrt{\frac{\mu_2 V_0}{\lambda_1(P)}} e^{-\frac{\varepsilon}{2}(t-t_0-\tau)}. \tag{23}$$

The leader-following systems (1) and (2) are exponential consensus with converge rate $\frac{\varepsilon}{2}$.

4   Numerical Example

In this section, we will provide a numerical example to show the feasibility of our results. Consider the leader-following MASs with four agents described by

$$\dot{x}_i(t) = Bx_i(t) + Cf(x_i(t)) + u_i(t), \tag{24}$$

where

$$B = \begin{pmatrix} -1 & 0 & 0 \\ 0 & -1 & 0 \\ 0 & 0 & -1 \end{pmatrix}, C = \begin{pmatrix} 0.25 & -0.2 & -0.2 \\ -0.2 & 0.1 & -0.4 \\ -0.2 & 0.5 & 0.1 \end{pmatrix},$$

with $f(x_i(t)) = (f_1(x_{i1}), f_2(x_{i2}), f_3(x_{i3}))^T$ and $f_i(y) = \frac{1}{2}(|y+1| + |y-1|), i = 1, 2, 3$. The corresponding topology construction is shown as in Fig. 1. Let $c = 4$ and $\tau_k = \tau = 0.02$. By Theorem 1 and the LMI Toolbox, there exist positive-definition matrices P and Ψ satisfying (9)–(11), hence the LFC of systems (24) can be achieved see in Fig. 2.

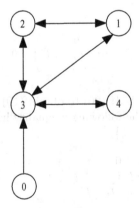

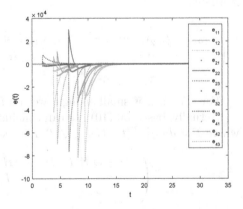

Fig. 1. The Network topology of MASs (24).

Fig. 2. The error state trajectory of leader-following MASs.

5   Conclusion and Further Work

In this paper, we employed a distributed ET impulsive controller to studied the LFC of nonlinear delayed MASs. Using an ET condition to determine the impulsive instants, our results enhanced the accuracy of general impulsive controller. Besides, considered the limited communication ability, time delays are concerned in the impulsive controller. Some sufficient conditions are obtained to assure the LFC for MASs. The effectiveness of our results is shown in a example. This paper used a matrix to consolidate all the systems to a form of integration. However, this method may increase the computation complexity. Hence, finding a efficient method to overcome the computation complexity is indispensable. Another interesting and urgent issue is to design more suitable ET condition to study the consensus problems for MASs.

Acknowledgments. This work was supported by the National Natural Science Foundation of China (Grant No. 61573102), the Natural Science Foundation of Jiangsu Province of China (Grant No. BK20170019), the Jiangsu Provincial Key Laboratory of Networked Collective Intelligence (Grant No. BM2017002), and the Postgraduate Research and Innovation Program of Jiangsu Province (Grant No. KYCX18_0052).

References

1. Ni, W., Cheng, D.: Leader-following consensus of multi-agent systems under fixed and switching topologies. Syst. Contr. Lett. **59**, 209–217 (2010)
2. Zhu, W., Cheng, D.: Leader-following consensus of second-order agents with multiple time-varying delays. Automatica **46**(12), 1994–1999 (2010)
3. Sun, F., Chen, J., Guan, Z.H., Ding, L., Li, T.: Leader-following finite-time consensus for multi-agent systems with jointly-reachable leader. Nonlinear Anal. Real World Appl. **13**(5), 2271–2284 (2012)
4. Song, Q., Liu, F., Cao, J., Yu, W.: M-matrix strategies for pinning-controlled leader-following consensus in multiagent systems with nonlinear dynamics. IEEE Trans. Cybern. **43**(6), 1688–1697 (2013)
5. Hong, Y., Chen, G., Bushnell, L.: Distributed observers design for leader-following control of multi-agent networks. Automatica **44**(3), 846–850 (2008)
6. Li, L., Ho, D.W., Lu, J.: A consensus recovery approach to nonlinear multi-agent system under node failure. Inform. Sci. **367**, 975–989 (2016)
7. Xu, W., Cao, J., Yu, W., Lu, J.: Leader-following consensus of non-linear multi-agent systems with jointly connected topology. IET Control Theor. Appl. **8**(6), 432–440 (2014)
8. Liu, B., Hill, J.: Impulsive consensus for complex dynamical networks with non-identical nodes and coupling time-delays. SIAM J. Contr. Optim. **49**(2), 315–338 (2011)
9. Liu, Z.W., Guan, Z.H., Shen, X., Feng, G.: Consensus of multi-agent networks with aperiodic sampled communication via impulsive algorithms using position-only measurements. IEEE Trans. Autom. Control **57**(10), 2639–2643 (2012)
10. He, W., Chen, G., Han, Q.L., Qian, F.: Network-based leader-following consensus of nonlinear multi-agent systems via distributed impulsive control. Inform. Sci. **380**, 145–158 (2017)

11. Lu, J., Ho, D.W., Cao, J.: A unified synchronization criterion for impulsive dynamical networks. Automatica **46**(7), 1215–1221 (2010)
12. Wang, N., Li, X., Lu, J., Alsaadi, F.E.: Unified synchronization criteria in an array of coupled neural networks with hybrid impulses. Neur. Netw. **101**, 25–32 (2018)
13. Wang, X., Li, C., Huang, T., Pan, X.: Impulsive control and synchronization of nonlinear system with impulse time window. Nonlinear Dyn. **78**(4), 2837–2845 (2014)
14. Donkers, M., Heemels, W.: Output-based event-triggered control with guaranteed $\mathcal{L}_\infty$-gain and improved and decentralized event-triggering. IEEE Trans. Autom. Control **57**(6), 1362–1376 (2012)
15. Li, H., Liao, X., Huang, T., Zhu, W.: Event-triggering sampling based leader-following consensus in second-order multi-agent systems. IEEE Trans. Autom. Control **60**(7), 1998–2003 (2015)
16. Xu, W., Ho, D.W.: Clustered event-triggered consensus analysis: an impulsive framework. IEEE Trans. Ind. Electron. **63**(11), 7133–7143 (2016)
17. Du, W., Sun Leung, S.Y., Tang, Y., Vasilakos, A.V.: Differential evolution with event-triggered impulsive control. IEEE Trans. Cybern. **47**(1), 244–257 (2015)
18. Zhu, W., Wang, D., Liu, L., Feng, G.: Event-based impulsive control of continuous-time dynamic systems and its application to synchronization of memristive neural networks. IEEE Trans. Neur. Netw. Learn. Syst. **29**(8), 3599–3609 (2018)
19. Tan, X., Cao, J., Li, X.: Consensus of leader-following multiagent systems: a distributed event-triggered impulsive control strategy. IEEE Trans. Cybern. **49**(3), 792–801 (2019)

Adaptive Distributed Observer for an Uncertain Leader over Directed Acyclic Graphs

Shimin Wang and Jie Huang$^{(\boxtimes)}$

Department of Mechanical and Automation Engineering,
The Chinese University of Hong Kong, Shatin, N.T., Hong Kong
{smwang,jhuang}@mae.cuhk.edu.hk

Abstract. Recently, an adaptive distributed observer for a class of uncertain leader systems was established over undirected connected graphs. In this paper, we further study the same problem over directed connected graphs. It is shown that, if the graph is acyclic, then an adaptive distributed observer exists in the sense that it can not only estimate the leader's state, but also the unknown parameters of the leader's system matrix exponentially.

Keywords: Distributed observer · Leader-following · Consensus · Uncertain leader · Multi-agent systems

1 Introduction

The distributed observer is an effective approach to dealing with various leader-following control problems of multi-agent systems. A distributed observer for a leader system is a distributed dynamic compensator that can provide for each follower the estimated state of the leader system. The distributed observer was first developed for solving the cooperative output regulation problem for linear multi-agent systems over static networks in [6] and over jointly connected switching networks in [7], respectively. In both [6] and [7], the distributed observer assumes that every follower knows the system matrix of the leader system. This assumption was removed in [1] where a so-called adaptive distributed observer for the leader system was proposed which estimates both the state, and the system matrix of the leader system. A distinct feature of the adaptive distributed observer in [1] is that only the children of the leader need to know system matrix of the leader. Thus, the adaptive distributed observer in [1] is more practical than the distributed observer in [6,7].

This work has been supported in part by the Research Grants Council of the Hong Kong Special Administration Region under grant No. 14201418 and in part by Projects of Major International (Regional) Joint Research Program NSFC (Grant no. 61720106011).

© Springer Nature Switzerland AG 2019
H. Lu et al. (Eds.): ISNN 2019, LNCS 11555, pp. 25–34, 2019.
https://doi.org/10.1007/978-3-030-22808-8_3

In practice, a leader system may contain some unknown parameters. Typically, a linear uncertain leader system takes the following form:

$$\dot{v} = S(\omega)v \tag{1}$$

where $v \in \mathbb{R}^m$ is the state of the leader system and $S(\omega) \in \mathbb{R}^{m \times m}$ relies on some unknown constant vector $\omega \in \mathbb{R}^l$.

Some efforts have been made on designing distributed observer for an uncertain leader system of the form (1) [5,9–11]. Specifically, reference [5] proposed a distributed dynamic compensator for an uncertain leader system and showed that this compensator can estimate the state of (1) asymptotically using off-policy reinforcement learning. However, [5] did not consider the convergence issue of the estimated unknown parameters of the matrix S to the actual value of the unknown parameters of the matrix S. [10] further developed an adaptive distributed observer for (1) that estimated both v and ω using the local information only and showed that the estimated parameters can converge to the actual value of the unknown parameters asymptotically provided the state of the leader system is persistently exciting. More recently, [11] further strengthened the result in [10] by showing that the adaptive distributed observer in [10] can actually estimate the unknown parameter vector ω exponentially. This strengthen result is the key for handling the cooperative output regulation problem for linear multi-agent systems with an uncertain leader system.

A common assumption in [5,9–11] is that the communication graph of the follower subsystems is undirected. This assumption limits the applications of the adaptive distributed observer. In this paper, we will further tackle if the assumption that the communication graph of the follower subsystems is undirected can be removed or can be relaxed. By using a quite different approach from the one in [10], we show that if the graph is connected and acyclic, then the distributed observer in [10,11] can also estimate the leader's state and the unknown parameters of the leader's system matrix exponentially. Thus, our result can at least handle systems whose communication graph does not contain loops.

Notation: $\otimes$ denotes the Kronecker product of matrices. A function $f : [t_0, \infty) \mapsto \mathbb{R}^{n \times m}$ is uniformly bounded if there exists a positive constant c such that

$$\|f(t)\| \leq c, \quad \forall t \geq t_0.$$

For $F_1, \cdots, F_k \in \mathbb{R}^{n \times m}$, let $\mathrm{col}\,(F_1, \cdots, F_k) = \left[F_1^T, \cdots, F_k^T\right]^T$. $\mathbb{1}_n = \mathrm{col}\,(1, \cdots, 1) \in \mathbb{R}^n$, $\forall f \in \mathbb{R}^n$, unless described otherwise, f_i denotes the ith component of f and let diag be such that

$$\mathrm{diag}\,(f) = \begin{bmatrix} f_1 & & \\ & \ddots & \\ & & f_n \end{bmatrix}. \tag{2}$$

For any $f \in \mathbb{R}^m$ with m an even positive integer, the matrix function $\phi(\cdot)$: $\mathbb{R}^m \mapsto \mathbb{R}^{\frac{m}{2} \times m}$ is defined as follows:

$$\phi(f) = \begin{bmatrix} -f_2 \ f_1 & \cdots & 0 & 0 \\ \vdots & \vdots & \ddots & \vdots & \vdots \\ 0 & 0 & \cdots & -f_m \ f_{m-1} \end{bmatrix}. \tag{3}$$

2   Some Technical Lemmas

As in [1], we view a multi-agent system of $N+1$ agents with (1) as the leader and the N subsystems as N followers. The network topology of the multi-agent system is described by a graph $\bar{\mathcal{G}} = (\bar{\mathcal{V}}, \bar{\mathcal{E}})$ with $\bar{\mathcal{V}} = \{0, \cdots, N\}$ and $(i, j) \in \bar{\mathcal{E}}$ if and only if $a_{ji} > 0$. Here the node 0 is associated with the leader system (1) and the node i, $i = 1, \ldots, N$, is associated with the ith follower of the multi agent system. We use $\bar{\mathcal{N}}_i$ to denote the neighbor set of note i. Let $\mathcal{G} = (\mathcal{V}, \mathcal{E})$ be the subgraph of $\bar{\mathcal{G}}$ where $\mathcal{V} = \{1, \cdots, N\}$ and $\mathcal{E}$ is obtained from $\bar{\mathcal{E}}$ by removing those edges of $\bar{\mathcal{E}}$ incident on the agent 0. We call $\mathcal{G}$ the communication graph of the follower subsystems.

We will first list some assumptions as follows.

Assumption 1. $\bar{\mathcal{G}}$ *contains a spanning tree with the node 0 as the root and $\mathcal{G}$ is acyclic graph.*

Since $\mathcal{G}$ is acyclic, by appropriately labeling the nodes of $\bar{\mathcal{G}}$, we can assume that the Laplacian matrix $\mathcal{L}$ of $\mathcal{G}$ is in the lower triangular matrix form. Let $\mathcal{D} = \text{block diag}\,(a_{10}, \cdots, a_{N0})$ and $H = \mathcal{L} + \mathcal{D}$. Then H is also a lower triangular matrix as follows:

$$H = \begin{bmatrix} h_{11} & 0 & \cdots & 0 \\ h_{21} & h_{22} & \cdots & 0 \\ \vdots & \vdots & \ddots & \vdots \\ h_{N1} & h_{N2} & \cdots & h_{NN} \end{bmatrix} \tag{4}$$

where $h_{ii} = \sum\limits_{j \neq i} a_{ij}$, $h_{ij} = -a_{ij}$ for $i \neq j$.

Assumption 2. *The matrix $S(\omega)$ takes the following form:*

$$S(\omega) = \text{diag}\,(\omega) \otimes a \tag{5}$$

where $a = \begin{bmatrix} 0 & 1 \\ -1 & 0 \end{bmatrix}$, $\omega = \text{col}\,(\omega_{01}, \cdots, \omega_{0l}) \in \mathbb{R}^l$, $2l = m$ and $\omega_{0k} > 0$, for $k = 1, \cdots, l$.

Remark 1. *It is assumed in [5, 9–11] that $\mathcal{G}$ is undirected. In Assumption 1, we have replaced this assumption with the assumption that $\mathcal{G}$ is acyclic. Assumption 2 is equivalent to saying that the matrix $S(\omega)$ is nonsingular and all of its eigenvalues are semi-simple with zero real part.*

Let us review some results in [8,11] as follows.

Definition 1 [8]. *A uniformly bounded piecewise continuous function f : $[0,+\infty) \mapsto \mathbb{R}^{n\times m}$ is said to be persistently exciting (PE) if there exist positive constants ϵ, t_0, T_0 such that,*

$$\frac{1}{T_0}\int_t^{t+T_0} f(s)f^T(s)ds \geq \epsilon I_n, \quad \forall t \geq t_0.$$

Lemma 1 [11]. *Suppose $f(t), g(t) : [0,+\infty) \mapsto \mathbb{R}^{n\times m}$ are uniformly bounded, and $\lim_{t\to\infty}(g(t) - f(t)) = 0$. Then, $f(t)$ is persistently exciting if and only $g(t)$ is.*

Lemma 2 [11]. *For any $z \in \mathbb{R}^l$ and $x, y \in \mathbb{R}^m$ with $m = 2l$,*

$$\phi(x)y = -\phi(y)x, \tag{6}$$
$$S(z)x = -\phi^T(x)z, \tag{7}$$

where $S(z) = \text{diag}(z) \otimes a$.

Before stating our main results, we first study the stability of the following system:

$$\dot{x} = (S(\omega) - \mu_2 c I_m)x + c\phi^T(y)\tilde{\theta} + z, \tag{8a}$$
$$\dot{\tilde{\theta}} = -\mu_3\phi(y)x, \tag{8b}$$

where μ_2, μ_3, c are any positive numbers.

Lemma 3. *Consider the system (8). Under Assumption 2, for any $x(0), \tilde{\theta}(0)$,*

1. if $y(t)$ and $z(t)$ are uniformly bounded and $\lim_{t\to\infty} z(t) = 0$ exponentially, then $x(t), \dot{x}(t), \dot{\tilde{\theta}}(t)$ and $\hat{\theta}(t)$ exist and are uniformly bounded,

$$\lim_{t\to\infty} x(t) = 0 \quad and \quad \lim_{t\to\infty} \dot{\tilde{\theta}}(t) = 0;$$

2. if $\lim_{t\to\infty} z(t) = 0$ exponentially, $y(t), \dot{z}(t)$ and $\dot{y}(t)$ are uniformly bounded and $\phi(y(t))$ is persistently exciting, then system (8) is exponentially stable.

Proof. (i) Consider the following function:

$$V = 0.5(x^T x + c\mu_3^{-1}\tilde{\theta}^T\tilde{\theta}). \tag{9}$$

Differentiating V along the trajectory of (8) gives

$$\dot{V} = x^T\dot{x} + c\mu_3^{-1}\tilde{\theta}^T\dot{\tilde{\theta}}$$
$$= x^T(S(\omega) - \mu_2 c I_m)x + cx^T\phi^T(y)\tilde{\theta}$$
$$+ x^T z + c\mu_3^{-1}\tilde{\theta}^T\dot{\tilde{\theta}}. \tag{10}$$

Using the fact that $S(\omega)$ is skew symmetric gives

$$
\begin{aligned}
\dot{V} &= -c\mu_2 x^T x + x^T z + c\tilde{\theta}^T \phi(y)x + c\mu_3^{-1}\tilde{\theta}^T \dot{\tilde{\theta}} \\
&= -c\mu_2 x^T x + x^T z + c\tilde{\theta}^T \left(\phi(y)x + \mu_3^{-1}\dot{\tilde{\theta}}\right) \\
&= -c\mu_2 x^T x + x^T z \\
&\leq -0.5c\mu_2\|x\|^2 + 0.25\varepsilon^{-1}\|z\|^2,
\end{aligned}
\tag{11}
$$

with $0 \leq \varepsilon \leq 0.5c\mu_2$. Thus, we have

$$
\dot{V} \leq 0.25\varepsilon^{-1}\|z\|^2.
\tag{12}
$$

Therefore, for any $t \geq 0$, $V(t) - V(0) \leq \int_0^t 0.25\varepsilon^{-1}\|z(\tau)\|^2 d\tau$. As $\lim_{t\to\infty} z(t) = 0$ exponentially, $\int_0^\infty \|z(\tau)\|^2 d\tau$ is bounded. Thus $V(t)$ is bounded over $t > 0$, which concludes that x and $\tilde{\theta}$ are uniformly bounded. Since y and z are uniformly bounded, from Eq. (8), $\dot{x}$ and $\dot{\tilde{\theta}}$ are uniformly bounded.

Let $W(t) = V(t) - \int_0^t x^T z d\tau$. As $\lim_{t\to\infty} z(t) = 0$ exponentially, we have $\int_0^\infty x^T z d\tau$ is bounded. Differentiating W along the trajectory (8) and using (11) gives

$$
\dot{W} = \dot{V} - x^T z = -c\mu_2 x^T x \leq 0.
\tag{13}
$$

Since $\ddot{W} = -2c\mu_2 x^T \dot{x}$, $\dot{W}(t)$ is uniformly continuous. By Barbalat's Lemma, we have $\lim_{t\to\infty} \dot{W}(t) = 0$, which implies $\lim_{t\to\infty} x(t) = 0$ from Eq. (13). From Eq. (8b), we have $\lim_{t\to\infty} \dot{\tilde{\theta}}(t) = 0$.

(ii) We first consider system (8) with $z = 0$, which can be rewritten as the following linear time varying system:

$$
\dot{Y} = B(t)Y,
\tag{14}
$$

where

$$
Y = \begin{bmatrix} x \\ \tilde{\theta} \end{bmatrix}, \quad B(t) = \begin{bmatrix} (S(\omega) - \mu_2 cI_m) & c\phi^T(y) \\ -\mu_3\phi(y) & 0 \end{bmatrix}.
$$

The rest of the proof is based on Lemma B.2.3 of [3], which can be found as Lemma 6 in the Appendix of this paper. Note that (14) is in the form (28) with

$$
A = (S(\omega) - \mu_2 cI_m), \quad P = I_m, \quad \Lambda = \mu_3 c^{-1}I_m, \quad \Omega(t) = c\phi(y(t)).
$$

We will verify that (14) satisfies all conditions of Lemma 6. Clearly Λ and P are positive definite matrices. Since all the eigenvalues of $S(\omega)$ have zero real parts, A is Hurwitz for any $\mu_2 > 0$. Next, it can be verified that $PA + A^T P = -2\mu_2 cI_m$. Since $y(t)$, $\dot{y}(t)$ are uniformly bounded and $\phi(y(t))$ is persistently exciting. Equation (14) satisfies all conditions of Lemma 6. Thus, system (8) with $z = 0$ is exponentially stable. Next note that system (8) can be written into the following form

$$\dot{Y} = B(t)Y + \begin{bmatrix} z \\ 0 \end{bmatrix}. \tag{15}$$

Let $\Phi(t, \tau)$ be the transition matrix of (14). Then, for any t_0, the solution of (15) satisfies

$$Y(t) = \Phi(t, t_0)Y(t_0) + \int_{t_0}^{t} \Phi(t, \tau) \begin{bmatrix} z(\tau) \\ 0 \end{bmatrix} d\tau \tag{16}$$

Since (14) is exponentially stable and $\lim_{t \to \infty} z(t) = 0$ exponentially, there exist some positive constants α_i and λ_i, $i = 1, 2$, such that, for any $t \geq \tau \geq 0$, $\|\Phi(t, \tau)\| \leq \alpha_1 e^{-\lambda_1(t-\tau)}$, and $\|z(t)\| \leq \alpha_2 e^{-\lambda_2(t-\tau)} \|z(\tau)\|$. Without loss of generality, we assume that $\lambda_1 \neq \lambda_2$. Simple calculation shows

$$\|Y(t)\| \leq \alpha_1 \|Y(t_0)\| e^{-\lambda_1(t-t_0)} + \frac{\|z(t_0)\| \alpha_1 \alpha_2}{\lambda_1 - \lambda_2} \left(e^{-\lambda_2(t-t_0)} - e^{-\lambda_1(t-t_0)} \right) \tag{17}$$

Thus, $\lim_{t \to \infty} Y(t) = 0$ exponentially.

3   Adaptive Distributed Observer

In what follows, we will remove the assumption that the subgraph $\mathcal{G}$ is undirected at the cost of assuming that the graph $\bar{\mathcal{G}}$ is acyclic. We first recall the distributed observer proposed in [10] as follows:

$$\dot{\xi}_i = S(\omega_i)\xi_i + \mu_2 \sum_{j=0}^{N} a_{ij}(\xi_j - \xi_i) \tag{18a}$$

$$\dot{\omega}_i = \mu_3 \phi \left(\sum_{j=0}^{N} a_{ij}(\xi_j - \xi_i) \right) \xi_i \tag{18b}$$

where, for $i = 1, \cdots, N$, $\xi_i \in \mathbb{R}^m$, $\xi_0 = v$, μ_2 and μ_3 are some positive numbers, $\omega_i \in \mathbb{R}^l$ is the estimator of ω, and $S(\omega_i) = (\text{diag}(\omega_i) \otimes a)$.

Remark 2. *It was shown in Lemma 4 of [11] that, under Assumption 2 and the assumption that the subgraph $\mathcal{G}$ is undirected and contains a spanning tree with node 0 as the root, the solution of (18) is such that*

$$\lim_{t \to \infty} (\xi_i(t) - v(t)) = 0 \ and \ \lim_{t \to \infty} (\omega_i(t) - \omega) = 0, \ i = 1, \cdots, N.$$

For $i = 1, \cdots, N$, let, $e_{vi} = \sum_{j=0}^{N} a_{ij}(\xi_j - \xi_i)$, $\tilde{\omega}_i = (\hat{\omega}_i - \omega)$, $\hat{v} = 1_N \otimes v$, $e_v = \text{col}(e_{v1}, \cdots, e_{vN})$, $\xi = \text{col}(\xi_1, \cdots, \xi_N)$, $\hat{\omega} = \text{col}(\omega_1, \cdots, \omega_N)$, $\tilde{\xi} =$

$(\xi - \hat{v})$, $S_d(\hat{\omega}) = \text{block diag}\,(S(\omega_1), \cdots, S(\omega_N))$. Then, we have the following relation:

$$e_v = -(H \otimes I_m)(\xi - \hat{v}).$$

System (18) can be put into the following compact system

$$\dot{\xi} = S_d(\hat{\omega})\xi - \mu_2(H \otimes I_m)(\xi - \hat{v}), \qquad (19a)$$

$$\dot{\hat{\omega}} = \mu_3 \phi_d(e_v)\xi, \qquad (19b)$$

where $\phi_d(e_v) = \text{block diag}\,(\phi(e_{v1}), \cdots, \phi(e_{vN}))$.

Under Assumption 1, by Theorem 2.5.3 of [4], there exists a positive definite diagonal matrix $D = \text{diag}\,(d_1, \cdots, d_N)$, such that $\bar{H} = DH + H^T D$ is positive definite. Let λ_h denote the minimum eigenvalue of $\bar{H}$ and $d_M = \max\{d_i, i = 1, \cdots, N\}$.

Lemma 4. *Consider systems (1) and (19a). Under Assumptions 1 and 2, for any $\xi(0)$ and $\omega(0)$, $\mu_2, \mu_3 > 0$, $\xi(t)$ and $\dot{\xi}(t)$ are uniformly bounded.*

Proof. We first consider Eq. (19a) with $v = 0$:

$$\dot{\xi} = A(t)\xi, \qquad (20)$$

where $A(t) = (S_d(\hat{\omega}) - \mu_2(H \otimes I_m))$. Define the following Lyapunov function for Eq. (20),

$$X_0 = \xi^T(D \otimes I_m)\xi \leq d_M\|\xi\|^2.$$

The time derivative of X_0 along the trajectory of (20) is

$$\begin{aligned}
\dot{X}_0 &= 2\xi^T(D \otimes I_m)\dot{\xi} \\
&= 2\xi^T(D \otimes I_m)S_d(\hat{\omega})\xi - 2\mu_2\xi^T(DH \otimes I_m)\xi \\
&= \sum_{i=1}^{N} 2d_i\xi_i^T S(\omega_i)\xi_i - \mu_2\xi^T((DH + H^T D) \otimes I_m)\xi.
\end{aligned}$$

Since, for, $i = 1, \cdots, N$, $S(\omega_i)$ is skew symmetric,

$$\dot{X}_0 = -\mu_2\xi^T(\bar{H} \otimes I_m)\xi \leq -\mu_2\lambda_h d_M^{-1} X_0. \qquad (21)$$

Thus, the origin of the system (20) is exponentially stable. Under Assumption 2, v is uniformly bounded for all $t \geq 0$. By Example 2.11 in [2], system (19a) is input-to-state stable with ξ as the state and $\mu_2\xi^T(\bar{H} \otimes I_m)\hat{v}$ as the input. Since $\hat{v}$ is uniformly bounded for all $t \geq 0$, ξ is also uniformly bounded for all $t \geq 0$.

Now differentiating (19a) gives

$$\ddot{\xi} = A(t)\dot{\xi} + S_d(\dot{\hat{\omega}})\xi + \mu_2(H \otimes I_m)\dot{\hat{v}}. \qquad (22)$$

Since system $\ddot{\xi} = A(t)\dot{\xi}$ is exponentially stable with $\dot{\xi}$ as the state, again, by Example 2.11 in [2], system (22) is input-to-state stable with $\dot{\xi}$ as the state and

$$S_d(\dot{\hat{\omega}})\xi + \mu_2(H \otimes I_m)\dot{\hat{v}}$$

as the input. Under Assumption 2, $\dot{v}$ is uniformly bounded. We have proved that ξ is uniformly bounded. From Eq. (19b), we have $\dot{\tilde{\omega}}$ is uniformly bounded. Thus, $S_d(\dot{\tilde{\omega}})\xi + \mu_2 (H \otimes I_m) \dot{\hat{v}}$ is uniformly bounded. Hence $\dot{\xi}$ is also uniformly bounded.

Under Assumption 1, the matrix H is in the form (4). Thus,

$$e_{vi} = \sum_{j=0}^{i-1} a_{ij}(\xi_j - \xi_i), i = 1, \cdots, N. \tag{23}$$

Equations (18) and (23) give

$$\dot{e}_{vi} = (S(\omega) - \mu_2 h_i I_m)e_{vi} - h_i S(\tilde{\omega}_i) \xi_i + m_i \tag{24a}$$

$$\dot{\tilde{\omega}}_i = \mu_3 \phi(e_{vi})\xi_i, \quad i = 1, \cdots, N, \tag{24b}$$

where $m_i = \sum_{j=0}^{i-1} a_{ij} (\mu_2 e_{vj} + S (\tilde{\omega}_j) \xi_j)$ and $h_i = \sum_{j=0}^{i-1} a_{ij}$.

By Lemma 2, we can rewrite Eq. (18) into the following form:

$$\dot{e}_{vi} = (S(\omega) - \mu_2 h_i I_m)e_{vi} + h_i \phi^T (\xi_i) \tilde{\omega}_i + m_i, \tag{25a}$$

$$\dot{\tilde{\omega}}_i = -\mu_3 \phi(\xi_i)e_{vi}, \quad i = 1, \cdots, N. \tag{25b}$$

Lemma 5. *Consider systems* (1) *and* (25). *Under Assumptions 1 and 2, suppose, for all* $k = 1, \cdots, l$, col $(v_{2k-1}(0), v_{2k}(0)) \neq 0$. *Then, for all* $\xi_i(0), \omega_i(0)$ *and* $\mu_2, \mu_3 > 0$, $\xi_i(t), \dot{\xi}_i(t)$ *and* $\omega_i(t), \dot{\omega}_i(t)$ *exist and are uniformly bounded and satisfy* $\lim_{t\to\infty} \dot{\xi}_i(t) = 0$ *and* $\lim_{t\to\infty} \tilde{\omega}_i(t) = 0$, *exponentially.*

Proof. When $i = 1$, Eq. (25) takes the following form:

$$\dot{e}_{v1} = (S(\omega) - \mu_2 h_1 I_m)e_{v1} + h_1 \phi^T (\xi_1) \tilde{\omega}_1, \tag{26a}$$

$$\dot{\tilde{\omega}}_1 = -\mu_3 \phi(\xi_1)e_{v1}, \tag{26b}$$

where $h_1 = a_{10}$. System (26) can be viewed as a special case of (8) with $z = 0$. By Lemma 4, ξ_1 is uniformly bounded for all $t > 0$. System (26) satisfies all the conditions in Case 1 of Lemma 3, we have $\lim_{t\to\infty} \dot{\tilde{\omega}}_1(t) = 0$ and $\lim_{t\to\infty} e_{v1}(t) = 0$ which implies $\lim_{t\to\infty} \dot{\tilde{\xi}}_1(t) = 0$. Thus, $\lim_{t\to\infty} (\phi (\xi_1(t)) - \phi (v(t))) = \lim_{t\to\infty} \phi(\tilde{\xi}_1(t)) = 0$.

Since, for all $k = 1, \cdots, l$, col $(v_{2k-1}(0), v_{2k}(0)) \neq 0$, under Assumptions 2, from Lemma 4 of [11], we have $\phi(v)$ is persistently exciting. By Lemma 1, $\phi(\xi_1)$ is persistently exciting. From Lemma 4, $\dot{\xi}_1$ is uniformly bounded. Since system (26) satisfies all conditions in Case 2 of Lemma 3 with $z = 0$, both

$$\lim_{t\to\infty} e_{v1}(t) = 0 \quad \text{and} \quad \lim_{t\to\infty} \tilde{\omega}_1(t) = 0$$

exponentially. Hence $\lim_{t\to\infty} \tilde{\xi}_1(t) = 0$ and $\lim_{t\to\infty} \tilde{\omega}_1(t) = 0$ exponentially.

Now suppose, for some n satisfying $1 \leq n-1 < N$, and, for all $i = 1, \cdots, n-1$,

$$\lim_{t \to \infty} \tilde{\omega}_i(t) = 0 \quad \text{and} \quad \lim_{t \to \infty} \tilde{\xi}_i(t) = 0$$

exponentially and $\dot{\tilde{\omega}}_i(t)$ is uniformly bounded. We prove

$$\lim_{t \to \infty} \tilde{\omega}_n(t) = 0 \quad \text{and} \quad \lim_{t \to \infty} \tilde{\xi}_n(t) = 0$$

exponentially and $\dot{\tilde{\omega}}_n(t)$ is uniformly bounded for all $t \geq 0$. From Eq. (25), we have the following equation

$$\dot{e}_{vn} = (S(\omega) - \mu_2 h_n I_m)e_{vn} + h_n \phi^T(\xi_n)\tilde{\omega}_n + m_n, \tag{27a}$$

$$\dot{\tilde{\omega}}_n = -\mu_3 \phi(\xi_n)e_{vn}. \tag{27b}$$

For $j = 1, \cdots, n-1$, $\lim_{t \to \infty} \tilde{\omega}_j(t) = 0$ and $\lim_{t \to \infty} \tilde{\xi}_j(t) = 0$ exponentially. Then, we have $\lim_{t \to \infty} m_n(t) = 0$ exponentially. Since, for $j = 1, \cdots, n-1$, $\tilde{\omega}_j(t)$ and $\dot{\tilde{\omega}}_j(t)$ are uniformly bounded for all $t \geq 0$ and, by Lemma 4, $\tilde{\xi}_j, \dot{\tilde{\xi}}_j$ are uniformly bounded for all $t > 0$. Also,

$$\dot{m}_n = \sum_{j=0}^{n-1} a_{nj}\left(\mu_2 \dot{e}_{vj} + S(\dot{\tilde{\omega}}_j)\tilde{\xi}_j + S(\tilde{\omega}_j)\dot{\tilde{\xi}}_j\right).$$

Then, we have $\dot{m}_n$ is uniformly bounded for all $t \geq 0$. Since system (27) satisfies all the conditions in Case 1 of Lemma 3, we have

$$\lim_{t \to \infty} \dot{\tilde{\omega}}_n(t) = 0 \quad \text{and} \quad \lim_{t \to \infty} e_{vn}(t) = 0.$$

Since

$$e_{vn} = \sum_{j=0}^{n-1} a_{nj}(\tilde{\xi}_j - \tilde{\xi}_n) = \sum_{j=0}^{n-1} a_{nj}\tilde{\xi}_j - h_n\tilde{\xi}_n$$

and, for $j = 1, \cdots, n-1$, $\lim_{t \to \infty} \tilde{\xi}_j(t) = 0$, we have $\lim_{t \to \infty} \tilde{\xi}_n(t) = 0$. Thus,

$$\lim_{t \to \infty} \left(\phi(\xi_n(t)) - \phi(v(t))\right) = \lim_{t \to \infty} \phi(\tilde{\xi}_n(t)) = 0.$$

Since, for all $k = 1, \cdots, l$, $\mathrm{col}\,(v_{2k-1}(0), v_{2k}(0)) \neq 0$, under Assumption 2, $\phi(v)$ is persistently exciting by Lemma 4 of [11]. Thus, by Lemma 1, $\phi(\xi_n)$ is persistently exciting. By Lemma 4, $\dot{\xi}_n$ is uniformly bounded for all $t > 0$. System (27) satisfies all the conditions in Case 2 of Lemma 3. Thus, both $\lim_{t \to \infty} \tilde{\xi}_n(t) = 0$ and $\lim_{t \to \infty} \tilde{\omega}_n(t) = 0$ exponentially.

4   Conclusion

In this paper, we have considered establishing adaptive distributed observer for an uncertain leader system over directed graphs. We have shown that such an observer exists if the graph is connected and acyclic. A natural future work is to further eliminate the assumption that the graph is acyclic.

Appendix

The following lemma is rephrased from Lemma B.2.3 of [3].

Lemma 6. *Consider the following linear time-varying system*

$$\dot{x} = Ax + \Omega^T(t)z, \quad x \in \mathbb{R}^n \tag{28a}$$

$$\dot{z} = -\Lambda\Omega(t)Px, \quad z \in \mathbb{R}^p \tag{28b}$$

where $A \in \mathbb{R}^{n \times n}$ is Hurwitz, $P \in \mathbb{R}^{n \times n}$ is a symmetric positive definite matrix satisfying $A^T P + PA = -Q$ with Q some symmetric positive definite matrix, and $\Lambda \in \mathbb{R}^{n \times n}$ is some symmetric positive definite matrix. If $\|\Omega(t)\|$ and $\|\dot{\Omega}(t)\|$ are uniformly bounded and $\Omega(t)$ is persistently exciting, then, the origin of system (28) is exponentially stable.

References

1. Cai, H., Huang, J.: The leader-following consensus for multiple uncertain Euler-Lagrange systems with an adaptive distributed observer. IEEE Trans. Autom. Control **61**(10), 3152–3157 (2016)
2. Chen, Z., Huang, J.: Stabilization and Regulation of Nonlinear Systems. Springer, Cham (2015). https://doi.org/10.1007/978-3-319-08834-1
3. Marino, R., Tomei, P.: Nonlinear Control Design: Geometric, Adaptive and Robust. Prentice Hall International (UK) Ltd., Upper Saddle River (1996)
4. Merino, D.I.: Topics in matrix analysis. Ph.D. dissertation, Johns Hopkins University (1992)
5. Modares, H., Nageshrao, S.P., Babuška, R., Lopes, G.A.D., Lewis, F.L.: Optimal model-free output synchronization of heterogeneous systems using off-policy reinforcement learning. Automatica **71**, 334–341 (2016)
6. Su, Y., Huang, J.: Cooperatve output regulation of linear multi-agent systems. IEEE Trans. Autom. Control **57**(4), 1062–1066 (2012)
7. Su, Y., Huang, J.: Cooperative output regulation with application to multi-agent consensus under switching network. IEEE Trans. Syst. Man Cybern. Part B Cybern. **42**(3), 864–875 (2012)
8. Slotine, J.J.E., Li, W.: Applied Nonlinear Control, vol. 199. Prentice Hall, Englewood Cliffs (1991)
9. Wu, Y., Lu, R., Shi, P., Su, H., Wu, Z.G.: Adaptive output synchronization of heterogeneous network with an uncertain leader. Automatica **76**, 183–192 (2017)
10. Wang, S., Huang, J.: Adaptive leader-following consensus for multiple Euler-Lagrange systems with an uncertain leader. IEEE Trans. Neural Netw. Learn. Syst. (2018). https://doi.org/10.1109/TNNLS.2018.2878463
11. Wang, S., Huang, J.: Cooperative output regulation of linear multi-agent systems subject to an uncertain leader system. In: 2018 15th IEEE International Conference on Information and Automation, 11–13 August, Fujian, China, pp. 57–62 (2018)

Regularization in DQN for Parameter-Varying Control Learning Tasks

Dazi Li[1(✉)], Chengjia Lei[1], Qibing Jin[1], and Min Han[2]

[1] College of Information Science and Technology,
Beijing University of Chemical Technology, Beijing 100029, China
lidz@mail.buct.edu.cn
[2] Dalian University of Technology, Dalian, China
minhan@dlut.edu.cn

Abstract. As an important technique of preventing overfitting, regularization is widely used in supervised learning. However, regularization has not been systematically studied in deep reinforcement learning (deep RL). In this paper, we study the generalization of deep Q-network (DQN), applying with mainstream regularization approaches, including l_1, l_2 and dropout. We pay attention on agent's performance not only in original environments, but also in parameter-varying environments which are variational but the same task type. Furthermore, the dropout is modified to make it more adaptive to DQN. Then, a new dropout is proposed to speed up the optimization of DQN. Experiments show that regularization helps deep RL achieve better performance in both original and parameter-varying environments when the number of samples is insufficient.

Keywords: Regularization · Deep RL · Control learning task

1 Introduction

Although neural networks are lauded for generalization capabilities [1], which means being able to approximate arbitrary functions, they are often troubled by overfitting problems, especially in deep networks [2]. In supervised learning, regularization, as a reliable technique preventing overfitting, has been widely practiced in the optimization of deep networks. Main idea of regularization is to control the solution's complexity in a large function space. A relatively simple model is smoother and less affected by the noise generated by uneven sample distribution than common ones [3]. Thus, in the classification task, regularization can impressively improve the generalization ability of the model for unseen samples [4].

As a general learning algorithm, deep reinforcement learning (deep RL) has made remarkable achievements on complex decision problems [5]. An RL agent entirely depends on experience-driven, obtaining an optimal policy through trial-and-error method in the process of continuous interaction with environments [6]. Regularization technique was also utilized to get a smooth policy network to improve generalization ability in new episodes [7]. At present, the mainstream RL evaluation method is to use the trained policy network to verify its performance in the original environment. However, if the agent just remembers the order of actions instead of learning decision

H. Lu et al. (Eds.): ISNN 2019, LNCS 11555, pp. 35–44, 2019.
https://doi.org/10.1007/978-3-030-22808-8_4

knowledge, it is also possible to get a good performance [8]. Thus, we are interested at whether the knowledge learned by deep RL can be generalized on parameter-varying environments.

In this paper, we investigate the generalization ability of deep RL and the impact of applying regularization with deep RL for parameter-varying control learning tasks. A parameter-varying task is training an agent in a default environment but evaluating the policy in variational environments. First, we employ l_1 regularization, l_2 regularization and dropout [9] in deep Q-network (DQN) [10] and train it in default environments. Cartpole task and mountain car task are chosen as our experimental environments and the pole length in cartpole task and the car weight in mountain car task are changed to create variational environments. Then we evaluate the learned policy in variational environments. In traditional control field, this kind of generalization ability can also be regarded as robustness. We show that most regularization approaches have an ability to help deep RL achieve better control performance and better generalization with limited learning samples. Besides, we adapt the dropout approach to DQN which makes it work well as other regularization approaches. Finally, a new dropout approach is proposed that enable the network work better and stable. Experimental results can be seen as a novel aspect for generalization in deep RL.

2 Background

2.1 Reinforcement Learning

RL is a universal algorithm that the agent aims to select good actions for maximizing accumulated rewards. The RL problem can be formalized as the Markov decision process (MDP), defined by a 5-tuple $\langle S, A, P, R, \gamma \rangle$, with the state space S, the action space A, the state-transition function p, the reward function R and the discounted factor γ [10]. At every time step t, the agent receives the state information $s_t \in S$ and then selects actions $a_t \in A$ that need to be executed according to the policy π. After that, the environment transfers to the next state $s_{t+1} \in S$ described by the state-transition function P and feeds reward signal r_t, generated by reward function R, to the agent. The goal of agent is to find an optimal policy π^* which enable agent to maximize long-term expected return $G_t \doteq \sum_{k=0}^{\infty} \gamma^k r_{t+k}$ with $\gamma \in [0, 1)$. γ determines the considered timespan or planning horizon [11].

DQN [12] is one of the most successful deep RL algorithm. DQN uses a multi-layer neural network to approximate the state-action pair value function, denoted as function $Q(s, a; \theta)$, where θ are weights of the network. To make a correct evaluation on state-action pairs, an object designed to be minimized as follows:

$$L^{DQN} = E_{\{s_t, a_t, r_t, s_{t+1}\} \sim U(\cdot)} \left[(r_t + \max_{a' \in A} Q(S_{t+1}, a'; \theta^-) - Q(s_t, a_t; \theta))^2 \right], \quad (1)$$

where $\{s_t, a_t, r_t, s_{t+1}\}$ are independently distributed samples, randomly being sampled from $U(\cdot)$, a fixed-capacity experience replay buffer. θ^- are target network parameters that copy the current network parameters according to a certain frequency k. There are two key techniques to ensure that DQN performs outstandingly and works reasonably.

One is experience replay, overcoming the problem caused by correlated data and non-stationary distribution. The other one is target network, which results in the well-defined optimization problem similar to supervised learning.

2.2 Regularization Approaches

In supervised learning problem, regularization is very important for improving generalization capability of networks. Suppose that we want to optimize an input-to-output parameterized mapping function $\hat{y}_i = f(X_i; \theta)$, where X_i denotes input data, $\hat{y}_i$ denotes predicted output, $f(\cdot)$ denotes mapping function and θ denotes parameters of the mapping function network. Overfitting problem makes the mapping function perform very badly on unseen data sets [13]. The main idea of regularization is to perform a series of operations on θ, so that the network model approaches smoothing and suppresses overfitting. Three popular regularization approaches are l_1 regularization, l_2 regularization and dropout.

l_1 and l_2 regularization achieve the purpose of improving generalization capability by adding a penalty term for network parameters to the optimization objective. As for l_1 regularization, the objective is:

$$\min_\theta \frac{1}{n} \sum_{i=1}^{n} L(y_i, \hat{y}_i) + \frac{\lambda_{l_1}}{2} ||\theta||_1, \tag{2}$$

where $L(\cdot)$ is a loss function which is differential and measures the distance between the actual output y_i and predicted output $\hat{y}_i$. The second term is the form of l_1 regularization and parameter λ_{l_1} is the penalty coefficient for l_1 regularization. As for l_2 regularization, the objective is:

$$\min_\theta \frac{1}{n} \sum_{i=1}^{n} L(y_i, \hat{y}_i) + \frac{\lambda_{l_2}}{2} ||\theta||_2^2, \tag{3}$$

where λ_{l_2} is the penalty coefficient for l_1 regularization and the second term is the form for l_2 regularization.

Dropout is based on the idea of monte-carlo model averaging and is a very efficient way of performing model averaging with neural networks. When using dropout to optimize network parameters, there is a certain probability $p_{dropout} \in [0, 1)$ randomly dropping some hidden units out during forward propagation, which makes the current network thinner than the original network. Researchers found that dropout regularization can be interpreted as a data augmentation approach in the input space without domain knowledge [14], which makes each hidden unit more robust and prevents them from co-adapting too much [15].

3 Regularization in DQN

In this paper, we explore the impact of regularization on DQN for parameter-varying control learning tasks, focusing on the following two aspects:

- The impact of different regularization approaches on the DQN in the original environment.
- The generalization capabilities of DQN with different regularization approaches in variational environments.

For general deep RL problems, policy evaluation is performed in the environment whose parameters of the environmental model is the same as that of training. Notice that in this case, the state-transition function P stays unchanged. Here, we pay attention to a new generalization perspective in deep RL. That is whether an optimized policy can generalize to parameter-varying environments that are slightly different from the original environments but do not affect the essence of the task. If we use MDP to describe this situation, the new state-transition function P' in the parameter-varying environment is slightly different from the state-transition function P in the original environment.

In DQN, we need to evaluate every state-action pair $Q(s_t, a_t; \theta)$ and choose the action with the max state-action value. As for an optimized DQN, the Bellman optimality equation for $Q_*(s_t, a_t; \theta)$ is

$$
\begin{aligned}
Q_*(s_t, a_t; \theta) &= E[r_t + \gamma \max_{a'} Q_*(s_{t+1}, a_t; \theta)] \\
&= \sum_{s_{t+1}, r_t} P(s_{t+1}, r_t | s_t, a_t)[r_t + \gamma \max_{a'} Q(s_{t+1}, a'; \theta)],
\end{aligned}
\tag{4}
$$

where $Q_*(\cdot)$ denotes the state-action pair value function under an optimal policy and $P(s_{t+1}, r_t | s_t, a_t)$ denotes the state-transition function that transfer the state s_t to the state s_{t+1}. In parameter-varying environments, $P(s_{t+1}, r_t | s_t, a_t)$ becomes $P'(s_{t+1}, r_t | s_t, a_t)$, which impact the evaluation for the optimal $Q(s_t, a_t; \theta)$. If the state action pair can be evaluated correctly no matter how the state transition function P changed, we think that the current network has a good generalization capability.

When applying l_1 regularization with DQN, we get the loss function as

$$
Loss = \frac{1}{n} \sum_{i=1}^{n} L(r_t + \max_{a' \in A} Q(s_{t+1}, a'; \theta^-), Q(s_t, a_t; \theta)) + \frac{\lambda_{l_1}}{2} ||\theta||_1,
\tag{5}
$$

where n is the size of a mini-batch.

When applying l_2 regularization with DQN, we get the loss function as

$$
Loss = \frac{1}{n} \sum_{i=1}^{n} L(r_t + \max_{a' \in A} Q(s_{t+1}, a'; \theta^-), Q(s_t, a_t; \theta)) + \frac{\lambda_{l_2}}{2} ||\theta||_2^2.
\tag{6}
$$

In supervised learning, dropout happens during every time when the network forward propagates to get predicted outputs. However, in DQN, there are three processes that perform forward propagation. First, the maximum Q value is calculated for the current state in order to choose a good action. The second is to use the target network to evaluate samples in mini-batch. The third is to use the current network to evaluate samples in mini-batch. We found that the dropout rate $p_{dropout}$ is very sensitive in DQN. If dropout is added in all three processes, the network optimization process will become extremely fragile. Therefore, we improve dropout approach to adapt itself to DQN. Unlike the dropout in supervised learning, the improved dropout approach only occurs when the current network evaluates samples from the mini-batch, while it does not occur at other times.

Base on the improved dropout, aiming at accelerating the optimization process and reducing fluctuating in performance during training, a new dropout approach is proposed. The main idea is that output of the subnet after random dropout should be consistent with the output of the original network. Therefore, the squared difference between subnet output and original network output is add to the loss function so we can get the loss function as

$$Loss = \frac{1}{n} \sum_{i=1}^{n} \left[L(r_t + \max_{a' \in A} Q(s_{t+1}, a'; \theta^-), Q(s_t, a_t; \theta)) + L(y_{sub}, y_{origin}) \right], \quad (7)$$

where y_{sub} denotes the output of the subnet obtained by randomly dropping unit out and y_{origin} denotes the output of the original network.

4 Problem Description and Evaluation Protocol

In control problems, there are two basic types of task specification [16]:

1. Avoidance problem. The goal is trying to keep the controlled object from falling into certain states. One of these typical problems is pole balancing.
2. Reaching goals problem. The goal is trying to reach the target state from the initial state as soon as possible. One of these typical problems is mountain car.

In this paper, we implement deep RL on these two control learning tasks, the cartpole task and the mountain-car task, which are supported by the OpenAI gym environment [17]. OpenAI gym environment is a popular testbed in deep RL domain. In order to explore the two aspects of regularization in deep RL, we modify one of the environmental parameters of each control task to generate a series of variational environments with the same task objectives but slightly different state-transition function P'. In the cartpole task, the length of the pole can be changed, which allows us to create a series of slightly different environments and test the performance under these conditions. In the mountain-car task, the weight of the car is also allowed to be changed for the same purpose.

In order to observe the role of regularization in deep RL, we redesign the experiment. First, set a default value for the adjustable parameter x. In cartpole task, the pole length x is set to 1.0 by default, and in mountain car task, the car weight x is set to 0.2 kg by default. After that, the agent optimizes in the default environment and ends up training after different time steps, which limits the sample size that agents have learned. If we finish learning after 10 thousand steps, it means that the agent has experienced a maximum of 10,000 * n samples and n is the size of a mini-batch.

Finally, we change the adjustable parameter in the environment and test the policy we obtained after different time steps to focus on the performance in parameter-varying environments.

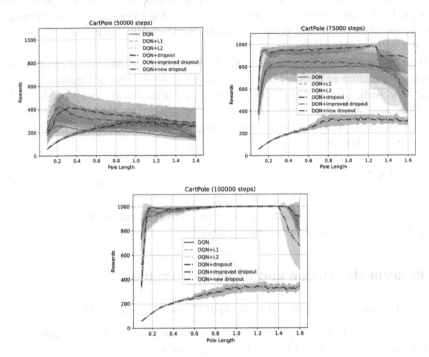

Fig. 1. Performance of agents trained with default pole length (1.0 unit) and evaluated the policy learned after 50 thousand steps, 75 thousand steps and 100 thousand steps with the pole length changing from 0.1 unit to 1.6 units. Notice that for each picture we train from the beginning. The x-axis is the pole length and the y-axis are the accumulated rewards received by the agent. The solid line is the mean value and the shadow is the standard deviation and each line is averaged over ten seeds. In each seed, the agent performs 100 times evaluations with the pole length and calculates the average value of these 100 times evaluations as the performance.

5   Experimental Results and Empirical Analysis

In cartpole task and mountain car task, state information is represented by several values. Thus, we construct a DQN with three hidden layers and there are 256 units on each hidden layer. The first two hidden layers are followed by rectified linear unit (ReLU) [18] layers, and the last hidden layer is the linear fully connected layer. We apply regularization approaches including l_1, l_2, dropout, improved dropout and new dropout proposed with DQN. In details, we separately apply these regularization approaches on all hidden layers in the network. Grid search is performed for the penalty coefficient $\lambda \in \{0.1, 0.01, 0.001, 0.0001, 0.00001\}$ and the dropout rate $p_{dropout} \in \{0.1, 0.01, 0.001, 0.0001\}$ on the default environment setting to find reasonable hyperparameters. We end up searching with $\lambda = 0.00001$ for both cartpole task and mountain car task and in cartpole task, $p_{dropout}$ is fixed at 0.01 while in mountain-car task, $p_{dropout}$ is fixed at 0.001. The capability of storing historical samples for experience replay is set to 10000 in both tasks.

In this paper, 6 sets of experiments are done and they are DQN, DQN with l_1 regularization, DQN with l_2 regularization, DQN with dropout, DQN with improved dropout and DQN with new dropout. In each set of experiments, we run twenty times random seeds and take the top ten groups with the highest average scores in evaluating for statistics and charting. Because it is possible for DQN algorithm rapidly fluctuating in performance during training, the statistical methods we utilized can effectively avoid the effects of fluctuations at the end of training.

5.1   The Cartpole Task

In cartpole task, we separately train 50 thousand steps, 75 thousand steps and 100 thousand steps for the six approaches, and evaluated policies after training. In this task, each step of the agent gets a reward of 1 until it failed or reached the maximum number of steps in an episode. During the training, the maximum number of steps in an episode is limited to 500. In policy evaluation, the maximum number of steps in an episode is limited to 1000. The experimental results are shown in Fig. 1. We can observe that when the training ends up at 50 thousand steps, most regularization approaches perform better with DQN, except for dropout. When the length of pole is set to 1.0 units, the DQN with regularization techniques perform better than DQN, the baseline. Furthermore, we can observe that the blue line quickly deteriorates when the pole length is reduced. At 75 thousand steps, except for the blue line, other line grows up and the line with regularization are still higher than the baseline. At 100 thousand steps, except for the blue one, other lines are basically coincident. The light blue line has the smallest performance fluctuation under both extreme pole length conditions.

Through the experimental results of cartpole task, it is found that DQN itself has a capability of generalization. Except for the dramatic change on the length of pole, the baseline does not crash down at most time. This means that the policy learned is able to generalize on parameter-varying environments. However, we can also observe that regularization is beneficial to DQN. In one hand, DQN applying with regularization approaches achieve better performance than the baseline under the same pole length condition. On the other hand, at 100 thousand steps, DQN with l_1 and l_2 regularization had almost no performance degradation with a very long pole although other methods crashed down quickly. Notice that when samples are insufficient, the new dropout we proposed help DQN get better performance than others at most time.

5.2   The Mountain Car Task

In mountain car task, we separately train 50 thousand steps, 75 thousand steps, 100 thousand steps and 200 thousand steps for the six approaches, and evaluated policies after training. In this task, each step of the agent gets a reward of -1 until it reached the goal. There is no limited number of steps in an episode. In policy evaluation, the maximum number of steps in an episode is limited to 1000. The experimental results are shown in Fig. 2. At 50 thousand steps, there are three lines higher than the baseline. Among these three lines, the yellow line is relatively stable under different car weight. As the number of training steps increases, all the lines except the blue line move up, and the performance is getting better and better. At 100 thousand steps, only DQN with improved dropout and DQN with new dropout perform better than the baseline. At 200 thousand, except the blue one, other lines are basically coincident. Among these 5 lines, when the weight of the car is lighter than about 0.27 kg, the purple line is slightly higher than the others. However, when the weight of the car is heavier than about 0.27 kg, the purple line deteriorates quickly.

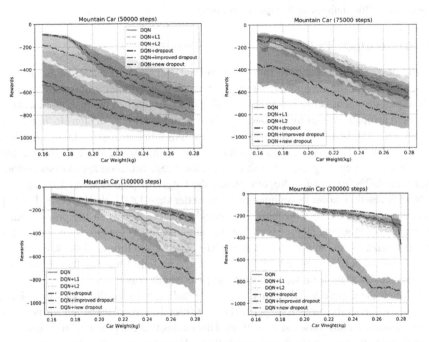

Fig. 2. Performance of agents trained with default car weight (2.0 kg) and evaluated the policy learned after 50 thousand, 75 thousand, 100 thousand and 200 thousand steps with the car weight changing from 0.16 kg to 0.26 kg. Notice that for each picture we train from the beginning. The x-axis is the car weight and the y-axis are the accumulated rewards received by the agent. The solid line is the mean value and the shadow is the standard deviation and each line is aver-aged over ten seeds. In each seed, the agent performs 100 times evaluations with the pole length and calculates the average value of these 100 times evaluations as the performance.

Through the experimental results of the mountain car task, we can observe that different regularization approaches have a different impact on DQN. In mountain car task, the basic rule is that the lighter the car, the easier it is to reach the target point. Notice that DQN with dropout behaved abnormally, so it is removed by default in comparison. With limited samples, DQN with l_1 show a better capability of generalization because at 50 thousand steps, when the weight of the car gradually increases, the performance is only slightly reduced. From the whole experiment, compared with the baseline, regularization based on dropout help DQN achieve better performance, whether in the default environment or variational environments. Furthermore, the new dropout we proposed helped DQN get better performance than baseline at all times except when the weight of the car becomes particularly heavy and the performance deteriorates rapidly.

6   Conclusion

In this paper, we systematically studied regularization approaches on deep RL. First, due to the difference between supervised learning and deep RL, we adapted dropout to DQN. Applying with the improved dropout, DQN worked well and reasonably. Base on it, a new dropout method was proposed which speeded up the optimization and enabled DQN get better control performance. Then, for cartpole task and mountain car task, a comparative test of DQN and its combination with five regularization approaches were conducted. In the case of relatively few samples, the role of regularization is obvious. On the one hand, most regularization approaches helped deep RL get a better control performance in the original environment. On the other hand, some regularization techniques also improved the generalization capability in parameter-varying environments and suppressed performance degradation. The more impressive regularization techniques are the improved dropout and the new dropout we proposed because their improvement in generalization capability is more obvious on deep RL. Finally, we believe our work is valuable in theoretical exploration on generalization of deep RL and it would be a milestone for the field if we were able to develop an algorithm that can generalize across among completely different tasks.

Acknowledgements. This work is supported by the National Natural Science Foundation of China (61873022, 61573052), the Beijing Natural Science Foundation (4182045), the China Postdoctoral Science Foundation (2018M640049) and the Fundamental Research Funds for the Central Universities (XK1802-4, ZY1839).

References

1. LeCun, Y., Bottou, L., Bengio, Y., Haffner, P.: Gradient-based learning applied to document recognition. Proc. IEEE **86**(11), 2278–2324 (1998)
2. Hardt, M., Recht, B., Singer, Y.: Train faster, generalize better: stability of stochastic gradient descent. In: 33rd International Conference on Machine Learning, pp. 1868–1877. IMLS Press, New York (2016)
3. Bishop, C.M.: Training with noise is equivalent to Tikhonov regularization. Neural Comput. **7**(1), 108–116 (1995)
4. Chandra, B., Sharma, R.K.: Adaptive noise schedule for denoising autoencoder. In: Loo, C. K., Yap, K.S., Wong, K.W., Teoh, A., Huang, K. (eds.) ICONIP 2014. LNCS, vol. 8834, pp. 535–542. Springer, Cham (2014). https://doi.org/10.1007/978-3-319-12637-1_67
5. Silver, D., et al.: Mastering the game of go without human knowledge. Nature **550**, 354–359 (2017)

6. Sutton, R.S., Barto, A.G.: Reinforcement Learning: An Introduction. MIT, Cambridge (1998)
7. Farahmand, A.M., Ghavamzadeh, M., Szepesvari, C., Mannor, S.: Regularized policy iteration. In: 22nd Annual Conference on Neural Information Processing Systems, pp. 441–448. Curran Associates Press, Vancouver (2009)
8. Henderson, P., Islam, R., Bachman, P., Pineau, J., Precup, D., Meger, D.: Deep reinforcement learning that matters. arXiv preprint arXiv:1709.06560 (2017)
9. Srivastava, N., Hinton, G., Krizhevsky, A., Sutskever, I., Salakhutdinov, R.: Dropout: a simple way to prevent neural networks from overfitting. JMLR 15(1), 1929–1958 (2014)
10. Strehl, A.L., Li, L., Littman, M.L.: Reinforcement learning in finite MDPs: PAC analysis. JMLR 10, 2413–2444 (2009)
11. Smith, S.L., Le, Q.V.: Understanding generalization and stochastic gradient descent. arXiv preprint arXiv:1710.06451v1 (2017)
12. Mnih, V., et al.: Human-level control through deep reinforcement learning. Nature 518, 529–533 (2015)
13. Whiteson, S., Tanner, B., Taylor, M.E., Stone, P.: Protecting against evaluation overfitting in empirical reinforcement learning. In: 2011 IEEE Symposium on Adaptive Dynamic Programming and Reinforcement Learning, Paris, pp. 120–127. IEEE Press (2011)
14. Bouthillier, X., Konda, K., Vincent, P., Memisevic, R.: Dropout as data augmentation. arXiv preprint arXiv:1506.08700 (2016)
15. Gal, Y., Ghahramani, Z.: Dropout as a Bayesian approximation: representing model uncertainty in deep learning. arXiv preprint arXiv:1506.02142 (2015)
16. Riedmiller, M.: Neural fitted Q iteration – first experiences with a data efficient neural reinforcement learning method. In: Gama, J., Camacho, R., Brazdil, P.B., Jorge, A.M., Torgo, L. (eds.) ECML 2005. LNCS, vol. 3720, pp. 317–328. Springer, Heidelberg (2005). https://doi.org/10.1007/11564096_32
17. Brockman, G., Cheung, V., Pettersson, L., Schneider, J., Zaremba, W.: OpenAI gym. arXiv preprint arXiv:1606.01540 (2016)
18. Hahnloser, R.H.R., Sarpeshkar, R., Mahowald, M.A., Douglas, R.J., Seung, H.S.: Digital selection and analogue amplification coexist in a cortex-inspired silicon circuit. Nature 405, 947–951 (2000)

A Gradient-Descent Neurodynamic Approach for Distributed Linear Programming

Xinrui Jiang[1], Sitian Qin[2(✉)], and Ping Guo[3]

[1] Department of Mathematics, Harbin Institute of Technology, Harbin, China
`jiangxrhit@163.com`
[2] Department of Mathematics, Harbin Institute of Technology, Weihai, China
`qinsitian@163.com`
[3] Laboratory of Image Processing and Pattern Recognition,
Beijing Normal University, Beijing, China
`pguo@ieee.org`

Abstract. In this paper, a gradient-descent neurodynamic approach is proposed for the distributed linear programming problem with affine equality constraints. It is rigorously proved that the state solution of the proposed gradient-descent approach with an arbitrary initial point reaches agreement and is convergent to an optimal solution of the considered optimization problem at the same time. In the end, some numerical experiments are conducted to verify the effectiveness of the proposed gradient-descent approach.

Keywords: Distributed linear programming ·
Gradient-descent neurodynamic approach · Reach agreement ·
Convergence to an optimal solution

1 Introduction

Optimization problems are ubiquitous in science and engineering field such as Traveling Salesman Problems [6] and economic resource allocation [1]. It has been a long history to study various optimization problems via numerous approaches [8,11,14,19]. As the complexity and scales of optimization problems increase, it becomes rather urgent to seek distributed approaches for solving complex and large-scale optimization problems. Thanks to the inspiring progresses in multi-agent systems [9,10,24], the study of distributed optimization has made great advances [3,5,7,17,18,22].

The research on asymptotic behaviors of dynamical systems [11,15,16,21] demonstrates that neurodynamic approaches exerts their great efficiency in real-time and parallel computing contrast to discrete-time algorithms [2]. As a consequence, recent years have witnessed abundant research results involving neurodynamic approaches for distributed optimization. Yi et al. [23] studied a distributed smooth convex optimization problem with inequality constraints and

© Springer Nature Switzerland AG 2019
H. Lu et al. (Eds.): ISNN 2019, LNCS 11555, pp. 45–53, 2019.
https://doi.org/10.1007/978-3-030-22808-8_5

presented a distributed gradient algorithm via Lagrangian method. In [25], Zhu et al. further proposed a projected primal-dual dynamic for a more general distributed nonsmooth convex optimization problem. However, due to the practical requirements, the neurodynamic approaches mentioned above may fail to solve distributed optimization problems over directed graphs. To break through this restrictions, Kia et al. [7] established a distributed coordination algorithm to solve the unconstrained distributed optimization problem over weight-balanced directed graphs. In [26], Zhu et al. constructed a novel neurodynamic approach for distributed convex optimization over weight-unbalanced directed graphs.

It is well known that gradient or subgradient descent algorithms are quite powerful for solving optimization problems [4,17]. Motivated by above research results, a gradient-descent neurodynamic approach based on Karush-Kuhn-Tucker condition is proposed to cope with the distributed linear optimization problem with affine equality constraints. The global existence of the proposed gradient-descent approach is firstly studied. By Lyapunov method, the consensus as well as convergence to an optimal solution of our gradient-descent approach is finally derived.

The rest of this paper is organized as follows. Several related preliminaries concerning graph theory are introduced in Sect. 2. In Sect. 3, we firstly formulate the considered distributed optimization problem and present a neurodynamic approach, and then study the asymptotic behaviors of the proposed gradient-descent approach. Some numerical examples are given to illustrate our results in Sect. 4, and Sect. 5 concludes this paper.

2   Preliminaries

In this section, we introduce some basic preliminaries concerning graph theory which are needed in this paper.

A directed graph is denoted by $\mathcal{G} = (\mathcal{V}, \mathcal{E}, \mathcal{A})$, where $\mathcal{V} = \{1, 2, \cdots, m\}$ is a node set, $\mathcal{E} \subset \mathcal{V} \times \mathcal{V}$ is an edge set and $\mathcal{A} = (a_{ij}) \in \mathbb{R}^{m \times m}$ is an weighted adjacency matrix. If $(i, j) \in \mathcal{E}$, then $a_{ij} > 0$ otherwise $a_{ij} = 0$. There is no self-connection in graph $\mathcal{G}$, that is $a_{ii} = 0$. Denote by $\mathcal{N}_i = \{j : a_{ij} > 0\}$ the neighbors set, $\mathcal{N}_i^2 = \{k : a_{ij} > 0 \text{ and } a_{jk} > 0\}$ the second-order neighbors set and $\mathcal{N}_i^- = \{l : a_{li} > 0\}$ the in-neighbors set of the ith node. Let $d_{out}^i = \sum_{j=1}^m a_{ij}$ be the outer degree of the node i. The Laplacian matrix of the graph $\mathcal{G}$ is defined as $L_m = D - \mathcal{A}$ with $D = \text{diag}\{d_{out}^1, d_{out}^2, \cdots, d_{out}^m\}$. A directed path between the node i_0 and i_k in graph $\mathcal{G}$ is a finite sequence of directed edges $(i_0, i_1), (i_2, i_3), \cdots, (i_{k-1}, i_k)$ with $i_l \in \mathcal{V}$ $(l = 0, 1, \cdots, k)$. A directed graph is said to be strongly connected if for any pair of distinct vertices i and j, there exists a directed path starting from i and reaching to j.

3   Problem Description and Convergence Analysis

In this paper, we consider a network of m agents over a strongly connected directed graph $\mathcal{G}$ to solve the following distributed linear programming cooperatively:

$$\min \quad f(x) = \sum_{i=1}^{m} e_i^T x \tag{1}$$
$$\text{s.t.} \quad A_i x = b_i, \ i = 1, 2, \cdots, m.$$

where $x \in \mathbb{R}^n$ is the decision variable, $A_i \in \mathbb{R}^{p_i \times n}$ and $b_i \in \mathbb{R}^{p_i}$. It is assumed that the distributed optimization problem (1) has at least an optimal solution.

Remark 1. In optimization problem (1), the matrices A_i $(i = 1, 2, \cdots, m)$ in constraints are not necessarily full row rank. Compared with the problems considered in [12,13,20], it is no need to equivalently convert A_i $(i = 1, 2, \cdots, m)$ into full row rank ones, which reduces computational loads.

Assumption 1. *The connection graph $\mathcal{G}$ is directed and strongly connected.*

Let $r = \sum_{i=1}^{m} r_i$, $p = \sum_{i=1}^{m} p_i$. Denote by $\mathbf{L} = L_m \otimes I_n$, $\mathbf{e} = (e_1^T, e_2^T, \cdots, e_m^T)^T \in \mathbb{R}^{mn}$, $\mathbf{b} = (b_1^T, b_2^T, \cdots, b_m^T)^T \in \mathbb{R}^r$, $\mathbf{A} = \text{blkdiag}\{A_1, A_2, \cdots, A_m\} \in \mathbb{R}^{r \times mn}$. Thus, according to Lemma 3.1 in [5], the following lemma can be derived.

Lemma 1. *Suppose Assumption 1 holds and let $\mathbf{x} = (x_1^T, x_2^T, \cdots, x_m^T)^T \in \mathbb{R}^{mn}$. Then the optimization problem (1) is equivalent to the following problem*

$$\min \quad f(x) = e^T x$$
$$\text{s.t.} \quad Ax = b, \tag{2}$$
$$\quad Lx = 0.$$

It is clear that the problem (1) is a convex optimization problem. Let $\mu = (\mu_1^T, \mu_2^T, \cdots, \mu_m^T)^T$, $\nu = (\nu_1^T, \nu_2^T, \cdots, \nu_m^T)^T$ with $\mu_i \in \mathbb{R}^{p_i}$ and $\nu_i \in \mathbb{R}^n$. The following lemma gives a necessary and sufficient condition of an optimal solution to the optimization problem (1).

Lemma 2. *x^* is an optimal solution to the optimization problem (2) if and only if there exist $\mu^* \in \mathbb{R}^p$ and $\nu^* \in \mathbb{R}^{mn}$ such that*

$$A^T \mu^* + L^T \nu^* + e = 0, \tag{3a}$$
$$Ax^* = b, \quad Lx^* = 0, \tag{3b}$$

For convenience, denote by $z = (x^T, \mu^T, \nu^T)^T$. Define an energy function V as follows.

$$V(z) = \|A^T \mu^* + L^T \nu^* + e\|^2 + \|Ax - b\|^2 + \|Lx\|^2. \tag{4}$$

We propose a neurodynamic approach to handle the distributed optimization problem (1):

$$\dot{x}_i = -A_i^T(A_i x_i - b_i) - \sum_{j \in \mathcal{N}_i}\left[a_{ij} \sum_{k \in \mathcal{N}_i} a_{ik}(x_i - x_k) - a_{ji} \sum_{k \in \mathcal{N}_j} a_{jk}(x_j - x_k) \right],$$

$$\dot{\mu}_i = -A_i\left[A_i^T \mu_i + e_i + \sum_{j \in \mathcal{N}_i}(a_{ij}\nu_i - a_{ji}\nu_j) \right],$$

$$\dot{\nu}_i = -\sum_{j \in \mathcal{N}_i} a_{ij}\left[e_i - e_j + A_i^T \mu_i - A_j^T \mu_j + \sum_{k \in \mathcal{N}_i}(a_{ik}\nu_i - a_{ki}\nu_k) - \sum_{l \in \mathcal{N}_j}(a_{jl}\nu_j - a_{lj}\nu_l) \right].$$
$$\tag{5}$$

Remark 2. It is worth noting that the ith agent is assumed to have access to its local information, neighbor information, second-order neighbor information, and in-neighbor information. Therefore, the gradient-descent approach (5) is distributed.

For simplicity, the distributed gradient-descent approach (5) can be further rewritten in the following compact forms:

$$\dot{z}(t) = -\nabla V(z(t)), \tag{6}$$

where V is defined in (4). Denote $\mathcal{C}$ the equilibrium point set of the gradient-descent approach (6), that is,

$$\mathcal{C} = \left\{ z \in \mathbb{R}^{mn+p+r} : \nabla V(z) = 0 \right\}.$$

Theorem 1. *The state solution of the gradient-descent approach (6) with an arbitrary initial point $z_0 \in \mathbb{R}^{mn+r+p}$ globally exists and is bounded.*

Proof. On the basis of the right side of (6) being continuous, we have that for any initial point $z_0 \in \mathbb{R}^{mn+r+p}$, there is a local solution $z(t)$ of the gradient-descent approach (6) defined on $[0, T_{max})$. Differentiating $\epsilon_0 V(z(t))$ with respect to t on $[0, T_{max})$, we have

$$\frac{dV(z(t))}{dt} = \langle \nabla V(z(t)), \dot{z}(t) \rangle = -\|\nabla V(z(t))\|^2 \leq 0. \tag{7}$$

So $V(z(t))$ is nonincreasing along the state solution of the gradient-descent approach (6). Due to $V(z(t)) \geq 0$, then it follows that $\lim\limits_{t \to T_{max}^-} V(z(t))$ exists, and denote by $\bar{V}$. That is,

$$\lim_{t \to T_{max}^-} V(z(t)) = \bar{V}. \tag{8}$$

Let

$$W(t) = V(z(t)) + \frac{1}{2}\|z(t) - z^*\|^2, \tag{9}$$

where z^* is an equilibrium point of the gradient-descent approach (6). Differentiating $W(t)$ on $[0, T_{max})$, owing to the fact that $\nabla V(z^*) = 0$ and the monotonicity of ∇V, one has

$$\begin{aligned}
\frac{dW(t)}{dt} &= \frac{dV(z(t))}{dt} + \langle z(t) - z^*, \dot{z}(t) \rangle \\
&= -\|\nabla V(z(t))\|^2 - \langle z(t) - z^*, \nabla V(z(t)) - \nabla V(z^*) \rangle \\
&\leq 0.
\end{aligned} \tag{10}$$

Integrating the both sides of (10) from 0 to t, for any $t \leq T_{max}$, it obtains

$$\frac{1}{2}\|z(t) - z^*\|^2 \leq W(t) \leq V(z(0)) + \frac{1}{2}\|z(0) - z^*\|^2, \tag{11}$$

which indicates $T_{max} = +\infty$ and $z(t)$ is bounded on $[0, +\infty)$. The proof is complete.

Theorem 2. *For any initial point $z_0 \in \mathbb{R}^{mn+p+r}$, the state solution $z(t)$ of gradient-descent approach (6) is convergent to an equilibrium point of gradient-descent approach (6).*

Proof. By Theorem 1, the state solution $z(t)$ is bounded. Thus, there is a sequence $\{t_n\}$ such that $\lim\limits_{n\to\infty} z(t_n) = \bar{z}$. Due to the continuity of ∇V and the gradient-descent approach (6), we have $\dot{z}(t)$ is also bounded. So there exists $M > 0$ such that equilibrium point of gradient-descent approach (6).

$$\|\dot{z}(t)\| \leq M. \tag{12}$$

Next, we prove that $\bar{z}$ is an equilibrium point of gradient-descent approach (6). If not, we have $\nabla V(\bar{z}) \neq 0$. Thus, there exists $m > 0$, $\delta > 0$, such that

$$\|\nabla V(z)\| > m, \quad \forall z \in B(\bar{z}, \delta). \tag{13}$$

On account of $\lim\limits_{n\to\infty} z(t_n) = \bar{z}$, there exists $N \in \mathbb{N}$ such that

$$z(t_n) \in B(\bar{z}, \frac{\delta}{2}), \quad \forall n \geq N. \tag{14}$$

For any $n \geq N$ and $t \in \left[t_n - \frac{\delta}{2M}, t_n + \frac{\delta}{2M}\right]$, it can be derived that

$$\|z(t) - \bar{z}\| \leq \|z(t) - z(t_n)\| + \|z(t_n) - \bar{z}\| \leq M \cdot \frac{\delta}{2M} + \frac{\delta}{2} \leq \delta. \tag{15}$$

As a result, by (7), (8), (13) and (15), it has

$$\lim_{t\to+\infty} V(z(t)) - V(z_0) = -\int_{t=0}^{+\infty} \|\nabla V(z(t))\|^2 dt$$

$$\geq -\sum_{n=N}^{\infty} \int_{t_n - \frac{\delta}{2M}}^{t_n + \frac{\delta}{2M}} m^2 dt \geq -\sum_{n=N}^{\infty} m^2 \cdot \frac{\delta}{M} = -\infty, \tag{16}$$

which results in a contradiction. Therefore, $\bar{z}$ is an equilibrium point of gradient-descent approach (6). Let

$$W_1(t) = V(z(t)) - V_\infty + \frac{1}{2}\|z(t) - \bar{z}\|^2. \tag{17}$$

Similar to (10) and (8), we also obtain

$$\frac{dW_1(t)}{dt} \leq 0, \tag{18}$$

and $\lim\limits_{t\to+\infty} W_1(t)$ exists. Finally, owing to $\lim\limits_{n\to\infty} W_1(t_n) = 0$, it is clear that

$$\lim_{t\to+\infty} W_1(t) = 0, \tag{19}$$

and (19) indicates

$$\lim_{t\to+\infty} z(t) = \bar{z}. \tag{20}$$

The proof is complete.

Remark 3. By Theorem 2, the state solution of gradient-descent approach (6) is convergent to an equilibrium point of gradient-descent approach (6). Combining Lemma 2 and the definition of V, an optimal solution to distributed optimization problem (2) can be obtained by the derived equilibrium point.

4   Simulations

Example 1. Consider a network of 3 agents interacting over a strongly connected directed graph to cooperatively minimize a global linear objective function with affine constraints:

$$\min \quad \sum_{i=1}^{3} e_i^{\mathrm{T}} x, \tag{21}$$

$$\text{s.t.} \quad 5x_1 + 11x_2 - x_3 + 3x_4 = 7,$$

where

$$x = (x_1, x_2, x_3, x_4)^{\mathrm{T}} \in \mathbb{R}^4, \; e_1 = (-1, -5, 1, 1)^{\mathrm{T}},$$
$$e_2 = (-2, 4, -2, -6)^{\mathrm{T}}, \quad e_3 = (-12, -32, 4, -4)^{\mathrm{T}}.$$

where $x = (x_1, x_2, x_3, x_4)^{\mathrm{T}}$. The considered connection directed graph in this example is shown in Fig. 1.

This optimization problem (21) can also be solved by the gradient-descent approach (6). The transient behaviors of the state solution of the gradient-descent approach (6) is depicted in Fig. 2, which implies that the output x of the gradient-descent approach (6) in each agent is convergent to same optimal solution $(0.7201, 0.6176, -0.0107, -1.1346)^{\mathrm{T}}$.

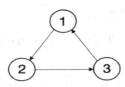

Fig. 1. The interaction topology of the 3-agent network in Example 1.

Example 2. Consider a network of 5 agents interacting over a strongly connected directed graph to minimize a global objective function with an affine equality constraint:

$$\min \quad \sum_{i=1}^{5} d_i^{\mathrm{T}} x, \tag{22}$$

$$\text{s.t.} \quad x_1 + x_2 + x_3 + x_4 + x_5 = 5,$$

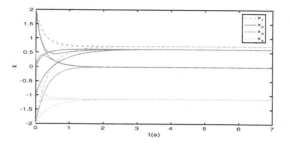

Fig. 2. Transient behaviors of the state of (6) in Example 1.

where
$$x = (x_1, x_2, x_3, x_4, x_5)^T \in \mathbb{R}^5, d_1 = (-10, -5, -10, 5, 10)^T,$$
$$d_2 = (-5, -10, 5, 10, -10)^T, d_3 = (-10, 5, 10, -10, -5)^T,$$
$$d_4 = (5, 10, -10, -5, -10)^T, d_5 = (10, -10, -5, -10, 5)^T.$$

In this example, the considered communication graph is described in Fig. 3. This optimization problem (22) can also be solved by the gradient-descent approach (6). The transient behaviors of the state solution of the gradient-descent approach (6) is exhibited in Fig. 4, which implies that the output x of the gradient-descent approach (6) reach consensus as well as converges to an optimal solution $(0.8800, 1.0800, 1.2200, 0.3800, 1.4400)^T$.

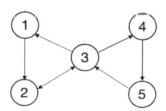

Fig. 3. The interaction topology of the 5-agent network in Example 2.

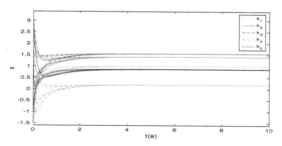

Fig. 4. Transient behaviors of the state of (6) in Example 2.

5 Conclusions

This paper presented a gradient-descent neurodynamic approach to cope with the distributed linear optimization problem. The state solution of the presented gradient-descent approach is proved to reach consensus and converges to an optimal solution via rigorously theoretical deduction. We obtain the related results by Lyapunov function theory, and give some numerical results in order to verify the obtained result.

Acknowledgments. This research is supported by the National Natural Science Foundation of China (61773136, 11471088) and the NSFC and CAS project in China with granted No. U1531242.

References

1. Balasubramanian, A., Levine, B.N., Venkataramani, A.: DTN routing as a resource allocation problem. ACM SIGCOMM Comput. Commun. Rev. **37**(4), 373–384 (2007)
2. Cochocki, A., Unbehauen, R.: Neural Networks for Optimization and Signal Processing. Wiley, New York (1993)
3. Deng, Z., Liang, S., Hong, Y.: Distributed continuous-time algorithms for resource allocation problems over weight-balanced digraphs. IEEE Trans. Cybern. **48**(11), 3116–3125 (2018)
4. Forti, M., Nistri, P., Quincampoix, M.: Convergence of neural networks for programming problems via a nonsmooth Lojasiewicz inequality. IEEE Trans. Neural Netw. Learn. Syst. **17**(6), 1471–1486 (2006)
5. Gharesifard, B., Cortes, J.: Distributed continuous-time convex optimization on weight-balanced digraphs. IEEE Trans. Autom. Control **59**(3), 781–786 (2012)
6. Hopfield, J., Tank, D.: Neural computation of decisions in optimization problems. Biol. Cybern. **52**(3), 141–152 (1985)
7. Kia, S.S., Cortes, J., Martinez, S.: Distributed convex optimization via continuous-time coordination algorithms with discrete-time communication. Automatica **55**, 254–264 (2015)
8. Li, G., Song, S., Wu, C., Du, Z.: A neural network model for non-smooth optimization over a compact convex subset. In: Wang, J., Yi, Z., Zurada, J.M., Lu, B.-L., Yin, H. (eds.) ISNN 2006. LNCS, vol. 3971, pp. 344–349. Springer, Heidelberg (2006). https://doi.org/10.1007/11759966_53
9. Li, Z., Liu, X., Wei, R., Xie, L.: Distributed tracking control for linear multiagent systems with a leader of bounded unknown input. IEEE Trans. Autom. Control **58**(2), 518–523 (2013)
10. Li, Z., Wei, R., Liu, X., Xie, L.: Distributed consensus of linear multi-agent systems with adaptive dynamic protocols. Automatica **49**(7), 1986–1995 (2013)
11. Liu, N., Qin, S.: A novel neurodynamic approach to constrained complex-variable pseudoconvex optimization. IEEE Trans. Cybern. (2018). https://doi.org/10.1109/TCYB.2018.2855724
12. Liu, N., Qin, S.: A neurodynamic approach to nonlinear optimization problems with affine equality and convex inequality constraints. Neural Netw. **109**, 147–158 (2019)

13. Liu, Q., Yang, S., Wang, J.: A collective neurodynamic approach to distributed constrained optimization. IEEE Trans. Neural Netw. Learn. Syst. **28**(8), 1747–1758 (2017)
14. Ma, K., et al.: Stochastic non-convex ordinal embedding with stabilized Barzilai-Borwein step size. arXiv:1711.06446
15. Min, H., Meng, Z., Zhang, Y.: Projective synchronization between two delayed networks of different sizes with nonidentical nodes and unknown parameters. Neurocomputing **171**, 605–614 (2016)
16. Min, H., Zhang, M., Qiu, T., Xu, M.: UCFTS: a unilateral coupling finite-time synchronization scheme for complex networks. IEEE Trans. Neural Netw. Learn. Syst. **30**(1), 255–268 (2019)
17. Nedic, A., Ozdaglar, A.: Distributed subgradient methods for multi-agent optimization. IEEE Trans. Autom. Control **54**(1), 48–61 (2009)
18. Nedic, A., Ozdaglar, A., Parrilo, P.A.: Constrained consensus and optimization in multi-agent networks. IEEE Trans. Autom. Control **55**(4), 922–938 (2010)
19. Qin, S., Le, X., Wang, J.: A neurodynamic optimization approach to bilevel quadratic programming. IEEE Trans. Neural Netw. Learn. Syst. **28**(11), 2580–2591 (2017)
20. Qin, S., Yang, X., Xue, X., Song, J.: A one-layer recurrent neural network for pseudoconvex optimization problems with equality and inequality constraints. IEEE Trans. Cybern. **47**(10), 3063–3074 (2017)
21. Wu, A., Zeng, Z.: Exponential stabilization of memristive neural networks with time delays. IEEE Trans. Neural Netw. Learn. Syst. **23**(12), 1919–1929 (2012)
22. Yang, S., Liu, Q., Wang, J.: A multi-agent system with a proportional-integral protocol for distributed constrained optimization. IEEE Trans. Autom. Control **62**(7), 3461–3467 (2017)
23. Yi, P., Hong, Y., Liu, F.: Distributed gradient algorithm for constrained optimization with application to load sharing in power systems. Syst. Control Lett. **83**(711), 45–52 (2015)
24. You, K., Xie, L.: Network topology and communication data rate for consensusability of discrete-time multi-agent systems. IEEE Trans. Autom. Control **56**(10), 2262–2275 (2011)
25. Zhu, Y., Yu, W., Wen, G., Chen, G.: Projected primal-dual dynamics for distributed constrained nonsmooth convex optimization. IEEE Trans. Cybern. (2018). https://doi.org/10.1109/TCYB.2018.2883095
26. Zhu, Y., Yu, W., Wen, G., Ren, W.: Continuous-time coordination algorithm for distributed convex optimization over weight-unbalanced directed networks. IEEE Trans. Circ. Syst. II Express Briefs (2018). https://doi.org/10.1109/TCSII.2018.2878250

An Asymptotically Stable Identifier Design for Unmanned Surface Vehicles Based on Neural Networks and Robust Integral Sign of the Error

Shengnan Gao, Lu Liu, Zhouhua Peng$^{(\boxtimes)}$, Dan Wang$^{(\boxtimes)}$, Nan Gu, and Yue Jiang

School of Marine Electrical Engineering, Dalian Maritime University,
Dalian 116026, China
{zhpeng,dwang}@dlmu.edu.cn

Abstract. In this paper, a robust identifier is developed for unmanned surface vehicles (USVs) subject to uncertain dynamics. The uncertain dynamics comes from parametric model uncertainty and external ocean disturbance. The identifier for USV is designed based on Robust Integral Sign of the Error (RISE) and neural networks. With the proposed identifier, asymptotic stability of the estimation errors can be proven in the presence of parametric model uncertainties and external ocean disturbances. The proposed method can be used in a variety of practical settings such as trajectory tracking and formation control of marine vehicles for achieving better performance.

Keywords: Neural networks · Unmanned surface vehicle · Derivative estimation · Robust identification

1 Introduction

One of the major challenges in motion control of autonomous marine vehicles is to identify the uncertain dynamics induced by parametric model uncertainty and external ocean disturbance [1–3]. Estimation of the model uncertainty and ocean disturbance is crucial for many practical applications [4–6], and a number of control methods are available such as sliding model control [7,8], neural network

L. Liu, Z. Peng, D. Wang—This work was supported in part by the National Natural Science Foundation of China under Grants 51579023, 61673081, the Innovative Talents in Universities of Liaoning Province under Grant LR2017014, High Level Talent Innovation and Entrepreneurship Program of Dalian under Grant 2016RQ036, China Postdoctoral Science Foundation 2019M650086, the National Key Research and Development Program of China under Grant 2016YFC0301500, the Traning Program for High-level Technical Talent in Transportation Industry under Grant 2018-030, the Fundamental Research Funds for the Central Universities under Grant 3132019101,3132019013, and Outstanding Youth Support Program of Dalian.

© Springer Nature Switzerland AG 2019
H. Lu et al. (Eds.): ISNN 2019, LNCS 11555, pp. 54–61, 2019.
https://doi.org/10.1007/978-3-030-22808-8_6

control [11,12,14,15,17–20], fuzzy control [9,10], and extended-state-observer-based control [16,21]. In particular, neural networks have received a significant growth for estimating unknown vehicle kinetics [12,14,15,17–20]. However, in these works, only uniform ultimate boundedness (UUB) of estimation errors can be achieved.

In this paper, an asymptotically stable identifier is developed for USVs subject to unknown dynamics which caused by parametric model uncertainty and external ocean disturbance. The identifier is designed based on RISE and neural networks. With the proposed identifier, asymptotic stability of the estimation errors are proven. The developed asymptotically stable identifier can be used in trajectory tracking, formation tracking and motion control of vehicles for achieving better performance and stability results. The main feature of the presented identifier is concluded as follows. Compared with the neural-network-based approximation methods in [12,14,15,17–20] where only UUB of estimation errors can be proven, the developed identifier can ensure the asymptotic stability of estimation errors based on neural networks and RISE.

2 Preliminaries and Problem Formulation

2.1 Neural Networks

A continuous function $g(\xi)$ can be approximated as

$$g(\xi) = W_g^T \beta(V_g^T \xi) + \varepsilon_g(\xi), \quad \xi \in \Omega, \tag{1}$$

where ξ is the input vector; W_g, V_g are the ideal wights of neural network satisfying $\|W_g\|_F \leq W_g^*$ and $\|V_g\|_F \leq V_g^*$ with $W_g^* > 0$ and $V_g^* > 0$; $\beta(V_g^T \xi)$ is the activation function satisfying $\|\beta(V_g^T \xi)\| \leq \beta^*$ with $\beta^* > 0$; $\varepsilon_g(V_g^T \xi)$ is the neural network approximation error satisfying $\|\varepsilon_g(V_g^T \xi)\| \leq \varepsilon_g^*$ with $\varepsilon_g^* > 0$; Ω is a compact set.

2.2 Problem Formulation

The dynamics of the USV can be described as [1]

$$\dot{\eta}_s = R(\psi_s)\nu_s, \tag{2}$$

$$M_s \dot{\nu}_s = \tau_s - C_s(\nu_s)\nu_s - D_s(\nu_s)\nu_s + g_s(\nu_s) + \tau_{ws}(t), \tag{3}$$

where $\eta_s = [x_s, y_s, \psi_s]$ denotes the position-yaw in earth-fixed; $\nu_s = [u_s, v_s, r_s]^T$ is the body-fixed velocity vector denoting surge velocity, sway velocity and angular velocity; $M_s = M_s^T$ is an inertial matrix; $C_s(\nu_s) = -C_s(\nu_s)$ denotes the centrifugal and coriolis matrix; $D_s(\nu_s)$ is a damping matrix;

$\tau_{ws} = [\tau_{wus}, \tau_{wvs}, \tau_{wrs}]^T$ is a vector of environmental forces satisfying $\|\tau_{ws}\| \le \|\tau_{ws}^*\|$ with $\tau_{ws}^* > 0$; $\tau_s = [\tau_{us}, \tau_{vs}, \tau_{rs}]^T$ is the control input; and

$$R(\psi_s) = \begin{bmatrix} \cos\psi_s & -\sin\psi_s & 0 \\ \sin\psi_s & \cos\psi_s & 0 \\ 0 & 0 & 1 \end{bmatrix}. \tag{4}$$

The *control objective* in this paper is to develop an asymptotically stable identifier for unmanned surface vehicles (USVs) subject to unknown dynamics.

3   Identifier Design and the Asymptotic Stability Analysis

3.1   Identifier Design

Rewrite (3) as

$$M_s \dot{\nu}_s = \tau_s + g(\nu_s) + \tau_{ws}(t), \tag{5}$$

where $g(\nu_s) = -C_s(\nu_s)\nu_s - D_s(\nu_s)\nu_s + g_s(\nu_s)$.

Lemma 1 [22,23]. The given constants ε_{kg}^* and τ_{ws}^* satisfy $\varepsilon_{kg}^* > 0$ and $\tau_{ws}^* > 0$ for $k = 1, 2, 3$, the ideal wight of neural network $W_g = [W_{1g}, W_{2g}, W_{3g}]$ satisfies $|W_{kg}| \le W_{kg}^*, (k = 1, 2, 3)$, and the continuous function $g(\nu_s)$ can be estimated by neural network as

$$g(\nu_s) = W_g^T \beta(V_g^T \nu_s) + \varepsilon_g, \tag{6}$$

where $\varepsilon_g = [\varepsilon_{1g}, \varepsilon_{2g}, \varepsilon_{3g}]^T$ with $|\varepsilon_{kg}| \le \varepsilon_{kg}^*, (k = 1, 2, 3)$, and $\tau_{ws} = [\tau_{wus}, \tau_{wvs}, \tau_{wrs}]^T$ with $\|\tau_{ws}\| \le \|\tau_{ws}^*\|$.

The asymptotically stable identifier design is based on the following assumptions.

Assumption 1. The state $\nu_s(t)$ and control input τ_s of system (3) are bounded, i.e., $\nu_s(t) \in \mathcal{L}_\infty$, $\tau_s \in \mathcal{L}_\infty$ and $\dot{\nu}_s(t) \in \mathcal{L}_\infty$.

Assumption 2. The neural network activation function $\beta(V_s^T \nu_s)$ and its derivative $\beta'(V_s^T \nu_s)$ are bounded.

$\hat{\nu}_s = [\hat{u}_s, \hat{v}_s, \hat{r}_s]^T$ denotes the estimation of ν_s, and then the following dynamic neural network identifier for (5) is proposed as [27]

$$\begin{cases} M_s \dot{\hat{\nu}}_s = \hat{W}_g^T \beta(\hat{V}_g^T \hat{\nu}_s) + \tau_s + \mu_s, \\ \mu_s \triangleq c_1 \tilde{\nu}_s(t) - c_1 \tilde{\nu}_s(0) + \varrho, \\ \tilde{\nu}_s(t) = \nu_s(t) - \hat{\nu}_s(t), \\ \dot{\varrho} = (c_1 c_2 + c_3)\tilde{\nu}_s + c_4 \mathrm{sgn}(\tilde{\nu}_s), \quad \varrho(0) = 0, \end{cases} \tag{7}$$

where μ_s is the RISE feedback term, $\mathrm{sgn}(\tilde{\nu}_s)$ is the vector signum function and c_1, c_2, c_3, c_4 are positive constants.

The dynamics of velocity estimation $\tilde{\nu}_s$ can be described as

$$M_s \dot{\tilde{\nu}}_s = W_g^T \beta(V_g^T \nu_s) - \hat{W}_g^T \beta(\hat{V}_g^T \hat{\nu}_s) + \varepsilon_g + \tau_{ws} - \mu_s. \tag{8}$$

3.2 Asymptotic Stability Analysis

To analyze the stability, the notion of a filtered identification error is introduced

$$e \triangleq M_s \dot{\tilde{\nu}}_s + c_2 M_s \tilde{\nu}_s. \tag{9}$$

Using (8), the derivative of (9) can be achieved as

$$\begin{aligned}
\dot{e} &= W_g^T \beta' V_g^T \dot{\nu}_s - \dot{\hat{W}}_g^T \hat{\beta} - \hat{W}_g^T \hat{\beta}' \hat{V}_g^T \dot{\nu}_s - \hat{W}_g^T \hat{\beta}' \dot{\hat{V}}_g^T \dot{\nu}_s \\
&\quad + \dot{\varepsilon}_f + \dot{\tau}_{ws} - c_1 e - c_3 \tilde{\nu}_s - c_4 \mathrm{sgn}(\tilde{\nu}_s) + c_2 \dot{\nu}_s.
\end{aligned} \tag{10}$$

Based on (10), $\hat{W}_g$ and $\hat{V}_g$ are updated as follows

$$\begin{cases}
\dot{\hat{W}}_g(t) = \mathrm{Proj}[\Gamma_W \hat{\beta}' \hat{V}_g^T \dot{\nu}_s \tilde{\nu}_s^T], \\
\dot{\hat{V}}_g(t) = \mathrm{Proj}[\Gamma_V \dot{\nu}_s \tilde{\nu}_s^T \hat{W}_g^T \hat{\beta}'],
\end{cases} \tag{11}$$

where Γ_W, Γ_V denote the positive constant adaptation gain matrices. From [24], there exist constants ϵ_1 and ϵ_2 satisfying

$$\|\hat{W}_g(t)\| \le W_q^* + \epsilon_1, \quad \|\hat{V}_q(t)\| \le V_g^* + \epsilon_2. \tag{12}$$

Rewrite (10) as

$$\dot{e} = \tilde{W}_1 + W_{B1} + \hat{W}_{B2} - c_1 e - c_3 \tilde{\nu}_s - c_4 \mathrm{sgn}(\tilde{\nu}_s). \tag{13}$$

$\tilde{W}_1, W_{B1}$ and $\hat{W}_{B2}$ are defined as

$$\begin{cases}
\tilde{W}_1 = c_2 \dot{\nu}_s \quad \dot{\hat{W}}_g^T \hat{\beta} - \hat{W}_g^T \hat{\beta}' \hat{V}_g^T \dot{\nu}_s + \frac{1}{2} W_g^T \hat{\beta}' \hat{V}_g^T \dot{\nu}_s + \frac{1}{2} \hat{W}_g^T \hat{\beta}' V_g^T \dot{\nu}_s, \\
W_{B1} = W_g^T \beta' V_g^T \dot{\nu}_s - \frac{1}{2} \hat{W}_g^T \hat{\beta}' \hat{V}_g^T \dot{\nu}_s - \frac{1}{2} \hat{W}_g^T \hat{\beta}' V_g^T \dot{\nu}_s + \dot{\varepsilon}_g + \dot{\tau}_{ws}, \\
\hat{W}_{B2} = \frac{1}{2} \hat{W}_g^T \hat{\beta}' \hat{V}_g^T \dot{\nu}_s + \frac{1}{2} \hat{W}_g^T \hat{\beta}' \hat{V}_g^T \dot{\nu}_s,
\end{cases} \tag{14}$$

where $W_B = W_{B1} + W_{B2}$, $\tilde{W}_g = W_g - \hat{W}_g$, and $\tilde{V}_g = V_g - \hat{V}_g$. Using Eq. (12) and Assumption 2 , we can get

$$\begin{cases}
\|\tilde{W}_1\| \quad \le \alpha_1(\|h\|)\|h\|, \ \|\tilde{W}_{B1}\| \le \lambda_1, \ \|\tilde{W}_{B2}\| \le \lambda_2, \\
\|\dot{W}_B\| \quad \le \lambda_3 + \lambda_4 \alpha_2(\|h\|)\|h\|, \\
\|\dot{\nu}_s^T \tilde{W}_{B2}\| \le \lambda_5 \|\tilde{\nu}_s\|^2 + \lambda_6 \|e\|^2,
\end{cases} \tag{15}$$

where $h = [\tilde{\nu}_s^T \ e^T]^T$, and α_1, α_2 are positive, globally invertible, non-decreasing functions, and λ_i $(i = 1, 2, ...)$ are positive constants. From [13], define a domain containing function $y(t) = 0$, where $y = [\tilde{\nu}_s^T e^T \sqrt{P_y} \sqrt{Q_y}]^T$ with $P_y(h, t)$ and $Q_y(\tilde{W}_g, \tilde{V}_g)$ satisfying

$$\begin{cases}
\dot{P}_y = -L_y, \ P_y(0) = c_4 \sum_{i=1}^n |\tilde{\nu}_{si}(0)| - \tilde{\nu}_s(0) W_B(0), \\
Q_y = \frac{1}{4} [tr(\tilde{W}_g^T \Gamma_W^{-1} \tilde{W}_g) + tr(\tilde{V}_g^T \Gamma_V^{-1} \tilde{V}_g)], \\
L_y = e^T (W_{B1} - c_4 \mathrm{sgn}(\tilde{\nu}_s)) + \dot{\nu}_s^T W_{B2} - c_5 \alpha_2(\|h\|)\|h\|\|\tilde{\nu}_s\|,
\end{cases} \tag{16}$$

and $tr(\cdot)$ is the trace of a matrix.

To ensure $P_y \geq 0$, c_4 and c_5 are chosen as follows [25]

$$c_4 \geq \max(\lambda_1 + \lambda_2, \lambda_1 + \frac{\lambda_3}{c_2}), \quad c_5 \geq \lambda_4. \tag{17}$$

Theorem 1. Based on the proposed neural network weight update laws (11), the presented identifier (7) is proved to be asymptotically stable. Besides, the presented identifier ensures the asymptotic stability of the velocity estimation, such that

$$\lim_{t \to \infty} ||\tilde{\nu}_s(t)|| = 0, \quad \lim_{t \to \infty} ||\dot{\tilde{\nu}}_s(t)|| = 0, \tag{18}$$

provided that

$$c_1 \geq \frac{\lambda_5}{c_2}, \quad c_1 \geq \lambda_6. \tag{19}$$

Proof. Construct the Lipschitz continuous definite function as

$$V_L = \frac{1}{2}e^T e + \frac{1}{2}c_3\tilde{\nu}_s^T\tilde{\nu}_s + P_y + Q_y, \tag{20}$$

which satisfies $U_1(w) \leq V_L(w) \leq U_2(w)$ with $U_1 = \frac{1}{2}\min(1,c_3)||w||^2$ and $U_2 = \frac{1}{2}\max(1,c_3)||w||^2$.

From (8), (11), (13) and (16), the closed-loop differential equations are defined as $\dot{w} = s(w,t)$ with $s(w,t)$ being the system error signals. By Filippov's theory [26], there exists $\dot{w} \in K[s](w,t)$ satisfying

$$\begin{cases} K[s] &= \bigcap_{\sigma>0}\bigcap_{\mu_s N=0} \overline{cos}(B(w,\sigma) - N, t), \\ B(w,\sigma) &= \{\nu_s|||w - \nu_s|| < \sigma\}, \end{cases} \tag{21}$$

where σ is a constant; $\bigcap_{\mu_s N=0}$ is the intersection over all sets N of Lebesgue measure zero; and $\overline{co}$ is convex closure.

From [13], to identify the stability of $\dot{w} = s(w,t)$, a Lyapunov stability theory is introduced. Define $\dot{V}_L$ is the derivative of V_L, and it is satisfying

$$\begin{cases} \dot{V}_L(w) \in^{a.e.} \overset{\cdot}{\tilde{V}}_L(w), \ t \in [t_0, t_f], \\ \overset{\cdot}{\tilde{V}}_L(w) = \bigcap_{\varsigma \in \partial V_L(w)} \varsigma^T K[\dot{e}^T \dot{\tilde{\nu}}_s^T \frac{1}{2}P_y^{-\frac{1}{2}}\dot{P}_y \frac{1}{2}Q_y^{-\frac{1}{2}}\dot{Q}_y]^T. \end{cases} \tag{22}$$

Using (13), (16) and (21), the derivative of V_L can be described as

$$\dot{V}_L \leq^{a.e.} -c_2c_3||\tilde{\nu}_s||^2 - c_1||e||^2 + \alpha_1(||h||)||h||||e||$$
$$+ \lambda_5||\tilde{\nu}_s||^2 + \lambda_6||e||^2 + c_5\alpha_2(||h||)||h||||\tilde{\nu}_s||.$$

Using $c_1 = c_{1a} + c_{1b}$, $c_3 = c_{3a} + c_{3b}$, $\dot{V}_L$ is

$$\dot{V}_L \leq^{a.e.} -(c_2c_{3a} - \lambda_5)||\tilde{\nu}||^2 - (c_{1a} - \lambda_6)||e||^2$$
$$+ \frac{\alpha_1(||h||)^2}{4c_{1b}}||h||^2 + \frac{c_5^2\alpha_2(||h||)^2}{4c_2c_{3b}}||h||^2. \tag{23}$$

Because of $c_1 \geq \frac{\lambda_5}{c_2}$ and $c_1 \geq \lambda_6$, for $\forall y \in \mathcal{D}_\infty$, we can get

$$\dot{V}_L \overset{a.e.}{\leq} -\rho||h||^2 + \frac{\alpha(||h||)^2}{4\eta}||h||^2$$

$$\overset{a.e.}{\leq} -U(w), \tag{24}$$

where

$$\begin{cases} \rho &= \min\{c_2 c_{3a} - \lambda_5, c_{1a} - \lambda_6\}, \\ \eta &= \min\{c_{1a}, c_2 c_{3b}/c_6^2\}, \\ \alpha(||h||^2) &= \alpha_1(||h||^2) + \alpha_2(||h||^2), \\ U(w) &= c||h||^2, \ c \geq 0, \\ \mathcal{D} &= \{w(t)| \ ||w|| \leq (2\sqrt{\rho\eta})\}. \end{cases} \tag{25}$$

Note that, from (24) and $U_1(w) \leq V_L(w) \leq U_2(w)$, we can get $V_L(w) \in \mathcal{L}_\infty$ in $\mathcal{D}$, therefore, $\tilde{\nu}_s(t), e(t) \in \mathcal{L}_\infty$ in $\mathcal{D}$. From (3) and Assumption 1, $\hat{\nu}_s(t) \in \mathcal{L}_\infty$ in $\mathcal{D}$. From (9) and the standard linear analysis, we can get $\dot{\tilde{\nu}}_s(t) \in \mathcal{L}_\infty$ in $\mathcal{D}$. From (11), $\dot{\hat{W}}_g(t) \in \mathcal{L}_\infty$; from Assumption 1, $\tau_s(t) \in \mathcal{L}_\infty$, $\hat{\beta}(t) \in \mathcal{L}_\infty$; from (7), $\mu_s(t) \in \mathcal{L}_\infty$ in $\mathcal{D}$. Based on the previous bounds, the $\dot{e}(t) \in \mathcal{L}_\infty$ in $\mathcal{D}$ and $U(w)$ is uniformly continuous in $\mathcal{D}$ can be easily obtained. Define a set χ which satisfies $\chi = \{w(t) \subset \mathcal{D}|U_2(w(t)) < \frac{1}{2}(\alpha^{-1}(2\sqrt{\rho\eta}))^2\}$, it reveals that we can increase the control gain η to contain any initial conditions, therefore, for $\forall w(0) \in \chi, t \to \infty$, we can obtain

$$c||h||^2 \to 0, \ ||\tilde{\nu}_s(t)|| \to 0, \ ||\dot{\tilde{\nu}}_s(t)|| \to 0, \ ||e|| \to 0. \tag{26}$$

As a consequence, the proposed identifier for USVs is asymptotically stable.

4   Conclusions

In this paper, an asymptotically stable identifier is developed for USVs subject to unknown dynamics which caused by parametric model uncertainty and external ocean disturbance. The identifier for USV is designed based on RISE and neural networks. Based on the proposed identifier, the estimation errors are proved to be asymptotically stable in the presence of parametric model uncertainties and external ocean disturbances. In the future, the developed asymptotically stable identifier can be used in trajectory tracking and formation tracking for achieving the desired formation tracking of a group of USVs.

References

1. Fossen, T.: Handbook of Marine Craft Hydrodynamics and Motion Control. Wiley, Hoboken (2011)
2. Peng, Z., Wang, J., Wang, J.: Constrained control of autonomous underwater vehicles based on command optimization and disturbance estimation. IEEE Trans. Ind. Electron. **66**(5), 3627–3635 (2019)

3. Peng, Z., Wang, J., Han, Q.: Path-following control of autonomous underwater vehicles subject to velocity and input constraints via neurodynamic optimization. IEEE Trans. Ind. Electron. (2019, in press). https://doi.org/10.1109/TIE.2018.2885726

4. Jin, K., Wang, H., Yi, H., Liu, J., Wang, J.: Key technologies and intelligence evolution of maritime UV. Chin. J. Ship Res. **13**(6), 1–8 (2018)

5. Li, F., Yi, H.: Application of USV maritime safety superition. Chin. J. Ship Res. **13**(6), 27–33 (2018)

6. Zhao, R., Xu, J., Xiang, X., Xu, G.: A review of path planning and coopreative control for MAUV systems. Chin. J. Ship Res. **13**(6), 58–65 (2018)

7. Ashrafiuon, H., Muske, K.R., McNinch, L.C., Soltan, R.A.: Sliding-mode tracking control of surface vessels. IEEE Trans. Ind. Electron. **55**(11), 4004–4012 (2008)

8. Cui, R., Zhang, X., Cui, D.: Adaptive sliding-mode attitude control for autonomous underwater vehicles with input nonlinearities. Ocean Eng. **123**, 45–54 (2016)

9. Xiang, X., Yu, C., Zhang, Q.: Robust fuzzy 3D path following for autonomous underwater vehicle subject to uncertainties. Comput. Oper. Res. **84**, 165–177 (2017)

10. Peng, Z., Wang, J., Wang, D.: Distributed maneuvering of autonomous surface vehicles based on neurodynamic optimization and fuzzy approximation. IEEE Trans. Control Syst. Technol. **26**(3), 1083–1090 (2018)

11. Chen, M., Ge, S.S., How, B.V.E., Choo, Y.S.: Robust adaptive position mooring control for marine vessels. IEEE Trans. Control Syst. Technol. **21**(2), 395–409 (2013)

12. Peng, Z., Wang, D.: Robust adaptive formation control of underactuated autonomous surface vehicles with uncertain dynamics. IET Control Theory Appl. **5**(12), 1378–1387 (2011)

13. Bhasin, S., Kamalapurkar, R., Johnson, M., Vamvoudakis, K.G., Lewis, F.L., Dixon, W.E.: A novel actor-critic-identifier architecture for approximate optimal control of uncertain nonlinear systems. Automatica **49**(1), 82–92 (2013)

14. Peng, Z., Wang, D., Chen, Z., Hu, X., Lan, W.: Adaptive dynamic surface control for formations of autonomous surface vehicles with uncertain dynamics. IEEE Trans. Control Syst. Technol. **21**(2), 513–520 (2013)

15. Zheng, Z., Sun, L.: Path following control for marine surface vessel with uncertainties and input saturation. Neurocomputing **177**, 158–167 (2016)

16. Liu, L., Wang, D., Peng, Z.: ESO-based line-of-sight guidance law for path following of underactuated marine surface vehicles with exact sideslip compensation. IEEE J. Oceanic Eng. **42**(2), 477–487 (2017)

17. Peng, Z., Wang, D., Shi, Y., Wang, H., Wang, W.: Containment control of networked autonomous underwater vehicles with model uncertainty and ocean disturbances guided by multiple leaders. Inf. Sci. **316**(20), 163–179 (2015)

18. Peng, Z., Wang, J., Wang, D.: Containment maneuvering of marine surface vehicles with multiple parameterized paths via spatial-temporal decoupling. IEEE ASME Trans. Mechatron. **22**(2), 1026–1036 (2017)

19. Peng, Z., Wang, J., Wang, D.: Distributed containment maneuvering of multiple marine vessels via neurodynamics-based output feedback. IEEE Trans. Ind. Electron. **64**(5), 3831–3839 (2017)

20. Peng, Z., Wang, D., Zhang, H., Sun, G.: Distributed neural network control for adaptive synchronization of uncertain dynamical multiagent systems. IEEE Trans. Neural Netw. Learn. Syst. **25**(8), 1508–1519 (2014)

21. Lei, Z.L., Guo, C.: Disturbance rejection control solution for ship steering system with uncertain time delay. Ocean Eng. **95**(1), 78–83 (2015)

22. Calise, A.J., Hovakimyan, N., Idan, M.: Adaptive output feedback control of non-linear systems using neural networks. Automatica **37**(12), 1201–1211 (2001)
23. Peng, Z., Wang, D., Wang, W., Liu, L.: Containment control of networked autonomous underwater vehicles: a predictor-based neural DSC design. ISA Trans. **59**, 160–171 (2015)
24. Lavretsky, E., Gibson, T.E.: Projection Operator in Adaptive Systems. arXiv:1112.4232 (2011)
25. Patre, P.M., MacKunis, W., Kaiser, K., Dixon, W.E.: Asymptotic tracking for uncertain dynamic systems via a multilayer neural network feedforward and RISE feedback control structure. IEEE Trans. Autom. Control **53**(9), 2180–2185 (2008)
26. Filippov, A.: Differential equations with discontinuous right-hand side. Am. Math. Soc. Trans. **42**(2), 199–231 (1964)
27. Xian, B., Dawson, D.M., Queiroz, M.S., Chen, J.: A continuous asymptotic tracking control strategy for uncertain nonlinear systems. IEEE Trans. Autom. Control **49**(7), 1206–1211 (2004)

Global Stabilization for Delayed Fuzzy Inertial Neural Networks

Qiang Xiao[1,2], Tingwen Huang[3], and Zhigang Zeng[1,2(✉)]

[1] School of Artificial Intelligence and Automation,
Huazhong University of Science and Technology, Wuhan 430074, China
{xq1620128,zgzeng}@hust.edu.cn
[2] Key Laboratory of Image Processing and Intelligent Control of Education
Ministry of China, Wuhan 430074, China
[3] Department of Mathematics, Texas A&M University at Qatar, Doha, Qatar
tingwen.huang@qatar.tamu.edu

Abstract. This paper studies the global stabilization problem for a class of fuzzy inertial neural networks (FINN) with time delays and deals with the FINN directly by non-reduced order method. By Lyapunov theory and some analytical techniques, some criteria of global asymptotic and exponential stabilization for the considered FINN are obtained. An example is given to show the effectiveness and validity of the theoretical results.

Keywords: Asymptotic and exponential stabilization ·
Inertial neural networks · Non-reduced order method · T-S fuzzy logic ·
Time delay

1 Introduction

In recent years, neural networks has received a great deal of attention due to its applications or potentials in associate memory, information science, bioinspired engineering, and signal processing [1–5]. In [5], Guo *et al.* gave results on global exponential synchronization of multiple coupled memristive neural networks via pinning control. As we know, stability of the addressed system is the basis of its application in many emerging technologies, therefore it is of both theoretical and practical significance to study the stability and stabilization problem for neural networks [6–8]. We also note that, fuzzy neural networks have been investigated for several years as they are closely related to complexity of neural networks. Among various fuzzy logics, especially the T-S fuzzy rules [9], is frequently used to analyze and synthesize nonlinear systems [10, 11].

Most of the previous work is mainly focused on first-order derivative of the states. However, it is also vital to investigate the dynamics of the networks with second-order terms, which is called an inertial term. It has been shown that the inertial term is a key tool to generate complicated bifurcations and chaos for artificial neural networks. Babcock and Westervelt mentioned in [12] that

© Springer Nature Switzerland AG 2019
H. Lu et al. (Eds.): ISNN 2019, LNCS 11555, pp. 62–69, 2019.
https://doi.org/10.1007/978-3-030-22808-8_7

the dynamics could be complex if the neuron couplings comprise of an inertial nature. Many authors have addressed inertial neural networks (INN) in a deep manner [13–21]. In [17], Xiao *et al.* studied the passivity and robust passivity for a class of memristive INNs by LMI techniques and matrix analysis. It is worthy of note that the second-order INNs are transformed into the first-order INNs by a linear transformation in most of the above works. However, it may be more meaningful to study the non-reduced order INNs directly. To the best of the authors' knowledge, just a few works have been done based on non-reduced order INNs [15]. In [15], the authors studied a class of delayed INNs without any order-reducing for the first time. As we can see, it remains much room for investigations of fuzzy INN (FINN) with non-reduced order approaches.

Motivated by the above statements, in this paper, we will study the stabilization problem for a class of delayed FINN by non-reduced order approach. Roughly stated, the main contributions of this paper are the following twofold. (i) The global asymptotic and exponential stabilization is investigated for the addressed FINN. Some criteria guaranteeing the stabilization of FINN are obtained; (ii) Different from using the linear transformation, this paper deals with the stabilization problem directly by non-reduced order approach, which has never been used for FINN.

The rest of this paper is organized as follows. In Sect. 2, some preliminaries are prepared, and the problem is also formulated. Then in Sect. 3, global stabilization problems, including global asymptotic stabilization and global exponential stabilization, are studied, respectively, followed by one numerical example in Sect. 4. Finally, conclusions are made in Sect. 5.

2   Preliminaries and Problem Formulation

Notations. $\mathbb{R}$ and $\mathbb{Z}$ are the real set and integer set; $\mathbf{I}_n$ is the index set $\{1, 2, \cdots, n\}$; $c_{pq}^{\sharp} = \max_r\{|c_{pq}^r|\}$, $d_{pq}^{\sharp} = \max_r\{|d_{pq}^r|\}$.

We consider a class of INNs described by:

$$\ddot{x}_p(t) = -a_p^{\natural}\dot{x}_p(t) - b_p^{\natural}x_p(t) + \sum_{q=1}^{n} c_{pq}f_q(x_q(t)) + \sum_{q=1}^{n} w_{pq}f_q(x_q(t - \tau_q)), \quad (1)$$

with the initial values $x_p(s) = \phi_p(s), \dot{x}_p(s) = \psi_p(s), -\tau \le s \le 0$, for $p \in \mathbf{I}_n$, where the second-order derivative of $x_p(t)$ is called an inertial term of INN (1), $x_p(t) \in \mathbb{R}$ is the state of neuron p at time t, τ_q is the time delay which satisfies $\max\{\tau_q\} = \tau$ for $q \in \mathbf{I}_n$, $a_p^{\natural} > 0$, $b_p^{\natural} > 0$, $f_q(\cdot)$ is the activation function, c_{pq} and d_{pq} represent connection weights without and with delays. $\phi_p(s), \psi_p(s)$ are bounded functions.

Introducing T-S fuzzy sets into INN (1), where the rth *rule* is as follows:
Plant Rule r:

IF $\zeta_1(t)$ is Σ_r^1 and $\cdots$ and $\zeta_q(t)$ is Σ_r^q

THEN $\ddot{x}_p(t) = -a_p^{\natural}\dot{x}_p(t) - b_p^{\natural}x_p(t) + \sum_{q=1}^{n} c_{pq}^r f_q(x_q(t)) + \sum_{q=1}^{n} w_{pq}^r f_q(x_q(t - \tau_q)),$

where $\zeta_l(t)(l \in I_q)$ are the premise variables, $\Sigma_r^l(r \in I_m, l \in I_q)$ are fuzzy sets and m is the number of **IF-THEN** rules.

Let $f_r(\zeta(t))$ be the normalized membership function, i.e., $f_r(\zeta(t)) = \frac{\pi_r(\zeta(t))}{\sum_{r=1}^{m} \pi_r(\zeta(t))}, r \in I_m$, where $\pi_r(\zeta(t)) = \Pi_{q=1}^{q} \Sigma_r^q(\zeta_q(t))$, $\Sigma_r^q(\zeta_q(t))$ is the grade of membership of $\zeta_q(t)$ in Σ_r^q. Then $f_\ell(\zeta(t)) \geq 0$ and $\sum_{\ell=1}^{m} f_\ell(\zeta(t)) = 1$.

After introducing fuzzy module, (1) can be represented by the following FINN

$$
\begin{aligned}
\ddot{x}_p(t) = & -a_p^\natural \dot{x}_p(t) - b_p^\natural x_p(t) + \sum_{q=1}^{n} \sum_{\ell=1}^{m} f_\ell(\zeta(t)) c_{pq}^r f_q(x_q(t)) \\
& + \sum_{q=1}^{n} \sum_{\ell=1}^{m} f_\ell(\zeta(t)) w_{pq}^r f_q(x_q(t - \tau_q)),
\end{aligned}
\tag{2}
$$

An assumption is made for the main results.

Assumption 1: $f_q(\cdot)$ is bounded and satisfies

$$
|f_q(\theta) - f_q(\vartheta)| \leq k_q |\theta - \vartheta|, f_q(0) = 0, \forall q \in I_n,
$$

for $\forall \theta, \vartheta \in \mathbb{R}$ and $\theta \neq \vartheta$, where constant $k_q > 0$. Obviously, $\mathbf{x}^0 = (0, 0, \cdots, 0)^T$ is an equilibrium of FINN (2).

In order to stabilize FINN (2) to $\mathbf{x}^0$, the error state feedback control term u_k is considered in this paper. Let

$$
u_p(t) = -a_p^\sharp \dot{x}_p(t) - b_p^\sharp x_p(t),
\tag{3}
$$

for any $p \in I_n$, where $a_p^\sharp$ and $b_p^\sharp$ are state feedback gains, then FINN (2) turns into

$$
\begin{aligned}
\ddot{x}_p(t) = & -a_p \dot{x}_p(t) - b_p x_p(t) + \sum_{q=1}^{n} \sum_{\ell=1}^{m} f_\ell(\zeta(t)) c_{pq}^r f_q(x_q(t)) \\
& + \sum_{q=1}^{n} \sum_{\ell=1}^{m} f_\ell(\zeta(t)) w_{pq}^r f_q(x_q(t - \tau_q)),
\end{aligned}
\tag{4}
$$

for any $p \in I_n$, where $a_p = a_p^\natural + a_p^\sharp$, $b_p = b_p^\natural + b_p^\sharp$.

Definition 1. FINN (1) is said to be globally asymptotically stabilized (GAS) to its equilibrium $\mathbf{x}^0$ under control protocol (3) provided $|x_p(t) - \mathbf{x}^0| \to 0$ as $t \to +\infty$ for any $p \in I_n$.

Definition 2. FINN (1) is said to be globally exponentially stabilized (GES) to its equilibrium $\mathbf{x}^0$ under control protocol (3) provided there exists positive constants r and $M = M(\phi, \psi)$ such that $|x_p(t) - \mathbf{x}^0| \leq Me^{rt}$ for any $t \geq 0$ and for any $p \in I_n$, ϕ, ψ is the initial value of FINN (1).

Lemma 1 [22]. Function $f(t)$ is defined on interval $[0, +\infty)$, if $f(t)$ is uniformly continuous and $\int_0^\infty f(\tau)d\tau$ exists and is bounded, then $\lim_{t \to \infty} f(t) = 0$.

3   GAS and GES of FINN

In this section, GAS and GES of FINN (2) are considered, respectively. For convenience, for any $p \in I_n$, we define

$$q_p = -2b_p + \sum_{q=1}^{n} \left(c_{pq}^{\sharp} l_q + 2c_{qp}^{\sharp} l_p + d_{pq}^{\sharp} l_q + 2d_{qp}^{\sharp} l_p \right), m_p = 4 - 2a_p - 2b_p$$

$$k_p = 2 - 2a_p + \sum_{q=1}^{n} \left(c_{pq}^{\sharp} + d_{pq}^{\sharp} \right) l_q, s = \max_p \left\{ q_p - \frac{m_p^2}{4k_p} \right\}.$$

Theorem 1. Assume that Assumption 1 holds. If

$$\max_p \{k_p\} < 0 \text{ and } s < 0, \tag{5}$$

then FINN (1) is globally asymptotically stabilized.

Proof. Consider the following Lyapunov candidate $V(t) = V_1(t) + V_2(t) + V_3(t)$, where

$$V_1(t) = \sum_{p=1}^{n} x_p^2(t), \ V_2(t) - \sum_{p=1}^{n} (x_p(t) \mid \dot{x}_p(t))^2,$$

$$V_3(t) = p \sum_{p=1}^{n} \sum_{q=1}^{n} 2d_{qp}^{\sharp} l_p \int_{t-\tau_p}^{t} x_p^2(s) \mathrm{d}s.$$

Taking the derivative of $V(t)$ along the trajectories of (4) gives

$$\dot{V}_1(t) = \sum_{p=1}^{n} 2x_p(t)\dot{x}_p(t), \dot{V}_2(t) = \sum_{p=1}^{n} 2(x_p(t) + \dot{x}_p(t))(\dot{x}_p(t) + \ddot{x}_p(t)),$$

$$\dot{V}_3(t) = \sum_{p=1}^{n} \sum_{q=1}^{n} 2d_{qp}^{\sharp} l_p x_p^2(t) - \sum_{p=1}^{n} \sum_{q=1}^{n} 2d_{qp}^{\sharp} l_p x_p^2(t - \tau_p).$$

By Assumption 1 and the fact that $2ab \leq a^2 + b^2$,

$$\sum_{p=1}^{n} 2x_p(t)(\dot{x}_p(t) + \ddot{x}_p(t))$$

$$\leq \sum_{p=1}^{n} 2(1 - a_p)x_p(t)\dot{x}_p(t) - \sum_{p=1}^{n} 2b_p x_p^2(t) + \sum_{p=1}^{n} \sum_{q=1}^{n} (c_{pq}^{\sharp} l_q$$

$$+ c_{qp}^{\sharp} l_p + d_{pq}^{\sharp} l_q)x_p^2(t) + \sum_{p=1}^{n} \sum_{q=1}^{n} d_{qp}^{\sharp} p l_p x_p^2(t - \tau_p)),$$

$$\sum_{p=1}^{n} 2\dot{x}_p(t)(\dot{x}_p(t) + \ddot{x}_p(t))$$

$$\leq \sum_{p=1}^{n} 2(1 - a_p)\dot{x}_p^2(t) - \sum_{p=1}^{n} 2b_p x_p(t)\dot{x}_p(t) + \sum_{p=1}^{n}\sum_{q=1}^{n}(c_{pq}^{\sharp}l_q$$

$$+ d_{pq}^{\sharp}l_q)\dot{x}_p^2(t) + \sum_{p=1}^{n}\sum_{q=1}^{n} c_{qp}^{\sharp}l_p x_p^2(t) + \sum_{p=1}^{n}\sum_{q=1}^{n} d_{qp}^{\sharp}l_p x_p^2(t - \tau_p)).$$

Combining them, we get

$$\dot{V}(t) \leq \sum_{p=1}^{n}\left(-2b_p + \sum_{q=1}^{n}(c_{pq}^{\sharp}l_q + 2c_{qp}^{\sharp}l_p + d_{pq}^{\sharp}l_q + 2d_{qp}^{\sharp}l_p)\right)x_p^2(t)$$

$$+ \sum_{p=1}^{n}\left(4 - 2a_p - 2b_p\right)x_p(t)\dot{x}_p(t) + \sum_{p=1}^{n}\left(2 - 2a_p + \sum_{q=1}^{n}\left(c_{pq}^{\sharp} + d_{pq}^{\sharp}\right)l_q\right)\dot{x}_p^2(t)$$

$$= \sum_{p=1}^{n} k_p\left(\dot{x}_p(t) + \frac{m_p}{2k_p}\right)^2 + \sum_{p=1}^{n}\left(q_p - \frac{m_p^2}{4k_p}\right)x_p^2(t).$$

Using (5), we get

$$\dot{V}(t) \leq sV_1(t) < 0, \tag{6}$$

therefore, $V(t) < V(0)$ for any $t \in \mathbb{R}^+$. It follows that $x_p(t)$ and $\dot{x}_p(t)$ are bounded for any $t \in \mathbb{R}^+$ and $p \in \mathbf{I}_n$. Hence, the derivative of $V_1(t)$ is bounded, implying that $V_1(t)$ is uniformly continuous. In addition, by (6), for any $t \in \mathbb{R}^+$, we have $\int_0^t V_1(\zeta)d\zeta \leq \frac{V(0)}{-s} < +\infty$, thus $\int_0^\infty V_1(\zeta)d\zeta \leq \frac{V(0)}{-s} < +\infty$, together with Lemma 1, we get $x_p(t) \to 0$ as $t \to +\infty$. By Definition 1, FINN (1) is globally asymptotically stabilized under the control protocol (3). ∎

Remark 2. Theorem 1 deals with the global asymptotic stabilization of FINN (1) under control protocol (3). One can see that Lemma 1 plays an important role in Theorem 1. Next, the global exponential stabilization is addressed.

Also, for any $p \in \mathbf{I}_n$, we define

$$\bar{q}_p = 2\lambda - 2b_p + \sum_{q=1}^{n}\left(2d_{qp}^{\sharp}l_p e^{\lambda\tau_p} + c_{pq}^{\sharp}l_q + 2c_{qp}^{\sharp}l_p + d_{pq}^{\sharp}l_q\right)\bar{m}_p = 4 + 2\lambda - 2a_p - 2b_p,$$

$$\bar{k}_p = 2 + \lambda - 2a_p + \sum_{q=1}^{n}\left(c_{pq}^{\sharp}l_q + d_{pq}^{\sharp}l_q\right), \bar{s} = \max_p\left\{q_p - \frac{m_p^2}{4k_p}\right\}.$$

Theorem 2. Assume that Assumption 1 holds. If there exists a positive constant λ such that $\max_p\{\bar{k}_p\} < 0$ and $\bar{s} < \lambda$, then FINN (1) is globally exponentially stabilized.

Proof. By the similar analysis, one gets $\dot{V}(t)e^{-\lambda t} \leq \bar{s}\sum_{p=1}^{n} x_p^2(t) = \bar{s}e^{-\lambda t}V_1(t)$, therefore, $\dot{V}(t) \leq \bar{s}V_1(t)$. Next, two cases are considered.

Case 1: $\bar{s} < 0$. Then we have $\dot{V}(t) < 0$, and hence $V(t) < V(0)$ for any $t \in \mathbb{R}^+$. It follows that $\sum_{p=1}^{n} x_p^2(t) = e^{-\lambda t}V_1(t) \leq e^{-\lambda t}V(t) \leq e^{-\lambda t}V(0)$.

Case 2: $\bar{s} \geq 0$. Then $\dot{V}(t) \leq \bar{s}V_1(t) \leq \bar{s}V(t)$, therefore, $V(t) \leq e^{\bar{s}t}V(0)$. So we have $\sum_{p=1}^{n} x_p^2(t) = e^{-\lambda t}V_1(t) \leq e^{-\lambda t}V(t) \leq e^{(\bar{s}-\lambda)t}V(0)$.

Together with Definition 2, FINN (1) is globally exponentially stabilized under control protocol (3). ∎

Remark 3. It is obvious that $\lambda > 0$ is a necessity of Theorem 2. If $\lambda = 0$, then we can only get the global asymptotically stabilization of FINN (1). Therefore, Theorem 2 could be viewed as a generalization of Theorem 1.

Remark 4. In Theorems 1 and 2, we deal with the stabilization problem for FINN directly by the non-reduced order approach, rather than the variable transformation method. As far as the authors' knowledge, very few paper has been considered the dynamics of FINN in a non-reduced order manner. Therefore, the results in this paper are novel.

4   An Example

Example 1. Consider a two-neuron delayed FINN with two fuzzy rules:
Plant Rule 1: IF $x_p(t)$ is $x_p(t) \leq 0$, THEN

$$\ddot{x}_p(t) - -u_p^\natural \dot{x}_p(t) - b_p^\natural x_p(t) + \sum_{q=1}^{2} c_{pq}^1 f_q(x_q(t)) + \sum_{q=1}^{2} d_{pq}^1 f_q(x_q(t-\tau_q)), \quad (7)$$

Plant Rule 2: IF $x_p(t)$ is $x_p(t) > 0$, THEN

$$\ddot{x}_p(t) = -a_p^\natural \dot{x}_p(t) - b_p^\natural x_p(t) + \sum_{q=1}^{2} c_{pq}^2 f_q(x_q(t)) + \sum_{q=1}^{2} d_{pq}^2 f_q(x_q(t-\tau_q)). \quad (8)$$

where $a_p^\natural = b_p^\natural = 0.01$ for any $p = 1,2$; $\tau_1 = 0.01$, $\tau_2 = 0.005$; $c_{11}^1 = 0.05$, $c_{11}^2 = 0.01$, $c_{12}^1 = 0.02$, $c_{12}^2 = 0.01$, $c_{21}^1 = 0.07$, $c_{21}^2 = 0.05$, $c_{22}^1 = 0.03$, $c_{22}^2 = 0.05$, $d_{11}^1 = 0.01$, $d_{11}^2 = 0.04$, $d_{12}^1 = 0.09$, $d_{12}^2 = 0.05$, $d_{21}^1 = 0.07$, $d_{21}^2 = 0.01$, $d_{22}^1 = 0.05$, $d_{22}^2 = 0.06$, and $f_1(x) = f_2(x) = (|x+1| - |x-1|)/2$.

The state trajectories of FINNs (7) and (8) are shown in Fig. 1(a)–(b), which shows FINNs (7)–(8) is unstable. Now the state feedback control protocol (3) is considered. Let $a_p^\sharp = b_p^\sharp = 1.49$ for any $p = 1,2$. Then it is easy to be verified that conditions of Theorem 1 are satisfied with $\max_p\{k_p\} = -0.750$ and $s = -0.9767$. Therefore, FINNs (7)–(8) is globally asymptotically stabilized, which is shown in Fig. 1(c)–(d).

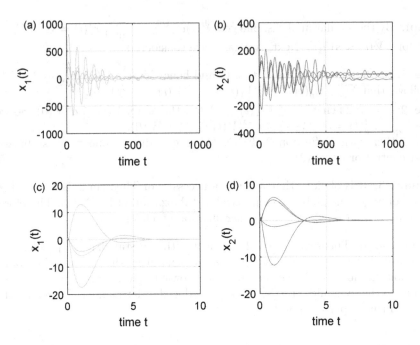

Fig. 1. The trajectories of state $x_1(k)$ and $x_2(k)$.

5   Conclusions

This paper considered the global asymptotic stabilization and global exponential stabilization problem for a class of delayed fuzzy inertial neural networks (FINN). Rather than using the reduced order approach, this paper utilizes the non-reduced order method. Some criteria for the stabilization of FINN are derived by Lyapunov method and some analytical techniques. The obtained results in this paper enrich and improve some previous ones. An example are given to illustrate the effectiveness and validity of the main results.

References

1. Liu, X., Zeng, Z., Wen, S.: Implementation of memristive neural networks with full-function Pavlov associative memory. IEEE Trans. Circuits Syst. I Reg. Papers **63**(9), 1454–1463 (2016)
2. Hayakawa, Y., Nakajima, K.: Design of the inverse function delayed neural networks for solving combinatorial optimization problems. IEEE Trans. Neural Netw. **21**, 224–237 (2010)
3. Cheng, G., Zhou, P., Han, J.: Learning rotation-invariant convolutional neural networks for object detection in VHR optical remote sensing images. IEEE Trans. Geosci. Remote Sens. **54**, 7405–7415 (2016)

4. Zhao, X., Shi, P., Zheng, X., Zhang, J.: Intelligent tracking control for a class of uncertain high-order nonlinear systems. IEEE Trans. Neural Netw. Learn. Syst. **27**, 1976–1982 (2016)

5. Guo, Z., Yang, S., Wang, J.: Global synchronization of memristive neural networks subject to random disturbances via distributed pinning control. Neural Netw. **84**, 67–79 (2016)

6. Yang, W., Yu, W., Cao, J., Alsaadi, F., Hayat, T.: Global exponential stability and lag synchronization for delayed memristive fuzzy Cohen-Grossberg BAM neural networks with impulses. Neural Netw. **98**, 122–153 (2018)

7. Zhou, Y., Li, C., Chen, L., Huang, T.: Global exponential stability of memristive Cohen-Grossberg neural networks with mixed delays and impulse time window. Neurocomputing **31**, 2384–2391 (2018)

8. Wen, S., Zeng, Z., Huang, T., Yu, X.: Noise cancellation of memristive neural networks. Neural Netw. **60**, 74–83 (2014)

9. Takagi, T., Sugeno, M.: Fuzzy identification of systems and its applications to modelling and control. IEEE Trans. Syst. Man Cybern. **15**, 116–132 (1985)

10. Zhao, X., Yin, Y., Niu, B., Zheng, X.: Stabilization for a class of switched nonlinear systems with novel average dwell time switching by T-S fuzzy modeling. IEEE Trans. Cybern. **46**, 1952–1957 (2016)

11. Li, Y., Liu, L., Feng, G.: Adaptive finite-time controller design for T-S fuzzy systems. IEEE Trans. Cybern. **47**(9), 2425–2436 (2017). https://doi.org/10.1109/TCYB.2017.2671902

12. Babcock, K., Westervelt, R.: Stability and dynamics of simple electronic neural networks with added inertia. Physica D **23**, 464–469 (1986)

13. Cao, J., Wan, Y.: Matrix measure strategies for stability and synchronization of inertial BAM network with time delays. Neural Netw. **53**, 165–172 (2014)

14. Tu, Z., Cao, J., Hayat, T.: Matrix measure based dissipativity analysis for inertial delayed uncertain neural networks. Neural Netw. **75**, 47–55 (2016)

15. Li, X., Li, X., Hu, C.: Some new results on stability and synchronization for delayed inertial neural networks based on non-reduced order method. Neural Netw. **96**, 91–100 (2017)

16. Zhang, G., Zeng, Z.: Exponential stability for a class of memristive neural networks with mixed time-varying delays. Appl. Math. Comput. **321**, 544–554 (2018)

17. Xiao, Q., Huang, Z., Zeng, Z.: Passivity analysis for memristor-based inertial neural networks with discrete and distributed delays. IEEE Trans. Syst. Man Cybern: Syst. **49**, 375–385 (2019)

18. Xiao, Q., Huang, T., Zeng, Z.: Passivity and passification of fuzzy memristive inertial neural networks on time scales. IEEE Trans. Fuzzy Syst. **26**, 3342–3355 (2018)

19. Xiao, Q., Huang, T., Zeng, Z.: Global exponential stability and synchronization for a class of generalized discrete-time inertial neural networks with time delays. IEEE Trans. Neural Netw. Learn. Syst. https://doi.org/10.1109/TNNLS.2018.2874982

20. Gong, S., Yang, S., Guo, Z., Huang, T.: Global exponential synchronization of inertial memristive neural networks with time-varying delay via nonlinear controller. Neural Netw. **102**, 138–148 (2018)

21. Guo, Z., Gong, S., Huang, T.: Finite-time synchronization of inertial memristive neural networks with time delay via delay-dependent control. Neurocomputing **293**, 100–107 (2018)

22. Popov, V.: Hyperstability of Control Systems. Springer, New York (1973)

Neural-Network-Based Modular Dynamic Surface Control for Surge Speed Tracking of an Unmanned Surface Vehicle Driven by a DC Motor

Chengcheng Meng[1], Lu Liu[1], Zhouhua Peng[1(✉)], Dan Wang[1(✉)],
Haoliang Wang[1], and Gang Sun[2]

[1] School of Marine Electrical Engineering, Dalian Maritime University,
Dalian 116026, People's Republic of China
zhpeng@dlmu.edu.cn, dwangdl@gmail.com
[2] School of Information Engineering, Jiangxi University of Science and Technology,
Ganzhou 341000, People's Republic of China
gs_sungang@126.com

Abstract. In this paper, a surge speed tracking problem for an unmanned surface vehicle (USV) driven by a direct current (DC) motor is investigated. The surge velocity tracking controller is developed based on a modular neural dynamic surface control (MNDSC) design method. Specifically, two neural-network-based identifiers are designed to identify the uncertainties existing in surge dynamics, propeller and DC motor. Then, a robust adaptive surge speed controller based on a dynamic surface control design method is designed by using the recovered unknown information from identifiers. The proposed surge velocity tracking controller is able to track any time-varying bounded velocity profiles. The stability of the closed-loop system is established based on cascade theory and input-to-state stability (ISS) theory. The effectiveness of the proposed surge speed controller for the USV is illustrated via simulations.

Keywords: Surge speed tracking · Neural networks ·
Dynamic surface control · DC motor · Modular design

This work was supported in part by the National Natural Science Foundation of China under Grants 51579023, 61673081, the Innovative Talents in Universities of Liaoning Province under Grant LR2017014, High Level Talent Innovation and Entrepreneurship Program of Dalian under Grant 2016RQ036, China Postdoctoral Science Foundation 2019M650086, the National Key Research and Development Program of China under Grant 2016YFC0301500, the Training Program for High-level Technical Talent in Transportation Industry under Grant 2018-030, the Fundamental Research Funds for the Central Universities under Grant 3132019101,3132019013, and Outstanding Youth Support Program of Dalian.

© Springer Nature Switzerland AG 2019
H. Lu et al. (Eds.): ISNN 2019, LNCS 11555, pp. 70–80, 2019.
https://doi.org/10.1007/978-3-030-22808-8_8

1   Introduction

In recent decades, unmanned surface vehicles (USVs) have gained increasing attention due to their wide applications in marine industry, military and scientific fields [1–4]. A large number of researchers are dedicated to motion control of USVs and numerous control methods are proposed [5–16].

Surge speed control is a fundamental problem in motion control of USVs. The speed of ship is not only related to the thrust provided by the propulsion system, but also to the shape of the hull, the size of the sea surface waves and the flowing speed of seawater. As for surge speed dynamics driven by a DC motor, it is difficult to attain good control performance by using conventional control methods such as PID control and traditional adaptive control [17]. With the development of modern control theory and artificial intelligence theory, fuzzy control, neural network and adaptive control combined with other control methods are used to control ship speed, exhibiting better control performance than conventional control methods [17–19]. In [17], three control algorithms are applied to velocity control of an USV, namely, classical PID control with gain scheduling, model reference adaptive control, and L1 adaptive control. Moreover, a comparative study of three approaches is presented to test the capability of speed stabilization. In [18], an L1 adaptive controller is designed to achieve surge speed control. In [19], a nominal PI controller is designed for speed tracking control by using the $\mathcal{H}_\infty$ output feedback technique. However, the previous studies on propulsion system and ship speed control have been carried out independently [17–20].

In this paper, we consider the surge speed control problem of USV subject to unknown models of DC motor, propeller, and surge dynamics. An MNDSC design method is adopted to design controller for USV to achieve the goal of surge speed tracking. Two identifiers based on NNs are developed to deal with the uncertainties due to the unknown model of surge dynamics, propeller, and DC motor. Then, a robust adaptive surge speed controller is designed based on a dynamic surface control design method. The proposed surge velocity tracking controller is capable of tracking any time-varying bounded velocity profiles in the presence of unknown surge dynamics and motor dynamics. Compared to the existing NNs based adaptive control method [21], the MNDSC method improves the speed of approaching unknown functions by NNs. Therefore, the proposed surge velocity tracking controller designed by the MNDSC has a satisfactory dynamic performance. The stability of close-loop system is proved by ISS theory and cascade theory. The simulation results indicate that the controller is able to achieve desired control performances not only in steady-state behavior but also in dynamic properties.

The remainder of this paper is organized as follows. The DC motor model, propeller model and surge dynamics are established in Sect. 2. Section 3 gives controller design procedure and stability analysis. Section 4 provides the simulation results to illustrate the efficacy of the proposed approach. Section 5 concludes this paper.

2   Problem Formulation

In this section, DC motor model, propeller model and surge speed dynamics are established and the motion equation for the system composed of them are provided.

2.1   DC Motor Model

According to the electromagnetic torque equation, armature voltage and motor torque balance equation in [22], the motion equation of DC motor can be obtained as

$$T_m \dot{\omega} + \omega = K_m u_a - K_c M_c, \tag{1}$$

where K_m, K_c are transfer coefficients of motor, T_m is electromechanical coefficient of motor, ω is motor rotating speed, u_a is the motor input voltage, and M_c is total load torque.

2.2   Propeller Model

According to [23], the thrust and torque of the propeller in open water can be expressed as (usually referred to as non-bounded forms)

$$P = (1 - r)K_P \rho D^4 n|n|, \tag{2}$$

$$T = K_T \rho D^5 n|n|, \tag{3}$$

where P and T are propeller thrust and torque, K_P and K_T are propeller thrust coefficient and torque coefficient, which are functions of the advance ratio J, i.e., $K_P = K_P(J)$, $K_T = K_T(J)$, r is the thrust reduction factor, ρ is the density of seawater, n is propeller shaft speed, and D is propeller diameter.

The advance ratio J is defined as [19]:

$$J = \frac{v_p}{nD}, \tag{4}$$

where v_p is actual forward speed of the propeller.

2.3   Surge Speed Dynamics

According to [24], a 3-DOF maneuvering model can be decoupled in a forward speed (surge) model and a sway-yaw subsystem for maneuvering. Therefore, the surge model can be written as follows

$$m\dot{u} - X_u u - X_{|u|u} u = \tau, \tag{5}$$

where m is the mass of ship, u is surge speed and τ is the sum of control and external forces on the ship.

According to (1), (2), (3) and (5), the dynamic equation can be rewritten in the following form:

$$\begin{cases} \dot{u} = f_1(u,\omega) + b_0\omega, \\ \dot{\omega} = f_2(\omega) + b_1 u_a, \end{cases} \qquad (6)$$

where $b_0 = 1/m$, $b_1 = K_m/T_m$, $f_1(u,\omega) = \{[(1-r)K_p\rho D^4|\omega|\omega/4\pi^2] + X_u u + X_{|u|u} u - \omega\}/m$ and $f_2(\omega) = (-\omega - K_c K_T \rho D^5|\omega|\omega/4\pi^2)/T_m$. The control objective of this paper is to develop a adaptive control law for a USV driven by a DC motor to track a time-varying velocity reference signal u_r.

3   Surge Tracking Controller Design

3.1   Identifier Design

In this subsection, two identifiers are developed to identify the uncertain nonlinear functions of the system (6).

Step 1: The unknown function $f_1(u,\omega)$ can be approximated by an NN as follows

$$f_1(u,\omega) = W_1^T \theta(u,\omega) + \epsilon_1, \quad u, \omega \in S, \qquad (7)$$

where $\theta(\cdot)$ is a known activation function, W_1 is an ideal NN weigh satisfying $\|W_1\|_F \leq W_1^*$ with W_1^* is a positive constant, $S \subset \Re$ is a compact set, and ϵ_1 is the function reconstruction error, there exists a positive constant ϵ_1^*, such that $\|\epsilon_1\| \leq \epsilon_1^*$.

A surge speed identifier is developed to design an update law for $\hat{W}_1$ as follows

$$\dot{\hat{u}} = \hat{W}_1^T \theta(u,\omega) + b_0\omega - (\zeta_1 + \mu_1)\tilde{u}, \qquad (8)$$

where $\zeta_1 \in R$ and $\mu_1 \in R$ are positive constants, $\hat{u}$ is an estimated value of u and $\tilde{u} = \hat{u} - u$.

The update law for $\hat{W}_1$ is designed as

$$\dot{\hat{W}}_1 = -\Gamma_1[\theta(u,\omega)\tilde{u} + k_1\hat{W}_1], \qquad (9)$$

where $\Gamma_1 \in R$ and $k_1 \in R$ are positive constants.

Step 2: The unknown function $f_2(\omega)$ can be approximated by an NN as follows

$$f_2(\omega) = W_2^T \theta(\omega) + \epsilon_2, \quad \omega \in S, \qquad (10)$$

where W_2 is an ideal NN weigh satisfying $\|W_2\|_F \leq W_2^*$ with W_2^* is a positive constant, and ϵ_2 is the function reconstruction error, there exists a positive constant ϵ_2^*, such that $\|\epsilon_2\| \leq \epsilon_2^*$.

A motor speed identifier is proposed to develop an update law for $\hat{W}_2$ as follows

$$\dot{\hat{\omega}} = \hat{W}_2^T \theta(\omega) + b_1 u_a - (\zeta_2 + \mu_2)\tilde{\omega}, \tag{11}$$

where $\zeta_2 \in R$ and $\mu_2 \in R$ are positive constants, $\hat{\omega}$ is an estimated value of ω and $\tilde{\omega} = \hat{\omega} - \omega$.

The update law for $\hat{W}_2$ is designed as

$$\dot{\hat{W}}_2 = -\Gamma_2[\theta(\omega)\tilde{\omega} + k_2\hat{W}_2], \tag{12}$$

where $k_2 \in R$ and Γ_2 are positive constants.

3.2   Controller Design

In this subsection, a dynamic surface control (DSC) technique is applied to design surge tracking controller for the USV driven by DC motor.

Step 1: First of all, define the first dynamic surface error $z_1 = u - u_r$, and the time derivative of z_1 along (6) is

$$\dot{z}_1 = f_1(u, \omega) + \omega - \dot{u}_r. \tag{13}$$

The virtual control law α_ω is designed as

$$\alpha_\omega = [-\zeta_1(u - u_r) + \dot{u}_r - \hat{W}_1^T \theta(u, \omega)]/b_0, \tag{14}$$

where u_r is the given surge speed.

Let α_ω pass through a first-order filter to get ω_d

$$\sigma_1 \dot{\omega}_d = \alpha_\omega - \omega_d, \alpha_\omega(0) = \omega_d(0), \tag{15}$$

where $\sigma_1 \in R$ is a positive constant, and ω_d is an estimate of α_ω.

Step 2: Similar to *Step 1*, let the second surface error be $z_2 = \omega - \omega_d$, and the time derivative of z_2 along (6) satisfies

$$\dot{z}_2 = f_2(\omega) + u_a - \dot{\omega}_d. \tag{16}$$

The practical control law u_a is designed as follows

$$u_a = [-\zeta_2(\omega - \omega_d) + \dot{\omega}_d - \hat{W}_2^T \theta(\omega)]/b_1, \tag{17}$$

where ζ_2 is a positive constant.

A frame diagram of the proposed surge tracking controller is displayed in Fig. 1.

4   Stability Analysis

Let $\hat{z}_1 = \hat{u} - u_r$, $\hat{z}_2 = \hat{\omega} - \omega_d$, $q_2 = \omega_d - \alpha_\omega$, $\tilde{W}_1 = \hat{W}_1 - W_1$, $\tilde{W}_2 = \hat{W}_2 - W_2$, and then, the error dynamics of $\hat{z}_1$, $\hat{z}_2$, $\tilde{u}$, $\tilde{\omega}$, $\tilde{W}_1$ and $\tilde{W}_2$ can be written as

$$\begin{cases} \dot{\hat{z}}_1 = -\zeta_1 \hat{z}_1 - \mu_1 \tilde{u} - \tilde{\omega} + \hat{z}_2 + q_2, \\ \dot{\hat{z}}_2 = -\zeta_2 \hat{z}_2 - \mu_2 \tilde{\omega}, \end{cases} \tag{18}$$

and

$$\begin{cases} \dot{\tilde{u}} = -(\zeta_1 + \mu_1)\tilde{u} + \tilde{W}_1^T \theta(u,\omega) - \epsilon_1, \\ \dot{\tilde{\omega}} = -(\zeta_2 + \mu_2)\tilde{\omega} + \tilde{W}_2^T \theta(\omega) - \epsilon_2, \\ \dot{\tilde{W}}_1 = -\Gamma_1[\theta(u,\omega)\tilde{u} + k_1 \hat{W}_1], \\ \dot{\tilde{W}}_2 = -\Gamma_2[\theta(\omega)\tilde{\omega} + k_2 \hat{W}_2]. \end{cases} \tag{19}$$

The stability of the subsystem (18) is provided by the lemma below:

Lemma 1. The subsystem (18) with states being $\hat{z}_1$ and $\hat{z}_2$, and inputs being $\tilde{u}$, $\tilde{\omega}$ and q_2 is ISS.

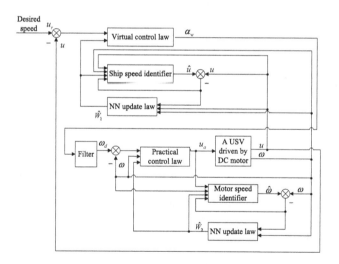

Fig. 1. A frame diagram of surge speed controller.

Proof. Firstly, select a Lyapunov function for the subsystem (18)

$$V_c = \frac{1}{2}\hat{z}_1^2 + \frac{1}{2\zeta_2 \mu_2}\hat{z}_2^2. \tag{20}$$

The derivative the Lyapunov function (20) to time is

$$\dot{V}_c = \hat{z}_1 \dot{\hat{z}}_1 + \frac{1}{\zeta_2 \mu_2} \hat{z}_2 \dot{\hat{z}}_2$$

$$= -\zeta_1 \hat{z}_1^2 - \mu_1 \tilde{u} \hat{z}_1 - \tilde{\omega} \hat{z}_1 + \hat{z}_1 \hat{z}_2 + \hat{z}_1 q_2 - \frac{1}{\mu_2} \hat{z}_2^2 - \frac{1}{\zeta_2} \tilde{\omega} \hat{z}_2$$

$$\leq -\zeta_1 \|\hat{z}_1\|^2 + \mu_1 \|\tilde{u}\|\|\hat{z}_1\| + \|\tilde{\omega}\|\|\hat{z}_1\| + \|\hat{z}_1\|\|\hat{z}_2\| + \|\hat{z}_1\|\|q_2\| \qquad (21)$$

$$- \frac{1}{\mu_2} \|\hat{z}_2\|^2 + \frac{1}{\zeta_2} \|\tilde{\omega}\|\|\hat{z}_2\|.$$

Let $G_1 = [\|\hat{z}_1\|, \|\hat{z}_2\|]^T$, $H_1 = [\mu_1 \|\tilde{u}\|, \frac{1}{\zeta_2}\|\tilde{\omega}\|, \|q_2\|]^T$, and $K = \begin{bmatrix} \zeta_1 & -1 \\ 0 & \frac{1}{\mu_2} \end{bmatrix}$, $c_1 = \lambda_{\min}(K)$, and then the aforementioned inequality (21) can be written as

$$\dot{V}_c \leq -c_1 \|G_1\|^2 + \|H_1\|\|G_1\|. \qquad (22)$$

If and only if $\|G_1\|$ satisfies $\|G_1\| \geq \frac{\|H_1\|}{c_1 \eta_1}$,

$$\dot{V}_c \leq -(1 - \eta_1) c_1 \|G_1\|^2, \qquad (23)$$

where $0 < \eta_1 < 1$.

We can draw the conclusion that the subsystem (18) is ISS. Letting $P_1 = \text{diag}(1, \frac{1}{\zeta_2 \mu_2})$, there exists a $\mathcal{KL}$ function $\beta_1(\cdot)$ and $\mathcal{K}_\infty$ functions $\Delta^{\tilde{u}}(\cdot)$, $\Delta^{\tilde{\omega}}(\cdot)$ and $\Delta^{q_2}(\cdot)$ such that

$$\|G_1(t)\| \leq \beta_1(\|G_1(t_0)\|, t - t_0) + \Delta^{\tilde{u}}(\|\tilde{u}\|) + \Delta^{\tilde{\omega}}(\|\tilde{\omega}\|) + \Delta^{q_2}(\|q_2\|), \qquad (24)$$

where $\Delta^{\tilde{u}}(s) = \frac{s \mu_1 \sqrt{\lambda_{\max}(P_1)}}{c_1 \eta_1 \sqrt{\lambda_{\min}(P_1)}}$, $\Delta^{\tilde{\omega}}(s) = \frac{s \sqrt{\lambda_{\max}(P_1)}}{c_1 \eta_1 \zeta_2 \sqrt{\lambda_{\min}(P_1)}}$, and $\Delta^{q_2}(s) = \frac{s \sqrt{\lambda_{\max}(P_1)}}{c_1 \eta_1 \sqrt{\lambda_{\min}(P_1)}}$.

The stability of the subsystem (19) is provided by the lemma below:

Lemma 2. The subsystem (19) with states being $\tilde{u}$, $\tilde{\omega}$, $\tilde{W}_1$ and $\tilde{W}_2$ and inputs being ϵ_1, ϵ_2, W_1 and W_2 is ISS.

Proof. Firstly, set up a Lyapunov function for subsystem (19)

$$V_e = \frac{1}{2}\tilde{u}^2 + \frac{1}{2}\tilde{\omega}^2 + \frac{1}{2}\Gamma_1^{-1}\tilde{W}_1^T \tilde{W}_1 + \frac{1}{2}\Gamma_1^{-1}\tilde{W}_2^T \tilde{W}_2. \qquad (25)$$

The derivative of the Lyapunov function (25) to time is

$$\dot{V}_e = \tilde{u}\dot{\tilde{u}} + \tilde{\omega}\dot{\tilde{\omega}} + \Gamma_1^{-1}\tilde{W}_1^T \dot{\tilde{W}}_1 + \Gamma_2^{-1}\tilde{W}_2^T \dot{\tilde{W}}_2,$$

$$= -(\zeta_1 + \mu_1)\tilde{u}^2 - \tilde{u}\epsilon_1 - k_1 \tilde{W}_1^2 - k_1 \tilde{W}_1^T W_1 - (\zeta_2 + \mu_2)\tilde{\omega}^2$$

$$- \tilde{\omega}\epsilon_2 - k_2 \tilde{W}_2^2 - k_2 \tilde{W}_2^T W_2,$$

$$\leq -(\zeta_1 + \mu_1)\|\tilde{u}\|^2 + \|\tilde{u}\|\|\epsilon_1\| - k_1\|\tilde{W}_1\|^2 + k_1\|\tilde{W}_1^T\|\|W_1\| \qquad (26)$$

$$- (\zeta_2 + \mu_2)\|\tilde{\omega}\|^2 + \|\tilde{\omega}\|\|\epsilon_2\| - k_2\|\tilde{W}_2\|^2 + k_2\|\tilde{W}_2^T\|\|W_2\|.$$

Let $G_2 = [\|\tilde{u}\|, \|\tilde{\omega}\|, \|\tilde{W}_1\|, \|\tilde{W}_2\|]^T$, $H_2 = [\|\epsilon_1\|, \|\epsilon_2\|, k\|W_1\|, k\|W_2\|]^T$, $k = \min(k_1, k_2)$, $c_2 = \min(\zeta_1 + \mu_1, \zeta_2 + \mu_2, k_1, k_2)$, and then the aforementioned inequality (26) can be rewritten as

$$\dot{V}_e \leq -c_2\|G_2\|^2 + \|H_2\|\|G_2\|. \tag{27}$$

If and only if $\|G_2\|$ satisfies $\|G_2\| \geq \frac{\|H_2\|}{c_2\eta_2}$,

$$\dot{V}_e \leq -(1 - \eta_2)c_2\|G_2\|^2, \tag{28}$$

where $0 < \eta_2 < 1$.

We can draw the conclusion that the subsystem (19) is ISS. Letting $P_2 = \mathrm{diag}(1, \Gamma_1^{-1}, \Gamma_2^{-1})$, there exists a $\mathcal{KL}$ function $\beta_2(\cdot)$ and $\mathcal{K}_\infty$ functions $\Delta^{\epsilon_1}(\cdot)$, $\Delta^{\epsilon_2}(\cdot)$, $\Delta^{W_1}(\cdot)$ and $\Delta^{W_2}(\cdot)$ such that

$$\|G_2(t)\| \leq \beta_2(\|G_2(t_0)\|, t - t_0) + \Delta^{\epsilon_1}(\|\epsilon_1\|) + \Delta^{\epsilon_2}(\|\epsilon_2\|) \tag{29}$$
$$+ \Delta^{W_1}(\|W_1\|) + \Delta^{W_2}(\|W_2\|),$$

where $\Delta^{\epsilon_1}(s) = \Delta^{\epsilon_2}(s) = \frac{s\sqrt{\lambda_{\max}(P_2)}}{c_2\eta_2\sqrt{\lambda_{\min}(P_2)}}$ and $\Delta^{W_1}(s) = \Delta^{W_2}(s) = \frac{sk\sqrt{\lambda_{\max}(P_2)}}{c_2\eta_2\sqrt{\lambda_{\min}(P_2)}}$.

The stability of the system cascaded by (18) and (19) is provided by the following theorem.

Theorem 1. The closed-loop system cascaded by (18) and (19) is ISS.

Proof. According to *Lemma 1* in [25], both the subsystems (18) and (19) are ISS, and then the system cascaded by (18) and (19) is ISS.

5   Simulation Verification

To verify the validity of the surge speed tracking controller, the simulation is established for the system described by (6). The system parameters are chosen as: $K_m = 6$, $K_c = 0.02$, $T_m = 3$, m $= 23.8\,\mathrm{kg}$, $X_u = -0.7225$, $X_{|u|u} = -1.3274$, $D = 0.05\,\mathrm{m}$, $\rho = 1025\,\mathrm{kg/m_3}$, $r = 0.9$. The controller parameters are chosen as: $\zeta_1 = \zeta_2 = 1$, $\sigma_1 = 0.01$, $\Gamma_1 = \Gamma_2 = 1000$, $k_1 = k_2 = 0.001$, $\mu_1 = \mu_2 = 10$. The activation function is chosen as: $\theta = (1 - e^{-x})/(1 + e^{-x})$.

Figures 2, 3, 4 and 5 show the simulation results. Figure 2 shows that the system response at a desired reference speed of $0.4\,\mathrm{m/s}$. It can be seen that the trajectory of surge speed rises quickly and smoothly and tracks the given speed well. Figure 3 depicts that the curve of the controller output, i.e., the DC motor input voltage. Figures 4 and 5 demonstrate that estimate performance of NNs and it can be observed that the outputs of NNs are able to fast approximate the uncertain nonlinear functions.

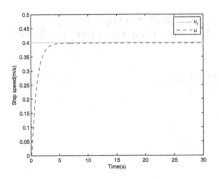

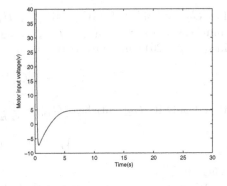

Fig. 2. The surge speed controller result.

Fig. 3. The output of controller.

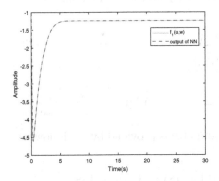

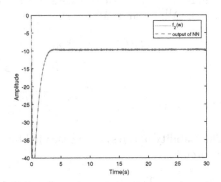

Fig. 4. The estimate performance of the first order NN.

Fig. 5. The estimate performance of the second order NN.

6   Conclusions

In this paper, a surge speed control method is presented for a USV driven by a DC motor. An MNDSC design scheme is used to develop the controller for the surge speed tracking. By utilizing neural-network-based identifiers, the uncertain nonlinear functions can be approximated. The surge speed tracking controller can track any time-varying bounded velocity profiles. The stability of the total closed-loop system is analyzed based on ISS theory and cascade theory. Finally, the simulation results illustrate the effectiveness and rationality of the designed controller.

References

1. Ramos, P.A., Neves, M.V., Pereira, F.L.: Mapping and initial dilution estimation of an ocean outfall plume using an autonomous underwater vehicle. Cont. Shelf Res. **27**(5), 583–593 (2007)
2. Bayat, B., Crasta, N., Crespi, A., Pascoal, A.M., Ijspeert, A.: Environmental monitoring using autonomous vehicles: a survey of recent searching techniques. Curr. Opin. Biotechnol. **45**, 76–84 (2017)
3. Jin, K., Wang, H., Yi, H., Liu, J., Wang, J.: Key technologies and intelligence evolution of maritime UV. Chin. J. Ship Res. **13**(6), 1–8 (2018)
4. Li, F., Yi, H.: Application of USV maritime safety supervision. Chin. J. Ship Res. **13**(6), 27–33 (2018)
5. Peng, Z., Wang, D., Chen, Z., Hu, X., Lan, W.: Adaptive dynamic surface control for formations of autonomous surface vehicles with uncertain dynamics. IEEE Trans. Control Syst. Technol. **21**(2), 513–520 (2013)
6. Peng, Z., Wang, D., Zhang, H., Sun, G.: Distributed neural network control for adaptive synchronization of uncertain dynamical multiagent systems. IEEE Trans. Neural Netw. Learn. Syst. **25**(8), 1508–1519 (2014)
7. Peng, Z., Wang, D., Shi, Y., Wang, H., Wang, W.: Containment control of networked autonomous underwater vehicles with model uncertainty and ocean disturbances guided by multiple leaders. Inf. Sci. **316**(20), 163–179 (2015)
8. Peng, Z., Wang, D., Wang, W., Liu, L.: Neural adaptive steering of an unmanned surface vehicle with measurement noises. Neurocomputing **186**, 228–234 (2016)
9. Liu, Z., Zhang, Y., Yu, Y., Yuan, C.: Unmanned surface vehicles: an overview of developments and challenges. Annu. Rev. Control **41**, 71–93 (2016)
10. Peng, Z., Wang, J., Wang, D.: Containment maneuvering of marine surface vehicles with multiple parameterized paths via spatial-temporal decoupling. IEEE/ASME Trans. Mechatron. **22**(2), 1026–1036 (2017)
11. Peng, Z., Wang, J., Wang, D.: Distributed containment maneuvering of multiple marine vessels via neurodynamics-based output feedback. IEEE Trans. Ind. Electron. **64**(5), 3831–3839 (2017)
12. Peng, Z., Wang, D., Wang, J.: Predictor-based neural dynamic surface control for uncertain nonlinear systems in strict-feedback form. IEEE Trans. Neural Netw. Learn. Syst. **28**, 2156–2167 (2017)
13. Peng, Z., Wang, J.: Output-feedback path-following control of autonomous underwater vehicles based on an extended state observer and projection neural networks. IEEE Trans. Syst. Man Cybern. Syst. **48**(4), 535–544 (2018)
14. Peng, Z., Wang, J., Wang, D.: Distributed maneuvering of autonomous surface vehicles based on neurodynamic optimization and fuzzy approximation. IEEE Trans. Control Syst. Technol. **26**(3), 1083–1090 (2018)
15. Peng, Z., Wang, J., Han, Q.: Path-following control of autonomous underwater vehicles subject to velocity and input constraints via neurodynamic optimization. IEEE Trans. Ind. Electron. 1 (2018)
16. Peng, Z., Wang, J., Wang, J.: Constrained control of autonomous underwater vehicles based on command optimization and disturbance estimation. IEEE Trans. Ind. Electron. **66**(5), 3627–3635 (2019)
17. Kragelund, S., Dobrokhodov, V., Monarrez, A., Hurban, M., Khol, C.: Adaptive speed control for autonomous surface vessels. In: MTS/IEEE Oceans - San Diego, pp. 1–10 (2013)

18. Svendsen, C.H., Hoick, N.O., Galeazzi, R., Blanke, M.: L1 adaptive manoeuvring control of unmanned high-speed water craft. IFAC Proc. Vol. **45**(27), 144–151 (2012)
19. Zhao, Z., Yang, Y., Zhou, J., Li, L., Yang, Q.: Adaptive fault-tolerant PI tracking control for ship propulsion system. ISA Trans. **80**, 279–285 (2018)
20. Fossen, T.I., Blanke, M.: Nonlinear output feedback control of underwater vehicle propellers using feedback form estimated axial flow velocity. IEEE J. Oceanic Eng. **25**(2), 241–255 (2000)
21. Wang, D., Huang, J.: Neural network-based adaptive dynamic surface control for a class of uncertain nonlinear systems in strict-feedback form. IEEE Trans. Neural Netw. **16**(1), 195–202 (2005)
22. Emhemed, A.A.A., Mamat, R.B.: Modelling and simulation for industrial DC motor using intelligent control. Procedia Eng. **41**, 420–425 (2012)
23. Barnitsas, M.M., Ray, D., Kinley, P.: KT, KQ and efficiency curves for the Wageningen B-series propellers (2012)
24. Fossen, T.I.: Handbook of Marine Craft Hydrodynamics and Motion Control. Wiley, Hoboken (2011)
25. Peng, Z., Wang, D., Wang, J.: Cooperative dynamic positioning of multiple marine offshore vessels: a modular design. IEEE/ASME Trans. Mechatron. **21**(3), 1210–1221 (2016)

Data-Based Approximate Policy Iteration for Optimal Course-Keeping Control of Marine Surface Vessels

Yuming Bai[1], Yifan Liu[1(✉)], Qihe Shan[1], Tieshan Li[1], and Yuzhen Lu[2]

[1] Navigation College, Dalian Maritime University, Dalian 116026, Liaoning, China
liuyf_613@163.com

[2] School of Science, Dalian Maritime University, Dalian 116026, Liaoning, China

Abstract. In this paper, the data-based approximate policy iteration (API) method is used for optimal course-keeping control with unknown ship model. When we deal with the nonlinear optimal control problem, the Hamilton Jacobi Bellman (HJB) equation, which is difficult to be solved analytically, needs to be tackled. Furthermore, because of numerous parameters to be determined and unknown nonlinear terms, it is usually difficult to establish the accurate mathematical model for ships. In order to overcome these difficulties, the API method, which can solve the problem of model-free system, is introduced for optimal course-keeping control of marine surface vessels. And the asymptotic stability of the closed-loop system can be guaranteed via Lyapunov analysis. Finally, a numerical example is provided to demonstrate the effectiveness of the control scheme.

Keywords: Optimal course-keeping control · Unknown ship model · Neural network · Approximate policy iteration

1 Introduction

With the development of shipping industry, ships are playing a more and more important role in world trade, and the navigation of ships has always been a major problem for sailors [1]. The energy consumption and the heading control accuracy are important indexes when the performance of ship course control system is evaluated. A large number of control methods for course-keeping control of ships have been reported [2–8], such as backstepping control, sliding model control, robust control, fuzzy logic control [10], PID control, neural network control [22], and so forth. However, with economic development, it is increasingly important to save energy as much as possible while ensuring the safety of ship navigation. Therefore, optimal control problem of ships becomes a hot topic and attracts extensive attentions of scholars nowadays.

The objective of optimal control is to find a control policy that drives the controlled system to a desired target in an optimal way, i.e., to minimize or maximize the performance index [9]. In recent years, many optimal control methods

© Springer Nature Switzerland AG 2019
H. Lu et al. (Eds.): ISNN 2019, LNCS 11555, pp. 81–92, 2019.
https://doi.org/10.1007/978-3-030-22808-8_9

have been introduced for ship control, and gradually become one of the research hotspots in control field. An estimator is developed in [10] based on the optimal command signal and the fuzzy system. By using the backstepping technique and disturbance observer method, an optimal control scheme is presented for the ship with external disturbances in [11]. In [12], a novel path-following control strategy for vessels with roll constraints is proposed with the combination of Kalman filter, disturbance observer, and robust constrained model predictive control method. However, most of these methods require the accurate mathematical model of ships.

In the real world, the ship motion is very complicated which includes 6 degrees of freedom and makes it difficult for system modeling. There are many remarkable achievements have been obtained for the simplification of the ship model [25,26]. For example, Abkowitz proposed the small perturbation model; the Manoeuvring Model Group (MMG) of Japan proposed the MMG mathematical model; Japanese scholar Nomoto simplified the third-order heading control model of ships. Using the ship rudder as the system input, the course as the output, establishing the second-order model of ship course-keeping. This work is of great significance. In real life, however, the model parameters are usually difficult to get when the mathematical model of ships is established. To solve this problem, many strategies have been proposed for the ship control with uncertain parameters. For example, a nonlinear robust controller is designed in [13] to improve the robust performance and tracking characteristics of ships in the presence of significant modeling uncertainties. A novel sliding mode control approach is proposed in [14] for robust tracking control of an underactuated surface vessel with parameter uncertainties. In [15], an online adaptive near-optimal controller is presented for linear and nonlinear systems with parameter uncertainty, which is also applied to course control of underactuated surface vessels.

Adaptive dynamic programming (ADP) is an optimization method that avoids solving the Hamilton Jacobi Bellman (HJB) equation, and this algorithm has developed rapidly in recent years [16]. In [17], ADP is introduced for optimal control of linear systems and nonlinear systems. For the infinite horizon optimal control of nonlinear systems, a novel discrete-time policy iteration approach is developed in [18]. With the utilization of the input-output data during the controller design process of the system, a novel data-driven adaptive iterative learning predictive control algorithm is presented in [19]. By developing a new data-based approximate policy iteration (API) algorithm, the model-free nonlinear optimal control problem is solved in [20].

In this paper, the API method is used to obtain the optimal course-keeping controller of ships by using the system input-output data, and the model-free problem of the ship is solved successfully.

The main contributions of this paper can be summarized as follows:

(1) The controller is designed based on the input-output data, rather than the explicit mathematical model of ships.

(2) By utilizing the input-output data, an optimal controller is developed for course-keeping control of ships, and the problem of unknown model of the ship is dealt with successfully.

(3) Based on the (API) method, a simulation is conducted on the YUPENG ship to verify the effectiveness of the control algorithm.

The rest of the paper is organized as follows. The problem formulation and some preliminaries are presented in Sects. 2 and 3. In Sect. 4, the critic neural network and actor neural network are introduced to approximate the ship performance index and the course-keeping controller, respectively. The application of data-based API method to course-keeping control of ships is developed in Sect. 5 and the simulation results are shown in Sect. 6. Finally, the conclusion of this paper is given in Sect. 7.

Notation. R is the set of real numbers. The superscript T represents the transpose of a matrix and I denotes the identify matrix of appropriate dimension. For a symmetric matrix A, $A > 0\,(\geq 0)$ means that the matrix A is a positive (semi-positive). $\|u\|_A^T = u^T A u$ for some real vector and symmetric matrix $A > 0\,(\geq 0)$ with appropriate dimension.

2 Problem Formulation

Consider the mathematical expression of the Nomoto model as follows [1]:

$$\ddot{\varphi} + \frac{1}{T}H(\dot{\varphi}) = \frac{K}{T}\delta \tag{1}$$

where φ and δ are the course and rudder angle of ships, respectively. K and T are model parameters. $H(\dot{\varphi}) = a_1\dot{\varphi} + a_2\dot{\varphi}^3 + a_3\dot{\varphi}^5 + \dots$, where $a_i(i = 1, 2, \dots)$ are nonlinear coefficients of ships and a_i is a constant.

Let $x_1 = \varphi$, $x_2 = \dot{\varphi}$ and $u = \delta$, system (1) is written as:

$$\begin{cases} \dot{x}_1 = x_2 \\ \dot{x}_2 = -\frac{1}{T}H(x_2) + \frac{K}{T}\delta \\ y = x_1 \end{cases} \tag{2}$$

where y is the output of the system.

Assume that the system (2) is controllable, and asymptotically stable on X, X $\subset$ R, where X is a compact set. In the real world, because of the variation of water depth, ship loading and speed, the dynamic parameters K and T will be changed. Accordingly, it is difficult to obtain the accurate mathematical model of the ship.

In order to minimize the energy consumption of ships, define the performance index as follow:

$$V(x_0) \triangleq \int_0^\infty (Q(x) + \|\delta(t)\|_R^2)\mathrm{d}t \tag{3}$$

where $\|\delta(t)\|_R^2 \triangleq \delta^T(t)R\delta(t)$, $R > 0$, and $Q(x)$ is a positive definite function.

In this paper, the optimal control is to find the optimal controller $\delta^*(x)$, such that the system (2) can track the desired trajectory stably, and the performance index (3) is minimized. i.e.

$$\delta(t) \overset{\triangle}{=} \delta^*(x) \overset{\triangle}{=} \arg \min_{\delta} V(x_0) \tag{4}$$

Remark 1. According to the different point of view, the mathematical model of ships has many different divisions. In the Nomoto model, the ship is regarded as a dynamic system, with ship control (rudder angle) and ship motion (ship heading or yaw rate) as input and output of system, respectively. Compared with other mathematical models of ships, the Nomoto model has a relatively simple structure and keeps accuracy high enough.

3   Preliminary Works

3.1   GHJB Equation

Definition 1 [21] **(GHJB equation).** Suppose that u is an admissible control, the function V satisfies the generalized Hamilton Jacobi Bellman (GHJB) equation, i.e.

$$\frac{\partial V}{\partial x}(f + gu) + l + \|u\|_R^2 = 0, \ V(0) = 0 \tag{5}$$

In order to achieve optimal control objective, the controller $u(x)$ minimize the pre-Hamiltonian equation, i.e.

$$
\begin{aligned}
u(x) &= \arg \min_{u \in \Omega} \left\{ \frac{\partial V^T}{\partial x}(f + gu) + l + \|u\|_R^2 \right\} \\
&= -\frac{1}{2} R^{-1} g^T \frac{\partial V}{\partial x}
\end{aligned}
\tag{6}
$$

The controller of u is given by the solution of the Eq. (5). It can be proved that when the process is iterated until $\hat{V}(x) \leq V(x)$ for each $x \in \Omega$, the value functions converges uniformly to the solution of the follow HJB equation: $\frac{\partial V^T}{\partial x}f + l - \frac{1}{4}\frac{\partial V^T}{\partial x}gR^{-1}g^T\frac{\partial V}{\partial x} = 0$, that is to say, $\lim_{i \to \infty} V^{(i)} = V^*$ and $\lim_{i \to \infty} u^{(i)} = u^*$.

3.2   API Algorithm

The explicit expression of the GHJB equation (5) is not available, because the dynamic parameters K, T and $H(\dot\varphi)$ in system (2) are completely unknown. Therefore, it is impossible to solve the GHJB equation (5) with model-based methods. The API algorithm using real system data instead of the explicit system information, so this method can be used for control problem of unknown ship model [20].

The system (3) can be rewritten as:

$$\dot{x} = -\frac{1}{T}H(x) + \frac{K}{T}\delta^{(i)} + \frac{K}{T}[\delta - \delta^{(i)}] \tag{7}$$

Combining Eqs. (5) and (6), the derivative of $V^{(i+1)}(x)$ along the state of system (7) is:

$$
\begin{aligned}
\frac{dV^{(i+1)}(x)}{dt} &= [\frac{\partial V^{(i+1)}}{\partial x}]^{\mathrm{T}}(-\frac{H}{T} + \frac{K}{T}\delta^{(i)}) + [\frac{\partial V^{(i+1)}}{\partial x}]^{\mathrm{T}}\frac{K}{T}[\delta - \delta^{(i)}] \\
&= -Q(x) - \left\|\delta^{(i)}\right\|_{\mathrm{R}}^{2} + 2[\delta^{(i+1)}]^{\mathrm{T}}R[\delta^{(i)} - \delta]
\end{aligned}
\tag{8}
$$

Integrating both sides of the above formula on the interval $[t, t+\varDelta t]$, i.e.,

$$
\begin{aligned}
&V^{(i+1)}(x(t)) - V^{(i+1)}(x(t+\varDelta t)) + 2\int_{t}^{t+\varDelta t}[\delta^{(i+1)}(x(\tau))]^{\mathrm{T}}R[\delta^{(i)}(x(\tau)) - \delta(\tau)]\mathrm{d}\tau \\
&= \int_{t}^{t+\varDelta t}[Q(x(\tau)) + \left\|\delta^{(i)}(x(\tau))\right\|_{R}^{2}]\mathrm{d}\tau
\end{aligned}
\tag{9}
$$

where $V^{(i+1)}(x)$ and $\delta^{(i+1)}(x)$ are the unknown cost function and undetermined controller, respectively. In Preliminary works, we need to solve the GHJB equation (5) to get performance index V and control policy δ. After the formula transformation, if there exist an initial control policy $\delta^{(0)}$, V and δ can be obtained by Eq. (9). Therefore, the parameter unknown problem of K, T and $H(\dot{\varphi})$ is sorted.

Lemma 1 [20]. When $(V^{(i+1)}(x), \delta^{(i+1)}(x))$ is the solution of Eqs. (5) and (6), it is also the solution of Eq. (9). In other words, Eq. (9) is equivalent to Eqs. (5) and (6).

The key step of the proof is proving $(V^{(i+1)}(x), \delta^{(i+1)}(x))$ is the unique solution of Eq. (9). As space is limited, the detailed proof procedure won't be described here. The proof is completed in [20].

Remark 2. Lemma 1 indicate that the controller $\delta^{(i+1)}$ which obtained by iterate Eq. (9) is equivalent to $\delta^{(i+1)}$ obtained by iterate Eqs. (5) and (6), which has been proven the convergence of the system in [21]. That means the API algorithm can be guaranteed to be convergent.

4 Ship Optimal Course-Keeping Actor-Critic Neural Network Structure

Because the accurate mathematical model of ships is unknown, it is impossible to obtain the performance index $V(x)$ and control policy $\delta(x)$ directly. To overcome this problem, the API method study $V^{(i+1)}(x)$ and $\delta^{(i+1)}(x)$ by Eq. (9). In this section, the critic and actor neural networks (NNs) are introduced to approximate the performance index $V^{(i+1)}(x)$ and controller $\delta^{(i+1)}(x)$ respectively. The frameworks of critic and actor NNs are constructed in light of the method of [23].

4.1 Critic Network

Critic NN is used to approximate the performance index $V^{(i+1)}(x)$. Suppose that the number of hide layer neurons in critic NN is N_c, and the weight vector of critic NN is ω_c, The activation function is $\phi(x) \triangleq [\phi_1(x) \ \ldots \ \phi_j(x) \ \ldots \ \phi_{N_c}(x)]^{\mathrm{T}}$, $j = 1, 2, \ldots, N_c$. So, the output of the critic NN is

$$\hat{V}^{(i)}(x) = \sum_{j}^{N_c} \omega_c^{(i)} \phi_j(x) = \phi^{\mathrm{T}}(x) \omega_c^{(i)} \tag{10}$$

for $\forall i = 0, 1, \ldots, N_c$, where $\omega_c^{(i)}$ means the weight vector ω_c with $i\text{-}th$ iteration.

4.2   Actor Network

Actor NN is used to approximate the controller $u^{(i+1)}(x)$. Suppose that the number of hide layer neurons in actor NN is N_a, and the weight vector of critic NN is ω_a, The activation function of the $n\text{-}th$ sub-actor NN for approximating control policy u_n is $\psi^n(x) \triangleq [\psi_1^n(x) \ \ldots \ \psi_k^n(x) \ \ldots \ \psi_{N_a}^n(x)]^{\mathrm{T}}$, $n = 1, \ldots, m$, where $k = 1, 2, \ldots, N_a$. So,the output of the $n\text{-}th$ sub-actor NN can be given by

$$\hat{\delta}^{(i)}(x) = \sum_{k}^{N_a} \omega_{a_n}^{(i)} \psi_k(x) = \psi^{\mathrm{T}}(x) \omega_{a_n}^{(i)} \tag{11}$$

for $\forall i = 0, 1, \ldots, N_a$, where $\omega_{a_n}^{(i)}$ means the weight vector ω_{a_n} with $i\text{-}th$ iteration.

4.3   Convergence Analysis

In the light of [20], the stability of the closed-loop system of the ship control system can be simplified as follows:

$$\dot{x} = -\frac{1}{T}H(x) + \frac{K}{T}\hat{\delta}^{(i)}(x) \tag{12}$$

The Lyapunov function is selected as $V^{(i)}(x)$. The estimation error of actor NN is $\varepsilon_\delta^{(i)}(x) \triangleq \hat{\delta}^{(i)}(x) - \delta^{(i)}(x)$, differentiating $V^{(i)}$ with respect to system (12) we have

$$\begin{aligned}
\dot{V}^{(i)} = & [\frac{\partial V^{(i)}}{\partial x}]^{\mathrm{T}}(-\frac{H}{T} + \frac{K}{T}\hat{\delta}^{(i)}) \\
= & [\frac{\partial V^{(i)}}{\partial x}]^{\mathrm{T}}[-\frac{H}{T} + \frac{K}{T}(\hat{\delta}^{(i)} + \varepsilon_\delta^{(i)})] + Q \\
& + \left\| \delta^{(i-1)} - \delta^{(i)} \right\|_R^2 - Q - \left\| \delta^{(i-1)} - \delta^{(i)} \right\|_R^2 \\
= & [\frac{\partial V^{(i)}}{\partial x}]^{\mathrm{T}}[-\frac{H}{T} + \frac{K}{T}\hat{\delta}^{(i)}] + Q + \left\| \delta^{(i-1)} - \delta^{(i)} \right\|_R^2 \\
& + [\frac{\partial V^{(i)}}{\partial x}]^{\mathrm{T}}[-\frac{H}{T} + \frac{K}{T} + \varepsilon_\delta^{(i)}] - Q - \left\| \delta^{(i-1)} - \delta^{(i)} \right\|_R^2
\end{aligned} \tag{13}$$

where the controller $\delta^{(i)}$ makes the Eq. (6) tenable. Substituting (5) into (13) and combining (5)

$$
\begin{aligned}
\dot{V}^{(i)} &= [\frac{\partial V^{(i)}}{\partial x}]^{\mathrm{T}}[-\frac{H}{T} + \frac{K}{T}(\delta^{(i-1)})] + Q + \left\|\delta^{(i-1)}\right\|_R^2 \\
&\quad + [\frac{\partial V^{(i)}}{\partial x}]^{\mathrm{T}}\frac{K}{T}\varepsilon_\delta^{(i)} - Q - \left\|\delta^{(i)}\right\|_R^2 - \left\|\delta^{(i-1)} - \delta^{(i)}\right\|_R^2 \\
&= [\frac{\partial V^{(i)}}{\partial x}]^{\mathrm{T}}\frac{K}{T}\varepsilon_\delta^{(i)} - Q - \left\|\delta^{(i)}\right\|_R^2 - \left\|\delta^{(i-1)} - \delta^{(i)}\right\|_R^2
\end{aligned}
\tag{14}
$$

According to Lyapunov theorem, the closed-loop system (12) is asymptotically stable when $\dot{V}^{(i)} < 0$. The appropriate activation function and N_a can be selected to ensure that the following formula is established

$$
Q(x) + \left\|\delta^{(i)}\right\|_R^2 + \left\|\delta^{(i-1)} - \delta^{(i)}\right\|_R^2 > [\frac{\partial V^{(i)}}{\partial x}]^{\mathrm{T}}\frac{K}{T}\varepsilon_\delta^{(i)}
\tag{15}
$$

where $\varepsilon_\delta^{(i)}$ is an arbitrarily small positive number so that the Eq. (14) satisfies $\dot{V}^{(i)} < 0$, i.e. system (12) is asymptotically stable.

5 The Optimal Course-Keeping Controller Design Based on API Algorithm

5.1 The Solution of Iterative Formula

In this section, the iteration formula of API algorithm is computed by input-output data of ships.

Using $\hat{V}^{(i+1)}$ and $\hat{\delta}^{(i+1)}$ instead of $V^{(i+1)}$ and $\delta^{(i+1)}$ in (11). Define $x'(t) \overset{\Delta}{=} x(t+\Delta t)$, Due to the approximation error of NN, adopt the residual error formula in [20] as follow:

$$
\begin{aligned}
&\sigma^{(i)}(x(t), \delta(t), x'(t)) \\
&\overset{\Delta}{=} [\phi(x(t)) - \phi(x'(t))]^{\mathrm{T}}\omega_c^{(i+1)} + 2\int_t^{t+\Delta t} [\delta^{(i)}(x(\tau)) - \delta(\tau)]^{\mathrm{T}}R\delta^{(i+1)}(x(\tau))d\tau \\
&\quad - \int_t^{t+\Delta t} Q(x(\tau))d\tau - \int_t^{t+\Delta t} \left\|\delta^{(i)}(x(\tau))\right\|_R^2 d\tau \\
&= [\phi(x(t)) - \phi(x'(t))]^{\mathrm{T}}\omega_c^{(i+1)} + 2\sum_{n_1=1}^{m}\sum_{n_2=1}^{m} r_{n_1,n_2}\int_t^{t+\Delta t} [(\psi^{n_1}(x(\tau)))^{\mathrm{T}}\omega_{a_{n_1}}^{(i)} \\
&\quad - \delta_{n_1}(\tau)](\psi^{n_2}(x(\tau)))^{\mathrm{T}}\omega_{a_{n_2}}^{(i+1)}d\tau - \int_t^{t+\Delta t} Q(x(\tau))d\tau \\
&\quad - \sum_{n_1=1}^{m}\sum_{n_2=1}^{m} r_{n_1,n_2}\int_t^{t+\Delta t} (\omega_{a_{n_1}}^{(i)})^{\mathrm{T}}\psi^{n_1}(x(\tau))(\psi^{n_2}(x(\tau)))^{\mathrm{T}}\omega_{a_{n_2}}^{(i)}d\tau
\end{aligned}
\tag{16}
$$

When $\sigma^{(i)}(x(t), \delta(t), x'(t))$ (for $\forall t \geq 0$) tends to zero, the undetermined NN weight vector $\omega^{(i+1)}$ can be computed.

Utilizing the least-square method, the iterative formula can be obtained as follow:

$$
\begin{aligned}
\omega^{(i+1)} = &\{ [[\phi(x(t)) - \phi(x'(t))]^{\mathrm{T}} \, 2 \int_t^{t+\Delta t} \delta_{n_1}(\tau)(\psi^{n_2}(x(\tau)))^{\mathrm{T}} \mathrm{d}\tau \,]^{\mathrm{T}} \\
&\times [[\phi(x(t)) - \phi(x'(t))]^{\mathrm{T}} \, 2 \int_t^{t+\Delta t} \delta_{n_1}(\tau)(\psi^{n_2}(x(\tau)))^{\mathrm{T}} \mathrm{d}\tau] \,\}^{-1} \\
&\times [[\phi(x(t)) - \phi(x'(t))]^{\mathrm{T}} 2 \int_t^{t+\Delta t} \delta_{n_1}(\tau)(\psi^{n_2}(x(\tau)))^{\mathrm{T}} \mathrm{d}\tau]^{\mathrm{T}} [\int_t^{t+\Delta t} Q(x(\tau)) \mathrm{d}\tau \\
&+ \sum_{n_1=1}^{m} \sum_{n_2=1}^{m} r_{n_1, n_2} (\omega_{a_{n_1}}^{(i)})^{\mathrm{T}} \int_t^{t+\Delta t} \psi^{n_1}(x(\tau))(\psi^{n_2}(x(\tau)))^{\mathrm{T}} \mathrm{d}\tau \, \omega_{a_{n_2}}^{(i)}]
\end{aligned}
\tag{17}
$$

5.2 The Step of the Optimal Course-Keeping Controller Design Based on API Algorithm

The controller of ship $\delta(x)$ is obtained by Eq. (11). In order to train the NNs, a sufficient number of ship input-output data are required. The ship maintains or changes its course and speed by using propeller thrust and rudder, which means the input data of ships include ship speed and rudder angle. They can be obtained by acoustic doppler current profiler (ADCP) and rudder angle sensor, respectively. The output data are ship course and they can be obtained by the rate sensor, gyro, numerical differentiation of the heading measurement or a state estimator [24].

The steps of the optimal course-keeping controller design are:

Step 1: Perform a large number of experiments on the target ship to get enough input-output data. For each different input signal u, the corresponding output data (x_k, δ_k, x_{k+1}) can be obtained. All of this data constitutes a sample set Ω_M.

Step 2: Select the appropriate initial weight vector of actor NN and critic NN.

Step 3: Let $i = 0$, utilize the input-output data obtained in Step 1, iterative calculate NN weight $\omega^{(i+1)}$ by Eq. (17).

Step 4: If $\|\omega^{(i+1)} - \omega^{(i)}\| \leq \xi$, where ξ is a small enough positive number, stop iteration, and $u^{(i+1)}$ is the final control policy of the ship system. Otherwise, return to Step 3 and continue iterating.

Step 5: Apply the optimal course-keeping controller to the ship model and get the output ship heading.

Remark 3. Through the steps of the controller design, we can learn that: (1) The real input-output data instead of the exact mathematical model of ships is used during the process of optimal course-keeping controller design. (2) The first part of the algorithm is collecting ship motion state and rudder angle input signal for data processing; The second part is iteratively computing the optimal controller of ships. Finally, the optimal controller is applied to model-free ship course-keeping control system.

6 Simulation

In this section, the API methods is applied to the YUPENG ship of Dalian Maritime University.

The research object of this paper is the model free system of ships. However, the system we chose has an accurate mathematical model in the simulation. That does not mean that we have to know the exact information of ship model. The system we chose in simulation just used to get the input-output data of ships. The information of system is not required in the process of algorithm implementation.

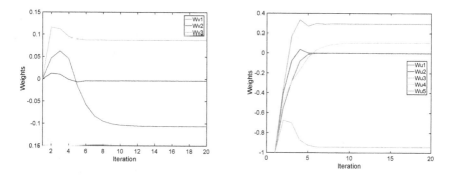

Fig. 1. The iteration of critic NN weights and actor NN weights.

In the simulation, we chose the ship Nomoto model as the control objectives. From system (3), the simulation model is written as:

$$\begin{cases} \dot{x}_1 = x_2 \\ \dot{x}_2 = -\frac{1}{T}(a_1 x_2 + a_2 x_2^3) + \frac{K}{T}u \end{cases} \tag{18}$$

where $K = 0.21, T = 107.78, a_1 = 13.14, a_2 = 16212.5$.

Choose $Q(x) = x^\mathrm{T}x$ and $R = I$ in Eq. (3). The value of the expectation error $\xi = 10^{-5}$. The activation functions of critic NN and actor NN are: $\phi(x) = [x_1^2 \ x_1 x_2 \ x_2^2]^\mathrm{T}$, $\psi(x) = [x_1 \ x_2 \ x_1^2 \ x_1 x_2 \ x_2^2]^\mathrm{T}$ with $N_c = 3$, $N_a = 5$, respectively. The initial weight vectors of critic NN and actor NN are: $\omega_c^{(0)} = [0 \ 0 \ 0]^\mathrm{T}$, and $\omega_a^{(0)} = [-1 \ -1 \ -1 \ -1 \ -1]^\mathrm{T}$, respectively.

From Fig. 1, we can learn that the weight vectors of critic NN and actor NN converge to ω_c^* and ω_a^* at the 20-*th* iteration, respectively. At the 20-*th* iteration

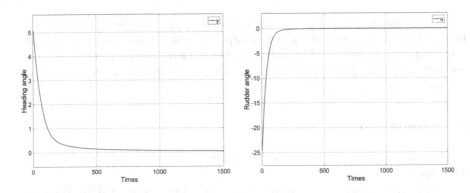

Fig. 2. The output and controller of ship model.

the weight $\omega_c^{(20)} = [-0.1062 \ -0.0041 \ 0.0870]^{\mathrm{T}}$, $\omega_a^{(20)} = [0 \ \ 0 \ -0.9428 \ 0.1035$
$0.2938]^{\mathrm{T}}$. Then, the optimal control policy is used to control the ship model (18), and its input signal (rudder angle) and output signal (ship heading) are shown in Fig. 2.

Figure 2 shows the course and rudder angle of the ship, respectively. And the initial value of the course and rudder angle are 5 and -30, respectively. We can learn from that the ship course control system is converges to zero along time.

7   Conclusions

The approximate policy iteration (API) method has been used to solve the optimal control problem of surface vessels with uncertain parameters, and the ship control system has proved to be stable. In the implementation of algorithm, the optimal controller calculates by real input-output data of ships instead of accurate mathematical model. The application on YUPENG ship has illustrated the effective performance of the API method. Further work of this paper would be focused on the controller design for ship course control problem.

Acknowledgments. This work was supported in part by the National Natural Science Foundation of China (Grant Nos. 61751202, 61751205, 61572540, 61803064); the Natural Science Foundation of Liaoning Province (20180550082, 20170540093); the Science & Technology Innovation Founds of Dalian (2018J11Y022); the Fundamental Research Funds for the Central Universities (3132018306); and the Fundamental Research Funds of Dalian Maritime University (3132018157).

References

1. Huang, Z., Zhuo, Y., Li, T.: Optimal tracking control of the ship course system with partially-unknown dynamics. In: 4th International Conference on Information, Cybernetics and Computational Social Systems (ICCSS), pp. 631–635. IEEE (2017)
2. Du, J., Guo, C., Yu, S., Zhang, X.: Adaptive autopilot design of time-varying uncertain ships with completely unknown control coefficient. IEEE J. Ocean. Eng. **32**(2), 346–352 (2007)
3. Wang, Y., Li, T., Philip Chen, C.L.: Adaptive terminal sliding mode control for formations of underactuated vessels. In: Sun, F., Liu, H., Hu, D. (eds.) ICCSIP 2016. CCIS, vol. 710, pp. 27–35. Springer, Singapore (2017). https://doi.org/10.1007/978-981-10-5230-9_3
4. Chen, M., Ge, S.S., How, B.V.E., Choo, Y.S.: Robust adaptive position mooring control for marine vessels. IEEE Trans. Control Syst. Technol. **21**(2), 395–409 (2017)
5. Wang, N., Er, M.J.: Direct adaptive fuzzy tracking control of marine vehicles with fully unknown parametric dynamics and uncertainties. IEEE Trans. Control Syst. Technol. **24**(5), 1845–1852 (2016)
6. Liu, S., Fang, L., Ge, Y.-M., Fu, H.-X.: GA-PID adaptive control research for ship course-keeping system. J. Syst. Simul. **19**(16), 3783–3786 (2007)
7. Zhang, K., Li, T., Li, Z., Philip Chen, C.L.: Neural network based dynamic surface second order sliding mode control for AUVs. In: Sun, F., Liu, H., Hu, D. (eds.) ICCSIP 2016. CCIS, vol. 710, pp. 417–424. Springer, Singapore (2017). https://doi.org/10.1007/978-981-10-5230-9_41
8. Liu, C., Li, T., Chen, C.L.P.: Leader-following consensus control for multiple marine vessels based on dynamic surface control and neural network. In: 4th International Conference on Information, Cybernetics and Computational Social Systems (ICCSS), pp. 160–165. IEEE (2017)
9. Lewis, F.L., Vrabie, D., Syrmos, V.L.: Optimal Control. Wiley, Hoboken (2012)
10. Peng, Z., Wang, J., Wang, D.: Distributed maneuvering of autonomous surface vehicles based on neurodynamic optimization and fuzzy approximation. IEEE Trans. Control Syst. Technol. **26**(3), 1083–1090 (2018)
11. Ejaz, M., Chen, M.: Optimal backstepping control for a ship using firefly optimization algorithm and disturbance observer. Trans. Inst. Meas. Control **40**(6), 1983–1998 (2018)
12. Zhang, J., Sun, T., Liu, Z.: Robust model predictive control for path-following of underactuated surface vessels with roll constraints. Ocean Eng. **143**, 125–132 (2017)
13. Khaled, N., Chalhoub, N.G.: A dynamic model and a robust controller for a fully-actuated marine surface vessel. J. Vib. Control **17**(6), 801–812 (2011)
14. Yu, R., Zhu, Q., Xia, G., Liu, Z.: Sliding mode tracking control of an underactuated surface vessel. IET Control Theory Appl. **6**(3), 461–466 (2012)
15. Zhang, Y., Li, S., Liu, X.: Adaptive near-optimal control of uncertain systems with application to underactuated surface vessels. IEEE Trans. Control Syst. Technol. **26**(4), 1204–1218 (2018)
16. Zhang, H., Zhang, X., Luo, Y., Yang, J.: An overview of research on adaptive dynamic programming. Acta Automatica Sin. **39**(4), 303–311 (2013)
17. Murray, J.J., Cox, C.J., Lendaris, G.G., Saeks, R.: Adaptive dynamic programming. IEEE Trans. Syst. Man Cybern. Part C-Appl. Rev. **32**(2), 140–153 (2002)

18. Liu, D., Wei, Q.: Policy iteration adaptive dynamic programming algorithm for discrete-time nonlinear systems. IEEE Trans. Neural Netw. Learn. Syst. **25**(3), 621–634 (2014)
19. Sun, H., Hou, Z.: A data-driven adaptive iterative learning predictive control for a class of discrete-time nonlinear systems. In: Proceedings of the 29th Chinese Control Conference (CCC), pp. 5871–5876. IEEE (2010)
20. Luo, B., Wu, H., Huang, T., Liu, D.: Data-based approximate policy iteration for affine nonlinear continuous-time optimal control design. Automatica **50**(12), 3281–3290 (2014)
21. Beard, R.W., Saridis, G.N., Wen, J.T.: Galerkin approximations of the generalized Hamilton-Jacobi-Bellman equation. Automatica **33**(12), 2159–2177 (1997)
22. Peng, Z., Wang, J.: Output-feedback path-following control of autonomous underwater vehicles based on an extended state observer and projection neural networks. IEEE Trans. Syst. Man Cybern. Syst. **48**(4), 535–544 (2018)
23. Wang, D., Mu, C., Liu, D.: Data-driven nonlinear near-optimal regulation based on iterative neural dynamic programming. Acta Automatica Sin. **43**(3), 366–375 (2017)
24. Fossen, T.I.: Guidance and Control of Ocean Vehicles. Wiley, New York (1994)
25. Abkowitz, M.A.: Lectures on Ship Hydrodynamics–Steering and Manoeuvrability (1964)
26. Kensaku, N., Kenshi, T., Keinosuke, H., Susumu, H.: On the steering qualities of ships. Int. Shipbuilding Prog. **4**(35), 354–370 (1957)

Intelligent Fuzzy Kinetic Control for an Under-Actuated Autonomous Surface Vehicle via Stochastic Gradient Descent

Yue Jiang, Lu Liu, Zhouhua Peng$^{(\boxtimes)}$, Dan Wang$^{(\boxtimes)}$, Nan Gu, and Shengnan Gao

School of Marine Electrical Engineering, Dalian Maritime University, Dalian 116026, People's Republic of China
zhpeng@dlmu.edu.cn, dwangdl@gmail.com

Abstract. In this paper, an intelligent fuzzy kinetic control scheme based on online identification is developed for an under-actuated autonomous surface vehicle in the presence of unknown uncertainties and disturbances from ocean environment. An adaptive fuzzy system is used to approximate the unknown dynamics in real time by using recorded input and output data of the vessel. To improve the learning performance, the parameters of the fuzzy system are updated based on a stochastic gradient descent approach and a predictor design. With the estimated dynamics from the fuzzy system, a robust kinetic controller is designed without any off-line learning. The proposed intelligent fuzzy control method can be applied at the kinetic level of various control scenarios, such as target tracking, path following and trajectory tracking.

Keywords: Intelligent fuzzy kinetic control ·
Stochastic gradient descent ·
Under-actuated autonomous surface vehicle · Unknown dynamics

1 Introduction

The kinetics of an autonomous surface vehicle (ASV) suffers from parametric uncertainties, unmodeled hydrodynamics and disturbances, which are generally unmeasurable [1–9]. A critical challenge in velocity control of an under-actuated

This work was supported in part by the National Natural Science Foundation of China under Grants 51579023, 61673081, the Innovative Talents in Universities of Liaoning Province under Grant LR2017014, High Level Talent Innovation and Entrepreneurship Program of Dalian under Grant 2016RQ036, China Postdoctoral Science Foundation 2019M650086, the National Key Research and Development Program of China under Grant 2016YFC0301500, the Training Program for High-level Technical Talent in Transportation Industry under Grant 2018-030, the Fundamental Research Funds for the Central Universities under Grant 3132019101, 3132019013, and Outstanding Youth Support Program of Dalian.

© Springer Nature Switzerland AG 2019
H. Lu et al. (Eds.): ISNN 2019, LNCS 11555, pp. 93–100, 2019.
https://doi.org/10.1007/978-3-030-22808-8_10

ASV lies in the identification and compensation of these unknown dynamics [10–13]. The fuzzy system, as an effective approximation approach, is wildly applied to identify the nonlinear dynamics, such as to deal with the unknown dynamics in steering system [14], to approximate the disturbance in output-feedback control [15], and to identify model uncertainties in tracking control [16]. However, the updated laws of these fuzzy control schemes are designed based on Lyapunov theory and the model reference adaptive control approach. The transient approximation performance of the fuzzy system is remained to be improved.

The stochastic gradient descent (SGD) approach, as a fast and reliable online learning technique, is widely used in system identification [17,18]. With the advantage of rapidity, robustness and extensibility, SGD takes the merit of training any model within a reasonable period of time [19]. It is applied to train the weights of radial basis function neural network (NN) in [17], and it is used in adaptive NN controller in [18]. However, the SGD algorithm is not used in a predictor-based adaptive fuzzy control scheme, where the key challenge is how to construct a loss function without using the unavailable dynamics.

This paper presents an intelligent fuzzy controller based on online identification for velocity control of an under-actuated ASV. An adaptive fuzzy system is used to identify the unknown dynamics of ASV, and a predictor is developed based on the fuzzy system. The updated law of system parameters is designed based on the SGD approach and the predictor errors, such that the transient learning performance of the fuzzy system is improved. Then, with the estimated terms from fuzzy system, robust fuzzy kinetic control laws are designed. The stability of proposed SGD-based adaptive fuzzy controller is proved to be input-to-state stable (ISS) via Lyapunov analysis, and the velocity tracking errors are uniformly ultimately bounded.

The paper is organized as follows. The preliminaries and problem formulation are introduced in Sect. 2. The design and analysis of SGD-based fuzzy kinetic controller are in Sect. 3. Conclusions are presented in Sect. 4.

2   Preliminaries and Problem Formulation

2.1   SGD-Based Fuzzy System

A fuzzy system which consists of a fuzzy rule base, a fuzzifier and a defuzzifier, has the capability to approximate a nonlinear function. Let F_i^l be the fuzzy sets for a state ξ_i, and its membership functions are given by $\mu_{F_i^l}(\xi_i)$ with $i = 1, \ldots, n$. The fuzzy rule base is composed of the If-Then inference rules [20].

R^l : If ξ_1 is F_1^l and ξ_2 is F_2^l and … and ξ_n is F_n^l, then g is G^l, $l = 1, 2, \ldots, N$.
Then, the fuzzy system can be expressed as

$$g(\xi) = \frac{\sum_{l=1}^{N} \bar{g}_l \prod_{i=1}^{n} \mu_{F_i^l}(\xi_i)}{\sum_{l=1}^{N} [\prod_{i=1}^{n} \mu_{F_i^l}(\xi_i)]}, \tag{1}$$

where $\bar{g}_l = \max_{g \in R} \mu_{G^l}(g)$ with membership functions $\mu_{G^l}(g)$; $\xi = [\xi_1, \xi_2, \ldots, \xi_n]^T$.

Let $\theta^T = [\theta_1, \theta_2, \ldots, \theta_N] = [\bar{g}_1, \bar{g}_2, \ldots, \bar{g}_N]$, $\phi(\xi) = [\phi_1(\xi), \phi_2(\xi), \ldots, \phi_N(\xi)]^T$ with

$$\phi_l = \frac{\prod_{i=1}^n \mu_{F_i^l}(\xi_i)}{\sum_{l=1}^N [\prod_{i=1}^n \mu_{F_i^l}(\xi_i)]}, \tag{2}$$

satisfying $\|\phi(\xi)\| \leq \phi^*$ with ϕ^* being a positive constant. Then, the fuzzy system can be represented as $g(\xi) = \theta^T \phi(\xi)$.

Lemma 1 [20]. For a given continuous function $f(\xi)$ defined in a compact set Ω_F, there exists a fuzzy system such that for any $\epsilon_e > 0$

$$\sup_{\xi \in \Omega_F} |f(\xi) - \theta^T \phi(\xi)| \leq \epsilon_e. \tag{3}$$

Then, define the fuzzy system approximation error $E = f(\xi) - \hat{\theta}^T \phi(\xi)$, where $\hat{\theta} \in \Re^N$ is the estimate of θ. To design the update law of $\hat{\theta}$, an SGD algorithm is applied here. This method calculates the gradient of the loss function $\frac{1}{2}E^2$ online repeatedly, such that a minimum of the loss function is achieved. Based on SGD algorithm [17], the parameter update rule of $\hat{\theta}_l$ is given by

$$\hat{\theta}_l(k+1) = \hat{\theta}_l(k) + \kappa E \phi_l, \tag{4}$$

where $l = 1, 2, \ldots, N$, $k \geq 0$, $k \in \aleph$, and the positive gain $\kappa \in \Re$ denotes the learning rate. The stability and convergence of SGD can be found in [21].

2.2 Problem Formulation

The motion of an under-actuated ASV can be described with kinematics [22]

$$\begin{cases} \dot{x}_t = u_t \cos \psi_t - v_t \sin \psi_t, \\ \dot{y}_t = u_t \sin \psi_t + v_t \cos \psi_t, \\ \dot{\psi}_t = r_t, \end{cases} \tag{5}$$

and kinetics

$$\begin{cases} m_u \dot{u} = f_u(u, v, r) + \tau_u + \tau_{du}(t), \\ m_v \dot{v} = f_v(u, v, r) + 0 + \tau_{dv}(t), \\ m_r \dot{r} = f_r(u, v, r) + \tau_r + \tau_{dr}(t), \end{cases} \tag{6}$$

where x, y, ψ represent the position of the ASV and its yaw angle; u, v, r represent the surge velocity, sway velocity and yaw rate; m_u, m_v and m_r denote mass effects; $f_u(\cdot), f_v(\cdot), f_r(\cdot)$ represent nonlinear uncertainties including hydrodynamic damping forces, Coriolis forces and unmodeled hydrodynamics; $\tau_{du}(t), \tau_{dv}(t), \tau_{dr}(t)$ are environmental disturbances; τ_u, τ_r denote control inputs.

The control objective is to develop a fuzzy kinetic controller for the ASV with kinetics (6) suffering from unknown dynamics and extraneous disturbance to track a desired velocity, such that

$$\lim_{t\to\infty} |u - u_d| \le \epsilon_u, \quad \lim_{t\to\infty} |r - r_d| \le \epsilon_r, \tag{7}$$

are satisfied for small constants ϵ_u and ϵ_r, where $u_d \in \Re$ and $r_d \in \Re$ are desired surge velocity and yaw rate, respectively.

3   Controller Design and Analysis

At first, kinetic tracking errors are defined as $u_e = u - u_d$ and $r_e = r - r_d$, and their time derivatives along with (6) are

$$\begin{cases} m_u \dot{u}_e = \sigma_u(\cdot) + \tau_u, \\ m_r \dot{r}_e = \sigma_r(\cdot) + \tau_r, \end{cases} \tag{8}$$

where $\sigma_u(\cdot) = f_u(u, v, r) + \tau_{du} - m_u \dot{u}_d$ and $\sigma_r(\cdot) = f_r(u, v, r) + \tau_{dr} - m_r \dot{r}_d$.

The time-varying disturbances τ_{du} and τ_{dr} are unknown, so $\sigma_u(\cdot)$ and $\sigma_r(\cdot)$ cannot be approximated by the fuzzy system directly. Hence, the inputs and outputs are used to approximate the unknown dynamics. From Lemma 1, $\sigma_u(\cdot)$ and $\sigma_r(\cdot)$ can be rewritten as

$$\begin{cases} \sigma_u = \theta_u^T \phi(\xi_u) + \varepsilon_u, \\ \sigma_r = \theta_r^T \phi(\xi_r) + \varepsilon_r, \end{cases} \tag{9}$$

where $\xi_u = [u_e(t), u_e(t - t_d), \tau_u]^T \in \Re^3$ and $\xi_r = [r_e(t), r_e(t - t_d), \tau_r]^T \in \Re^3$ with the sample period t_d; $\theta_u \in \Re^N$ and $\theta_r \in \Re^N$ satisfying $||\theta_u|| \le \theta_u^*$ and $||\theta_r|| \le \theta_r^*$ with positive constants θ_u^*, θ_r^*; $|\varepsilon_u| \le \varepsilon_u^*$ and $|\varepsilon_r| \le \varepsilon_r^*$ with positive constants ε_u^*, ε_r^*.

To achieve the fast learning of unknown functions, a predictor for (8) is designed based on the fuzzy system as

$$\begin{cases} m_u \dot{\hat{u}}_e = -k_1(\hat{u}_e - u_e) + \hat{\theta}_u^T \phi(\xi_u) + \tau_u, \\ m_r \dot{\hat{r}}_e = -k_2(\hat{r}_e - r_e) + \hat{\theta}_r^T \phi(\xi_r) + \tau_r, \end{cases} \tag{10}$$

where k_1, k_2, are positive constants; $\hat{u}_e$ and $\hat{r}_e$ are the estimates of u_e and r_e; $\hat{\theta}_u = [\hat{\theta}_{u1}, \hat{\theta}_{u2}, \ldots, \hat{\theta}_{uN}]^T$ and $\hat{\theta}_r = [\hat{\theta}_{r1}, \hat{\theta}_{r2}, \ldots, \hat{\theta}_{rN}]^T$ are the estimates of θ_u and θ_r. The weights $\hat{\theta}_{ul}(t)$ and $\hat{\theta}_{rl}(t)$ are defined by

$$\begin{cases} \hat{\theta}_{ul}(t) = (t - k)\big(\hat{\theta}_{ul}(k + 1) - \hat{\theta}_{ul}(k)\big) + \hat{\theta}_{ul}(k), \\ \hat{\theta}_{rl}(t) = (t - k)\big(\hat{\theta}_{rl}(k + 1) - \hat{\theta}_{rl}(k)\big) + \hat{\theta}_{rl}(k), \end{cases} \tag{11}$$

where $l = 1, 2, \ldots, N$.

Based on the SGD algorithm and the predictor errors, the iterative equation of $\hat{\theta}_{ul}$ and $\hat{\theta}_{rl}$ in (11) are given by

$$
\begin{cases}
\hat{\theta}_{ul}(k+1) = \hat{\theta}_{ul}(k) - \kappa_u E_u \phi_l, \\
\hat{\theta}_{rl}(k+1) = \hat{\theta}_{rl}(k) - \kappa_r E_r \phi_l,
\end{cases}
\tag{12}
$$

where $k \geq 0$, $k \in \mathbb{N}$, and $\kappa_u \in \Re$, $\kappa_r \in \Re$ are positive learning rates. E_u and E_r are defined by

$$
E_u = -k_1 \tilde{u}_e - m_u \tilde{u}_{er}^d, \quad E_r = -k_2 \tilde{r}_e - m_r \tilde{r}_{er}^d,
\tag{13}
$$

where $\tilde{u}_e = \hat{u}_e - u_e$, $\tilde{r}_e = \hat{r}_e - r_e$; $\tilde{u}_{er}^d$, $\tilde{r}_{er}^d$ are estimates of $\dot{\tilde{u}}_e$, $\dot{\tilde{r}}_e$ from a tracking differentiator (TD) designed as

$$
\begin{cases}
\dot{\tilde{u}}_{er} = \tilde{u}_{er}^d, \quad \dot{\tilde{u}}_{er}^d = -\kappa_1^2(\tilde{u}_{er} - \tilde{u}_e) - 2\kappa_1 \tilde{u}_{er}^d, \\
\dot{\tilde{r}}_{er} = \tilde{r}_{er}^d, \quad \dot{\tilde{r}}_{er}^d = -\kappa_2^2(\tilde{r}_{er} - \tilde{r}_e) - 2\kappa_2 \tilde{r}_{er}^d,
\end{cases}
\tag{14}
$$

with $\kappa_1, \kappa_2 \in \Re$ being positive gains, and $\tilde{u}_{er}, \tilde{r}_{er} \in \Re$ being estimates of $\tilde{u}_e, \tilde{r}_e$, respectively. According to the property of TD in [23], there exist positive constants $\gamma_{ud}^*, \gamma_{rd}^*$ such that

$$
\tilde{u}_{er}^d - \dot{\tilde{u}}_e = \gamma_{ud}, |\gamma_{ud}| \leq \gamma_{ud}^*, \quad \tilde{r}_{er}^d - \dot{\tilde{r}}_e = \gamma_{rd}, |\gamma_{rd}| \leq \gamma_{rd}^*.
\tag{15}
$$

From (8), (10), (13) and (15), it leads to

$$
\begin{cases}
E_u = \sigma_u(\cdot) - \hat{\theta}_u^T \phi(\xi_u) - m_u \gamma_{ud}, \\
E_r = \sigma_r(\cdot) - \hat{\theta}_r^T \phi(\xi_r) - m_r \gamma_{rd}.
\end{cases}
\tag{16}
$$

According to the convergence of SGD and (15), one has

$$
\sigma_u(\cdot) - m_u \gamma_{ud} = \hat{\theta}_u^T \phi(\xi_u) + \varepsilon_{\sigma u}, \quad \sigma_r(\cdot) - m_r \gamma_{rd} = \hat{\theta}_r^T \phi(\xi_r) + \varepsilon_{\sigma r},
\tag{17}
$$

where $\varepsilon_{\sigma u}, \varepsilon_{\sigma r} \in \Re$ satisfy $|\varepsilon_{\sigma u}| \leq \varepsilon_{\sigma u}^*$, $|\varepsilon_{\sigma r}| \leq \varepsilon_{\sigma r}^*$ with positive constants $\varepsilon_{\sigma u}^*$, $\varepsilon_{\sigma r}^*$. Let $\tilde{\theta}_u = \hat{\theta}_u - \theta_u$ and $\tilde{\theta}_r = \hat{\theta}_r - \theta_r$. From (9) and (17), the bounds of $\tilde{\theta}_u^T \phi(\xi_u)$ and $\tilde{\theta}_r^T \phi(\xi_r)$ are given by

$$
\begin{cases}
|\tilde{\theta}_u^T \phi(\xi_u)| = |m_u \gamma_{ud} + \varepsilon_{\sigma u} + \varepsilon_u| \leq m_u \gamma_{ud}^* + \varepsilon_{\sigma u}^* + \varepsilon_u^* \triangleq \varepsilon_1^*, \\
|\tilde{\theta}_r^T \phi(\xi_r)| = |m_r \gamma_{rd} + \varepsilon_{\sigma r} + \varepsilon_r| \leq m_r \gamma_{rd}^* + \varepsilon_{\sigma r}^* + \varepsilon_r^* \triangleq \varepsilon_2^*.
\end{cases}
\tag{18}
$$

Based on the estimated terms from fuzzy system, two control laws are proposed for (10)

$$
\begin{cases}
\tau_u = -k_u \hat{u}_e - \hat{\theta}_u^T \phi(\xi_u), \\
\tau_r = -k_r \hat{r}_e - \hat{\theta}_r^T \phi(\xi_r),
\end{cases}
\tag{19}
$$

where k_u, k_r, are positive constants.

Taking the time derivatives of $\tilde{u}_e$, $\tilde{r}_e$ along (8) and (10), one has

$$
\begin{cases}
m_u \dot{\tilde{u}}_e = -k_1 \tilde{u}_e + \tilde{\theta}_u^T \phi(\xi_u) - \varepsilon_u, \\
m_r \dot{\tilde{r}}_e = -k_2 \tilde{r}_e + \tilde{\theta}_r^T \phi(\xi_r) - \varepsilon_r.
\end{cases}
\tag{20}
$$

The stability of predictor error subsystem (20) is presented as follows.

Lemma 2. The subsystem (20) with the states $\tilde{u}_e$, $\tilde{r}_e$ and the inputs $\tilde{\theta}_u^T\phi(\xi_u)$, $\tilde{\theta}_r^T\phi(\xi_r)$, ε_u, and ε_r is ISS.

Proof. The Lyapunov function is defined as follows

$$V_1 = \frac{1}{2}\{m_u\tilde{u}_e^2 + m_r\tilde{r}_e^2\}. \tag{21}$$

The time derivative of V_1 along (20) satisfies

$$\dot{V}_1 = -X_1^T K_1 X_1 + l_1^T X_1 \le -\lambda_{\min}(K_1)\|X_1\|^2 + \|l_1\|\|X_1\|, \tag{22}$$

where $X_1 = [\tilde{u}_e, \tilde{r}_e]^T$, $K_1 = \text{diag}\{k_1, k_2\}$, and $l_1 = [\tilde{\theta}_u^T\phi(\xi_u) - \varepsilon_u, \tilde{\theta}_u^T\phi(\xi_u) - \varepsilon_r]^T$. Since $\|X_1\| \ge \frac{2}{\lambda_{\min}(K_1)}(|\tilde{\theta}_u^T\phi(\xi_u)| + |\tilde{\theta}_u^T\phi(\xi_u)| + |\varepsilon_u| + |\varepsilon_r|)$ renders $\dot{V}_1 \le -\frac{1}{2}\lambda_{\min}(K_1)\|X_1\|^2$, the subsystem (20) is ISS, and

$$\|X_1(t)\| \le \max\{\varpi_1(\|X_1(0)\|, t), \beta_{11}(|\tilde{\theta}_u^T\phi(\xi_u)|) + \beta_{12}(|\tilde{\theta}_r^T\phi(\xi_r)|) + \beta_{13}(|\varepsilon_u|)$$
$$+ \beta_{14}(|\varepsilon_r|)\}, \tag{23}$$

where ϖ_1 is a $\mathcal{KL}$ function and $\beta_{11} = \beta_{12} = \beta_{13} = \beta_{14} = \sqrt{\frac{\lambda_{\max}(S_1)}{\lambda_{\min}(S_1)}}\frac{2s}{\lambda_{\min}(K_1)}$, with $S_1 = \text{diag}\{m_u, m_r\}$. The proof is completed.

Then, substituting (19) into (10), one gets the controller error subsystem as follows

$$\begin{cases} m_u\dot{\hat{u}}_e = -k_u\hat{u}_e - k_1\tilde{u}_e, \\ m_r\dot{\hat{r}}_e = -k_r\hat{r}_e - k_2\tilde{r}_e. \end{cases} \tag{24}$$

The following lemma presents the stability of subsystem (24).

Lemma 3. The velocity tracking error subsystem (24) with the states being $\hat{u}_e$, $\hat{r}_e$ and the inputs being $\tilde{u}_e$, $\tilde{r}_e$ is ISS.

Proof. Construct the Lyapunov function as follows

$$V_2 = \frac{1}{2}\{m_u\hat{u}_e^2 + m_r\hat{r}_e^2\}, \tag{25}$$

whose time derivative along (24) satisfies

$$\dot{V}_2 = -X_2^T K_2 X_2 + l_2^T X_2 \le -\lambda_{\min}(K_2)\|X_2\|^2 + \|l_2\|\|X_2\|, \tag{26}$$

with $X_2 = [\hat{u}_e, \hat{r}_e]^T$, $K_2 = \text{diag}\{k_u, k_r\}$, and $l_2 = [k_1\tilde{u}_e, k_2\tilde{r}_e]^T$.

Since $\|X_2\| \ge \frac{2}{\lambda_{\min}(K_2)}(|\tilde{u}_e| + |\tilde{r}_e|)$ renders $\dot{V}_2 \le -\frac{1}{2}\lambda_{\min}(K_2)\|X_2\|^2$, the subsystem (24) is ISS, and

$$\|X_2(t)\| \le \max\{\varpi_2(\|X_2(0)\|, t), \beta_{21}(|\tilde{u}_e|) + \beta_{22}(|\tilde{r}_e|)\}, \tag{27}$$

where ϖ_2 is a $\mathcal{KL}$ function and

$$\beta_{21} = \sqrt{\frac{\lambda_{\max}(S_1)}{\lambda_{\min}(S_1)}}\frac{2k_1 s}{\lambda_{\min}(K_2)}, \quad \beta_{22} = \sqrt{\frac{\lambda_{\max}(S_1)}{\lambda_{\min}(S_1)}}\frac{2k_2 s}{\lambda_{\min}(K_2)}. \tag{28}$$

At last, following theorem states the stability of the cascade system formed by the predictor error subsystem and the velocity tracking error subsystem.

Theorem 1. The cascade system formed by subsystem (20) and subsystem (24) is ISS. Besides, all error signals in the closed-loop system are uniformly ultimately bounded.

Proof. It is shown in Lemmas 2 and 3 that subsystem (20) with states being $\tilde{u}_e$, $\tilde{r}_e$ and inputs being $\tilde{\theta}_u^T \phi(\xi_u)$, $\tilde{\theta}_r^T \phi(\xi_r)$, ε_u, ε_r is ISS; subsystem (24) with states being $\hat{u}_e$, $\hat{r}_e$ and inputs being $\tilde{u}_e$, $\tilde{r}_e$ is ISS. By Lemma C.4 in [24], it follows that the cascade system formed by (20) and (24) with states being $\hat{u}_e$, $\hat{r}_e$, $\tilde{u}_e$, $\tilde{r}_e$ and inputs being $\tilde{\theta}_u^T \phi(\xi_u)$, $\tilde{\theta}_r^T \phi(\xi_r)$, ε_u, ε_r is ISS, and there exists class $\mathcal{KL}$ function ϖ and class $\mathcal{K}$ function β, such that

$$\|X(t)\| \leq \max \left\{ \varpi(\|X(0)\|, t), \beta(\|l\|) \right\}, \tag{29}$$

where $X = [\hat{u}_e, \hat{r}_e, \tilde{u}_e, \tilde{r}_e]^T$ and $l = [\tilde{\theta}_u^T \phi(\xi_u), \tilde{\theta}_r^T \phi(\xi_r), \varepsilon_u, \varepsilon_r]^T$.

Since $\tilde{\theta}_u^T \phi(\xi_u)$, $\tilde{\theta}_r^T \phi(\xi_r)$, ε_u, ε_r are bounded by ε_1^*, ε_2^*, ε_u^*, ε_r^* respectively, the errors $\hat{u}_e$, $\hat{r}_e$, $\tilde{u}_e$, $\tilde{r}_e$ are uniformly ultimately bounded. Moreover, since $|u_e| = |\hat{u}_e - \tilde{u}_e| \leq |\hat{u}_e| + |\tilde{u}_e|$ and $|r_e| = |\hat{r}_e - \tilde{r}_e| \leq |\hat{r}_e| + |\tilde{r}_e|$, the kinetic errors u_e, r_e are uniformly ultimately bounded, such that the control objective (7) is achieved.

4 Conclusions

In this paper, an intelligent fuzzy velocity control scheme is proposed for an under-actuator ASV suffering from unknown dynamics and environmental disturbances. The parameters of the fuzzy system are updated based on an SGD approach such that the unknown dynamics and disturbances of the ASV can be approximated online. Moreover, a predictor is developed based on the fuzzy system, which optimizes the learning performance of the fuzzy system. Then, fuzzy adaptive control laws are designed based on the estimated dynamics. The errors in the closed-loop kinetic control system are proved to be uniformly ultimately bounded with Lyapunov analysis.

References

1. Peng, Z., Wang, D., Wang, J.: Predictor-based neural dynamic surface control for uncertain nonlinear systems in strict-feedback form. IEEE Trans. Neural Netw. Learn. Syst. **18**(9), 2156–2167 (2017)
2. Peng, Z., Wang, J., Wang, D.: Distributed maneuvering of autonomous surface vehicles based on neurodynamic optimization and fuzzy approximation. IEEE Trans. Control Syst. Technol. **26**(3), 1083–1090 (2018)
3. Peng, Z., Wang, J., Wang, D.: Distributed containment maneuvering of multiple marine vessels via neurodynamics-based output feedback. IEEE Trans. Control Syst. Technol. **64**(5), 3831–3839 (2017)
4. Cui, R., Ge, S.S., How, B.V.E., Choo, Y.S.: Leader-follower formation control of underactuated autonomous underwater vehicles. Ocean Eng. **37**(7), 1491–1502 (2010)

5. Xiang, X., Yu, C., Zhang, Q.: Robust fuzzy 3D path following for autonomous underwater vehicle subject to uncertainties. Comput. Oper. Res. **84**, 165–177 (2017)
6. Shi, Y., Shen, C., Buckham, B.: Integrated path planning and tracking control of an AUV: a unified receding horizon optimization approach. IEEE/ASME Trans. Mechatron. **22**(3), 1163–1173 (2017)
7. Jin, K., Wang, H., Yi, H., Liu, J., Wang, J.: Key technologies and intelligence evolution of maritime UV. Chin. J. Ship Res. **13**(6), 1–8 (2018)
8. Li, F., Yi, H.: Application of USV to maritime safety supervision. Chin. J. Ship Res. **13**(6), 27–33 (2018)
9. Zhao, R., Xu, J., Xiang, X., Xu, G.: A review of path planning and cooperative control for MAUV systems. Chin. J. Ship Res. **13**(6), 58–65 (2018)
10. Peng, Z., Wang, J.: Output-feedback path-following control of autonomous underwater vehicles based on an extended state observer and projection neural network. IEEE Trans. Syst. Man Cybern. Part A Syst. Hum. **48**(4), 535–544 (2018)
11. Peng, Z., Wang, J., Wang, J.: Constrained control of autonomous underwater vehicles based on command optimization and disturbance estimation. IEEE Trans. Ind. Electron. **66**(5), 3627–3635 (2019)
12. Peng, Z., Wang, J., Han, Q.: Path-following control of autonomous underwater vehicles subject to velocity and input constraints via neurodynamic optimization. IEEE Trans. Ind. Electron. (2019). https://doi.org/10.1109/TIE.2018.2885726
13. Liu, L., Wang, D., Peng, Z.: State recovery and disturbance estimation of unmanned surface vehicles based on nonlinear extended state observers. Ocean Eng. **171**, 625–632 (2018). https://doi.org/10.1016/j.oceaneng.2018.11.008
14. Yang, Y., Zhou, C., Ren, J.: Model reference adaptive robust fuzzy control for ship steering autopilot with uncertain nonlinear systems. Appl. Soft Comput. **3**(4), 305–316 (2003)
15. Li, Y., Tong, S.: Adaptive fuzzy output-feedback stabilization control for a class of switched nonstrict-feedback nonlinear systems. IEEE Trans. Cybern. **47**(7), 1007–1016 (2017)
16. Hou, X., Zou, A., Tan, M.: Adaptive control of an electrically driven nonholonomic mobile robot via backstepping and fuzzy approach. IEEE Trans. Control Syst. Technol. **17**(4), 803–815 (2009)
17. Bottou, L.: Stochastic gradient learning in neural networks. Proc. Neuronîmes **91**(8), 12 (1991)
18. Yang, X., Zheng, X., Gao, H.: SGD-based adaptive NN control design for uncertain nonlinear systems. IEEE Trans. Neural Netw. Learn. Syst. **99**, 1–13 (2018). https://doi.org/10.1109/TNNLS.2018.2790479
19. Hardt, M., Recht, B., Singer, Y.: Train faster, generalize better: stability of stochastic gradient descent. In: Proceedings of 33rd International Conference on Extreme Learning Machines, pp. 1225–1234 (2016)
20. Wang, L.: Adaptive Fuzzy Systems and Control: Design and Stability Analysis. Prentice Hall, Upper Saddle River (1994)
21. Poggio, T., Voinea, S., Rosasco, L.: Online learning, stability, and stochastic gradient descent. arXiv:1105.4701 (2011)
22. Fossen, T.: Handbook of Marine Craft Hydrodynamics and Motion Control. Wiley, Hoboken (2011)
23. Guo, B., Zhao, Z.: On convergence of tracking differentiator. Int. J. Control **84**(4), 693–701 (2011)
24. Krstic, M., Kokotovic, P., Kanellakopoulos, I.: Nonlinear and Adaptive Control Design. Wiley, Hoboken (1995)

A Novel Second-Order Consensus Control in Multi-agent Dynamical Systems

Boshan Chen[1], Jiejie Chen[2(✉)], Zhigang Zeng[3], and Ping Jiang[4]

[1] College of Mathematics and Statistics, Hubei Normal University,
Huangshi 435002, China
[2] College of Computer Science and Technnology, Hubei Normal University,
Huangshi 435002, China
chenjiejie118@gmail.com
[3] School of Artificial Intelligence and Automation,
Huazhong University of Science and Technology, Wuhan 430074, China
[4] Computer School, Hubei Polytechnic University, Huangshi 435002, China

Abstract. In this paper, a new type protocol is proposed. Second-order consensus in multi-agent dynamical systems with this protocol is studied using a new analytical method. A necessary and sufficient condition for reaching consensus of the system with the this protocol is obtained, which depending on the spectrum of the Laplacian matrix and the control parameter setting. Meanwhile, a simple and practical criterion of sampling period is given in the ordinary case. Finally, two simulation examples are given to verify and illustrate the theoretical analysis.

Keywords: Second-order consensus · Multi-agent systems · Sampling period

1 Introduction

The consensus problem has a long history in the computer science especially for the field of distributed computing. The idea for consensus originated from statistical consensus theory in [1]. The consensus means that a group of autonomous agents converge to a common state under some appropriate protocols. In recent decade, the consensus control problem in a multi-agent systems networks (MASN) is getting a lot of attention because their potential and practical applications in many aspects, such as sensor networks, unmanned air vehicle formations, robotic teams, and underwater vehicles (see [2–7]). The recent works can be found in [8–21] and references therein. Theoretically, the consensus problem in MASN is equivalent to the stability problem of some system. Formally, the consensus problem in MASN can be divided into two broad categories: one is MASN with dynamic model (leader-following consensus) [22–29], and the other is MASN without dynamic model (leaderless consensus) [30–40]. On other hand, it plays a key role to design some appropriate protocols (control or algorithm) in the consensus in MASN. So far, various types of protocols have been proposed

© Springer Nature Switzerland AG 2019
H. Lu et al. (Eds.): ISNN 2019, LNCS 11555, pp. 101–110, 2019.
https://doi.org/10.1007/978-3-030-22808-8_11

on the consensus in MASN, such as continuous protocols, sampled protocols, hybrid protocols and so on.

Since most of the original physical models in the real world are second-order differential systems, so it is important to study second-order control problems. The second order control problem is essentially different from the first order control, and its controller types are more selective than the first order control, which means that its theories are richer in content. In last decade, the second-order consensus problem has come to receiving particular attention. Firstly, a necessary and sufficient condition was proved [30], where the second-order dynamics are governed by the position and velocity terms of the agents. Since then, many researchers have focused on investigating second-order consensus in multi-agent systems with various types of protocols. By using zero-order holds or direct discretization, some necessary and sufficient conditions were established for multi-agent systems with sampled-data control [31–33]. On the other hand, for second-order continuous-time multi-agent systems, the consensus problem under a time-varying topology and sampled data control was addressed in [35]. In addition, via the position and velocity sampled data protocols, continuous position and sampled velocity data protocols, continuous position protocols with delay and sampled position protocols with delay, the necessary and sufficient conditions developed for reaching consensus in [36–39], respectively. A containment control problems for networked multi-agent systems is investigated in [40]. It should be noted that all these aforementioned works were considered based on the basis solution matrix of the second order system.

Although sufficient and necessary conditions are obtained, they are complicated and difficult to apply in the ordinary case. The main contributions of this paper lies in the following two aspects: (i) Second-order consensus in multi-agent dynamical systems with this protocol is studied. A necessary and sufficient condition for reaching consensus of the system with the this protocol is obtained, which depending on the spectrum of the Laplacian matrix and the control parameter setting. Meanwhile, a simple and practical criterion of sampling period is given in the ordinary case. (ii) A new method is used to study linear consensus in multi-agent dynamical systems, which does not need the calculation of the fundamental solution matrix of the common subsystem.

2   Preliminaries

In this section, some preliminaries are introduced. The commonly studied second-order consensus protocol is described as follows:

$$\ddot{x}_i(t) = u_i(t), i \in \mathcal{N} \tag{1}$$

where $\mathcal{N} = \{1, 2, \cdots, N\}$, $x_i(t) \in \mathbb{R}$ and $u_i(t) \in \mathbb{R}$ represent the ith agent's state and control input, respectively.

A new consensus protocol $u_i(t)$ with current position state, sampled position state data and velocity state is considered as follows:

$$
\begin{aligned}
u_i(t) = &-\alpha \sum_{j=1}^{N} c_{ij}(x_j(t) - x_i(t)) \\
&+\beta \sum_{l=1}^{N} c_{il}[(\dot{x}_l(t_k) - \dot{x}_i(t_k)) + (x_l(t_k) - x_i(t_k))], i \in \mathcal{N}
\end{aligned}
\tag{2}
$$

for $t \in [t_k, t_{k+1})$, where $c_{ij} \geq 0$ for $i \neq j$ and $i, j \in \mathcal{N}$, t_k are the sampling instants satisfying $0 = t_0 < t_1 < \cdots < t_k < \cdots$, and α and β are the coupling strengths. For simplicity, assume that $t_{k+1} - t_k = T$, where $T > 0$ is the sampling period.

Remark 2.1. In [36–39], some protocols are designed respectively. However, no work has been found on the consensus protocols (2) for second-order continuous-time multi-agent systems.

Under the protocol (2), we have the following second closed-loop system:

$$
\ddot{x}_i(t) = \alpha \sum_{j=1}^{N} \ell_{ij}x_j(t) - \beta \sum_{l=1}^{N} \ell_{il}[\dot{x}_l(t_k) + x_l(t_k)], i \in \mathcal{N},
\tag{3}
$$

for $t \in [t_k, t_{k+1})$, where $L = (\ell_{ij})_{n \times n}$ denote the Laplacian matrix of the graph $\mathcal{G}$, that is $\ell_{ii} = \sum_{j \neq i} c_{ij}$ and $\ell_{ij} = -c_{ij}$ for $i \neq j$.

Definition 2.1. Second-order consensus in multi-agent system (MAS) (3) is said to be achieved if, for any initial conditions,

$$
\lim_{t \to \infty} |x_i(t) - x_j(t)| = 0, \text{ and } \lim_{t \to \infty} |\dot{x}_i(t) - x_j(t)| = 0,
$$

for any $i, j \in \mathcal{N}$.

3 Main Results

MAS (3) can be written as

$$
\ddot{x}(t) = \alpha L x(t) - \beta L(\dot{x}(t_k) + x(t_k)),
\tag{4}
$$

for $t \in [t_k, t_{k+1})$, where $x(t) = (x_1(t), x_2(t), \cdots, x_N(t))^T$. For Laplacian matrix L, there exists a nonsingular matrix P such that $L = PJP^{-1}$, where J is the Jordan form associated with the Laplacian matrix L. Then, from MAS (4) one has

$$
\ddot{y}(t) = \alpha J y(t) - \beta J(\dot{y}(t_k) + y(t_k)),
\tag{5}
$$

for $t \in [t_k, t_{k+1})$, where $y(t) = P^{-1}x(t)$.

Since $\mathcal{G}$ is directed, some eigenvalues of L can be complex. Let

$$
J = \text{diag}\{J_1, J_2, \cdots, J_r\}
$$

where J_i $(i = 1, 2, \cdots r)$ are Jordan block of L. It is well known that 0 is a simple eigenvalue of the Laplacian matrix L if the network $\mathcal{G}$ contains a directed spanning tree. Let $y(t) = (y_1(t), y_2(t), \cdots, y_N(t))^T$. If $i_0 \in \mathcal{N}$ is root of the directed spanning tree, then

$$\ddot{y}_{i_0}(t) = 0 \tag{6}$$

for $t \in [t_k, t_{k+1})$. Without loss of generality, assume $i_0 = 1$.

Lemma 3.1. Suppose that the network $\mathcal{G}$ contains a directed spanning tree. Then, the second consensus in MAS (3) can be reached if and only if the system (5) is asymptotically stable, i.e.,

$$\lim_{t \to \infty} y_i(t) = 0, \lim_{t \to \infty} \dot{y}_i(t) = 0 \tag{7}$$

for any $i = 2, \cdots, N$.

Lemma 3.2. Suppose that the network $\mathcal{G}$ contains a directed spanning tree. Then, the second consensus in MAS (3) can be reached if and only if the following r systems are asymptotically stable:

$$\ddot{z}_i(t) = \alpha \lambda_i z_i(t) - \beta \lambda_i (\dot{z}_i(t_k) + z_i(t_k)), i = 1, 2, \cdots, r. \tag{8}$$

for $t \in [t_k, t_{k+1})$, where $\lambda_i \in \sigma(L) \setminus \{0\}$.

Introducing new variables $v_{i1}(t) = z_i(t), v_{i2}(t) = \dot{z}_i(t)$, and let $v_i(t) = (v_{i1}(t), v_{i2}(t))^T, i \in \mathcal{N}$. The system (8) can be further written as:

$$\dot{v}_i(t) = (A + \alpha \lambda_i B)v_i(t) - \beta \lambda_i C v_i(t_k), i \in \mathcal{N}, \tag{9}$$

for $t \in [t_k, t_{k+1})$ and $i = 1, \cdots, r-1$ where $A, B \in \mathbb{R}^{n \times n}$,

$$A = \begin{pmatrix} 0 & 1 \\ 0 & 0 \end{pmatrix}, B = \begin{pmatrix} 0 & 0 \\ 1 & 0 \end{pmatrix}, C = \begin{pmatrix} 0 & 0 \\ 1 & 1 \end{pmatrix}.$$

The following basic assumption is required in our results.

Assumption 1. $0 < \alpha < \beta$ and $\frac{\beta}{\sqrt{\alpha}} \Re(\sqrt{\lambda}) + \Im_{\min} > 1$ for all $\lambda \in \sigma(L) \setminus \{0\}$, where $\Im_{\min} = \min_{\lambda \in \sigma(L) \setminus \{0\}} \{|\Im(\lambda)|\}$.

Remark 3.1. Assumption 1 is a necessary condition even in the case of the consensus protocol.

Under Assumption 1, one has

$$\det(A + \alpha \lambda_i B - \beta \lambda_i C) = \lambda_i(\beta - \alpha) \neq 0, \tag{10}$$

and

$$\det(A - \lambda_i \alpha B - \beta \lambda_i I_2) = \lambda_i^2 \beta^2 - \lambda_i \alpha \neq 0 \tag{11}$$

for any $\lambda_i \in \sigma(L) \setminus \{0\}$.

Let $w_i(t) = v_i(t) + (A + \alpha\lambda_i B - \beta\lambda_i I_2)^{-1}(\beta\lambda_i C - \beta\lambda_i I_2)v_i(t_k)$, then

$$\dot{w}_i(t) = (A + \alpha\lambda_i B)w_i(t) - \beta\lambda_i w_i(t_k), t \in [t_k, t_{k+1}) \tag{12}$$

where $\lambda_i \in \sigma(L) \setminus \{0\}$, and $i = 1, 2, \cdots, r$.

Lemma 3.3. Under Assumption 1, the system (12) is asymptotically stable if and only if the system (12) is asymptotically stable.

By using the structure of $J(A - \alpha\lambda_i B)$, one has the following lemma.

Lemma 3.4. Consider the following systems:

$$\dot{\hat{w}}_{ij}(t) = \pm\sqrt{\lambda_i \alpha}\hat{w}_{ij}(t) - \beta\lambda_i \hat{w}_{ij}(t_k), \tag{13}$$

for $i = 1, 2, \cdots, r$, where $\lambda_i \in \sigma(L)\setminus\{0\}$. Then, the system (12) is asymptotically stable if and only if the system (13) is asymptotically stable. From Lemmas 3.2, 3.3 and 3.4, we have the following lemma.

Lemma 3.5. Suppose that the network $\mathcal{G}$ contains a directed spanning tree. Under Assumption 1, the second consensus in MAS (3) can be reached if and only if the system (13) is asymptotically stable.

Theorem 3.1. Suppose that the network $\mathcal{G}$ contains a directed spanning tree. Under Assumption 1, the second-order consensus in MAS (3) if and only if

$$\left|1 + \left(e^{\pm\sqrt{\lambda\alpha}T} - 1\right)\left(1 \mp \frac{\beta}{\sqrt{\alpha}}\sqrt{\lambda}\right)\right| < 1, \tag{14}$$

for any $\lambda \in \sigma(L) \setminus \{0\}$.

Remark 3.2. A sufficient and necessary condition (14) is established in Theorem 3.1 to achieve the second-order consensus in MAS (3). In general, it is still a challenging issue that how to devise the system parameters that satisfy the condition (14). But, we can consider some particular cases or some simple and convenient sufficient conditions.

Theorem 3.2. Suppose that the network $\mathcal{G}$ contains a directed spanning tree. Under Assumption 1, the second-order consensus in MAS (3) can be reached if T is sufficiently small. In addition, Let $\rho = |1 \mp \frac{\beta}{\sqrt{\alpha}}\sqrt{\lambda}|$,

$$
\begin{aligned}
T_\lambda^+ &= \frac{1}{\Re(\sqrt{\lambda})\sqrt{\alpha}} \\
&\quad \times \ln\left[\frac{2}{\rho^2}\left[\frac{\beta}{\sqrt{\alpha}}(\Re(\sqrt{\lambda}) + |\Im(\sqrt{\lambda})|) - 1\right] + 1\right] \\
T_\lambda^- &= \frac{1}{-\Re(\sqrt{\lambda})\sqrt{\alpha}}, \\
&\quad \times \ln\left[\frac{-2}{\rho^2}\left[\frac{\beta}{\sqrt{\alpha}}(\Re(\sqrt{\lambda}) - \Im(\sqrt{\lambda})) + 1\right] + 1\right], \\
T_{\min}^* &= \min_{\lambda\in\sigma(L)\setminus\{0\}}\{T_\lambda^+, T_\lambda^-; T_\lambda^- \in \mathbb{R}^+\}, \\
T_{\max}^* &= \max_{\lambda\in\sigma(L)\setminus\{0\}}\{T_\lambda^+, T_\lambda^-; T_\lambda^- \in \mathbb{R}^+\}.
\end{aligned}
\tag{15}
$$

Then the second-order consensus in MAS (3) can be reached if $0 < T < T^* - o(T\Im_{\max})$ and the second-order consensus in MAS (3) can not be reached if $T > T_{\max}^* + o(T\Im_{\max})$, where $\Im_{\max} = \max_{\lambda\in\sigma(L)\setminus\{0\}}\{|\Im(\lambda)|\}$.

Corollary 3.1. Suppose that the network $\mathcal{G}$ contains a directed spanning tree, all the eigenvalues of its Laplacian matrix are real and $\alpha > 0$. Under Assumption 1, the second-order consensus in MAS (3) can be reached if and only if $\beta > \sqrt{\frac{\alpha}{\lambda_{\min}}}$ and

$$T < T^* = \min\{T^+, T^-; T^- \in \mathbb{R}^+\}, \tag{16}$$

where $\lambda_{\min} = \min_{1 \le i \le r-1}\{\lambda_i\}$.

$$T^{\pm} = \min_{1 \le i \le r-1} \frac{1}{\pm\sqrt{\lambda_i\alpha}} \ln\left(\frac{\pm 2\sqrt{\lambda_i\alpha}}{\beta\lambda_i \mp \sqrt{\lambda_i\alpha}} + 1\right). \tag{17}$$

Remark 3.3. The criteria are fairly simple and broad in Theorem 3.2 and Corollary 3.1. When the network $\mathcal{G}$ has an undirected topology T^* can be a threshold of the sampling period T. In general, $T^*_{\min}$ can be an approximate threshold of the sampling period T.

4   Examples and Simulations

In this section, we take two examples to further test the previous theoretical analysis.

Example 4.1. Consider the problem of second-order consensus in an MAS (3) with a directed topology, where $\alpha = 1$, $\beta = 1.6$ and

$$L = \begin{pmatrix} 1 & -1 & 0 & 0 \\ 0 & 1 & -1 & 0 \\ 0 & 0 & 1 & -1 \\ -1 & 0 & 0 & 1 \end{pmatrix}.$$

By computing, one has $\lambda_1 = 0$, $\lambda_{2,3} = 1 \pm i$ and $\lambda_4 = 2$. $T^*_{\min} = 0.6712$ and $T^*_{\max} = 1.1885$. Therefore, the second-order consensus in an MAS (3) can be reached for some $T < T^* - o(T)$. The states and the consensus error with $T = 0.5$ and $T = 0.55$ are shown in Figs. 1 and 2.

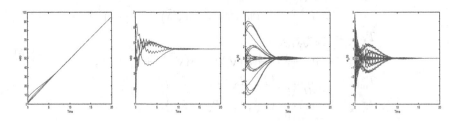

Fig. 1. The consensus can be reached when $T = 0.32$ in Example 4.2.

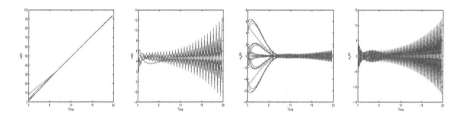

Fig. 2. The consensus can not be reached when $T = 0.34$ in Example 4.2.

Example 4.2. Consider the problem of second-order consensus in an MAS (3) with an undirected topology, where $\alpha = 1$, $\beta = 2$ and

$$L = \begin{pmatrix} 3 & -1 & -1 & -1 & 0 \\ -1 & 2 & -1 & 0 & 0 \\ -1 & -1 & 2 & 0 & 0 \\ -1 & 0 & 0 & 2 & -1 \\ 0 & 0 & 0 & -1 & 1 \end{pmatrix}.$$

By computing, one has $\lambda_1 = 0$, $\lambda_2 = 0.5188$, $\lambda_3 = 2.3111$, $\lambda_4 = 3$ and $\lambda_5 = 4.1701$. $T^* = 0.3319$. Therefore, the second-order consensus in an MAS (3) can be reached for some $T < 0.3319$. The states $(x(t), V(t))$ and the consensus error $(e_x(t), e_V(t))$ with $T = 0.32$ and $T = 0.34$ are shown in Figs. 3 and 4.

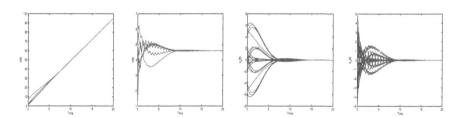

Fig. 3. The consensus can be reached when $T = 0.32$ in Example 4.2.

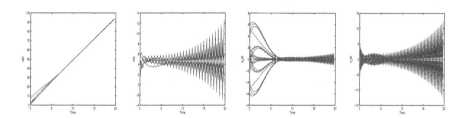

Fig. 4. The consensus can not be reached when $T = 0.34$ in Example 4.2.

5 Conclusion

In this paper, a new type protocol is proposed. Second-order consensus in multi-agent dynamical systems with this protocol is studied using a new analytical method. A necessary and sufficient condition for reaching consensus of the system with the this protocol is obtained, which depending on the spectrum of the Laplacian matrix and the control. Meanwhile, a simple and practical criterion of sampling period is given in the ordinary case. In particular, used method and obtained results does not need the calculation of the fundamental solution matrix of the common subsystem. There are many ways to promote this article, such as the synchronization problem of nonlinear MAS, the containment control problem, the event-triggered control problem and so on. These will be considered in future.

Acknowledgements. The work is supported by the Natural Science Foundation of China under Grant 61603129, 61841301, the Natural Science Foundation of Hubei Province under Grant 2016CFC734.

References

1. DeGroot, M.H.: Reaching a consensus. J. Am. Stat. Assoc. **69**(345), 118–121 (1974)
2. McLain, T., Chandler, P., Rasmussen, S., Pachter, M.: Cooperative control of UAV rendezvous. In: American Control Conference, pp. 2309–2314. IEEE (2001)
3. Lin, Z., Francis, B., Maggiore, M.: Necessary and sufficient graphical conditions for formation control of unicycles. Trans. Autom. Control **50**(1), 121–127 (2005)
4. Fax, J., Murray, R.: Information flow and cooperative control of vehicle formations. IEEE Trans. Autom. Control **49**(9), 1465–1476 (2004)
5. Cortés, J., Bullo, F.: Coordination and geometric optimization via distributed dynamical systems. SIAM J. Control Optim. **44**(5), 1543–1574 (2006)
6. Wu, C.W., Chua, L.O.: On a conjecture regarding the synchronization in an array of linearly coupled dynamical systems. IEEE Trans. Circuits Syst. I Regular Paper **43**(2), 161–165 (1996)
7. Leblanc, H.J., Koutsoukos, X.: Resilient first-order consensus and weakly stable, higher order synchronization of continuous-time networked multi-agent systems. IEEE Trans. Control Netw. Syst. **5**(3), 1219–1231 (2017)
8. Ge, X., Han, Q., Ding, D., Zhang, X., Ning, B.: A survey on recent advances in distributed sampled-data cooperative control of multi-agent systems. Neurocomputing **275**, 1684–1701 (2018)
9. Li, H., Zhu, Y., Wang, J.: Consensus of nonlinear second-order multi-agent systems with mixed time-delays and intermittent communications. Neurocomputing **251**(16), 115–126 (2017)
10. Zheng, Y., Ma, J., Wang, L.: Consensus of hybrid multi-agent systems. IEEE Trans. Neural Netw. Learn. Syst. **29**(4), 1359–1365 (2018)
11. Liu, X., Zhang, K., Xie, W.C.: Consensus of multi-agent systems via hybrid impulsive protocols with time-delay. Nonlinear Anal. Hybrid Syst. **30**, 134–146 (2018)
12. Hou, W., Fu, M., Zhang, H.: Consensus conditions for general second-order multi-agent systems with communication delay. Automatic **75**, 293–298 (2017)

13. Hu, J., Cao, J., Yu, J.: Consensus of nonlinear multi-agent systems with observer-based protocols. Syst. Control Lett. **72**, 71–79 (2014)
14. He, W., Chen, G., Han, Q.L.: Network-based leader-following consensus of nonlinear multi-agent systems via distributed impulsive control. Inf. Sci. **380**, 145–158 (2017)
15. Liu, X., Zhang, K., Xie, W.C.: Consensus seeking in multi-agent systems via hybrid protocols with: impulse delays. Nonlinear Anal. Hybrid Syst. **25**, 90–98 (2017)
16. Liu, H., Xie, G., Yu, M.: Necessary and sufficient conditions for containment control of fractional-order multi-agent systems. Neurocomputing **323**, 86–95 (2019)
17. Yu, Z., Jiang, H., Hu, C., Yu, J.: Necessary and sufficient conditions for consensus of fractional-order multiagent systems via sampled-data control. IEEE Trans. Syst. Man Cybern. **47**(8), 1892–1901 (2017)
18. Wang, F., Yang, Y.: Leader-following consensus of nonlinear fractional-order multi-agent systems via event-triggered control. Int. J. Syst. Sci. **48**(3), 571–577 (2017)
19. Ma, X., Sun, F., Li, H., He, B.: The consensus region design and analysis of fractional-order multi-agent systems. Int. J. Syst. Sci. **48**(3), 629–636 (2017)
20. Zhu, W., Li, W., Zhou, P., Yang, C.: Consensus of fractional-order multi-agent systems with linear models via observer-type protocol. Neurocomputing **230**(3), 60–65 (2017)
21. Chen, Y., Wen, G., Peng, Z., Rahmani, A.: Consensus of fractional-order multi-agent system via sampled-data event-triggered control. J. Franklin Inst. (2018). https://doi.org/10.1016/j.jfranklin.01.043
22. Song, Q., Cao, J., Yu, W.: Second-order leader-following consensus for nonlinear multi-agent systems via pinning control. Syst. Control Lett. **59**(9), 553–562 (2010)
23. Song, Q., Liu, F., Cao, J., Yu, W.: M-matrix strategies for pinning-controlled leader-following consensus in multiagent systems with nonlinear dynamics. IEEE Trans. Cybern. **43**(6), 1688–1697 (2013)
24. Wen, G., Duan, Z., Yu, W., Chen, G.: Consensus in multi-agent systems with communication constraints. Int. J. Robust Nonlinear Control **22**(2), 170–182 (2012)
25. You, K., Li, Z., Xie, L.: Consensus condition for linear multi-agent systems over randomly switching topologies. Automatica **49**(10), 3125–3132 (2013)
26. Liu, H., Cheng, L., Tan, M., Hou, Z.G.: Containment control of continuous-time linear multi-agent systems with aperiodic sampling. Automatica **57**, 78–84 (2015)
27. Li, Z., Wei, R., Liu, X.: Distributed containment control of multi-agent systems with general linear dynamics in the presence of multiple leaders. Int. J. Robust Nonlinear Control **23**(5), 534–547 (2013)
28. Qin, J., Gao, H.: A sufficient condition for convergence of sampled-data consensus for double-integrator dynamics with nonuniform and time-varying communication delays. IEEE Trans. Autom. Control **57**(9), 2417–2422 (2012)
29. Wu, Y., Su, H., Shi, P., Shu, Z., Wu, Z.G.: Consensus of multiagent systems using aperiodic sampled-data control. IEEE Trans. Cybern. **46**(9), 2132–2143 (2016)
30. Yu, W., Chen, G., Cao, M.: Some necessary and sufficient conditions for second-order consensus in multi-agent dynamical systems. Automatica **46**(6), 1089–1095 (2010)
31. Cao, Y., Ren, W.: Multi-vehicle coordination for double-integrator dynamics under fixed undirected/directed interaction in a sampled-data setting. Int. J. Robust Nonlinear Control **20**(9), 987–1000 (2010)
32. Gao, Y., Wang, L., Xie, G., Wu, B.: Consensus of multi-agent systems based on sampled-data control. Int. J. Robust Nonlinear Control **82**(12), 2193–2205 (2009)

33. Liu, H., Xie, G., Wang, L.: Necessary and sufficient conditions for solving consensus problems of double-integrator dynamics via sampled control. Int. J. Robust Nonlinear Control **20**(15), 1706–1722 (2010)
34. Ma, Q., Xu, S., Lewis, F.: Second-order consensus for directed multiagent systems with sampled data. Int. J. Robust Nonlinear Control **24**(16), 2560–2573 (2014)
35. Gao, Y., Wang, L.: Sampled-data based consensus of continuous time multiagent systems with time-varying topology. IEEE Trans. Autom. Control **56**(5), 1226–1231 (2011)
36. Yu, W., Zhou, L., Yu, X., Lü, J., Lu, R.: Consensus in multi-agent systems with second-order dynamics and sampled data. IEEE Trans. Ind. Inform. **9**(4), 2137–2146 (2013)
37. Yu, W., Zheng, W.X., Chen, G., Ren, W., Cao, J.: Second-order consensus in multi-agent dynamical systems with sampled position data. Automatica **47**(7), 1496–1503 (2011)
38. Yu, W., Chen, G., Cao, M., Ren, W.: Delay-induced consensus and quasiconsensus in multi-agent dynamical systems. IEEE Trans. Circuits Syst. I Regular Paper **60**(10), 679–2687 (2013)
39. Huang, N., Duan, Z., Chen, G.: Some necessary and sufficient conditions for consensus of second-order multi-agent systems with sampled position data. Automatica **63**, 148–155 (2016)
40. Liu, H., Xie, G., Wang, L.: Necessary and sufficient conditions for containment control of networked multi-agent systems. Automatica **48**(7), 1415–1422 (2012)

Adaptive Backstepping Dynamic Surface Control Design of a Class of Uncertain Non-lower Triangular Nonlinear Systems

Gang Sun[1,2(✉)], Mingxin Wang[1], and Sheng Wang[1]

[1] School of Mathematical Science and Energy Engineering,
Hunan Institute of Technology, No. 18, Henghua Road, Zhuhui District,
Hengyang 421002, People's Republic of China
gs_sungang@126.com, wangmx54@126.com

[2] School of Information Engineering, Jiangxi University of Science and Technology,
No. 86, Hongqi Avenue, Ganzhou 341000, People's Republic of China

Abstract. An adaptive backstepping control method is developed for a class of uncertain non-lower triangular nonlinear systems by combining techniques of neural network online approximation and dynamic surface control. In the design, adaptive backstepping technique is employed to establish virtual control laws and actual control law recursively. The unknown functions contained in control laws are replaced by neural network online approximators. And dynamic surface control technique is used to eliminate the problem of circular structure of the controller. The results of stability analysis show that all the closed-loop system signals are guaranteed to be uniformly ultimately bounded, and the steady-state tracking error can be made to converge to an arbitrarily small neighborhood of zero by choosing control parameters appropriately. The effectiveness of the proposed approach is demonstrated via a numerical simulation example.

Keywords: Adaptive backstepping · Dynamic surface control (DSC) ·
Neural network (NN) ·
Uncertain non-lower triangular nonlinear systems

1 Introduction

Adaptive backstepping has been a powerful method for synthesizing controllers for uncertain nonlinear systems with lower triangular forms [1,2]. In [3,4], adaptive backstepping control design was developed for lower triangular nonlinear systems with linearly parameterized uncertainties. For nonlinear systems with

This work is supported by National Natural Science Foundation of China under Grant 61463019, by Natural Science Foundation of Hunan Province under Grant 2019JJ40062, by Research Foundation of Education Bureau of Hunan Province under Grant 17C0434, and by Research Foundation of Hunan Institute of Technology under Grant HQ14011.

H. Lu et al. (Eds.): ISNN 2019, LNCS 11555, pp. 111–119, 2019.
https://doi.org/10.1007/978-3-030-22808-8_12

uncertainties that cannot be linearly parameterized or are completely unknown, online approximation based adaptive backstepping control approaches were proposed, such as NN control [5–12], fuzzy control [13–18], and wavelet control [19–22]. A drawback of "complexity of controller design" exists in traditional backstepping based adaptive control design methods. To reduce the complexity of control design, techniques of DSC [23–28], minimal learning parameter [26,29,30], single neural network online approximation [31–34], and so on were incorporated into backstepping based adaptive control design framework.

Aforementioned methods all focus on control design of uncertain nonlinear systems with a lower triangular form. While the uncertain lower triangular nonlinear systems have been much studied via adaptive backstepping design, relatively fewer results are available in literatures for the control of uncertain nonlinear systems with a non-lower triangular form. One of the main difficulties in control of non-lower triangular nonlinear systems is a circular construction of controller which is caused by the structure feature of the class of nonlinear systems and the differentiations of virtual control laws in backstepping design procedure. In fact, for some lower triangular nonlinear systems with pure-feedback forms, the circular structure problem also exists in backstepping based control design. In [10], by using input-to-state stability modular approach, the circularity problem was solved in adaptive backstepping control of non-affine pure-feedback nonlinear systems. In [27], an adaptive backstepping DSC design approach was proposed for a class of pure-feedback nonlinear systems. In the work, DSC technique was used to eliminate the circular structure problem.

In this paper, we will consider the problem of adaptive control design for a class of uncertain nonlinear systems with a non-lower triangular form. It should be pointed out that the aforementioned adaptive backstepping based control design methods are not valid for the class of uncertain non-lower nonlinear systems because the circular structure problem cannot be solved. Inspired by the approach of solving circular structure problem in [27], we will develop an NN online approximation based adaptive backstepping DSC design method. For an nth-order uncertain non-lower triangular nonlinear system, the controller design is divided into n steps. Firstly, by replacing an unknown function with an NN online approximator, a full state feedback virtual control law is designed to stabilize the first i subsystems at intermediate step i, $i = 1, 2, ..., n - 1$. Then, employing a first-order filter to handle the virtual control signal, and let the filter output enter into next step of the design. By doing this, the circular structure problem can be eliminated effectively. At the last step (step n), a full state feedback actual control law is given to stabilize all n subsystems. Stability results of the closed-loop system are given. It is showed that all the closed-loop system signals are uniformly ultimately bounded, and the tracking error of the system can converge to a small neighborhood of zero via appropriately choosing control parameters. A simulation example is given to demonstrate the effectiveness of the approach.

The rest of the paper is organized as follows. Section 2 presents the problem formulation. And the brief description of NN online approximator is also given in this section. In Sect. 3, the NN based adaptive backstepping DSC design procedure is presented. Stability results of the closed-loop control system are given in Sect. 4. Numerical simulation verification and conclusion of the work are provided in Sects. 5 and 6 respectively.

2 Problem Description and Preliminaries

2.1 Problem Description

Consider a class of uncertain nonlinear dynamical systems as follows,

$$
\begin{cases}
\dot{x}_i = x_{i+1} + f_i(\bar{x}_n), & i = 1, 2, ..., n - 1, \\
\dot{x}_n = u + f_n(\bar{x}_n), \\
y = x_1,
\end{cases}
\tag{1}
$$

where $\bar{x}_n = (x_1, x_2, ..., x_n)$, $x_i \in R$ are system state variables, $i = 1, 2, ..., n$; $u \in R$ and $y \in R$ are input and output of the system, respectively; $f_i(\bar{x}_n)$ are unknown continuous nonlinear functions, $i = 1, 2, ..., n$.

The control objective is, by designing an adaptive controller for the class of nonlinear systems (1), the system output y can follow a given reference signal y_r, and all the close-loop system signals remain uniformly ultimately bounded.

Assumption 1. *The reference signal y_r is sufficiently smooth.*

2.2 Neural Network Description

A brief introduction to Gaussian radial basis function neural network online approximator is given as follows,

$$
F_{NN}(\bar{x}_n) = \theta^T \xi(\bar{x}_n),
\tag{2}
$$

where

$\bar{x}_n = (x_1, x_2, ..., x_n)$	input vector of the network;
$F_{NN}(\bar{x}_n) = \theta^T \xi(\bar{x}_n)$	output of the network;
$\theta = (\theta_1, \theta_2, ..., \theta_N)^T$	weight vector of the network;
$\xi(\bar{x}_n) = (\rho_1(\bar{x}_n), \rho_2(\bar{x}_n), ..., \rho_N(\bar{x}_n))^T$	vector valued function;
$\rho_i(\bar{x}_n) = \mu e^{-\|x - \varsigma_i\|^2 / \sigma^2}$	Gaussian kernel functions, $i = 1, 2, ..., N$;
N	number of neurons of the network;
ς_i	centers of kernel functions, $i = 1, 2, ..., N$;
σ	width of kernel functions;
μ	amplification factor of kernel functions.

Remark 1. The neural network given above has the ability of approximating any continuous function [35,36]. That is, for a given continuous function, by appropriately choosing the number of neurons and parameters of the network (centers, width, and amplification factor), there is an ideal weights vector, such that the output of the network can approximate the given function with the network reconstruction error no more than a specified upper bound, i.e.,

$$f(\bar{x}_n) = \theta^T \xi(\bar{x}_n) + \varepsilon, \tag{3}$$

where $|\varepsilon| \leq \varepsilon^*$, and ε^* is upper bound of network reconstruction error.

Remark 2. Being unknown about $f(\bar{x}_n)$, the ideal weights vector of the network need to be estimated online in the operation process of the control system. In the controller design of Sect. 3, $\hat{\theta}$ denotes the estimation of θ, and an updated law will established to modify its value.

3   Adaptive Backstepping DSC Design

In this section, an NN online approximation based adaptive backstepping DSC design is established for the class of uncertain non-lower triangular nonlinear systems (1). The procedure of the controller design is as follows.

Step 1: Consider the first error surface $S_1 = x_1 - y_r$ (i.e. the system tracking error). The derivative of S_1 is

$$\dot{S}_1 = \dot{x}_1 - \dot{y}_r = x_2 + f_1(\bar{x}_n) - \dot{y}_r. \tag{4}$$

By regarding x_2 as control input of subsystem (4) and replacing unknown function $f_1(\bar{x}_n)$ with an NN online approximator, then a virtual control law can be designed as follows,

$$\alpha_2 = -k_1 S_1 - \hat{\theta}_1^T \xi_1(\bar{x}_n) + \dot{y}_r, \tag{5}$$

where $k_1 > 0$ is a control parameter. The updated law of $\hat{\theta}_1$ is as follows,

$$\dot{\hat{\theta}}_1 = \Gamma_1 \left(\xi_1(\bar{x}_n)S_1 - \eta_1\hat{\theta}_1 \right), \tag{6}$$

where Γ_1 is a real matrix satisfies $\Gamma_1 = \Gamma_1^T > 0$, and η_1 is a real scalar satisfies $\eta_1 > 0$.

A first-order filter is introduced to process virtual control signal α_2 (That is to use the DSC technique),

$$\tau_2 \dot{z}_2 + z_2 = \alpha_2, \tag{7}$$

where z_2 is the output of the filter, and τ_2 is a filter parameter.

Remark 3. x_n is contained in network $\hat{\theta}_1^T \xi_1(\bar{x}_n)$, thus the derivative of α_2 will contain actual control input u. If α_2 is passed into x_2-subsystem directly, the established virtual control law in next step will depend on u. Thus, the problem of circular dependence of control laws will arise. By letting z_2 enter into x_2-subsystem, virtual control law would not depend on u directly in next step.

Step i(i = 2, 3, ..., n − 1): Consider the *i*th error surface $S_i = x_i − z_i$. The derivative of S_i is

$$\dot{S}_i = \dot{x}_i − \dot{z}_i = x_{i+1} + f_i(\bar{x}_n) − \dot{z}_i. \tag{8}$$

By regarding x_{i+1} as control input of subsystem (8) and replacing unknown function $f_i(\bar{x}_n)$ with an NN online approximator, then a virtual control law can be designed as follows,

$$\alpha_{i+1} = −k_i S_i − \hat{\theta}_i^T \xi_i(\bar{x}_n) + \dot{z}_i, \tag{9}$$

where $k_i > 0$ is a control parameter. The updated law of $\hat{\theta}_i$ is as follows,

$$\dot{\hat{\theta}}_i = \Gamma_i \left(\xi_i(\bar{x}_n)S_i − \eta_i \hat{\theta}_i \right), \tag{10}$$

where Γ_i is a real matrix satisfies $\Gamma_i = \Gamma_i^T > 0$, and η_i is a real scalar satisfies $\eta_i > 0$.

A first-order filter is introduced to process virtual control signal α_{i+1},

$$\tau_{i+1} \dot{z}_{i+1} + z_{i+1} = \alpha_{i+1}, \tag{11}$$

where z_{i+1} is the output of the filter, and τ_{i+1} is a filter parameter.

Step n: The last error surface is $S_n = x_n − z_n$. The derivative of S_n is

$$\dot{S}_n = \dot{x}_n − \dot{z}_n = u + f_n(\bar{x}_n) − \dot{z}_n. \tag{12}$$

By replacing unknown function $f_n(\bar{x}_n)$ with an NN online approximator, then an actual control law can be designed as follows,

$$u = −k_n S_n − \hat{\theta}_n^T \xi_n(\bar{x}_n) + \dot{z}_n, \tag{13}$$

where $k_n > 0$ is a control parameter. The updated law of $\hat{\theta}_n$ is as follows,

$$\dot{\hat{\theta}}_n = \Gamma_n \left(\xi_n(\bar{x}_n)S_n − \eta_n \hat{\theta}_n \right), \tag{14}$$

where Γ_n is a real matrix satisfies $\Gamma_n = \Gamma_n^T > 0$, and η_n is a real scalar satisfies $\eta_n > 0$.

4   Stability of the Closed-Loop Systems

The stability of the control system is described by the following theorem.

Theorem 1. *Consider the closed-loop control system consisting of system (1), control laws (5), (9), (13), adaptive laws of NN online approximators (6), (10), (14), and first-order filters (7), (11). For any bounded initial states of the system, all signals of the closed-loop system can be guaranteed uniformly ultimately bounded, and the error of system output tracking the reference signal can be made arbitrarily small by appropriately choosing control parameters.*

The proof is omitted due to the space limit.

5   Numerical Simulation

Consider a third-order uncertain nonlinear system in the form of Eq. (1). Then, according to the control design procedure given in Sect. 3, the controller of the system contains three control laws, three NN online approximators, and two first-order filters.

For simulating, it is assumed that $f_1(\bar{x}_3) = x_1^2 + x_2 + x_3$, $f_2(\bar{x}_3) = x_1 + x_2^2 + 2x_3$, $f_3(\bar{x}_3) = x_1 x_2 + x_3^2$, $y_r = \sin t$, and the initial states of the system are $(x_1(0), x_2(0), x_3(0))^T = (0.8, 0, 0)^T$.

In simulation, control parameters are chosen as $k_1 = 3$, $k_2 = 3$, $k_3 = 3$. For three NN online approximator, centers of kernel functions are chosen as $\{\varsigma_i | i = 1, 2, ..., 45\} = \{-1, -0.5, 0, 0.5, 1\} \times \{-0.5, 0, 0.5\} \times \{-0.5, 0, 0.5\}$, and other parameters are $\sigma = 10$, $\mu = 1$, $\Gamma = 1$, $\eta = 0.2$, $\hat{\theta}_1(0) = \hat{\theta}_2(0) = \hat{\theta}_3(0) = 0$.

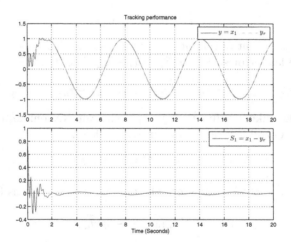

Fig. 1. Performance of the system output tracking the reference signal.

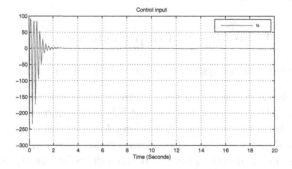

Fig. 2. Input signal of the control system.

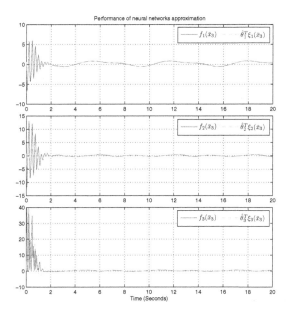

Fig. 3. Performance of neural networks approximating unknown functions.

For two first-order filters, filtering parameters are $\tau_2 = 0.1$, $\tau_3 = 0.05$, and $z_2(0) = z_3(0) = 0$.

Figures 1, 2 and 3 demonstrate the simulation results of this case. From Fig. 1, it can be seen that the system output y follows the reference signal y_r precisely. The maximum steady state tracking error is less than 3%. The actual control input signal is given in Fig. 2. The performance of three NN online approximators are given in Fig. 3.

6 Conclusion

An NN online approximation based adaptive backstepping DSC method is proposed for a class of uncertain non-lower triangular nonlinear systems. In the control design, NN online approximators are used to compensate the uncertainties of the nonlinear systems. And DSC technique is introduced to eliminate the problem of "circular design". By doing this, the recursive design of adaptive backstepping is feasible. The result of stability shows that all signals in the closed-loop control system can be guaranteed to be uniformly ultimately bounded, and the steady state tracking error converge to a small neighborhood of zero. A numerical simulation example demonstrates the effectiveness of the approach.

References

1. Krstic, M., Kanellakopoulos, I., Kokotovic, P.V.: Nonlinear and Adaptive Control Design. Wiley, New York (1995)
2. Zhou, J., Wen, C.Y.: Adaptive Backstepping Control of Uncertain Systems: Nonsmooth Nonlinearities, Interactions or Time-Variations. Springer, Heidelberg (2008). https://doi.org/10.1007/978-3-540-77807-3
3. Kanellakopoulos, I., Kokotovic, P.V., Morse, A.S.: Systematic design of adaptive controllers for feedback linearizable systems. IEEE Trans. Autom. Control 36(11), 1241–1253 (1991)
4. Seto, D., Annaswamy, A.M., Baillieul, J.: Adaptive control of nonlinear systems with a triangular structure. IEEE Trans. Autom. Control 39(7), 1411–1428 (1994)
5. Polycarpou, M.M.: Stable adaptive neural scheme for nonlinear systems. IEEE Trans. Autom. Control 41(3), 447–451 (1996)
6. Kwan, C., Lewis, F.L., Dawson, D.M.: Robust neural-network control of rigid-link electrically driven robots. IEEE Trans. Neural Netw. 9(4), 581–588 (1998)
7. Ge, S.S., Hang, C.C., Zhang, T.: A direct method for robust adaptive nonlinear control with guaranteed transient performance. Syst. Control Lett. 37(5), 275–284 (1999)
8. Kwan, C., Lewis, F.L.: Robust backstepping control of nonlinear systems using neural networks. IEEE Trans. Syst. Man Cybern. Part A Syst. Hum. 30(6), 753–766 (2000)
9. Wang, D., Huang, J.: Adaptive neural network control for a class of uncertain nonlinear systems in pure-feedback form. Automatica 38(8), 1365–1372 (2002)
10. Wang, C., Hill, D.J., Ge, S.S., Chen, G.R.: An ISS-modular approach for adaptive neural control of pure-feedback systems. Automatica 42(5), 723–731 (2006)
11. Chen, W.S., Jiao, L.C., Li, J., Li, R.H.: Adaptive NN backstepping output-feedback control for stochastic nonlinear strict-feedback systems with time-varying delays. IEEE Trans. Syst. Man Cybern. Part B Cybern. 40(3), 939–950 (2010)
12. Chen, M., Ge, S.S., How, B.V.E., Choo, Y.S.: Robust adaptive position mooring control for marine vessels. IEEE Trans. Control Syst. Technol. 21(2), 395–409 (2013)
13. Chen, B., Liu, X.P., Tong, S.C.: Adaptive fuzzy output tracking control of MIMO nonlinear uncertain systems. IEEE Trans. Fuzzy Syst. 15(2), 287–300 (2007)
14. Zou, A.M., Hou, Z.G., Tan, M.: Adaptive control of a class of nonlinear pure-feedback systems using fuzzy backstepping approach. IEEE Trans. Fuzzy Syst. 16(4), 886–897 (2008)
15. Lin, F.J., Shieh, P.H., Chou, P.H.: Robust adaptive backstepping motion control of linear ultrasonic motors using fuzzy neural network. IEEE Trans. Fuzzy Syst. 16(3), 676–692 (2008)
16. Tong, S.C., He, X.L., Zhang, H.G.: A Combined backstepping and small-gain approach to robust adaptive fuzzy output feedback control. IEEE Trans. Fuzzy Syst. 17(5), 1059–1069 (2009)
17. Tong, S.C., Wang, T., Li, Y.M., Chen, B.: A combined backstepping and stochastic small-gain approach to robust adaptive fuzzy output feedback control. IEEE Trans. Fuzzy Syst. 21(2), 314–327 (2013)
18. Li, Y.M., Tong, S.C., Li, T.S., Jing, X.J.: Adaptive fuzzy control of uncertain stochastic nonlinear systems with unknown deadzone using small-gain approach. Fuzzy Sets Syst. 235(1), 1–24 (2014)

19. Wai, R.J., Chang, H.H.: Backstepping wavelet neural network control for indirect field-oriented induction motor drive. IEEE Trans. Neural Netw. **15**(2), 367–382 (2004)
20. Hsu, C.F., Lin, C.M., Lee, T.T.: Wavelet adaptive backstepping control for a class of nonlinear systems. IEEE Trans. Neural Netw. **17**(5), 1175–1183 (2006)
21. Kulkarni, A., Purwar, S.: Wavelet based adaptive backstepping controller for a class of nonregular systems with input constraints. Expert Syst. Appl. **36**(3), 6686–6696 (2009)
22. Chen, C.H., Hsu, C.F.: Recurrent wavelet neural backstepping controller design with a smooth compensator. Neural Comput. Appl. **19**(7), 1089–1100 (2010)
23. Swaroop, D., Gerdes, J.C., Yip, P.P., Hedrick, J.K.: Dynamic surface control of nonlinear systems. In: American Control Conference, Albuquerque, New Mexico, pp. 3028–3034 (1997)
24. Wang, D., Huang, J.: Neural network based adaptive dynamic surface control for nonlinear systems in strict-feedback form. IEEE Trans. Neural Netw. **16**(1), 195–202 (2005)
25. Zhang, T.P., Ge, S.S.: Adaptive dynamic surface control of nonlinear systems with unknown deadzone in pure feedback form. Automatica **44**(7), 1895–1903 (2008)
26. Li, T.S., Wang, D., Feng, G., Tong, S.C.: A DSC approach to robust adaptive NN tracking control for strict-feedback nonlinear systems. IEEE Trans. Syst. Man Cybern. Part B Cybern. **40**(3), 915–927 (2010)
27. Wang, D.: Neural network-based adaptive dynamic surface control of uncertain nonlinear pure-feedback systems. Int. J. Robust Nonlinear Control **21**(5), 527–541 (2011)
28. Sun, G., Wang, D., Li, X.Q., Peng, Z.H.: A DSC approach to adaptive neural network tracking control for pure-feedback nonlinear systems. Appl. Math. Comput. **219**(11), 6224–6235 (2013)
29. Yang, Y.S., Ren, J.S.: Adaptive fuzzy robust tracking controller design via small gain approach and its application. IEEE Trans. Fuzzy Syst. **11**(6), 783–795 (2003)
30. Yang, Y.S., Feng, G., Ren, J.S.: A combined backstepping and small-gain approach to robust adaptive Fuzzy control for strict-feedback nonlinear systems. IEEE Trans. Syst. Man Cybern. Part A Syst. Hum. **34**(3), 406–420 (2004)
31. Sun, G., Wang, D., Peng, Z.H.: Adaptive control based on single neural network approximation for non-linear pure-feedback systems. IET Control Theor. Appl. **6**(15), 2387–2396 (2012)
32. Sun, G., Wang, D., Wang, M.X.: Robust adaptive neural network control of a class of uncertain strict-feedback nonlinear systems with unknown dead-zone and disturbances. Neurocomputing **145**(18), 221–229 (2014)
33. Wang, X., Li, T.S., Chen, C.L.P., Lin, B.: Adaptive robust control based on single neural network approximation for a class of uncertain strict-feedback discrete-time nonlinear systems. Neurocomputing **138**(11), 325–331 (2014)
34. Wang, M.X., Sun, G.: Single neural network approximation-based adaptive control of a class of uncertain strict-feedback nonlinear systems. In: Proceedings of the 6th International Conference on Information Science and Technology, pp. 280–284, IEEE Press, Dalian (2016)
35. Park, J., Sandberg, I.W.: Universal approximation using radial-basis-function networks. Neural Comput. **3**(2), 246–257 (1991)
36. Sanner, R.M., Slotine, J.E.: Gaussian networks for direct adaptive control. IEEE Trans. Neural Netw. **3**(6), 837–863 (1992)

Neurodynamics-Based Receding Horizon Control of an HVAC System

Jiasen Wang[1,3(✉)], Jun Wang[1,2,3], and Shenshen Gu[4]

[1] Department of Computer Science, City University of Hong Kong,
Kowloon, Hong Kong
jiasewang2-c@my.cityu.edu.hk, jwang.cs@cityu.edu.hk
[2] School of Data Science, City University of Hong Kong, Kowloon, Hong Kong
[3] Shenzhen Research Institute, City University of Hong Kong, Shenzhen, China
[4] School of Mechatronic Engineering and Automation, Shanghai University,
Shanghai, China
gushenshen@shu.edu.cn

Abstract. This paper addresses receding horizon control of a heating, ventilation, and air-conditioning (HVAC) system based on neurodynamic optimization. The receding horizon control problem for the HVAC system is formulated as sequential quadratic programs, which are solved by using a neurodynamic optimization model. Simulation results on temperature setpoint regulation of the HVAC system are discussed to substantiate the efficacy of the approach.

Keywords: HVAC · Receding horizon control ·
Neurodynamic optimization

1 Introduction

Heating, ventilation, and air-conditioning (HVAC) of buildings or rooms are widely demanded [23]. Control of HVAC systems is of great engineering value [1, 3,24]. A good control approach saves energy and renders comfort simultaneously.

Many methods are available for controlling HVAC systems. In [3], a learning-based model predictive control approach is used to reduce electricity/energy consumption in an HVAC system. In [5,6], heuristic algorithms are used for energy management of HVAC systems. Reinforcement learning technique is used in [39] for controlling HVAC systems. An open-loop demand response control algorithm is introduced in [8] for commercial HVAC systems. Model predictive control technique is used in [21] for temperature controlling in a real building.

This work was supported in part by the Research Grants Council of the Hong Kong Special Administrative Region of China, under Grants 11208517 and 11202318, in part by the National Natural Science Foundation of China under grants 61673330 and 61876105, and in part by International Partnership Program of Chinese Academy of Sciences under Grant GJHZ1849.

© Springer Nature Switzerland AG 2019
H. Lu et al. (Eds.): ISNN 2019, LNCS 11555, pp. 120–128, 2019.
https://doi.org/10.1007/978-3-030-22808-8_13

Neural network based model predictive control approaches are used in [2,12] for controlling HVAC systems in commercial/residential buildings. Model predictive controllers are presented in [7] for energy savings and comfort optimization in buildings. A model predictive control algorithm with a specifically tailored cost function is used in [9] for temperature control of buildings.

Neurodynamics-based receding horizon control (also known as model predictive control) approaches are able to handle constrained systems [19,34,35,37,38]. HVAC systems have various physical constraints [3,24]. Thus neurodynamics-based receding horizon control approaches are suitable for controlling HVAC systems.

In this paper, a neurodynamics-based receding horizon control approach is presented to an HVAC system. The receding horizon control technique is used to regulate the temperature. The control problem is formulated as sequential quadratic programs, and solved by using a neurodynamic model. In contrast to, for example, neural dynamic surface control approach in [20], the proposed approach is able to handle constraints and optimize performances.

2   Problem Formulations

2.1   An HVAC System

Consider an HVAC model [24]:

$$\dot{x}_1 = \frac{1}{V_{he}}[(T_0 - x_2)u_3 + (x_2 - x_1)u_2 + \frac{u_1}{\rho c_p}],$$

$$\dot{x}_2 = \frac{1}{V_{ts}}[(x_1 - x_2)u_2 + \frac{q_l}{\rho c_p}], \tag{1}$$

where x_1 is the air temperature of the heat exchanger, x_2 is the thermal space temperature, V_{he} is the effective heat exchanger volume, V_{ts} is the effective thermal space volume, T_0 is the temperature of the fresh air (usually refers to the outdoor air), u_1 is the heat input to the heat exchanger, u_2 is the volumetric airflow rate, u_2/u_3 is the flow-rate ratio, ρ is the density of the air, c_p is the constant heat of air, q_l is the thermal load on the room.

By using Euler method, system dynamic equation (1) is discretized as

$$x_1(k+1) = x_1(k) + T_s\zeta_1(k) := x_1(k) +$$
$$\frac{T_s}{V_{he}}[(T_0 - x_2(k))u_3(k) + (x_2(k) - x_1(k))u_2(k) + \frac{u_1(k)}{\rho c_p}],$$
$$x_2(k+1) = x_2(k) + T_s\zeta_2(k) := x_2(k) + \frac{T_s}{V_{ts}}[(x_1(k) - x_2(k))u_2(k) + \frac{Q_l}{\rho c_p}], \tag{2}$$

where T_s is a sampling interval.

2.2 Receding Horizon Control

Let T_r denote the reference to the thermal space temperature. It is supposed that T_r are piecewise constants; i.e., only regulation problems are considered. The control objective is to make x_2 track T_r by using control variables u_1, u_2, and u_3, and meanwhile the control variables and states should not violate a series of physical constraints.

The receding horizon control problem is formulated as an optimization problem at time k as follows:

$$\min_{u(t|k)} \sum_{t=k}^{k+N-1} \left\{ \|[x_1(t+1|k) - T_r, x_2(t+1|k) - T_r]^T\|_Q^2 + \right.$$

$$\left. \|[\zeta_1(t|k), \zeta_2(t|k)]^T\|_R^2 \right\} + \|[x_1(k+N|k) - T_r, x_2(k+N|k) - T_r]^T\|_P^2, \quad (3a)$$

$$\text{s.t. } x_i(t+1|k) = x_i(t|k) + T_s\zeta_i(t|k), \ i = 1, 2, \ t = k, ..., k+N-1 \quad (3b)$$

$$\|[x_1(k+N|k) - T_r, x_2(k+N|k) - T_r]^T\|_O \leq \alpha, \quad (3c)$$

$$u_1(t|k) \in [-4000, 4000], \ t = k, ..., k+N-1 \quad (3d)$$

$$u_2(t|k), u_3(t|k) \in [0.0354, 2], \ u_2(t|k) \geq u_3(t|k), \ t = k, ..., k+N-1 \quad (3e)$$

$$x_1(t|k) \in [5, 35], \ x_2(t|k) \in [10, 30], \ t = k+1, ..., k+N, \quad (3f)$$

where $v(t|k)$ denotes the prediction of variable v at time t with known information $v(k)$, N is the prediction horizon, O, P, Q, R are positive definite weighting matrices, α is a positive parameter. The objective function (3a) is a trade-off between transient temperature tracking error and terminal temperature tracking error by using matrices P, Q, and R. (3b) are the predicted form of system state equations in (2). (3c) is an elliptical terminal state constraint. (3d)–(3f), introduced in [24], are polyhedral constraints of states and control variables. In particular, the constraint $u_3(t|k) \geq 0.0354$ in (3e) is used to render comfort and hygiene.

Problem (3) is rewritten into a compact form as follows:

$$\min \ f(\xi) \text{ s.t. } g(\xi) \leq 0, \ h(\xi) = 0, \quad (4)$$

where ξ is a vector containing all the decision variables in problem (3), $g(\xi)$ is a concatenated vector containing (3c)–(3f), $h(\xi)$ is a concatenated vector containing all the predicted state equations in (3b).

Sequential quadratic programming [4,18] can be used for solving problem (4). A slack variable s is introduced such that $g(\xi)+s = 0$ [18]. Two direction vectors d_1 and d_2 are introduced with $d_1 = \text{col}(d_{11}, d_{12})$, $d_2 = \text{col}(d_{21}, d_{22})$, where $\text{col}(\cdot)$ denotes column concatenation of vectors or matrices. Vectors d_{11} and d_{21} are used to move the decision variable ξ toward its local optimal solution; vectors d_{12} and d_{22} are used to move the slack variable s toward its local optimal solution. Problem (4) can be approximately solved by sequentially solving the following quadratic programs [4]:

$$\min_{d_1} \ d_1^T A_l A_l^T d_1 + 2[h^T(\xi_l), (g(\xi_l) + s_l)^T]A_l^T d_1$$

$$\text{s.t. } \|d_1\| \leq br, \ -d_{12} - \tau/2 \leq 0, \quad (5)$$

$$\min_{d_2} \frac{1}{2} d_2^T G_l d_2 + \nabla f^T(\xi_l) d_{21} - \mu e^T d_{22} + (G_l d_1)^T d_2$$

$$\text{s.t.} \quad \|d_2\|^2 \le r^2 - \|d_1\|^2, \quad -d_{22} \le \tau e + d_{12},$$

$$\nabla h^T(\xi_l) d_{21} = 0, \quad \nabla g^T(\xi_l) d_{21} + S_l d_{22} = 0,$$

(6)

where l denotes iteration index, $A_l = \mathrm{col}([\nabla h(\xi_l), \nabla g(\xi_l)], [0, S_l])$, ξ_l is the decision vector at the lth iteration, $S_l = \mathrm{diag}(s_l)$, s_l is the slack variable at the lth iteration, $G_l = \mathrm{diag}(\nabla_x^2 L_l, S_l \Sigma_l S_l)$, $L_l = f(\xi_l) - \mu e^T \ln s_l + y_l^T h(\xi_l) + z_l^T(g(\xi_l) + s_l)$, $\Sigma_l = \mu \mathrm{diag}(s_{l1}^{-2}, s_{l2}^{-2}, ...)$, r, τ, b are positive scalars, $e = [1, ..., 1]^T$.

As in [4], a merit function δ is defined as $\delta(\xi, s, \nu) = f(\xi) - \mu e^T \ln(s) + \nu \|[h^T, (g+s)^T]\|$; an evaluation function V is defined as $V(\xi, s, \mu) = \|[\nabla f(\xi) + \nabla h(\xi) y + \nabla g(\xi) z, Sz - \mu e, h(\xi), g(\xi) + s]\|_\infty$. A sequential optimization procedure for solving (4)/(3) is presented as follows [4]:

- Initialization: Set initial values for scalars r, μ, ϵ_{tol}, ϵ_μ, τ, and for vector ξ_0; Set positive elements for the slack variable s_0.
- Repeat until $V(\xi_l, s_l, 0) \le \epsilon_{tol}$ or reaching the maximum number of iterations:
  - Repeat until $V(\xi_l, s_l, \mu) \le \epsilon_\mu$ or reaching the maximum number of iterations: (i) Solve problem (5) to obtain solution d_1. (ii) Solve $(A_l^T A_l) \mathrm{col}(y, z) = A_l^T \mathrm{col}(-\nabla f(\xi_l), \mu e)$ to get solutions y_l and z_l. (iii) Solve problem (6) to get solution d_2. (iv) $d_\xi = d_{11} + d_{21}$, $d_s - S_l(d_{12} + d_{22})$. If $\delta(\xi_l + d_\xi, s_l + d_s, \nu) < \delta(\xi_l, s_l, \nu)$, then $\xi_{l+1} = \xi_l + d_\xi$, $s_{l+1} = s_l + d_s$, $r = (1 + \theta)r$; otherwise $\xi_{l+1} = \xi_l$, $s_{l+1} = s_l$, $r = (1 - \theta)r$. $l = l + 1$.
  - $\mu = \varrho\mu$, $\epsilon_\mu = \varrho\epsilon_\mu$, $r = \max(5r, 1)$.

To speedup the convergence, the solution of problem (3) at time $k - 1$ is shifted one step forward to initialize problem (3) at time k.

When the solution of problem (3) is obtained, $u(k|k)$ is used to control system (2) at time k. At time $k + 1$, the optimization and control processes are repeated again. The iterative control procedure based on optimization on a finite horizon is where the term "receding horizon control" comes from.

3 Receding Horizon Control Based on Neurodynamic Optimization

Problems (5) and (6) are quadratic programs. Many neurodynamic models are available for solving quadratic programs [10, 11, 13–17, 22, 25–33, 36, 40]. In particular, for solving nonlinear inequality constrained program $\{\min f(x), \text{ s.t. } g(x) \le 0\}$, a neurodynamic model is introduced in [16] with the following dynamic equation:

$$\epsilon \frac{dx}{dt} = -\nabla f(x) - \sigma \nabla g(x) \varphi_{[0,1]}(g(x)),$$

(7)

where ϵ is a positive time constant, σ is a positive scalar, $\nabla g(x) = [\nabla g_1(x), \nabla g_2(x), ...]$, $\varphi_{[0,1]}(g(x)) = [\varphi_{[0,1]}(g_1(x)), \varphi_{[0,1]}(g_2(x)), ...]^T$,

$$\varphi_{[0,1]}(g_i(x)) = \begin{cases} 1 & g_i(x) > 1, \\ g_i(x) & 0 \le g_i(x) \le 1, \\ 0 & g_i(x) < 0. \end{cases}$$

4   Simulation Results

Consider a temperature setpoint regulation problem. Specifically, regulating thermal space temperature x_2 to $T_r = 16\,^{\circ}\mathrm{C}$ is specified. The parameters are set as follows: (i) The fresh air temperature T_0 is $15\,^{\circ}\mathrm{C}$. As in [24], $V_{he} = 25.5\,\mathrm{m}^3$, $V_{ts} = 255\,\mathrm{m}^3$, $c_p = 1005\,\mathrm{J/kg\,^{\circ}C}$, $\rho = 1.19\,\mathrm{kg/m}^3$, $q_l = 0\,\mathrm{W}$. (ii) As in [4], $r = 1$, $\tau = 0.995$, $\epsilon_\mu = 0.1$, $\varrho = 0.2$, $b = 0.8$. (iii) $Q = R = \mathrm{diag}(5,5)$, $P = \mathrm{diag}(50,50)$, $O = \mathrm{diag}(1,1)$, $\epsilon = 10^{-2}$, $T_s = 10\,\mathrm{s}$, $\alpha = 1$, $N = 20$. Extra bound constraints $x_1(t|k) \le T_r$ $(t = k+1, ..., k+N)$ are added in the optimization problem (3) to eliminate the overshooting of x_1.

The simulation results are shown in Figs. 1, 2, 3, 4 and 5. Figure 1 depicts a snapshot of the transient neuronal states. Figure 2 depicts the transient objective values. Figure 3 depicts the transient behaviors of the control variable u_1.

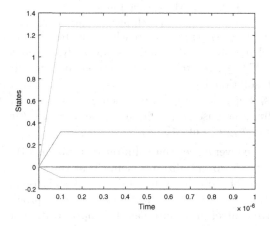

Fig. 1. A snapshot of the transient neuronal states.

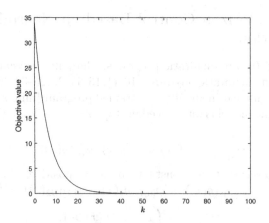

Fig. 2. Transient objective values.

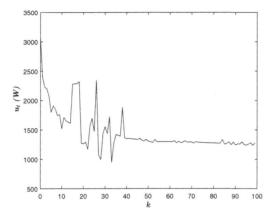

Fig. 3. Transient behaviors of u_1.

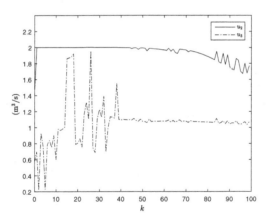

Fig. 4. Transient behaviors of u_2 and u_3.

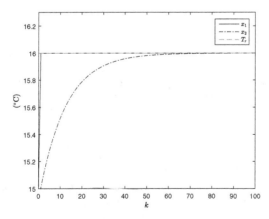

Fig. 5. Transient behaviors of the states and the reference.

Figure 4 depicts the transient behaviors of the control variables u_2 and u_3. It shows that u_2 and u_3 do not violate their bounds as specified in problem (3), rendering comfort to the occupants. Figure 5 depicts the transient behaviors of states x_1 and x_2. It shows that the thermal space temperature x_2 is regulated to the desired temperature $T_r = 16\ °C$, and the states do not violate their bounds as specified in problem (3).

5 Conclusions

In this paper, neurodynamics-based receding horizon control of an HVAC system is addressed. The control problem on a discrete HVAC system is formulated as a nonlinear optimization problem, which are further formulated as sequential quadratic programs. A neurodynamic model is used for solving the quadratic programs. Simulation results on a temperature setpoint regulation problem are reported to substantiate the efficacy of the control approach.

References

1. Afram, A., Janabi-Sharifi, F.: Theory and applications of HVAC control systems - a review of model predictive control (MPC). Build. Environ. **72**, 343–355 (2014)
2. Afram, A., Janabi-Sharifi, F., Fung, A.S., Raahemifar, K.: Artificial neural network (ANN) based model predictive control (MPC) and optimization of HVAC systems: a state of the art review and case study of a residential HVAC system. Energy Build. **141**, 96–113 (2017)
3. Aswani, A., Master, N., Taneja, J., Culler, D., Tomlin, C.: Reducing transient and steady state electricity consumption in HVAC using learning-based model-predictive control. Proc. IEEE **100**(1), 240–253 (2012)
4. Byrd, R., Hribar, M., Nocedal, J.: An interior point algorithm for large-scale nonlinear programming. SIAM J. Optim. **9**(4), 877–900 (1999)
5. Fong, K.F., Hanby, V.I., Chow, T.T.: System optimization for HVAC energy management using the robust evolutionary algorithm. Appl. Therm. Eng. **29**(11), 2327–2334 (2009)
6. Fong, K.F., Yuen, S.Y., Chow, C.K., Leung, S.W.: Energy management and design of centralized air-conditioning systems through the non-revisiting strategy for heuristic optimization methods. Appl. Energy **87**(11), 3494–3506 (2010)
7. Freire, R.Z., Oliveira, G.H., Mendes, N.: Predictive controllers for thermal comfort optimization and energy savings. Energy Build. **40**(7), 1353–1365 (2008)
8. Goddard, G., Klose, J., Backhaus, S.: Model development and identification for fast demand response in commercial HVAC systems. IEEE Trans. Smart Grid **5**(4), 2084–2092 (2014)
9. Hazyuk, I., Ghiaus, C., Penhouet, D.: Optimal temperature control of intermittently heated buildings using model predictive control: Part II - control algorithm. Build. Environ. **51**, 388–394 (2012)
10. Hu, X., Wang, J.: Design of general projection neural networks for solving monotone linear variational inequalities and linear and quadratic optimization problems. IEEE Trans. Syst. Man Cybern. Part B Cybern. **37**(5), 1414–1421 (2007)

11. Hu, X., Wang, J.: An improved dual neural network for solving a class of quadratic programming problems and its k-winners-take-all application. IEEE Trans. Neural Netw. **19**(12), 2022–2031 (2008)
12. Huang, H., Chen, L., Hu, E.: A neural network-based multi-zone modelling approach for predictive control system design in commercial buildings. Energy Build. **97**, 86–97 (2015)
13. Li, G., Yan, Z., Wang, J.: A one-layer recurrent neural network for constrained nonconvex optimization. Neural Netw. **61**, 10–21 (2015)
14. Liang, X., Wang, J.: A recurrent neural network for nonlinear optimization with a continuously differentiable objective function and bound constraints. IEEE Trans. Neural Netw. **11**(6), 1251–1262 (2000)
15. Liu, Q., Wang, J.: A one-layer recurrent neural network with a discontinuous hard-limiting activation function for quadratic programming. IEEE Trans. Neural Netw. **19**(4), 558–570 (2008)
16. Liu, Q., Wang, J.: A one-layer recurrent neural network for constrained nonsmooth optimization. IEEE Trans. Syst. Man Cybern. Part B Cybern. **40**(5), 1323–1333 (2011)
17. Liu, S., Wang, J.: A simplified dual neural network for quadratic programming with its KWTA application. IEEE Trans. Neural Netw. **17**(6), 1500–1510 (2006)
18. Nocedal, J., Wright, S.J.: Numerical Optimization. Springer, New York (2006). https://doi.org/10.1007/978-0-387-40065-5
19. Pan, Y., Wang, J.: Model predictive control of unknown nonlinear dynamical systems based on recurrent neural networks. IEEE Trans. Ind. Electron. **59**(8), 3089–3101 (2012)
20. Peng, Z., Wang, D., Wang, J.: Predictor-based neural dynamic surface control for uncertain nonlinear systems in strict-feedback form. IEEE Trans. Neural Netw. Learn. Syst. **28**(9), 2156–2167 (2017)
21. Prívara, S., Široký, J., Ferkl, L., Cigler, J.: Model predictive control of a building heating system: the first experience. Energy Build. **43**(2–3), 564–572 (2011)
22. Qin, S., Le, X., Wang, J.: A neurodynamic optimization approach to bilevel quadratic programming. IEEE Trans. Neural Netw. Learn. Syst. **28**(11), 2580–2591 (2017)
23. Rousseau, P., Mathews, E.: Needs and trends in integrated building and HVAC thermal design tools. Build. Environ. **28**(4), 439–452 (1993)
24. Teeter, J., Chow, M.Y.: Application of functional link neural network to HVAC thermal dynamic system identification. IEEE Trans. Ind. Electron. **45**(1), 170–176 (1998)
25. Wang, J.: A deterministic annealing neural network for convex programming. Neural Netw. **7**(4), 629–641 (1994)
26. Xia, Y., Feng, G., Wang, J.: A recurrent neural network with exponential convergence for solving convex quadratic program and related linear piecewise equations. Neural Netw. **17**(7), 1003–1015 (2004)
27. Xia, Y., Feng, G., Wang, J.: A novel recurrent neural network for solving nonlinear optimization problems with inequality constraints. IEEE Trans. Neural Netw. **19**(8), 1340–1353 (2008)
28. Xia, Y., Leung, H., Wang, J.: A projection neural network and its application to constrained optimization problems. IEEE Trans. Circuits Syst. Part I **49**(4), 447–458 (2002)
29. Xia, Y., Wang, J.: A general methodology for designing globally convergent optimization neural networks. IEEE Trans. Neural Netw. **9**(6), 1331–1343 (1998)

30. Xia, Y., Wang, J.: Global exponential stability of recurrent neural networks for solving optimization and related problems. IEEE Trans. Neural Netw. **11**(4), 1017–1022 (2000)
31. Xia, Y., Wang, J.: A general projection neural network for solving monotone variational inequalities and related optimization problems. IEEE Trans. Neural Netw. **15**(2), 318–328 (2004)
32. Xia, Y., Wang, J.: A recurrent neural network for nonlinear convex optimization subject to nonlinear inequality constraints. IEEE Trans. Circuits Syst. Part I **51**(7), 1385–1394 (2004)
33. Xia, Y., Wang, J.: A bi-projection neural network for solving constrained quadratic optimization problems. IEEE Trans. Neural Netw. Learn. Syst. **27**(2), 214–224 (2016)
34. Yan, Z., Le, X., Wang, J.: Tube-based robust model predictive control of nonlinear systems via collective neurodynamic optimization. IEEE Trans. Ind. Electron. **63**(7), 4377–4386 (2016)
35. Yan, Z., Wang, J.: Model predictive control of nonlinear systems with unmodeled dynamics based on feedforward and recurrent neural networks. IEEE Trans. Ind. Inform. **8**(4), 746–756 (2012)
36. Yan, Z., Fan, J., Wang, J.: A collective neurodynamic approach to constrained global optimization. IEEE Trans. Neural Netw. Learn. Syst. **28**(5), 1206–1215 (2017)
37. Yan, Z., Wang, J.: Robust model predictive control of nonlinear systems with unmodeled dynamics and bounded uncertainties based on neural networks. IEEE Trans. Neural Netw. Learn. Syst. **25**(3), 457–469 (2014)
38. Yan, Z., Wang, J.: Nonlinear model predictive control based on collective neurodynamic optimization. IEEE Trans. Neural Netw. Learn. Syst. **26**(4), 840–850 (2015)
39. Yang, L., Nagy, Z., Goffin, P., Schlueter, A.: Reinforcement learning for optimal control of low exergy buildings. Appl. Energy **156**, 577–586 (2015)
40. Zhang, Y., Wang, J.: A dual neural network for convex quadratic programming subject to linear equality and inequality constraints. Phys. Lett. A **298**, 271–278 (2002)

Robust Adaptive Resonant Damping Control of Nanopositioning

Wei Wei, Pengfei Xia, Zaiwen Liu, and Min Zuo[✉]

School of Computer and Information Engineering,
Beijing Technology and Business University,
FuCheng Road 11, Beijing 100048, China
weiweizdh@126.com, a1232015@163.com, {liuzw,zuomin}@btbu.edu.cn

Abstract. Hysteresis and vibration are main factors in reducing the accuracy and the speed of a nanopositioner. In order to improve the positioning accuracy and the speed, a robust adaptive resonant damping control is proposed for trajectory tracking of a nanopositioner system driven by a piezoelectric actuator. Radial basis function neural network is employed to approximate unknown nonlinearities in the controller so as to reduce the dependence on model information. Linear and hysteresis model are establish based on nanopositioning stage measured data. The proposed control strategy is verified by simulation using an identified model.

Keywords: Piezoelectric actuators · Robust adaptive ·
Resonant damping · Radial basis function · Hysteresis

1 Introduction

Recently, nanopositioning systems driven by piezoelectric actuators (PEAs) are widely utilized in various of precision manufacturing equipments, such as scanning probe microscopy, micro-manipulators, and nanofabrication equipments [1–3]. The key part of a nanopositioning stage is the PEA for its superior virtues of high resolution, fast response, and large execution force [4]. However, the PEAs possess two major demerits that limit the accuracy and speed of a nanopositioner: (i) hysteresis nonlinearity of piezoelectric ceramic materials, and (ii) lightly damped resonant modes due to the mechanical structure [5].

Hysteresis is the key problem in decreasing the positioning accuracy. Hysteresis nonlinearity produces positioning error up to 15% of the motion range [6] and can even result in a unstable system. In order to address the hysteresis, many effective efforts have been made. Inverse-based control is a commonly

This work is supported by National Natural Science Foundation (NNSF) of China under Grant 61403006, 61873005; Key program of Beijing Municipal Education Commission KZ201810011012; and Support Project of High-level Teachers in Beijing Municipal Universities in the Period of 13th Five-year Plan CIT&TCD201704044.

H. Lu et al. (Eds.): ISNN 2019, LNCS 11555, pp. 129–138, 2019.
https://doi.org/10.1007/978-3-030-22808-8_14

used method [7]. Among them, the feedforward control based on the inverse hysteresis models, to some extent, is effective. However, when the hysteresis model does not match the system dynamics, the inverse-based approaches do not take effect. Generally, rather than reconstructing the inverse hysteresis model, feedback control, such as PID [8], is utilized to mitigate the hysteresis. In this paper, in order to improve the tracking performance in presence of hysteresis, robust adaptive (RA) control, combined with sliding mode control and adaptive control methods, is presented. Desired transient and steady state performances are guaranteed in the presence of disturbances and hysteresis. Radial basis function (RBF) neural network is employed to approximate unknown nonlinearities in a robust adaptive controller. Recently, related neural network modeling [9] and control [10] has been successfully applied for PEAs.

Resonant modes is the main issues that hinder the operation speed of the PEAs. It limits the operating frequency to less than $1/10$–$1/100$ of the first resonant frequency of a stage [11]. Although it is possible to construct a sufficiently stiff stage to augment the operating speed, its maximum motion range is limited to a few micrometers and the driving force is limited [12]. Therefore, it is necessary to exploit a desired control approach to suppress the vibration effects. In order to address this issue, a number of effective damping control approaches, such as shunt damping [13], positive position feedback [14], positive speed and position feedback [15], resonant control [16] and robustness control [17] has been proposed. However, high order controllers and accurate system models are necessary [18]. Integral resonant control (IRC) has been demonstrated to be a simple method in damping vibrations for nanoscale positioning [19]. However, IRC is limited in dealing with hysteresis and disturbances in the case of voltage driven control.

In order to improve the control performance of IRC, a robust adaptive control algorithm is designed to overcome hysteresis and disturbances. By combining the benefits of RA and IRC control, a robust adaptive resonance damping control is presented to reduce the resonant modes, improve the operation speed and positioning accuracy in nanopositioning.

2   System Description and Modeling

Dynamical model of a piezoelectric-driven nanopositioning stage can be described as [20]

$$\ddot{x}(t) + a_0\dot{x}(t) + a_1x(t) = b_0v_{in}(t) - b'H(q). \tag{1}$$

where a_0, a_1, b_0, b' denote system parameters, v_{in} represents the input voltage, $H(q)$ describes the hysteresis effect, and $x(t)$ is the position of the positioner. Let $y = x = x_1$, system (1) is and can be rewritten as

$$\begin{aligned}\dot{x}_1 &= \dot{x}_2 \\ \dot{x}_2 &= f(x) + g(x)u + w(t). \\ y &= x_1\end{aligned} \tag{2}$$

where $\mathbf{x} = [x_1, x_2]^T, y, u = v_{in}$ indicate state variables, system output and input, respectively; $f(x)$ and $g(x)$ as unknown nonlinear functions; $w(t)$ is the bounded disturbances, we have a known constant $w_0 > 0$, i.e., $|w(t)| \leq w_0$.

Without losing generality, assume that $g(x) > 0$. One has a known smooth function $\bar{g}(x)$ satisfies $|g(x)| \leq \bar{g}(x)$. In this paper, $\bar{g}(x)$ is taken as a constant.

Define vector $\mathbf{x}_d, \mathbf{e}$ and an extended error s as $\mathbf{x}_d = [y_d, \dot{y}_d]^T, \mathbf{e} = [e, \dot{e}]^T, s = [\lambda, 1]\mathbf{e} = \lambda e + \dot{e}$. Here $\lambda > 0$ is chosen such that polynomial $s + \lambda$ is Hurwitz. The time derivative of s can be written as

$$\dot{s} = f(x) + v + g(x)u + w(t). \tag{3}$$

where $v = -\ddot{y}_d + \lambda\dot{e}$.

2.1   Linear Model

The linear dynamics model of nanopositioning stage is identified by the input and output data

$$G(s) = \frac{9.207 \times 10^5}{s^2 + 332.4s + 1.296 \times 10^6}. \tag{4}$$

By comparing the coefficients of (1) and (5), parameters in dynamics model (1) are $a_1 = 332.4, a_0 = 1.296 \times 10^6$, and $b_0 = 9.207 \times 10^5$, respectively.

2.2   Prandtl-Ishlinskii Hysteresis Model

Prandtl-Ishlinskii (PI) model was chosen to establish the nonlinear dynamics. As a basic element of the PI model, rate independent backlash operator can be implemented as [21]

$$\begin{aligned} y_b(t) &= H_r[v_{in}, y_{b0}](t) = \max\{v_{in}(t) - r, \min\{v_{in}(t) + r, y_b(t - T)\}\} \\ y_b(0) &= \max\{v_{in}(0) - r, \min\{v_{in}(0) + r, y_{b0}\}\} \end{aligned}. \tag{5}$$

where $v_{in} = 21.12\sin(\pi t) + 18.59$ (V) is chosen as input signal, y_b represents the output of the backlash operator, r denotes the set threshold, and T indicates the sampling period. In general, the initial value of y_{b0} is nonzero.

PI hysteresis model consists of the sum of a number of rate independent backlash operators with different thresholds and weight values [22]. In this paper, for simple calculation, the threshold is set to be a fixed value, i.e., $r = 1.5$, and the weight is recognized by the measured data of the nanopositioning stage.

$$y_h(t) = \sum_{i=0}^{n} w_h^{T^i} H_r^i[v_{in}, y_{b0}](t), \tag{6}$$

where y_h represents the output of the hysteresis model. Considering the range of measured data, 19 operators have been chosen [21]. We have $n = 18$. Measured data and model output are shown in Fig. 1. Modeling accuracy indices, including root mean square error (RMSE), mean absolute error (MAE), and maximum absolute error (MAXE), have been presented in Table 1.

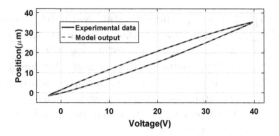

Fig. 1. Hysteresis loop generated by the identified model.

Table 1. Indexes of hysteresis model

	MAXE	MAE	RMSE
Value	0.6460	0.0709	0.1032

3   Robust Adaptive Resonant Damping Control

In order to improve positioning accuracy and eliminates oscillation, a compound control method is proposed.

3.1   Integral Resonant Control

The integral resonant control (IRC) changes the zero-pole interleaving of the colocated system by adding a constant as feed-through term [23]. The unique phase response of such a system enables a simple integral feedback controller to raise substantial damping. The block diagram of a typical IRC scheme is presented in Fig. 2 [24]. i indicates the output of the controller, d represents the feed-through term, K_d refers to the damping gain of IRC, $w(t)$ denotes disturbances. Linear dynamics can be rewritten as

$$G(s) = \frac{\gamma^2}{s^2 + 2\zeta\omega_p s + \omega_p^2}. \tag{7}$$

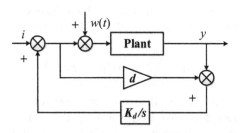

Fig. 2. Block diagram for the IRC damping controller.

According to Eq. (5) identified in Sect. 2.1, $\zeta = 1.282 \times 10^{-4}$ represents the damping coefficient, $\omega_p = 1138$ is the natural frequency, $\gamma = 959.5$. According to Ref. [23] $d = -2\dfrac{\gamma^2}{\omega_p^2} = -1.42$, and $K_d = 1000$.

3.2 Robust Adaptive Control by RBF Neural Network

Theorems 1. Ignoring the disturbance, desired feedback control (DFC) law is designed as

$$u^* = -\frac{1}{g(x)}(f(x) + v) - \left(\frac{1}{\varepsilon g(x)} + \frac{1}{\varepsilon g^2(x)} - \frac{\dot{g}(x)}{2g^2(x)}\right)s. \tag{8}$$

where $\varepsilon > 0$ as a adjustable parameter, one has $\lim\limits_{t\to\infty} \|e(t)\| = 0$.

Proof. Substituting DFC control law (8) u^* into Eq. (3) and letting $w(t) = 0$, we have

$$\dot{s} = -\left(\frac{1}{\varepsilon} + \frac{1}{\varepsilon g(x)}\right)s + \frac{\dot{g}(x)}{2g(x)}s. \tag{9}$$

Selecting a Lyapunov function as $V - s^2/2g(x)$, so

$$\dot{V} = -\frac{1}{g(x)}s\dot{s} - \frac{\dot{g}(x)}{2g^2(x)}s^2 = -\left(\frac{1}{\varepsilon g(x)} + \frac{1}{\varepsilon g^2(x)}\right)s^2. \tag{10}$$

Since $g(x) > 0$, then $\dot{V} \leq 0$, it implies $\lim\limits_{t\to\infty} |s| = 0$ [24]. Then, $\lim\limits_{t\to\infty} \|e(t)\| = 0$.

From (10), it is indicated that the smaller the adjustment parameter ε is, the more negative $\dot{V}$ will be. Therefore, the convergence rate relies on ε. From (8), we can see that u^* can be rewritten as a function of $\mathbf{x}, v, s$ and ε. Let

$$\mathbf{z} = [\mathbf{x}^T, s, \frac{s}{\varepsilon}, v]^T \in \Omega_z \in R^5. \tag{11}$$

where $\Omega_z = \{(\mathbf{x}^T, s, \frac{s}{\varepsilon}, v) | x \in \Omega; x_d \in \Omega_d; v = -\ddot{y}_d + \lambda\dot{e}; s = [\lambda \, 1]\mathbf{e}\}$.

Due to nonlinear functions $f(x)$ and $g(x)$ are unknown, u^* is unavailable for controller. Hence, RBF neural network (NN) [25] is applied for approximating the unknown u^*. From (9), we can see that elements s and s/ε are in different scopes when ε is small. Therefore, s/ε is also an available input for RBF NN to improve the approximation accuracy.

There is an ideal NN weight vector $\mathbf{W}^*$, and adaptive neural deal with unknown hysteresis be introduced in [26]

$$u^*(\mathbf{z}) = \mathbf{W}^{*T}\mathbf{h}(\mathbf{z}) + \mu_l, \quad \forall \mathbf{z} \in \Omega_\mathbf{z}. \tag{12}$$

where $\mathbf{h}(\mathbf{z})$ is radial basis function vector; μ_l indicates the approximation error of NN, satisfied $|\mu_l| < \mu_0$, and

$$\mathbf{W}^* = \arg \min_{\mathbf{W}\in R}\left\{\sup_{\mathbf{z}\in\Omega_z} |\mathbf{W}^{*T}\mathbf{h}(\mathbf{z}) - u^*(\mathbf{z})|\right\}$$

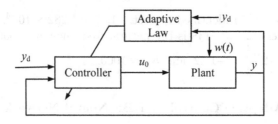

Fig. 3. Block diagram of the robust adaptive control.

Closed-loop block diagram is presented in Fig. 3. Define $\widehat{\mathbf{W}}$ as an estimate NN weight of $\mathbf{W}^*$. The output of RBF NN u_0 as direct robust adaptive controller, one has

$$u_0 = \widehat{\mathbf{W}}^T \mathbf{h}(\mathbf{z}), \tag{13}$$

and the NN weights updating law as [27]

$$\dot{\widehat{\mathbf{W}}} = -\mathbf{\Gamma}(\mathbf{h}(\mathbf{z})s + \sigma\widehat{\mathbf{W}}). \tag{14}$$

where $\mathbf{\Gamma} = \mathbf{\Gamma}^T > 0$ represents an adaptation gain matrix and $\sigma > 0$ denotes a constant.

3.3   Robust Adaptive Resonant Damping Design

Combining IRC and robust adaptive control strategy, a robust adaptive control is designed to improve the positioning accuracy in existence of disturbances and uncertain nonlinearities. IRC is designed to damp the resonant modes of the PEAs and eliminate oscillations. The compound control diagram is presented in Fig. 4.

$$i = u_0 \frac{K_t}{s}. \tag{15}$$

where K_t is the adjustable gain of the integral controller.

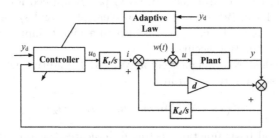

Fig. 4. Robust adaptive resonant damping control scheme.

4   Simulation Results

In this section, the presented control strategy is actualized on the system model established in Sect. 2. For the comparison, three control approaches are designed: (1) the proportional integral (PI) control; (2) the robust adaptive (RA) control realized by a RBF neural network; (3) the robust adaptive resonant damping control realized by the combination of the robust adaptive and IRC (RA+IRC). Control parameters are given in Table 2.

The structure of a RBF NN is selected to be 5-9-1, and center point vector values of Gaussian function are selected by the range of input $\mathbf{z}$. Parameters c_i and b_i can be chosen to be $[-2, -1.5, -1, 0, 0.5, 1, 1.5, 2] \times 10^6$ and 5×10^6. The sampling time is selected to be $2 \times 10^{-4}s$. Disturbance signals are described as follows

$$d(t) = \begin{cases} 0.5 & T_s/4 \leq t < T_s/2 \\ 0.25 \sin(\frac{2\pi}{T_s} + \frac{\pi}{2}) + 0.25 & T_s/2 \leq t < T_s \end{cases} . \tag{16}$$

where defined T_s as the simulation time.

Table 2. Parameters of the three controllers

Controllers	λ	Γ	ε	σ	$K_t(P)$	$K_d(I)$
PI	-	-	-	-	0.5	100
RA	100	$1 \times 10^{-7}I_9$	0.1	1×10^{-7}	-	-
RA+IRC	100	$1 \times 10^{-7}I_9$	0.1	1×10^{-7}	5	1000

4.1   Step Response

Firstly, the set-point positioning performances and the disturbances rejection performances have been confirmed. Simulation results are depicted in Fig. 5.

Figure 5(a) presents that the response of RA+IRC is faster than RA and PI. It is worth noting that, root mean square (RMS) of control efforts generated by PI, RA, and RA+IRC control are 16.7362, 16.8162, and 13.9892, respectively. Thus, RA+IRC requires significantly small control efforts to get desired performance. RMSE, MAE, and ITAE values are listed in Table 3.

4.2   Triangular Response

In this section, the motion tracking performances for a 20-Hz triangular trajectory have been tested. Figure 6(a) reveals that RA produces the best tracking accuracy with almost no tracking error. The cost of high precision, as can be seen from Fig. 6(b), is the obvious oscillates of the control effort. Meanwhile, RA+IRC has a smooth control effort, but positioning accuracy is sacrificed.

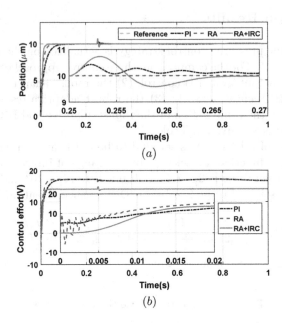

Fig. 5. Step responses and control efforts. (a) Position tracking results. (b) Control efforts.

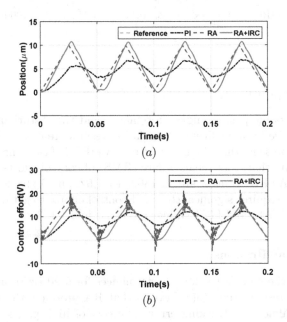

Fig. 6. Triangular responses and control efforts. (a) Position tracking results. (b) Control efforts.

Table 3. Parameters of the three controllers

Input	Performance	PI	RA	RA+IRC
Step wave	RMSE	0.8710	0.7241	0.8823
	MAE	0.1876	0.1060	0.1070
	ITAE	0.0122	0.0023	0.0023
Tria wave	RMSE	2.3014	0.0599	1.2590
	MAE	1.9911	0.0523	1.1008
	ITAE	0.0378	0.0010	0.0228
	Vibration	Small	Large	Small

5 Conclusion

A novel RA+IRC control has been designed and verified in this paper. The performance of RA+IRC has been validated by simulations. Numerical results demonstrate that RA scheme can improve the robustness of the positioning system to hysteresis and disturbance. The robust adaptive resonant damping control can reduce the oscillations, but to a certain extent, the positioning accuracy is sacrificed. Future work is to improve the positioning accuracy of the robust adaptive resonant damping control.

References

1. Tuma, T., Sebastian, A., Lygeros, J.: The four pillars of nanopositioning for scanning probe microscopy: the position sensor, the scanning device, the feedback controller, and the reference trajectory. IEEE Control Syst. Mag. **33**(6), 68–85 (2013)
2. Li, Y., Xu, Q.: A totally decoupled piezo-driven XYZ flexure parallel micropositioning stage for micro/nan-manipulation. IEEE Trans. Autom. Sci. Eng. **8**(2), 265–279 (2011)
3. Wang, F., Liang, C., Tian, Y.: A flexure-based kinematically decoupled micropositioning stage with a centimeter range dedicated to micro/nano manufacturing. IEEE/ASME Trans. Mechatron. **21**(2), 1055–1062 (2016)
4. Devasia, S., Eleftheriou, E., Moheimani, S.O.: A survey of control issues in nanopositioning. IEEE Trans. Control Syst. Technol. **15**(5), 802–823 (2007)
5. Gu, G.Y., Zhu, L.M., Su, C.Y.: Integral resonant damping for high-bandwidth control of piezoceramic stack actuators with asymmetric hysteresis nonlinearity. Mechatronics **24**(4), 367–375 (2014)
6. Ge, P., Jouaneh, M.: Tracking control of a piezoceramic actuator. IEEE Trans. Control Syst. Technol. **4**(3), 209–216 (1996)
7. Iyer, R.V., Tan, X.: Control of hysteretic systems through inverse compensation. IEEE Control Syst. **29**(1), 83–99 (2009)
8. Jayawardhana, B., Logemann, H., Ryan, E.P.: PID control of second-order systems with hysteresis. Int. J. Control **81**(8), 1331–1342 (2008)
9. Liu, W., Cheng, L., Hou, Z.G.: An inversion-free predictive controller for piezoelectric actuators based on a dynamic linearized neural network model. IEEE/ASME Trans. Mechatron. **21**(1), 214–226 (2016)

10. Cheng, L., Liu, W., Yang, C.: A neural-network-based controller for piezoelectric-actuated stick-slip devices. IEEE Trans. Ind. Electron. **65**(3), 2598–2607 (2018)

11. Clayton, G.M., Tien, S., Leang, K.K.: A review of feedforward control approaches in nanopositioning for high-speed SPM. J. Dyn. Syst. Meas. Control **131**(6), 061101 (2009)

12. Lee, C., Salapaka, S.M.: Robust broadband nanopositioning: fundamental trade-offs, analysis, and design in a two-degree-of-freedom control framework. Nanotechnology **20**(3), 035501 (2008)

13. Moheimani, S.O.: A survey of recent innovations in vibration damping and control using shunted piezoelectric transducers. IEEE Trans. Control Syst. Technol. **11**(4), 482–494 (2003)

14. Fanson, J.L., Caughey, T.K.: Positive position feedback control for large space structures. AIAA J. **28**(4), 717–724 (1990)

15. Bhikkaji, B., Ratnam, M., Fleming, A.J.: High-performance control of piezoelectric tube scanners. IEEE Trans. Control Syst. Technol. **15**(5), 853–866 (2007)

16. Pota, H.R., Moheimani, S.O., Smith, M.: Resonant controllers for smart structures. Smart Mater. Struct. **11**(1), 1–8 (2002)

17. Salapaka, S., Sebastian, A., Cleveland, J.P.: High bandwidth nano-positioner: a robust control approach. Rev. Sci. Instrum. **73**(9), 3232–3241 (2002)

18. Wei, W., Li, D., Zuo, M.: Compound active disturbance rejection control for resonance damping and tracking of nanopositioning. In: Proceedings of the 33rd China Control Conference, Nanjing, pp. 5906–5909. IEEE Press (2014)

19. Fleming, A.J., Aphale, S.S., Moheimani, S.O.: A new method for robust damping and tracking control of scanning probe microscope positioning stages. IEEE Trans. Nanotechnol. **9**(4), 438–448 (2010)

20. Gu, G.Y., Zhu, L.M., Su, C.Y.: Modeling and control of piezo-actuated nanopositioning stages: a survey. IEEE Trans. Autom. Sci. Eng. **13**(1), 313–332 (2016)

21. Wei, W., Xia, P., Zuo, M.: On disturbance rejection of piezo-actuated nanopositioner. In: 7th IEEE Data Driven Control and Learning Systems Conference, Enshi, pp. 688–692. IEEE Press (2018)

22. Ghafarired, H., Rezaei, S.: Observer-based sliding mode control with adaptive perturbation estimation for microposition actuators. Precis. Eng. **35**(2), 271–281 (2011)

23. Namavar, M., Fleming, A.J., Aleyaasin, M.: An analytical approach to integral resonant control of second-order systems. IEEE/ASME Trans. Mechatron. **19**(2), 651–659 (2014)

24. Narendra, K.S., Lin, Y.H., Valavani, L.S.: Stable adaptive controller design, Part II: Proof of stability. IEEE Trans. Autom. Control **25**(3), 440–448 (1980)

25. Liu, J., Lu, Y.: Adaptive RBF neural network control of robot with actuator non-linearities. J. Control Theory Appl. **8**(2), 249–256 (2010)

26. Chen, M., Ge, S.S.: Adaptive neural output feedback control of uncertain nonlinear systems with unknown hysteresis using disturbance observer. IEEE Trans. Ind. Electron. **62**(12), 7706–7716 (2015)

27. Ioannou, P.A., Sun, J.: Robust Adaptive Control. PTR Prentice-Hall, Upper Saddle River (1996)

Active Optimal Fault-Tolerant Control Method for Multi-fault Concurrent Modular Manipulator Based on Adaptive Dynamic Programming

Bing Li[1], Fan Zhou[1], Bo Dong[1], Yucheng Liu[1], Fu Liu[2], Huiqiu Lu[2], and Yuanchun Li[1(✉)]

[1] Department of Control Science and Engineering,
Changchun University of Technology, Changchun, China
bingli_c@foxmail.com, liyc@ccut.edu.cn
[2] Department of Control Science and Engineering,
Jilin University, Changchun, China

Abstract. In this paper, a novel active optimal fault-tolerant control (FTC) scheme is designed based on adaptive dynamic programming (ADP) for modular manipulator when sensor and actuator faults are concurrency. Firstly, the sensor fault is transformed into the pseudo-actuator fault by constructing a nonlinear transformation with diffeomorphism theory. Secondly, the faults estimated by the adaptive fault observer are applied to establish an improved performance index function. Next, the online policy iteration (PI) algorithm is used to solve the Hamilton-Jacobi-Bellman (HJB) equation via establishing a critic neural network. The optimal fault-tolerant controller is proved to be uniformly ultimately bounded (UUB) based on Lyapunov stable theory. Finally, the effectiveness of the proposed multi-fault-tolerant control algorithm is verified by simulation results.

Keywords: Adaptive dynamic programming · Modular manipulators · Optimal fault-tolerant control · Critic neural network · Policy iteration

1 Introduction

With the rapid development of robotics, modular manipulators are widely used in the real world such as space exploration, smart factories, and high-risk operations owing to the structural flexibility [1]. However, due to the complex and harsh working environment of the modular manipulator, the actuator and the sensor inevitably occur fault and cannot be artificially interfered. To ensure the safety and reliability of the modular manipulator system, it is an urgent need to design a control system that can tolerate multi-fault during operation.

Active FTC (AFTC) is greatly attracting researcher's attention owing to they can cope with unexpected faults. Researchers have achieved outstanding results in the field of AFTC recently. Niemann et al. [2] proposed a fault-tolerant control approach based on the model, Xu et al. [3] designed a dynamic surface control technology based

© Springer Nature Switzerland AG 2019
H. Lu et al. (Eds.): ISNN 2019, LNCS 11555, pp. 139–150, 2019.
https://doi.org/10.1007/978-3-030-22808-8_15

adaptive fuzzy fault-tolerant control, Yoo et al. [4] presents actuator fault detection based adaptive accommodation control of flexible-joint robots. However, in the above literature, the research on AFTC is only for single fault. Actually, the actuator and sensor concurrency faults are more general and practical. In this regard, Rotondo et al. [5] proposed a virtual actuator and sensor method for FTC of linear variable parameter systems. Sami and Patton [6] proposed AFTC for nonlinear systems when actuator and sensor faults are concurrency. Under the premise of ensuring the stability and accuracy of the modular manipulator trajectory tracking control, achieving the optimal control performance and power consumption is one of the most theoretical research topics in the field of robotics. However, there are few fault-tolerant control methods based on optimal control scheme.

In recent years, the optimal control policies based on ADP have received extensive attentions [7–10]. Since the modular manipulator is a nonlinear system, the difficulty of the optimal fault-tolerant strategy is the solution of the HJB equation. ADP has a unique advantage in solving the HJB equations for nonlinear systems. Some efforts have been made to modular manipulator based on ADP. Zhao et al. [11] develop an ADP based decentralized approximate optimal tracking control scheme for reconfigurable manipulators. Xia et al. [12] proposed an ADP based FTC scheme for reconfigurable manipulators with actuator failures.

Aiming at this research situation, an ADP based multiple FTC method for modular manipulators with simultaneous actuator and sensor faults is proposed in this paper. Firstly, using the diffeomorphism theory, the sensor fault is converted into a pseudo-actuator fault. Then a proper performance index function is constructed via the fault which is estimated through the adaptive fault observer. Besides, the improved HJB equation is settled by a critic neural network. At the same time, the parameter can be automatically updated by gradient descent method. Based on the Lyapunov stability theory, it is proved the closed-loop system with multiple faults is uniformly ultimately bounded (UUB). Finally, the proposed control method is verified by 2-DOF modular manipulators with two different configurations with concurrent actuator and sensor failures. The proposed control policy not only guaranteed the fault system to be UUB but also meet the minimum performance index function.

2   Problem Formulation

Consider the n-degree-of-freedom modular manipulator dynamics model as follows:

$$M(q)\ddot{q} + C(q,\dot{q})\dot{q} + G(q) = u \tag{1}$$

Where $q \in R^n, \dot{q} \in R^n$ and $\ddot{q} \in R^n$ are the actual joint position, velocity and acceleration, $M(q) \in R^{n \times n}$ represents the inertia matrix term, $C(q,\dot{q}) \in R^{n \times n}$ represents the Coriolis and centrifugal force term, $G(q) \in R^n$ represents the gravity term, and $u \in R^n$ is the joint motor output torque.

Equation (1) can be expressed as

$$S : \begin{cases} \dot{x} = Ax + B[f(x) + g(x)u] \\ y = x_1 \end{cases} \tag{2}$$

where $x = [x_1, x_2]^T = [q, \dot{q}]^T$, $A = \begin{bmatrix} 0 & 1 \\ 0 & 0 \end{bmatrix}$, $B = \begin{bmatrix} 0 \\ 1 \end{bmatrix}$ and $g(x) = M^{-1}(q)$, $f(x) = M^{-1}(q)[-C(q, \dot{q})\dot{q} - G(q)]$ both are nonlinear local Lipschitz continuous functions.

Considering the situation where the system suffering actuator and the sensor failures, the faulty dynamic model is shown below:

$$S_f : \begin{cases} \dot{x} = Ax + B[f(x) + g(x)(u + f_a)] \\ y = x_1 + f_s \end{cases} \tag{3}$$

where f_a is the actuator fault of the system S_f, f_s is the sensor fault of system S_f.

Introduce a simple filter as

$$\dot{x}_a = -ax_a + b(x_1 + f_s) \tag{4}$$

where a and b are constants and $a > 0$, $b \neq 0$. So the extended system dynamics can be formulated as

$$\begin{cases} \dot{x}_1 = x_2 \\ \dot{x}_2 = f(x) + g(x)(u + f_a) \\ \dot{x}_a = -ax_a + bx_1 + bf_s \\ y = x_a \end{cases} \tag{5}$$

3 Fault Tolerant Controller Design

3.1 Performance Index Function and HJB Equation Establishment

Assumption 1. The desired joint angle q_d, the desired joint velocity $\dot{q}_d$ and the desired joint acceleration $\ddot{q}_d$ are bounded, besides $f(x)$, $g(x)$ are also bounded.

Assumption 2. The actuator sensor faults of the system have the unknown upper bound as $\|f_a\| \leq \varepsilon_1$, $\|f_s\| \leq \varepsilon_2$ where ε_1 and ε_2 are positive constants.

The established performance index function is as follows

$$J(e(\tau)) = \int_0^\infty \left(\rho_1 \hat{f}_a^T(\tau)\hat{f}_a(\tau) + \rho_2 \hat{f}_s^T(\tau)\hat{f}_s(\tau) + \gamma(\nabla J^*(e(\tau)))^2 \right. \\ \left. + r(e(\tau), u(e(\tau))) \right) d\tau \tag{6}$$

where $e = x - x_d$ is the tracking error, $r(e, u) = e^T Q e + u^T R u$ is the effect function, $r(0, 0) = 0$ and $r(e, u) > 0$, in which Q, R are the positive definite matrices. $\hat{f}_a(\tau)$ is the observed value of the actuator failure, $\hat{f}_s(\tau)$ is the observed value of the sensor failure and $\gamma > 0$, $\rho_1 > 0$, $\rho_2 > 0$ are constants.

Assume that the given expected trajectory is as follows

$$\dot{x}_d = f(x_d) + g(x_d)u_d \tag{7}$$

u_d is the nominal control rate, the tracking error is guaranteed to be zero,

$$u_d = g^+(x_d)(\dot{x}_d - f(x_d)) \tag{8}$$

where $g^+(\cdot)$ is the generalized inverse of $g(\cdot)$.

The time derivative of e is

$$
\begin{aligned}
\dot{e} &= f(x) - f(x_d) + g(x)(u + f_a) - g(x_d)u_d \\
&= f_e + g(x)u - g(x_d)u + g(x_d)u - g(x_d)u_d + g(x)f_a \\
&= f_e + [g(x) - g(x_d)]u + g(x_d)u_e + g(x)f_a
\end{aligned} \tag{9}
$$

where $f_e = f(x) - f(x_d)$, $u = u_d + u_e$, the performance index function is rewritten as

$$
\begin{aligned}
J(e(\tau)) = \int_{u_e \in \psi(\Omega)}^{\infty} & \left(\rho_1 \hat{f}_a^T(\tau)\hat{f}_a(\tau) + \rho_2 \hat{f}_s^T(\tau)\hat{f}_s(\tau) + \gamma(\nabla J^*(e(\tau)))^2 \right. \\
& \left. + r(e(\tau), u_e(e(\tau)))\right)d\tau
\end{aligned} \tag{10}
$$

where $r(e, u_e) = e^T Q e + u_e^T R u_e$ is the effect function and $r(e, u_e) > 0$, $\psi(\Omega)$ is a set of permissible control sequences.

Definition 1. For the system (1), $\mu(e)$ is defined as a control policy, denoted by $\mu(e) \in \psi(\Omega)$, if $\mu(e)$ is continuous on Ω, $\mu(0) = 0$, $\mu(e) = u(e)$ stabilizes on Ω and $J(e)$ is finite, $\forall e \in \Omega$.

For the system with the admissible control set and the improved performance index function (10), it is needed that find a control policy $u_e(e) \in \mu(e)$, in addition, the performance index function (10) is minimized.

If the performance index function

$$
\begin{aligned}
V(e(\tau)) = \int_0^{\infty} & \left(\rho_1 \hat{f}_a^T(\tau)\hat{f}_a(\tau) + \rho_2 \hat{f}_s^T(\tau)\hat{f}_s(\tau) + \gamma(\nabla V^*(e(\tau)))^2 \right. \\
& \left. + r(e(\tau), u_e(e(\tau)))\right)d\tau
\end{aligned} \tag{11}
$$

is continuously differentiable, then the infinitesimal form of Eq. (11) is the so-called Lyapunov equation as

$$
\begin{aligned}
0 = & \rho_1 \hat{f}_a^T(\tau)\hat{f}_a(\tau) + \rho_2 \hat{f}_s^T(\tau)\hat{f}_s(\tau) + \gamma(\nabla V^*(e(\tau)))^2 + r(e, u_e) \\
& + (\nabla V(e))^T \times (f_e + [g(x) - g(x_d)]u + g(x_d)u_e)
\end{aligned} \tag{12}
$$

where $\nabla V(e) = \frac{\partial V(e)}{\partial e}$, $V(0) = 0$ and $r(0,0) = 0$.

Define the Hamiltonian function as

$$
\begin{aligned}
H(e, u_e, \nabla V(e)) &= \rho_1 \hat{f}_a^T(\tau)\hat{f}_a(\tau) + \rho_2 \hat{f}_s^T(\tau)\hat{f}_s(\tau) + \gamma(\nabla V^*(e(\tau)))^2 \\
&+ r(e, u_e) + (\nabla V(e))^T (f_e + [g(x) - g(x_d)]u + g(x_d)u_e)
\end{aligned}
\tag{13}
$$

Define the optimal performance indicator function as

$$
J^*(e) = \min_{u_e} \int_0^\infty r(e(\tau), u_e(e(\tau)))d\tau
\tag{14}
$$

and which also satisfies

$$
0 = \min_{u_e} H(e, u_e, \nabla J^*(e))
\tag{15}
$$

where $\nabla J^*(e) = \frac{\partial J^*(e)}{\partial e}$. Therefore, the optimal multi-fault-tolerant control strategy is obtained,

$$
u_e^* = -\frac{1}{2}R^{-1}g^T(x)\nabla J^*(e)
\tag{16}
$$

Combine Eqs. (12) and (15), one obtains

$$
\begin{aligned}
&(\nabla J(e))^T (f_e + [g(x) - g(x_d)]u + g(x_d)u_e) \\
&= -e^T Q e - u_e^T R u_e - \rho_1 \hat{f}_a^T \hat{f}_a - \rho_2 \hat{f}_s^T \hat{f}_s - \gamma(\nabla J^*(e))^2
\end{aligned}
\tag{17}
$$

3.2 Adaptive Fault Observer Design

Assumption 3. The estimated error of the actuator and sensor failure have an unknown upper bound as $\left\|f_a - \hat{f}_a\right\| \leq \varepsilon_3$, $\left\|f_s - \hat{f}_s\right\| \leq \varepsilon_4$.

Design the following adaptive faults observer form

$$
\begin{cases}
\dot{\hat{x}} = f(x) + g(x)\left(u + \hat{f}_a\right) + \alpha_1(x - \hat{x}) \\
\dot{\hat{x}}_a = -a\hat{x}_a + b\hat{x}_1 + b\hat{f}_s + \alpha_2(x_a - \hat{x}_a)
\end{cases}
\tag{18}
$$

where $\hat{x}$ is the observed value of the system state vector x, the definition $\hat{f}_a$ is the estimated value of the actuator fault f_a, the definition $\hat{f}_s$ is the estimated value of the actuator fault f_s, α_1 and α_2 are the positive definite observer gain. The observed values of the sensor and actuator failures are updated according to the following adaptive update rate.

$$
\dot{\hat{f}}_a = \alpha_3 g^T(x)e_x
\tag{19}
$$

$$\dot{\hat{f}}_s = b\alpha_4 e_{xa} \tag{20}$$

where $e_x = \begin{bmatrix} e_{x1} \\ e_{x2} \end{bmatrix} = \begin{bmatrix} x_1 - \hat{x}_1 \\ x_2 - \hat{x}_2 \end{bmatrix}$, $e_{xa} = x_a - \hat{x}_a$, the actuator observation error as $e_a = f_a - \hat{f}_a$, and the sensor observation error as $e_s = f_s - \hat{f}_s$, α_3 and α_4 are the positive definite matrix.

3.3   Construction of the Critic Network

The approximate performance indicator function can be expressed as

$$V(e) = W_c^T \sigma_c(e) + \varepsilon_c \tag{21}$$

where $W_c \in R^k$ is the ideal weight vector, k is the number of neurons in the hidden-layer, $\sigma_c(e)$ is the activation function, ε_c is the approximation error of the neural network, and the gradient of $V(e)$ is approximated by the neural network: $\nabla V(e) = (\nabla\sigma_c(e))^T W_c + \nabla\varepsilon_c$.

Therefore, the Hamiltonian function is expressed as

$$\begin{aligned}
H(e, u_e, W_c) &= \rho_1 \hat{f}_a^T(\tau)\hat{f}_a(\tau) + \rho_2 \hat{f}_s^T(\tau)\hat{f}_s(\tau) \\
&\quad + \gamma\big((\nabla\sigma_c(e))^T W_c + \nabla\varepsilon_c\big)^T \times \big((\nabla\sigma_c(e))^T W_c + \nabla\varepsilon_c\big) \\
&\quad + r(e, u_e) + \big(W_c^T \nabla\sigma_c(e)\big)\dot{e} = -\nabla\varepsilon_c^T \dot{e} = e_{cH}
\end{aligned} \tag{22}$$

where e_{cH} is the evaluation error of the evaluation network. Define $\hat{W}_c$ to be the estimated value of W_c, then evaluate the actual output of the network as $\hat{V}(e) = \hat{W}_c^T \sigma_c(e)$, then the gradient of $\hat{V}(e)$ is $\nabla\hat{V}(e) = (\nabla\sigma_c(e))^T \hat{W}_c$. So the approximate Hamiltonian function is expressed as

$$\begin{aligned}
\hat{H}(e, u_e, \hat{W}_c) &= \rho_1 \hat{f}_a^T(\tau)\hat{f}_a(\tau) + \rho_2 \hat{f}_s^T(\tau)\hat{f}_s(\tau) + \gamma\big((\nabla\sigma_c(e))^T \hat{W}_c\big)^T \\
&\quad \times \big((\nabla\sigma_c(e))^T \hat{W}_c\big) + r(e, u_e) + \big(\hat{W}_c^T \nabla\sigma_c(e)\big)\dot{e} = e_c
\end{aligned} \tag{23}$$

Define $\vartheta = \nabla\sigma_c(e)\dot{e}$, the minimum performance criterion in the training process is $E_c = \frac{1}{2}e_c^T e_c$, the weight is updated by the gradient descent method $\dot{\hat{W}}_c = -\alpha_c e_c \vartheta$, where $\alpha_c > 0$ is the adaptive gain of the evaluation network.

Define the weight estimation error as

$$\tilde{W}_c = W_c - \hat{W}_c \tag{24}$$

Combine (22), (23) and (24), we can obtain

$$e_c = e_{cH} - \tilde{W}_c^T \vartheta \tag{25}$$

The weight approximation error is updated as

$$\dot{\tilde{W}}_c = -\dot{\hat{W}}_c = \alpha_c \left(e_{cH} - \tilde{W}_c^T \vartheta \right) \vartheta \tag{26}$$

The ideal optimal feedback fault-tolerant control strategy can be obtained

$$u_e = -\frac{1}{2} R^{-1} g^T(x) \left((\nabla \sigma_c(e))^T W_c + \nabla \varepsilon_c \right) \tag{27}$$

and it can be approximated as

$$\hat{u}_e = -\frac{1}{2} R^{-1} g^T(x) (\nabla \sigma_c(e))^T \hat{W}_c \tag{28}$$

3.4   Online Policy Iteration Algorithm

The online PI algorithm is used to approximate the HJB and obtain the optimal feedback control strategy u_e^*, the specific method is shown below:

Step 1: Parameter initialization, select a small normal number ε, select $i = 0$ as the number of iterations and, $V^{(1)} = 0$, start from the initial control strategy $u_e^{(0)}$ to iterate;

Step 2: Combine the $u_e^{(i)}$ to update $J^{(i+1)}$ by the following equation

$$0 = \rho_1 \hat{f}_a^T(\tau) \hat{f}_a(\tau) + \rho_2 \hat{f}_s^T(\tau) \hat{f}_s(\tau) + \gamma \left(\nabla V^{(i+1)}(e(\tau)) \right)^2 + r\left(e, u_e^{(i)} \right) + \left(\nabla V^{(i+1)}(e) \right)^T$$
$$\times \left(f_e + [g(x) - g(x_d)]u + g(x_d)u_e^{(i)} \right)$$

Step 3: Control strategy $u_e^{(i+1)}$ is updated by $u_e^{(i+1)} = -\frac{1}{2} R^{-1} g^T(x) \nabla V^{(i+1)}(e)$.

Step 4: If $i > 0$ and $\left| V^{(i+1)}(e) - V^{(i)}(e) \right| \le \varepsilon$, the operation is stopped, and one can obtain the approximate optimal control strategy, otherwise $i = i + 1$, then the process returns to step 2.

3.5   Stability Analysis

It can be seen from the above analysis that the optimal control strategy consists of two parts: the nominal control rate u_d and the optimal feedback control rate u_e, so the control rate can be described as

$$u = u_d + \hat{u}_e \tag{29}$$

where u_d can be directly obtained by the formula (8).

Assumption 4. The nominal controller u_d is bounded and $\|u_d\| \leq \eta$, where η is an unknown positive constant.

Theorem 1. For the multi-fault-tolerant control problem of the modular manipulator system with Assumptions 1–4, if the solution based on the HJB equation exists, the control strategy can ensure that the tracking error of the system is UUB in the case of multiple faults.

Proof. Choose the Lyapunov function as

$$V_1(t) = \frac{1}{2}e^T e + J^*(e) \tag{30}$$

Combine (17), the time derivative of (30) is

$$\begin{aligned}
\dot{V}_1(t) &= e^T \dot{e} + (\nabla J^*(e))^T \dot{e} \\
&= e^T g(x) f_a + e^T [f_e + [g(x) - g(x_d)]u + g(x_d)u_e] \\
&\quad + (\nabla J^*(e))^T g(x) f_a - e^T Q e - u_e^T R u_e - \rho_1 \hat{f}_a^T \hat{f}_a \\
&\quad - \rho_2 \hat{f}_s^T \hat{f}_s - \gamma (\nabla J^*(e))^2
\end{aligned} \tag{31}$$

Since $f(z)$ is a Lipschitz function, $L_f > 0$, so that $\|f_e\| \leq L_f\|e\|$, $\|g(x)\| \leq D_g$, $\|g(x_d)\| \leq D_{gd}$, and thus $\|g(x) - g(x_d)\| \leq \Delta D$. Using triangular inequalities, and considering the Assumptions 2, 3 and 4, the above equation is rewritten as

$$\begin{aligned}
\dot{V}_1(t) &\leq - \left(\psi_1\|e\| - \Delta D\eta - \frac{\varepsilon_l}{\|e\|} \right)\|e\| - \psi_2\|u_e\|^2 - \rho_2\hat{f}_s^T\hat{f}_s \\
&\quad - \left(\rho_1 - K_g^2 \right)\hat{f}_a^T\hat{f}_a - \left(\gamma - \frac{1}{2} \right)(\nabla J^*(e))^2
\end{aligned} \tag{32}$$

where $\psi_1 = \lambda_{\min}(Q) - L_f - \frac{1}{2}D_{gd}^2 - \frac{1}{2}$, $\psi_2 = \lambda_{\min}(R) - \frac{1}{2}D_{gd}^2 - \frac{1}{2}\Delta D^2 - \frac{1}{2}$, $\varepsilon_l = D_g^2$ $(2\varepsilon_1 + \varepsilon_3)\varepsilon_3$.

We find that $\dot{V}_1(t) \leq 0$ when e lies outside the compact set $\Omega_3 = \left\{ e : \|e\| \leq \sqrt{\frac{2\Delta D^2\eta^2 + 4\psi_1\varepsilon_l}{4\psi_1^2}} \right\}$, and the following conditions need to be met $\lambda_{\min}(Q) \geq L_f + \frac{1}{2}D_{gd}^2 + 1$, $\lambda_{\min}(R) \geq \frac{1}{2}D_{gd}^2 + \frac{1}{2}\Delta D^2 + \frac{1}{2}$, $\rho_1 \geq D_g^2$, $\rho_2 > 0$, $\gamma \geq \frac{1}{2}$.

Therefore, the closed-loop system is UUB. This completes the proof.

4 Simulation Research

In order to verify the effectiveness of the proposed active optimal multi-fault-tolerant control method, two-degree-of-freedom manipulator models with different configurations are selected for simulation verification.

The system's control torque $u = [u_1, u_2]^T$, the starting position and angular velocity are $q_1(0) = q_2(0) = 0.5$, $\dot{q}_1(0) = \dot{q}_2(0) = 0$. The critic NN is chosen as 4-10-1 with 4 input neurons, 10 hidden neurons and 1 output neuron, and defined the NN weight vector $\hat{w} = [\hat{w}_1, \hat{w}_2, \ldots, \hat{w}_9, \hat{w}_{10}]^T$, with the initial value $\hat{w} = [20, 30, 40, 20, 30, 40, 50, 40, 50, 55]^T$. The desired trajectories of the modular manipulator configuration a and configuration b is shown below

$$q_{a1d} = 0.4\sin(0.3t) - 0.1\cos(0.5t) \qquad q_{b1d} = 0.2\cos(0.5t) + 0.25\sin(0.4t)$$
$$q_{a2d} = 0.3\cos(0.6t) + 0.6\sin(0.2t) \qquad q_{b2d} = 0.3\cos(0.2t) - 0.4\sin(0.6t)$$

Considering configuration a, an actuator fault $f_a = 2\sin(2 \times q_1)$ is added to the joint 1 at $t = 20$ s, and a sensor fault $f_s = -0.5q_1$ is added to the joint 2 at $t = 40$ s. Considering configuration b, an actuator fault $f_a = 3 \times \sin(0.2t) + \cos(1t)$ is added to the joint 1at $t = 30$ s, and a sensor fault $f_s = 2$ is added to the joint 2 at $t = 40$ s.

The simulation results of the configuration a are shown in Figs. 1, 2, 3 and 4. Figures 1 and 2 are the actuator fault and sensor fault estimation curves respectively. It can be seen that the fault observer designed can accurate estimate the faults. Figure 3 is a trajectory tracking curve of the configuration a. It can be observed that the actual trajectories follow the desired ones well though after the actuator failure occurred in the 20 s and the sensor failure occurred in the 40 s. Figure 4 shows the trajectories of tracking error.

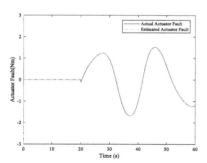

Fig. 1. The estimation of actuator fault.

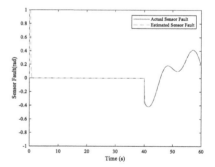

Fig. 2. The estimation of sensor fault.

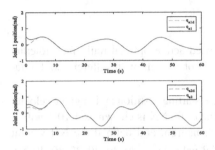

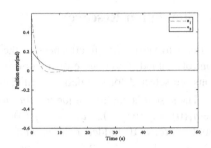

Fig. 3. Tracking curves for configuration *a*.

Fig. 4. Tracking error curves for configuration *a*.

We further verify the effectiveness of our multi-fault-tolerant control algorithm without changing the controller parameters, the simulation results of the configuration *b* as shown Figs. 5, 6, 7 and 8, Figs. 5 and 6 are the actuator fault and sensor fault estimation curves respectively. It can be seen from the figures both the actuator and the sensor fault observer accurately estimate fault information online. Figure 7 is a trajectory tracking curve of the configuration *b*. It can be observed that the actual trajectories follow the desired ones well though after the actuator failure occurred in the 30 s and the sensor failure occurred in the 40 s. Figure 8 shows the trajectories of tracking error. The simulation results show that when the sensor and actuator of the modular manipulator system are faulty, the control method designed is used for different configurations, which proves the effectiveness of the control method.

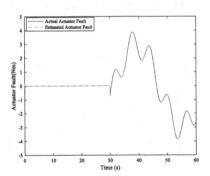

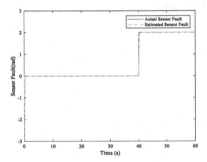

Fig. 5. The estimation of actuator fault.

Fig. 6. The estimation of sensor fault.

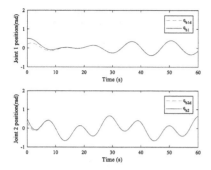 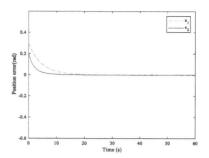

Fig. 7. Tracking curves for configuration b.   **Fig. 8.** Tracking error curves for configuration b.

5 Conclusion

This paper proposes an active optimal multi-fault-tolerant control method for modular manipulators based on ADP. An adaptive fault observer is utilized to build an improved performance index function to transform the multiple FTC problems into optimal control. Next, an evaluation neural network is established to handle the HJB equation, and the online PI algorithm is used to solve the optimal feedback control rate. In addition, the Lyapunov theory is utilized to prove that the closed-loop system is UUB. Finally, one validated the effectiveness of our designed fault-tolerant control method by selecting two-degree-of-freedom manipulators in two configurations.

Acknowledgement. This work is supported by the National Natural Science Foundation of China (Grant nos. 61374051, 61773075 and 61703055) and the Scientific Technological Development Plan Project in Jilin Province of China (Grant nos. 20170204067GX, 20160520013JH and 20160414033GH).

References

1. Paredis, C.: A rapidly deployable manipulator system. Robot. Auton. Syst. **21**(3), 289–304 (1997)
2. Niemann, H.: A model-based approach to fault-tolerant control. Int. J. Appl. Math. Comput. Sci. **22**(1), 67–86 (2012)
3. Xu, Y., Tong, S., Li, Y.: Adaptive fuzzy fault-tolerant control of static var compensator based on dynamic surface control technique. Nonlinear Dyn. **73**(3), 2013–2023 (2013)
4. Yoo, S.J.: Actuator fault detection and adaptive accommodation control of flexible-joint robots. IET Control Theory Appl. **6**(10), 1497–1507 (2012)
5. Rotondo, D., Nejjari, F., Puig, V.: A virtual actuator and sensor approach for fault tolerant control of LPV systems. J. Process Control **24**(3), 203–222 (2014)
6. Sami, M., Patton, R.J.: Active fault tolerant control for nonlinear systems with simultaneous actuator and sensor faults. Int. J. Control Autom. Syst. **11**(6), 1149–1161 (2013)

7. Wang, D., Liu, D.: Policy iteration algorithm for online design of robust control for a class of continuous-time nonlinear systems. IEEE Trans. Autom. Sci. Eng. **11**(2), 627–632 (2014)
8. Chang, S., Lee, J.Y.: An online fault tolerant actor-critic neuro-control for a class of nonlinear systems using neural network HJB approach. Int. J. Control Autom. Syst. **13**(2), 311–318 (2015)
9. Fan, Q.: Adaptive fault-tolerant control for affine nonlinear systems based on approximate dynamic programming. IET Control Theory Appl. **10**(6), 655–663 (2016)
10. Wang, Z., Liu, L., Wu, W., Zhang, H.: Optimal fault-tolerant control for discrete-time nonlinear strict-feedback systems based on adaptive critic design. IEEE Trans. Neural Netw. Learn. Syst. **29**(6), 2179–2191 (2018)
11. Zhao, B., Li, Y.: Model-free adaptive dynamic programming based near-optimal decentralized tracking control of reconfigurable manipulators. Int. J. Control Autom. Syst. **16**(2), 478–490 (2018)
12. Xia, H., Zhao, B., Zhou, F., Dong, B., Liu, G., Li, Y.: Fault tolerant control for reconfigurable manipulators based on adaptive dynamic programming with an improved performance index function. In: 2016 12th World Congress on Intelligent Control and Automation IEEE (2016)

Signal Processing, Industrial Application, and Data Generation

Analysing Epileptic EEG Signals Based on Improved Transition Network

Yang Li[1,2], Yao Guo[1,2], Qingfang Meng[1,2(✉)], Zaiguo Zhang[3],
Peng Wu[1,2], and Hanyong Zhang[1,2]

[1] School of Information Science and Engineering, University of Jinan,
Jinan 250022, China
ise_mengqf@ujn.edu.cn
[2] Shandong Provincial Key Laboratory of Network Based Intelligent
Computing, Jinan 250022, China
[3] CET Shandong Electronics Co., Ltd., Jinan 250101, China

Abstract. The epileptic automatic detection was very significance in clinical. The nonlinear time series analysis method based on complex network theory provided a new perspective understand the dynamics of nonlinear time series. In this paper, we proposed a new epileptic seizure detection method based on statistical properties of improved transition network. First, we improved the transition network and electroencephalogram (EEG) signal was constructed into the improved transition network. Then, based on the statistical characteristics of improved transition network, the mathematical expectation of node distribution in a network was extracted as the classification feature. Finally, the performance of the algorithm was evaluated by classifying the epileptic EEG dataset. Experimental results showed that the classification accuracy of proposed algorithm is 97%.

Keywords: Complex network · Improved transition network ·
Mathematical expectation · Seizure detection · EEG

1 Introduction

Epilepsy is a kind of chronic neurological disease of the brain which has a wide influence and is not infectious. Clinically, epilepsy was diagnosed by electroencephalogram (EEG). However, there are many problems in the prevention, diagnosis, treatment and other aspects of epilepsy at present. For example, the hospital lack specialists in epilepsy, the clinical manifestations of epilepsy are very complex, seizure types are diverse and epilepsy misdiagnosis and mistreatment is very common. Therefore, the study of automatic classification algorithm with high performance is of great clinical significance.

The two main parts of automatic epilepsy detection algorithm are feature extraction method and classifier. The features extracted by feature extraction method should be able to clearly represent the fundamental differences between the different EEG signals. Yuan et al. [1] and Kumar et al. [2] studied approximate entropy to characterize EEGs. Khoa et al. [3] utilized Lyapunov spectrum to filter noise and detect epilepsy. Empirical

© Springer Nature Switzerland AG 2019
H. Lu et al. (Eds.): ISNN 2019, LNCS 11555, pp. 153–161, 2019.
https://doi.org/10.1007/978-3-030-22808-8_16

mode decomposition (EMD) [4], fast Fourier transform (FFT) [5] and wavelet transform (WT) [6] were introduced into epileptic signal processing to extract the time domain, frequency domain and time-frequency domain features from the processed signals. Artificial neural network (ANN), decision tree classifier, extreme learning machine (ELM) and other classifiers combined with features are also important research directions in epilepsy detection algorithms.

The transformation of nonlinear time series into complex networks opens a new perspective for the study of time series. Zhang and Small [7] constructed complex networks from pseudo-periodic time series and found that time series with different dynamic characteristics have different topological structures. Lacasa et al. [8] presented a method called visibility graph that can transform arbitrary time series into complex networks. Sun et al. [9] proposed transition network construction algorithm, by which the dynamics behind the time series are encoded into the network structure. Many scholars had improved the algorithm of visibility graph and proximity network [11, 12] and used complex network to study the automatic detection and classification algorithms of epileptic EEG [13, 14]. However, due to the disadvantages of information loss and low efficiency in the construction of the transition network, few scholars use the transition network to study epileptic EEG.

In this study, we make improvements from two aspects for the transition network construction algorithm. First, the transition network construction algorithm combines the proximity network method to process the original data. Second, our new method reduces the number of nodes appearing in the network, thus decreasing the redundant information and computing complexity. According to our improved transition network construction algorithm, EEG signals are transformed into complex network. Then mathematical expectation of node distribution in a network is extracted as features. The classification results of our improved method reach 97%, which means that our method can effectively detect epilepsy automatically.

2 Method

Figure 1 shows a summary of the improved transition network analysis of nonlinear time series. The most important part is the construction of complex network.

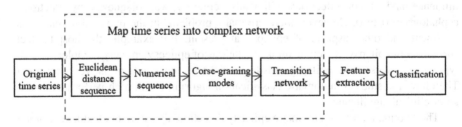

Fig. 1. The process of the improved transition network analysis of time series.

2.1   Construction of the Improved Transition Network from Time Series

For time series as $\{x_1, x_2, \ldots, x_i, \ldots, x_N\}(i \in [1, N])$, according to the local maximum value, the original time series is divided into several non-overlapping cycles, and the cycle set is expressed as $\{c_1, c_2, \ldots, c_n\}$. Then, we calculate the Euclidean distance between every adjacent c_i. Because each c_i has a different length, the Euclidean distance is modified to:

$$d_{ij} = min_{l=0,1,\ldots,l_j-l_i} \left(\frac{1}{l_i} \sqrt{\sum_{k=1}^{l_i} \left(c_i(k) - c_j(k+l) \right)^2} \right) \tag{1}$$

where l_i and l_j are respectively the length of c_i and c_j, the $c_i(k)$ and $c_j(k)$ are respectively the k_{th} point of c_i and c_j (supposing $l_i \le l_j$). A new sequence is obtained by calculating the Euclidean distance between each pair of adjacent cycles, denoted as $\{d_{12}, d_{23}, \ldots, d_{(n-1)n}\}$.

For every d_{ij}, according coarse gaining method, we denote the new numerical sequence $\{s_i\}$:

$$s_i = \begin{cases} 4, & Q_3 < d_{ij} \\ 3, & Q_2 < d_{ij} < Q_3 \\ 2, & Q_1 < d_{ij} < Q_2 \\ 1, & d_{ij} < Q_1 \end{cases} \tag{2}$$

where Q_1, Q_2 and Q_3 are the threshold values. Coarse graining processing discards small details, decomposes the original time series data into finite subintervals, and uses different characters to represent each subinterval.

After the coarse graining processing, the Euclidean distance sequence is transformed into the new numerical sequence $\{s_i\}$, which is defined as:

$$S = (s_1, s_2, s_3, \ldots, s_{n-1})(s_i \in (1, 2, 3, 4)) \tag{3}$$

Based on the generated numerical sequence $\{s_i\}$ above, we set the appropriate dimension m and delay time τ to divide the numerical sequence into coarse gaining modes. The resulting mode can be denoted as $\{CG_i\}, i = 1, 2, \ldots, (n-1-m)/\tau+1$. Each CG_i is a vector of length m.

As the sliding window moves, many coarse gaining modes that have the same vector are created. In the original transition network construction algorithm. Each coarse gaining mode is a network node. There are a lot of redundant information in the network. and the computational complexity is high.

In our study, the different coarse graining modes consider as network nodes and denoted to be v_i, $i = 1, 2, \ldots, M, M < (n-1-m)/\tau+1$. Therefore, we can study the characteristics of different time series by studying the modes that fluctuate with time.

The edge weight between node v_i and node v_j is expressed as:

$$w_{v_i v_j} = count \tag{4}$$

where the coarse graining modes represented by node v_i and node v_j is CG_{v_i} and CG_{v_j}. $count$ denotes the transformation frequency of CG_{v_i} transferred to CG_{v_j} in the $\{CG_i\}$.

The complex network constructed by the improved transition network algorithm is a directed weighted network. The connection directions between nodes are set in the order in which coarse graining modes appear when moving with sliding windows.

If the mapping numerical sequence of the time series is $\{s_i\}_{i=1,2,...,15} = \{321432321414321\}$. Setting m is 4 and τ is 1. Figure 2 shows the network structure of improved transition network. Although coarse graining modes $\{3214\}$ and $\{1432\}$ appear twice, they only appear once as nodes in the network. Edge weight between nodes are all 1.

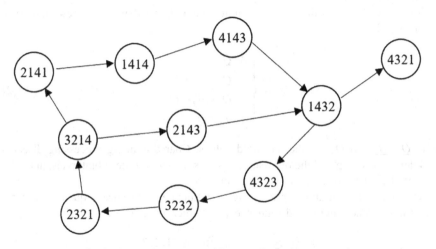

Fig. 2. Network structure of improved transition network.

2.2   Feature Extraction Method Based on Improved Transition Network

Weighted degree of node is the basic statistic in complex network. In directed and weighted networks, the weighted degree of the node i is denoted as the total weights of all the edges that connected to node i, and it can be calculated as follows:

$$k_i = \sum_{j=1}^{n} a_{ij} = k_{in}^i + k_{out}^i \tag{5}$$

where k_{out}^i and k_{in}^i are the out-strength and in-strength of node i. n is the number of the nodes in complex network.

Based on the structural characteristics of improved transition network, mathematical expectation of node distribution in a network is proposed to evaluate and measure the structural complexity of improved transition network.

According to the weighted degree of node in a network, the probability set of weighted degree is constructed and expressed as:

$$Q = [q_1, q_2 \ldots q_i \ldots q_n] \tag{6}$$

where, the q_i can be calculated using the following formula:

$$q_i = \frac{k_i}{\sum_{i=1}^{n} k_i} \tag{7}$$

The mathematical expectation of node distribution in a network is calculated based on the probability set Q, defined as:

$$E = \sum_{i=1}^{n} q_i k_i \tag{8}$$

3   Experimental Result and Analysis

In this paper, the EEG data is obtained from Department of Epileptology, University of Bonn, Germany. This public database has five sets. The interictal periods (set D) and ictal periods (set E) are utilized for evaluating the performances of the extracted feature and the proposed detection algorithm. Each set has 100 single-channel EEG data lasting 23.6 s. The EEG signals are recorded using 128 channel amplifiers and sampled at a rate of 173.61 Hz. Every signal has 4097 points.

The improved transition network algorithm is applied to the EEG database to extract the mathematical expectation of node distribution in a network.

Table 1. The classification results of the proposed method.

Method	Data length	Dimension of modes	Sensitivity	Specificity	Accuracy
Improved transition network	2048	3	98%	91%	94.5%
	2048	4	96%	93%	94.5%
	2048	5	95%	94%	94.5%
	2048	6	96%	93%	94.5%
	4096	3	97%	96%	96.5%
	4096	4	98%	96%	97%
	4096	5	98%	96%	97%
	4096	6	98%	96%	97%

In the process of complex network construction, the selection of parameters determines the characteristics of the final network, so it will affect the classification accuracy. In performance evaluation process, the interictal EEG and the ictal EEG are regarded as the positive case and the negative case. In order to find out the best performance of the classification method based on improved transition network, the dimensions of the coarse graining modes are 3, 4, 5, and 6, respectively, the delay time τ is 1 and the data length are 2048 and 4096 points, respectively. Q_1, Q_2 and Q_3 are respectively the lower quartile, median and upper quartile corresponding to the Euclidean distance sequence generated by 100 sample signals in the interictal periods.

The detailed classification results are shown in Table 1. As can be seen from Table 1, data length is the most important factor affecting classification accuracy. Only when the data reaches a certain length can the difference of network topological characteristics represented by different data sets be more obvious. As the data length is 4096, 97% accuracy can be achieved.

When the accuracy is the same, we hope to achieve higher sensitivity and specificity at the same time. When the data length is 2048 and m is 5, the sensitivity and specificity reach the same value. But when m is changed in a certain range, both specificity and sensitivity are acceptable. Especially when the length of data reaches 4096, specificity and sensitivity will not be affected. The dimension of the coarse graining modes in this method is not suitable to choose too large value. If the dimension is too large, the number of nodes and the computational complexity in the network will increase.

The mathematical expectation of node distribution with data length of 4096 and m of 5 is shown in Fig. 3. Each '+' presents the ictal signal and each '*' presents the interictal signal. The solid line in the figure is the threshold for the best classification. As can be seen from Fig. 3, the feature values of interictal signals are more concentrated and the values of the ictal signals are very sparse.

In order to more vividly describe the difference of the mathematical expectation of node distribution between interictal signals and ictal signals, Fig. 4 show the box plots for signals using mathematical expectation of node distribution in network. There are significant numerical differences between interictal signals and ictal signals.

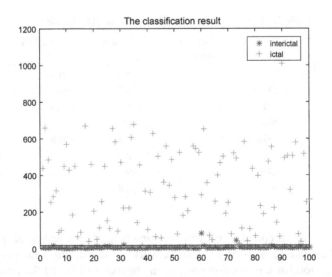

Fig. 3. The classification result with data length of 4096 and m of 5.

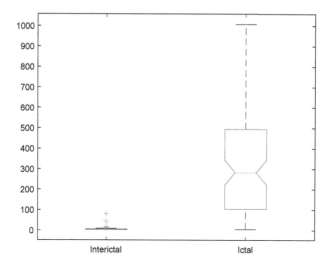

Fig. 4. The box plots with data length of 4096 and m of 5.

Table 2. Classification accuracy of several detection algorithms.

Method	Data length	Feature	Acc.
WVG [13]	1024	Modularity and weighted degree	93.25%
RQA [16]	1024	DET	90.5%
Sample entropy [15]	2048	***	91%
DWT [2]	4096	Approximate entropy	95%
Proposed method	4096	Mathematical expectation	97%

Table 2 shows the accuracy of epileptic EEG single feature classification based on improved transition network and other methods, which use same data sets. It shows that the improved transition network algorithm based on the mathematical expectation of node distribution in a network achieves the highest classification accuracy compared with other established detection algorithms. Our method provides a new idea for improving the algorithm of complex network construction.

4 Conclusions

In this paper, an improved transition network construction algorithm is proposed for automatic seizure detection and classification. Firstly, we combine the proximity network construction algorithm with the original transition network construction algorithm to process EEG signals, which makes it easier to obtain the statistical characteristics of epileptic data. Then, in order to reduce the computational complexity, we make different coarse graining modes as nodes in the network instead of each coarse graining mode as node, and the edge weight is determined by the transition frequency. Finally, the mathematical expectation of node distribution is extracted to verify the classification performance.

According to the experimental results, we conclude that the method proposed in this paper can effectively distinguish the interictal signals and ictal signals. With the increase of epilepsy patients, we hope that our method can provide a good idea for the automatic detection of epilepsy.

Acknowledgments. This work was supported by the National Natural Science Foundation of China (Grant No. 61671220, 61701192, 61640218), the Natural Science Foundation of Shandong Province, China (Grant No. ZR2017QF004).

References

1. Yuan, Q., Zhou, W., Li, S., Cai, D.: Epileptic EEG classification based on extreme learning machine and nonlinear features. Epilepsy Res. **96**(1–2), 29–38 (2011)
2. Kumar, Y., Dewal, M.L., Anand, R.S.: Epileptic seizure detection using DWT based fuzzy approximate entropy and support vector machine. Neurocomputing **133**, 271–279 (2014)
3. Khoa, T.Q.D., Thi Minh Huong, N., Toi, V.V.: Detecting epileptic seizure from scalp EEG using Lyapunov spectrum. Comput. Math. Methods Med. **2012**, 11 (2012)
4. Pachori, R.B., Bajaj, V.: Analysis of normal and epileptic seizure EEG signals using empirical mode decomposition. Comput. Methods Programs Biomed. **104**(3), 373–381 (2011)
5. Polat, K., Güneş, S.: Classification of epileptiform EEG using a hybrid system based on decision tree classifier and fast Fourier transform. Appl. Math. Comput. **187**(2), 1017–1026 (2007)
6. Patnaik, L.M., Manyam, O.K.: Epileptic EEG detection using neural networks and post-classification. Comput. Methods Programs Biomed. **91**(2), 100–109 (2008)
7. Zhang, J., Small, M.: Complex network from pseudoperiodic time series: topology versus dynamics. Phys. Rev. Lett. **96**, 238701 (2006)
8. Lacasa, L., Luque, B., Ballesteros, F., Luque, J., Nuno, J.C.: From time series to complex networks: the visibility graph. PNAS **105**(13), 4972–4975 (2008)
9. Sun, X., Small, M., Zhao, Y., Xue, X.: Characterizing system dynamics with a weighted and directed network constructed from time series data. Chaos Interdiscip. J. Nonlinear Sci. **24**(2), 024402 (2014)
10. Luque, B., Lacasa, L., Ballesteros, F., Luque, L.: Horizontal visibility graphs: exact results for random time series. Phys. Rev. E **80**(4), 046103 (2009)
11. Bezsudnov, I.V., Snarskii, A.A.: From the time series to the complex networks: the parametric natural visibility graph. Phys. A Stat. Mech. Appl. **414**, 53–60 (2014)
12. Li, X., Sun, M., Gao, C., Han, D., Wang, M.: The parametric modified limited penetrable visibility graph for constructing complex networks from time series. Phys. A Stat. Mech. Appl. **492**, 1097–1106 (2018)
13. Supriya, S., Siuly, S., Wang, H., Cao, J., Zhang, Y.: Weighted visibility graph with complex network features in the detection of epilepsy. IEEE Access **4**, 6554–6566 (2016)
14. Gao, Z.K., Jin, N.D.: A directed weighted complex network for characterizing chaotic dynamics from time series. Nonlinear Anal. Real World Appl. **13**(2), 947–952 (2012)

15. Lake, D.E., Richman, J.S., Griffin, M.P., Moorman, J.R.: Sample entropy analysis of neonatal heart rate variability. Am. J. Physiol. Regul. Integr. Comp. Physiol. **283**, 789–797 (2002)
16. Meng, Q.F., Chen, S.S., Chen, Y.H., Feng, Z.Q.: Automatic detection of epileptic EEG based on recurrence quantification analysis and SVM. Acta Phys. Sin. **63**(5), 050506 (2014)

Neural Network Based Modeling of Hysteresis in Smart Material Based Sensors

Yonghong Tan[1], Ruili Dong[2(✉)], and Hong He[1]

[1] College of Mechanical and Electrical Engineering,
Shanghai Normal University, Shanghai, China
tany@shnu.edu.cn
[2] College of Information Science and Technology, Donghua University,
Shanghai, China
ruilidong@dhu.edu.cn

Abstract. Hysteresis is a nonlinear phenomenon which is involved with dynamics, non-smoothness and multi-valued mapping. It usually exists in elastic materials, smart materials, and energy-storage materials. For describing the characteristic of hysteresis, a basis function based neural network model is proposed in this paper. In this method, the multi-valued mapping of hysteresis is transferred into a one-to-one mapping with an expanded input space involving the input variable and a constructed hysteretic auxiliary function. Thus, the neural network can be employed to approximate the characteristic of hysteresis. Finally, the method is used to the modeling of hysteresis in a smart material based sensor.

Keywords: Hysteresis · Expanded input space · Neural network · Modeling

1 Introduction

It is known that hysteresis usually exists in many smart materials such as piezoceramic, ionic polymer-metal composite (IPMC) and electromagnetic materials etc. [1–3]. Hysteresis is a nonlinear phenomenon with dynamics, non-smoothness and multi-valued mapping. The existence of hysteresis usually affects the performance of actuators or sensors made by these smart materials, e.g. dynamic performance and positioning accuracy etc. Accurate modeling of hysteresis in smart material based actuators or smart material based sensors is important for the design of a model based compensator to reduce the effect of hysteresis.

It has been found out that modeling of hysteresis is a challenge due to its features of multi-valued mapping and non-smoothness. Up till now, there have been some models proposed to describe the hysteresis phenomenon such as Preisach [5], Prandtl-Ishlinskii (PI) [6] and Bouc-Wen models [4]. However, both Preisach model and PI model are having the structure with a sum of weighted hysteretic operators and used for description of rate-independent hysteresis. Moreover, they usually employ lots of hysteretic operators even more than hundred for modeling of hysteresis. Additionally, Bouc-Wen model is usually used to describe the hysteresis between restoring force and displacement. It can also be used to describe the rate-dependent hysteretic behavior.

© Springer Nature Switzerland AG 2019
H. Lu et al. (Eds.): ISNN 2019, LNCS 11555, pp. 162–172, 2019.
https://doi.org/10.1007/978-3-030-22808-8_17

However, the identification of a Bouc-Wen model needs to use a method of nonlinear optimization [4] which may encounter the obstacle that the optimization may trap into local minima. Moreover, the parameters needed to be estimated in the Bouc-Wen model are involved in a group of non-smooth components, the gradients of the hysteresis output with respect to the estimated parameters may not exist at non-smooth points of the system. Furthermore, it is noted that the conventional identification methods cannot be utilized to model hysteresis behavior, directly, due to its characteristic of multi-valued mapping.

For modeling of hysteresis, neural networks are potential measures due to their well-known capability of universal approximation. However, the neural networks are applicable only for modeling the systems with one-to-one mapping between the input and output [7, 8]. Those neural networks will be unable to model the hysteresis which has the features of non-smoothness and multi-valued mapping between the input and output. Hence, using neural networks for modeling hysteresis directly becomes a challenge.

For modeling the hysteresis with non-smoothness and multi-valued mapping, Refs. [7, 8] proposed the so-called expanded input space based neural network hysteresis models. In these models, the sigmoidal functions are used as the active functions of neurons. For the simplification of modeling procedure, in this paper, a simple basis function based neural network is proposed for modeling of hysteresis. In this method, an expanded input space is built to transform the multi-valued mapping between input and output of hysteresis to a one-to-one mapping. Then, a basis function based neural network is applied to modeling of hysteresis on the expanded input space. The proposed basis function has a simple structure and can be trained conveniently just by least-square-type algorithm.

For test the performance of the proposed modeling method, the experimental results of modeling for an IPMC sensor are presented.

2 Hysteresis in Smart Material Based Sensors

Ionic polymer-metal composite (IPMC) is a kind of electroactive polymer (EAP) material, which is also called as artificial muscle. IPMC can generate electric signal correlated with its mechanical deformation or displacement. Based on this property, it can be used as a sensor to measure displacement and deformation of flexible mechanism. It is known that IPMC is usually composed of three layers, namely, an ion-exchange polymer membrane sandwiched in between noble metal electrodes. The negatively charged anions, in the polymer membrane, covalently fixed to polymer chains are balanced by positively charged moving cations. When an externally mechanical force to make IPMC deformed, the moving cations will be redistributed. In this case, IPMC will produce a detectable electric signal (e.g. voltage or current) which is associated with the externally mechanical excitation [2, 9]. This phenomenon enables IPMC to be used as a sensor to measure the displacement, vibration, or deformation of a load. Conversely, based on its inverse electroactive feature, IPMC can also be used as an actuator when voltage is implemented on it.

Moreover, IPMC can be used as a sensor to measure mechanical displacement in humid or flexible case for IPMC has the property that generates electrical signal corresponding to its mechanical deformation.

Just like most smart materials, hysteresis is also inherent in IPMC materials. Figure 1 has shown the experimental result having shown that the hysteresis in IPMC sensor has dynamic drift and rate-dependent characteristic. As hysteresis is a nonsmooth function with multi-valued mapping, the conventional neural networks may fail to model it properly due to the conventional neural networks can only be applied to modeling of smooth systems with one-to-one mapping [7, 8].

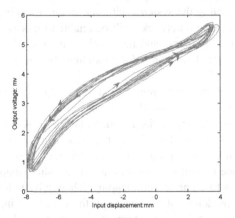

Fig. 1. Rate-dependent behavior of hysteresis with drift in IPMC sensor.

Therefore, in this paper, an important task is to transform the multi-valued mapping of hysteresis to a one-to-one mapping. In order to realize the task to transform the multi-valued mapping between the input and output of hysteresis to a single-valued mapping, an expanded input space will be constructed using the system input and a hysteresis auxiliary function [10]. The hysteretic auxiliary function introduced into the input space is served as an additional coordinate to construct an expanded input space. Actually, the function of the hysteretic auxiliary function is used as an "imagine" of hysteresis that can extract the movement tendency of hysteresis, such as ascending, descending and turning point etc.

3   Expanded Input Space

From the well known Preisach formula to describe hysteresis [10], we have

$$
H[u](k) = \iint_S \mu(\alpha, \beta) \gamma_{\alpha,\beta}[u](k) d\alpha d\beta
$$

$$
= \iint_{S_{+1}} \mu(\alpha, \beta) d\alpha d\beta - \iint_{S_{-1}} \mu(\alpha, \beta) d\alpha d\beta
$$

(1)

where $H[u](k)$ is the output of hysteresis, $\mu(\alpha, \beta)$ is the weighting function and $\gamma_{\alpha,\beta}[u](k)$ is the hysteretic operator. Then, it can be decomposed as

$$
\begin{aligned}
H_+[u](k) &= H_-[u_p] + \Delta H_+[u - u_p], \Delta u > 0 \\
H_-[u](k) &= H_+[u_p] - \Delta H_-[u_p - u], \Delta u < 0
\end{aligned}
\tag{2}
$$

where u_p is the local extreme of the input, $H_+[u_p]$ and $H_-[u_p]$ represent the dominant maximum and minimum of the output of the hysteretic model, while $\Delta H_+[u - u_p]$ and $\Delta H_-[u_p - u]$ are the incremental output of the model as $u(k)$ monotonically increases or decreases from the extreme. It is known that the hysteresis described in (1) has the following feature:

For two different time instants k_1 and k_2, if $u(k_1) = u(k_2)$, $H[u](k_1) \neq H[u](k_2)$ due to the different input extremes. Hence, suppose that an auxiliary function satisfies $f[u](k_1) \neq f[u](k_2)$ can be found. Then, we introduce it to the expanded input space to uniquely determine the output of hysteresis on $(u(k), f[u](k))$.

Definition 1: Define the hysteresis auxiliary function as

$$
f(k) = f(u_{ex}) + |u(k) - u_{ex}| f(|u(k) - u_{ex}|),
\tag{3}
$$

where u_{ex} is the local extreme of the input, and $f(x) \geq 0$ is a smooth and monotonic function not dependent on time k, and $f(u_{ex})$ is the current local extreme of the auxiliary function.

Remark: Based on Definition 1, it is known the procedure to construct an auxiliary function, i.e. (i) selecting a piecewise function with the structure shown as in Eq. (3), which includes local extreme of input and the output of function corresponding to the local extreme of input. (ii) each segment of piecewise function is a smooth and monotonic function not dependent on time k. (iii) the switch of the piecewise function is triggered at the extreme of input.

Then, we have

Definition 2: Define

$$
U = \{u(k)\}
\tag{4}
$$

as the input set. Then, the expanded input space is defined as

$$
E = \{u(k), f(k)\}
\tag{5}
$$

Afterward, we have

Theorem: Suppose $u(k) \in U$ and $f(k)$ is the hysteresis auxiliary function defined by (3) and

$$\Phi = \{H(k)\} \qquad (6)$$

is the output set of hysteresis. If there exists a continuous mapping Γ, such that

$$E| \xrightarrow{\Gamma} \Phi \qquad (7)$$

Then, Γ is a one-to-one mapping.

Proof: Please refer to Appendix.

4   Basis Function Based Neural Network

On the proposed expanded input space, the multi-valued mapping of hysteresis can be transformed to a one-to-one mapping.

Subsequently, a basis function based neural network (BFBNN) on the constructed expanded input space is used to describe the hysteresis characteristic of IPMC sensor. The advantage of using BFBNN is that it has a simpler model structure determined based on the parsimony principle, which does not require a large number of neural basis functions, hysteretic operators, or not rely on empirical skills by comparing with the modeling methods provided by Refs. [10, 11] and [12]. Thus, the corresponding model on expanded input space is defined by

$$y(k) = g[\mathbf{u}(k-1), \mathbf{f}(k-1), \mathbf{y}(k-1)] \qquad (8)$$

where $g(\cdot)$ is the mapping between the input space and the output of system, $\mathbf{u}(k-1) = [u(k-1), \cdots, u(k-n_u)]^T$ and $\mathbf{y}(k-1) = [y(k-1), \cdots, y(k-n_y)]^T$ are the input vector and output vector, respectively; $\mathbf{f}(k-1) = [f(k-1), \cdots, f(k-n_f)]^T$ is the output vector of hysteretic auxiliary function. n_u, n_f and n_y are the lags for sequences $\{u\}$, $\{f\}$ and $\{y\}$, respectively. Then, (8) can also be described by:

$$y(k) = \sum_{i=1}^{p} \theta_i z_i(k), \qquad (9)$$

where θ_i is the ith weighting factor, $z_i(k)$ is the ith basis function of the neural network while p is the number of basis function. Figure 2 illustrates the corresponding structure of the basis function based neural network.

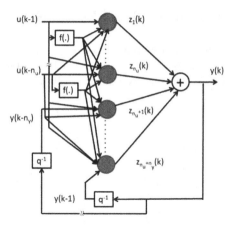

Fig. 2. Structure of basis function based neural network.

The candidates of basis function can be $sin(u)$, $cos(u)$; $1/(1 + exp(-x))$; $exp(-(x - m)^2/\rho^2)$; or $u^i y^j$. The selection of the proper basis function depends upon the specific requirement of application. In this paper, we use $u^i y^j$ as the basis function just for the simple structure of the neural model. Thus, Eq. (8) can be represented by

$$
y(k) = \sum_{i_1=1}^{p} \hat{\theta}_{i_1} z_{i_1}(k) + \sum_{i_1=1}^{p} \sum_{i_2=i_1}^{p} \hat{\theta}_{i_1 i_2} z_{i_1}(k) z_{i_2}(k) + \cdots
$$
$$
+ \sum_{i_1=1}^{p} \cdots \sum_{i_q=i_{q-1}}^{p} \hat{\theta}_{i_1 \cdots i_q} z_{i_1}(k) \cdots z_{i_q}(k) + \hat{\theta}_{n_1+1},
$$
(10)

where $\hat{\theta}_{n_1+1}$ is the estimated bias, $\hat{\boldsymbol{\theta}} = [\hat{\theta}_1, \cdots, \hat{\theta}_{n_1}, \hat{\theta}_{n_1+1}]^T$ is the estimated coefficient vector, $n_1 = (p+q)!/(p!q!) - 1$, and q is the order of model. Define the variable vector as

$$
\mathbf{h}(k) = [z_{i_1}(k), \cdots, z_p(k), z_1(k) \cdot z_1(k), \cdots, z_p(k) \cdot z_p(k),
$$
$$
\cdots, \underbrace{z_1(k) \cdots z_1(k)}_{q}, \ldots, \underbrace{z_p(k) \cdots z_p(k)}_{q}, 1]^T.
$$
(11)

Then, the BFBNN model can be denoted as:

$$
y(k) = \mathbf{h}^T(k)\hat{\boldsymbol{\theta}}(k-1) + e(k),
$$
(12)

where $e(k)$ is the modeling residual. Hence, the problem to model the dynamic drift and hysteresis becomes the training of the BFBNN model.

5 Training of Basis Function Based Neural Network

As the expanded input space based BFBNN has a very simple structure shown as in Eq. (12), the least square algorithm can be used for network training. Then, the corresponding recursive training algorithm is shown as follows:

$$e(k) = y(k) - \mathbf{h}^T(k)\widehat{\boldsymbol{\theta}}(k-1), \tag{13}$$

$$\widehat{\boldsymbol{\theta}}(k) = \widehat{\boldsymbol{\theta}}(k-1) + \mathbf{K}(k)e(k), \tag{14}$$

$$S_e(k) = \lambda(k)\mathbf{h}^T(k)\mathbf{P}(k-1)\mathbf{h}(k) + \mu(k)\widehat{\sum}(k), \tag{15}$$

$$\mathbf{K}(k) = \lambda(k)\mathbf{P}(k-1)\mathbf{h}(k)S_e^{-1}(k), \tag{16}$$

$$\mathbf{P}(k) = (1/\mu(k))[\mathbf{P}(k-1) - \mathbf{K}(k)\mathbf{h}^T(k)\mathbf{P}(k-1)], \tag{17}$$

and

$$\widehat{\sum}(k) = \widehat{\sum}(k-1) + \rho(k)[\hat{\sigma}^2(k) - \widehat{\sum}(k-1)], \tag{18}$$

where $\rho(k) \leq 1$ is the convergence factor, $\mu(k) = \frac{\rho(k-1)}{\rho(k)}[1 - \rho(k)]$ is the forgetting factor, $\hat{\sigma}^2(k) = \hat{\sigma}^2(k-1) + e^2(k)/k$, $\mathbf{P}(0) = \eta\mathbf{I}$, $\mathbf{I}$ is an identity matrix, $0 < \eta < \infty$, and $\lambda(k)$ is a switch coefficient and defined as

$$\lambda(k) = \begin{cases} 0, & |\varepsilon(k)| \leq |r| \\ \frac{\beta(\hat{\sigma}^2(k) - r^2)}{\mathbf{h}^T(k)\mathbf{P}(k-1)\mathbf{h}(k)r^2} & otherwise \end{cases}, \tag{19}$$

where $0 < \beta \leq \mu(k)\widehat{\sum}(k)$.

6 Experimental Results

In this experiment, the sampling frequency is chosen as 1000 Hz. Also, the proposed BFBNN model on expanded input space is applied to the modeling of hysteresis behavior of IPMC sensor. To determine the architecture of the model, the criterion shown in the following is defined to evaluate what structure can properly describe the behavior of the IPMC sensor.

$$C(n) = \frac{1}{n}\sum_{k=1}^{n}(y(k) - \hat{y}(k))^2. \tag{20}$$

Then, $IC(n)$ and $TC(n)$ are used to denote the values of $C(n)$ in training and model validation procedures, respectively. Afterwards, the proposed training algorithm is employed to estimate the coefficients of the model with the corresponding model architecture. Table 1 presents the corresponding values $IC(n)$ and $TC(n)$ of different model structures, so the structure of the proposed model can be selected as $n_y = 0$, $n_u = 9$, $n_f = 9$, when $q = 2$, and in this case, $TC(n)$, the criterion of model validation reaches the smallest value. For comparison, a classical nonlinear auto-regressive and moving-average with exogenous input (NARMAX) is also used to describe the characteristics of the IPMC sensor. The structure parameters of the NARMAX model are chosen as $n_y = 0$, $q = 2$ and $n_u = 9$, respectively.

Table 1. Determination of model structure of IPMC sensor.

$n_y = 0$	$n_u = 7$ $n_f = 7$	$n_u = 8$ $n_f = 8$	$n_u = 9$ $n_f = 9$	$n_u = 10$ $n_f = 10$
$IC(n)$	1.41×10^{-4}	1.01×10^{-5}	1.58×10^{-6}	1.23×10^{-7}
$TC(n)$	1.45×10^{-4}	1.20×10^{-5}	4.68×10^{-6}	1.29×10^{-5}

Figure 3 shows the comparison of model validation between the proposed model and the BFBNN model not on the expanded input space. To show the detail of the model validation results, the data from 0 s to 0.098 s are removed to avoid illustrating the influence of initial values. In Fig. 3(a), the dotted line denotes the output of the proposed model while the dot and dash line represents the output of classical NARMAX model. In Fig. 3(b), the solid line and dotted line denote modeling errors of the proposed model and classical NARMAX model, respectively. Note that the modeling error of the classical NARMAX model illustrates larger fluctuation and the maximum absolute model error of the classical NARMAX model is 0.14 mV, while that of the proposed modeling method is 0.014 mV, respectively. Moreover, Fig. 4 shows the comparison of output versus input curves between the proposed modeling strategy and the classical NARMAX modeling result. Obviously, the proposed model is more suitable to describe the performance of IPMC sensor.

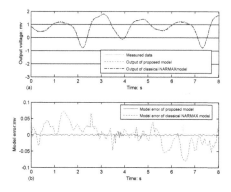

Fig. 3. Comparison of model validation results between the proposed model and the classical NARMAX model.

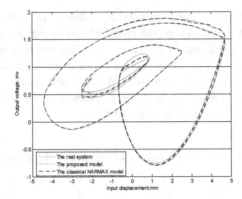

Fig. 4. Comparison of the input-output curves between the proposed model and the classical NARMAX model (the proposed model: dotted line; the classical NARMAX model: dashed line).

7 Conclusions

In this paper, a BFBNN model on the expanded input space to describe the hysteretic and dynamic drifting behavior of IPMC sensor is developed. A method for constructing a proper hysteresis auxiliary function for expanded input space is also presented. Then, a training algorithm is presented to train the BFBNN model of IPMC sensor.

In the modeling experiment, it has shown that the proposed method can achieve better modeling result of hysteresis and dynamic drift in IPMC sensor than a classical NARMAX model.

Acknowledgment. The work presented in this paper has been funded by the National Science Foundation of China under Grants 61671303 and 61571302, the Open Fund of the Key Laboratory of Nano-Devices and Applications, Chinese Academy of Sciences under Grant 18ZS06, the Shanghai Pujiang Program under Grant 18PJ1400100, the Natural Science Foundation of Shanghai under Grant 16ZR1446700, and the project of the Science and Technology Commission of Shanghai under Grant 18070503000.

Appendix

The proof of theorem is as follows:

For any $H(k) \in \Phi$, there, at least, exists a $u(k) \in U$. Thus, $E| \xrightarrow{\Gamma} \Phi$ is a surjective mapping. Figure 5 illustrates the case of surjective mapping of the expanded input space. Now, we prove $E| \xrightarrow{\Gamma} \Phi$ is also a one-to-one mapping.

Case 1: For two different time instants k_1 and k_2, assume $u(k_1) = u(k_2)$ are not extremes, but the output of hysteresis $H(k_1) \neq H(k_2)$ for the effect of the different extremes related to the output of hysteresis. In this case, we can choose a proper $f(k)$

satisfies $f(k_1) \neq f(k_2)$. Then, $E| \xrightarrow{\Gamma} \Phi$ is an injective mapping. To demonstrate the mapping of Case 1, Fig. 6 is presented.

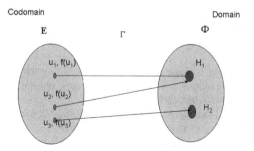

Fig. 5. The surjection of expanded input space.

Case 2: If $u(k_1) = u(k_2)$ and $H(k_1) = H(k_2)$ where $u(k_1)$ and $u(k_2)$ are extremes, based on the characteristic of hysteresis [13], it leads to $H(k_1) = H(k_2)$. In this case, it also results in $f(k_1) = f(k_2)$. In this situation, $E| \xrightarrow{\Gamma} \Phi$ is also an injective mapping. The corresponding description of Case 2 is illustrated in Fig. 7.

In terms of what we discussed, $E| \xrightarrow{\Gamma} \Phi$ is not only a surjective but also an injective mapping. Therefore, it is a bijection, i.e. Γ is a one-to-one mapping.

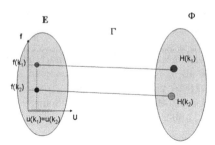

Fig. 6. The injective mapping of expanded input space (Case 1).

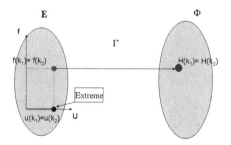

Fig. 7. The injective mapping of expanded input space (Case 2).

References

1. Hu, H.: Compensation of hysteresis in piezoceramic actuators and control of nanopositioning system. Ph.D. thesis of University of Toronto, Canada (2003)
2. Shahinpoor, M., Kim, K.: Ionic polymer-metal composites. Part I. Fundamentals. Smart Mater. Struct. **10**, 819–833 (2001)
3. Mayergoyz, D.: Dynamic Preisach models of hysteresis. IEEE Trans. Magnetics **24**(6), 2925–2927 (1988)
4. Awrejcewicz, J., Dzyubak, L., Lamarque, C.: Modelling of hysteresis using Masing–Bouc-Wen's framework and search of conditions for the chaotic responses. Commun. Nonlinear Sci. Numer. Simul. **13**, 939–958 (2008)
5. Yu, Y., Naganathan, N., Dukkipati, R.: Preisach modeling of hysteresis for piezoceramic actuator system. Mech. Mach. Theory **37**(1), 49–59 (2002)
6. Su, C., Wang, Q., Chen, X.: Adaptive variable structure control of a class of nonlinear systems with unknown Prandtl-Ishlinskii hysteresis. IEEE Trans. Autom. Control **50**(12), 2069–2074 (2005)
7. Dong, R., Tan, Y., Chen, H., Xie, Y.: A neural networks based model for rate-dependent hysteresis for piezoceramic actuators. Sens. Actuators A **143**(2), 370–376 (2008)
8. Deng, L., Tan, Y.: Diagonal recurrent neural network with modified backlash operators for modeling of rate-dependent hysteresis in piezoelectric actuators. Sens. Actuators A **148**(1), 259–270 (2008)
9. Chen, X., Zhu, G., Yang, X., Hung, D., Tan, X.: Model-based estimation of flow characteristics using an ionic polymer-metal composite beam. IEEE/ASME Trans. Mechatron. **18**(3), 932–943 (2013)
10. Zhao, X., Tan, Y.: Neural network based identification of Preisach-type hysteresis in piezoelectric actuator using hysteretic operator. Sens. Actuators A **126**, 306–311 (2006)
11. Nam, D., Ahn, K.: Identification of an ionic polymer metal composite actuator employing Preisach type fuzzy NARX model and Particle Swarm Optimization. Sens. Actuators A **183**, 105–114 (2012)
12. Li, Z., Hao, L.: The identification of discrete Preisach model based on IPMC. In: Proceedings of the 2008 IEEE International Conference on Robotics and Biomimetics, Bangkok, Thailand, vol. 21–26, pp. 751–755 (2009)
13. Mayergoyz, I.D.: Mathematical Models of Hysteresis. Springer, New York (1991). https://doi.org/10.1007/978-1-4612-3028-1

Approximation of Edwards-Anderson Spin-Glass Model Density of States

Magomed Y. Malsagov, Iakov M. Karandashev[(✉)],
and Boris V. Kryzhanovsky

SRISA RAS, Center of Optical Neural Technologies, Moscow, Russia
{malsagov,karandashev,kryzhanov}@niisi.ras.ru

Abstract. We have investigated the density of states of the Edwards-Anderson model and derived an approximation formula which agrees well with the results of numerical experiments. It is important that the formula can well approximate not only the density of states, but also its first and second derivatives, which are most valuable for obtaining the critical parameters of the system. The evaluations can be further used for examining the behavior of 2D Ising models at different temperatures, particularly for tackling Bayesian inference problems and learning algorithms.

Keywords: Edwards-Anderson model · Spin glass · Global state · Density of states

1 Introduction

The finding of the statistical sum is the key goal of statistical physics and informatics. A few conceptual models permit exact solutions [1]. Among all the models the 2D Ising model [2] is most illustrative. Though simple, it proved very important for investigating critical phenomena. Having made much contribution to the development of the spin glass theory, the Edwards-Anderson model [3] and Sherrington-Kirkpatrick model [4] are also worth pointing out. However, the models that allow exact solutions are very few. That is why numerical methods are used to examine the behavior of complex systems. We would like to mark two most advanced methods. First, it is the Monte-Carlo approach [5, 6] allowing us to determine and examine the critical parameters of a system [7, 8]. The capabilities and accuracy of the method are scrutinized in papers [9, 10]. Unfortunately, the method requires a large amount of computation and doesn't permit us to find the free energy directly. Based on the ideas formulated in [11, 12], the second approach has been recently implemented as a fast algorithm [13, 14]. The point of the approach is that the finding of the free energy is reduced to the computation of the determinant of a particular matrix. The algorithm is popular because the free energy is computed to machine accuracy with concurrent determination of ground-state energy E_0 and configuration S_0.

We investigate the distribution of states of a finite spin system. Its interaction matrix is of Edwards-Anderson type and has a quadratic-functional Hamiltonian (1). This sort of functional is often used in machine learning and image processing. Since

© Springer Nature Switzerland AG 2019
H. Lu et al. (Eds.): ISNN 2019, LNCS 11555, pp. 173–179, 2019.
https://doi.org/10.1007/978-3-030-22808-8_18

quantities $s_i = \pm 1$ can indicate the affiliation to either of two pixel classes in a pattern (background/object) or the activity of neurons in a Bayes neural network, the use of the approach in such systems can reveal new algorithms for machine learning.

2 The Density of States

Let us consider a system which is described by Hamiltonian:

$$E = -\frac{1}{2N} \sum_{i \neq j}^{N} J_{ij} s_i s_j \tag{1}$$

This is a set of N Ising spins $s_i = \pm 1$ ($i = 1, 2, \ldots, N$) located at the nodes of a planar grid, its nodes being numbered by index i.

We deal with a planar Edwards-Anderson model (EA model) where spins are positioned at the nodes of a two-dimensional grid. Each spin has non-zero connections with four nearest neighbors; couple interactions J_{ij} are normally distributed ($\bar{J} = 0$, $\sigma_J = 1$)

Our interest is the free energy of the system:

$$f = -\frac{1}{N} \ln Z, \ \ Z = \sum_S e^{-N\beta E(S)}, \tag{2}$$

where statistical sum Z is defined as a sum over all possible configurations S, and $\beta = 1/kT$ is the inverse temperature. The knowledge of the free energy allows us to calculate the key measurable parameters of the system:

$$U = \frac{\partial f}{\partial \beta}, \ \ \sigma^2 = -\frac{\partial^2 f}{\partial \beta^2}, \ \ C = -\beta^2 \frac{\partial^2 f}{\partial \beta^2}, \tag{3}$$

where internal energy $U = \langle E \rangle$ is the ensemble average at a given β, $\sigma^2 = \langle E^2 \rangle - \langle E \rangle^2$ is the energy variance, and $C = \beta^2 \sigma^2$ is the specific heat. Papers [15–17] give thorough description of how quantities (3) depend on the grid dimension and how they change when the matrix with a constant non-zero mean suffers from slight noise. Now we concentrate on the analysis of the density of states of EA model when the interaction matrix elements comply with a normal distribution and have a non-zero mean. Let $D(E)$ be the number of states with energy E. Then the statistical sum can be represented as $Z = \sum_E D(E) e^{-N\beta E}$. Going from summation to integration, we get the following expression which is accurate to an insignificant constant:

$$Z \sim \int_{-\infty}^{\infty} e^{N[\Psi(E) - \beta E]} dE \tag{4}$$

where $\Psi(E) = \ln D(E)/N$. If we evaluate integral with the saddle point method, we get $Z \sim \exp[-Nf(\beta)]$, where

$$f(\beta) = \beta E - \Psi(E), \quad \frac{d\Psi(E)}{dE} = \beta \tag{5}$$

The first of expressions (5) defines the free energy, the second determines E at the saddle point where the derivative of $\Psi(E) - \beta E$ becomes zero.

Papers [13, 14] offer an algorithm for accurate computation of the free energy of planar grids. The algorithm allowed us to compute function $f = f(\beta)$ and its derivatives. That in turn permits us to investigate the energy distribution $D(E) = \exp[N\Psi(E)]$. Indeed, it isn't difficult to derive from Eq. (5) the spectral function:

$$\Psi(E) = \beta E - f(\beta), E = \frac{df}{d\beta} \tag{6}$$

and its derivatives

$$\frac{d\Psi}{dE} = \beta, \quad \frac{d^2\Psi}{dE^2} = \left(\frac{d^2f}{d\beta^2}\right)^{-1} \tag{7}$$

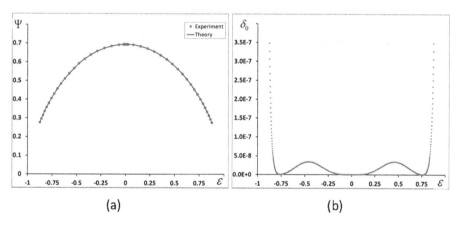

(a) (b)

Fig. 1. (a) The density of states; the solid line represents the analytical expression, marks stand for experimental data (only one in ten points is depicted). (b) The square of the difference between theoretical and experimental data at each point.

Note that $\Psi(E)$ is the entropy up to a constant, and Eq. (6) is the well-known Lagrange transformation which is also applicable to the analysis of state distribution of finite-dimension models [18, 19]. It follows from these equations that when β changes from 0 to ∞, E changes from 0 to E_0, and we have a couple of values of E and $\Psi(E)$ for each β. Thereby we define the form of function $\Psi(E)$ and its derivatives. Figures 1–3 give the plots of function $\Psi(E)$ built around experimental data. In the figures the values plotted as abscissas present parameter $\varepsilon = E/|E_0|$.

3 Approximating the Experimental Data

It is almost impossible to find the density of states of a system because the number of its states grows with dimensionality exponentially. The exact analytical expressions are also impossible to get. In the previous paragraph, using free energy $f = f(\beta)$, we determined density function $\Psi(E)$ that can be computed experimentally.

Examining distribution $\Psi(E)$, we have come to the approximation formula:

$$\Psi(\varepsilon) = \ln 2 - a[(1 - b\varepsilon)\ln(1 - c\varepsilon) + (1 + b\varepsilon)\ln(1 + c\varepsilon)]. \tag{8}$$

where

$$\varepsilon = E/|E_0| \tag{9}$$

Varying free parameters a, b, and c, it is possible to determine their optimal values:

$$a = 0.32, \ b = 1.19, \ c = 0.94. \tag{10}$$

which give the best agreement with experimental data.

Figure 1a shows the agreement between experimental (marks) and formula (8) (solid line). It is seen that (8) describes $\Psi(E)$ excellently. The mean square deviation of approximation (Fig. 1b) is $\langle \delta \rangle \approx 3.4 \cdot 10^{-8}$ throughout the range. The worst agreement is observed at the ends of the distribution because there are few states near the global minimum and they are hard to be discovered experimentally. To visualize the approximation error, the plot of error δ_0 is presented in Fig. 1b. This error is the square of the difference between theoretical (8) and experimental data at each point:

$$\delta_0 = \left[\Psi(E) - \Psi_{\text{exp}}(E)\right]^2 \tag{11}$$

Good agreement with experimental data also results if we take the derivative of expression (8):

$$\frac{\partial \Psi}{\partial E} = -\frac{a}{|E_0|}\left[b\ln\frac{1 + c\varepsilon}{1 - c\varepsilon} + \frac{2c(b - c)\varepsilon}{1 - c^2\varepsilon^2}\right] \tag{12}$$

It is seen from Fig. 2a that expression (12) provides excellent description of the behavior of the first derivative of the density function without having to adjust parameters or add new ones. The approximation error (Fig. 2b) doesn't exceed $2 \cdot 10^{-5}$ within $[-0.85; 0.85]$ interval, and is one to three orders of magnitude smaller in the middle of the distribution.

Figure 2b shows the plot of error δ_1, which is the squared difference between theoretical (12) and experimental data at each point:

$$\delta_1 = \left[\frac{d\Psi}{dE} - \frac{d\Psi_{\text{exp}}}{dE}\right]^2 \tag{13}$$

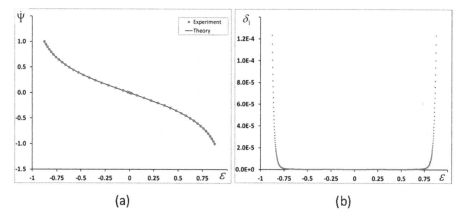

(a)   (b)

Fig. 2. (a) the first derivative of the density function; the solid line indicates the theory, marks stand for experimental data (only one in ten points is shown). (b) the squared difference between theoretical (12) and experimental data at each point.

A similar result comes out with the second derivative:

$$\frac{\partial^2 \Psi}{\partial E^2} = -\frac{2ac}{E_0^2} \cdot \frac{2b - c - c^3 \varepsilon^2}{\left(1 - c^2 \varepsilon^2\right)^2}. \tag{14}$$

In Fig. 3b the effect of deficiency of experimental data near the global minimum is quite noticeable. Expression (13) provides good agreement in the middle portion of the distribution, disagreement growing rapidly at the ends of the distribution.

Figure 3b presents the plot of error δ_2, which is the squared difference between theoretical (14 and experimental data at each point:

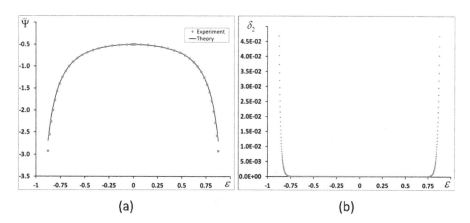

(a)   (b)

Fig. 3. (a) the second derivative of the density function: the solid line indicates the theory, marks stand for experimental data (only one in ten points is shown). (b) the squared difference between theoretical (12) and experimental data at each point.

$$\delta_2 = \left[\frac{d^2\Psi}{dE^2} - \frac{d^2\Psi_{exp}}{dE^2} \right]^2 \tag{15}$$

Nevertheless, expressions (8) and (9) are excellent to describe not only the density function itself (6), but also its first and second derivatives (7). It is important that experiments showed that with given normalization of Hamiltonian (1) the change of dimensionality doesn't change the form of the density function and the corresponding plots coalesce. For this reason expression (8) can be used to describe the EA model on a planar grid of any dimensionality.

4   Conclusion

Having used the experimental data of accurate computations of the free energy, we could build the function of the density of states for the EA model on a planar grid. We obtained the simple expression (8) that defines the density function. More importantly, the approximation proved so good that it allowed us to derive the approximations for the first and second derivatives of the density function without having to use additional adjustments. The latter fact is very important: research [21] shows that even a slight deviation of the density of states can bring about radical changes of thermodynamical properties of the system. For instance, the relative difference between the densities of states of two-dimensional models of Ising and Bethe lattice is no greater than 1%. However, the two models exhibit widely different properties: in the Ising model the heat capacity has a divergence at the critical point, and in the Bethe model it demonstrates a finite jump.

In the future the findings of [20] may allow us to connect the free parameters of the approximation to the statistical characteristics of the grid-element distribution. We hope that the simple approximation formulae we have obtained will enable us to develop new algorithms for computer-aided image processing and tackle other planar neural-net-based Bayes-inference problems.

References

1. Baxter, R.J.: Exactly Solved Models in Statistical Mechanics. Academic Press, London (1982)
2. Onsager, L.: Crystal Statistics. I. A two-dimensional model with an order–disorder transition. Phys. Rev. **65**, 117–149 (1944)
3. Edwards, S.F., Anderson, P.W.: Theory of spin glasses. J. Phys. F: Metal Phys. **5**, 965–974 (1975)
4. Sherrington, D., Kirkpatrick, P.: Solvable model of a spin-glass. Phys. Rev. Lett. **35**(26), 1792–1796 (1975)
5. Metropolis, N., Ulam, S.: The Monte Carlo Method. J. of Am. Stat. Assoc. **44**(247), 335–341 (1949)
6. Fishman, G.S.: Monte Carlo: Concepts, Algorithms, and Applications. Springer Series in Operations Research and Financial Engineering, 1st edn. Springer, New York (1996). https://doi.org/10.1007/978-1-4757-2553-7

7. Bielajew, A.F.: Fundamentals of the Monte Carlo method for neutral and charged particle transport. The University of Michigan, Dep. of Nuclear Engineering and Radiological Sciences (2001)

8. Foulkes, W.M.C., Mitas, L., Needs, R.J., Rajagopal, G.: Quantum Monte Carlo simulations of solids. Rev. Mod. Phys. **73**(1), 33–83 (2001)

9. Binder, K.: Finite Size scaling analysis of Iising model block distribution functions. Z. Phys. B **43**, 119–140 (1981)

10. Binder, K., Luijten, E.: Monte Carlo tests of renormalization-group predictions for critical phenomena in Ising models. Phys. Rep. **344**, 179–253 (2001)

11. Kasteleyn, P.: Dimer statistics and phase transitions. J. Math. Phys. **4**(2), 287–293 (1963)

12. Fisher, M.: On the dimer solution of planar Ising models. J. Math. Phys. **7**(10), 1776–1781 (1966)

13. Karandashev, Ya.M., Malsagov, M.Yu.: Polynomial algorithm for exact calculation of partition function for binary spin model on planar graphs. Opt. Mem. Neural Networks (Inf. Opt.) **26**(2), 87–95 (2017)

14. Schraudolph, N., Kamenetsky, D.: Efficient exact inference in planar Ising models. In: NIPS (2008). https://arxiv.org/abs/0810.4401

15. Kryzhanovsky, B.V., Malsagov, M.Yu., Karandashev, I.M.: Investigation of finite-size 2D Ising model with a noisy matrix of spin-spin interactions. Entropy **20**(8), 585 (2018)

16. Kryzhanovsky, B.V., Karandashev, I.M., Malsagov, M.Yu.: Dependence of critical temperature on dispersion of connections in 2D grid. In: Huang, T., Lv, J., Sun, C., Tuzikov, Alexander V. (eds.) ISNN 2018. LNCS, vol. 10878, pp. 695–702. Springer, Cham (2018). https://doi.org/10.1007/978-3-319-92537-0_79

17. Karandashev, I.M., Kryzhanovsky, B.V., Malsagov, M.Yu.: Spectral characteristics of a finite 2D Ising model. Opt. Mem. Neural Networks (Inf. Opt.) **27**(3), 147–151 (2018)

18. Häggkvist, R., Rosengren, A., Andrén, D., Kundrotas, P., Lundow, P.H., Markström, K.: Computation of the Ising partition function for 2-dimensional square grids. Phys. Rev. E **69**, 046104 (2004)

19. Beale, P.D.: Exact distribution of energies in the two-dimensional Ising model. Phys. Rev. Lett. **76**, 78–81 (1996)

20. Litinskii, L., Kryzhanovsky, B.: Spectral density and calculation of free energy. Physica A: Stat. Mech. Appl. **510**, 702–712 (2018)

21. Kryzhanovsky, B.V.: Features of the Spectral Density of a Spin System. Dokl. Math. **97**(2), 188–192 (2018)

Generalization of Linked Canonical Polyadic Tensor Decomposition for Group Analysis

Xiulin Wang[1,2], Chi Zhang[1], Tapani Ristaniemi[2], and Fengyu Cong[1,2(✉)]

[1] School of Biomedical Engineering, Faculty of Electronic Information and Electrical Engineering, Dalian University of Technology, Dalian 116024, China
xiulin.wang@foxmail.com, {chizhang,cong}@dlut.edu.cn
[2] Faculty of Information Technology, University of Jyväskylä, 40100 Jyväskylä, Finland
tapani.e.ristaniemi@jyu.fi

Abstract. Real-world data are often linked with each other since they share some common characteristics. The mutual linking can be seen as a core driving force of group analysis. This study proposes a generalized linked canonical polyadic tensor decomposition (GLCPTD) model that is well suited to exploiting the linking nature in multi-block tensor analysis. To address GLCPTD model, an efficient algorithm based on hierarchical alternating least squa res (HALS) method is proposed, termed as GLCPTD-HALS algorithm. The proposed algorithm enables the simultaneous extraction of common components, individual components and core tensors from tensor blocks. Simulation experiments of synthetic EEG data analysis and image reconstruction and denoising were conducted to demonstrate the superior performance of the proposed generalized model and its realization.

Keywords: Linked tensor decomposition ·
Hierarchical alternating least squares · Canonical polyadic ·
Simultaneous extraction

1 Introduction

Linked tensor decomposition (LTD) is an emerging technique for group analysis in recent years, specially designed for simultaneous analysis of multi-block tensor data. It has been successfully applied in the fields of neuroscience [1], multi-dimensional harmonic retrieval [2], array signal processing [3] and metabolic physiology [4].

Linked tensor decomposition can be seen as an extension of tensor decomposition applied to single-block tensor [5–7] in multi-block data analysis, e.g., analysis of electrophysiological (EEG) data collected from different subjects under a certain stimulus, which can be naturally linked together for sharing the similar brain activities [1]. LTD method can take full advantage of such linking/coupling

© Springer Nature Switzerland AG 2019
H. Lu et al. (Eds.): ISNN 2019, LNCS 11555, pp. 180–189, 2019.
https://doi.org/10.1007/978-3-030-22808-8_19

information among data blocks to improve the decomposition identifiability [3]. In addition, LTD method has its advantage in imposing constraints on particular modes or components compared to its matrix counterpart [9,10]. Any combination of constraints including independence, sparsity, smoothness and non-negativity can be added more easily and flexibly [11]. Moreover, imposing specific constraints on different modes or components would contribute to obtaining more reasonable decomposition solutions with convincing interpretations [6,8,11]. For example, the constraint of non-negativity is applied in the processing of ERP data with time-frequency representation [6]. Furthermore, tensor decomposition is superior to two-way matrix factorization such as solution uniqueness and component identification in some cases [12]. To unfold some of the modes in matrix factorization will inevitably loss the potential interactions under the multiway structure [13]. Therefore, it is reliable to take the high-order characteristics of tensors into consideration in data analysis.

With the LTD model, simultaneous extraction of common components, individual components and core tensors can be obtained. The notion 'linked' is based on the assumption that different data blocks share the same or highly correlated components while retaining individual information [14]. In group data analysis, e.g. face images collected from different subjects with the same expression [14], or EEG data collected from different participants under the same stimulus [8], all subjects may share the similar or even identical information, which can be regarded as linking factors among tensors. However, individual characteristics will exist in particular subjects at the same time, which may lead to inconsistent number of components for tensors. Obviously, this inconsistency does not match the linked canonical polyadic tensor decomposition (LCPTD) model in [14]. Therefore, this study aims to develop a more generalized and flexible model with inconsistent component number for linked tensor decomposition. To obtain the solution of the new model, we propose a generalized linked canonical polyadic tensor decomposition algorithm based on HALS strategy [7], which is termed as GLCPTD-HALS algorithm. The experiment results show that the generalized model is more practical in multi-block data analysis, and its realization can achieve better performance.

This paper is organized as follows. Section 2 introduces LCPTD model and its generalization. In Sect. 3, GLCPTD-HALS algorithm is proposed. In Sect. 4, simulation experiments are conducted to verify the performance of proposed algorithm. The last section summarizes this paper.

2   Problem Formulation

In this section, we mainly introduce the linked canonical polyadic tensor decomposition (LCPTD) model [14] and its generalization. CP model [15] is also called parallel factor analysis (PARAFAC) [16] and canonical composition (CANDE-COMP) [17]. CP decomposition (CPD) can decompose a tensor into a minimal number of rank-1 tensors, and the minimum number R is termed as the rank of a tensor. It can achieve good unique identification under some mild conditions without any special constraints. Please refer to [18] for a detailed description of standard notations and basic tensor operations.

2.1   Review of LCPTD Model

To deal with multi-block tensors with coupling information, researchers in [14] proposed a model of simultaneous decomposition, namely LCPTD model, which is defined as follows:

$$
\underline{\boldsymbol{X}}^{(s)} \approx \hat{\underline{\boldsymbol{X}}}^{(s)} = \sum_{r=1}^{R} \lambda_r^{(s)} \boldsymbol{u}_r^{(1,s)} \circ \boldsymbol{u}_r^{(2,s)} \circ \cdots \circ \boldsymbol{u}_r^{(N,s)}
$$

$$
= \left[\!\left[\underline{\boldsymbol{G}}^{(s)}; \boldsymbol{U}^{(1,s)}, \boldsymbol{U}^{(2,s)}, \cdots, \boldsymbol{U}^{(N,s)} \right]\!\right], \tag{1}
$$

where $\underline{\boldsymbol{X}}^{(s)} \in \Re^{I_1 \times I_2 \times \cdots I_N}$ and $\hat{\underline{\boldsymbol{X}}}^{(s)} \in \Re^{I_1 \times I_2 \times \cdots I_N}$ denote the original and estimated tensors, respectively. $\boldsymbol{U}^{(n,s)} = \left[\boldsymbol{u}_1^{(n,s)}, \boldsymbol{u}_2^{(n,s)}, \cdots, \boldsymbol{u}_R^{(n,s)} \right] \in \Re^{I_n \times R}$ denotes the n-mode factor matrix of sth tensor. S, R, N are denoted as the number, rank and order of tensors, respectively. $\underline{\boldsymbol{G}}^{(s)} \in \Re^{R \times R \times \cdots R}$ denotes the sth core tensor with non-zero entries only on the super-diagonal. $\lambda_r^{(s)}$ is the $(r, r, ..., r)$th element of $\underline{\boldsymbol{G}}^{(s)}$. The LCPTD model assumes that each factor matrix $\boldsymbol{U}^{(n,s)} = \left[\boldsymbol{U}_C^{(n)} \ \boldsymbol{U}_I^{(n,s)} \right] \in \Re^{I_n \times R}$ consists of two parts: $\boldsymbol{U}_C^{(n)} \in \Re^{I_n \times L_n}$, $0 \leq L_n \leq R$ and $\boldsymbol{U}_I^{(n,s)} \in \Re^{I_n \times (R-L_n)}$. The former shared by all tensor blocks represents the coupling (same or highly correlated) information, whereas the latter corresponds to the individual characteristics of each tensor block.

2.2   Generalization of LCPTD Model

Even though multiple data blocks are collected under the same condition, individual characteristics will exist in the particular blocks due to the individual differences. These characteristics may lead to inconsistent number of components for tensors. Obviously, this inconsistency does not match the LCPTD model. Therefore, we extend the LCPTD model to the generalized case with different component number $R^{(s)}$, termed as GLCPTD, which is defined as:

$$
\underline{\boldsymbol{X}}^{(s)} \approx \hat{\underline{\boldsymbol{X}}}^{(s)} = \sum_{r=1}^{R^{(s)}} \lambda_r^{(s)} \boldsymbol{u}_r^{(1,s)} \circ \boldsymbol{u}_r^{(2,s)} \circ \cdots \circ \boldsymbol{u}_r^{(N,s)}
$$

$$
= \left[\!\left[\underline{\boldsymbol{G}}^{(s)}; \boldsymbol{U}^{(1,s)}, \boldsymbol{U}^{(2,s)}, \cdots, \boldsymbol{U}^{(N,s)} \right]\!\right]. \tag{2}
$$

The generalized LCPTD model still assumes that each factor matrix $\boldsymbol{U}^{(n,s)} = \left[\boldsymbol{U}_C^{(n)} \ \boldsymbol{U}_I^{(n,s)} \right] \in \Re^{I_n \times R^{(s)}}$ consists of two parts: $\boldsymbol{U}_C^{(n)} \in \Re^{I_n \times L_n}$, $0 \leq L_n \leq min (R^{(s)})$ and $\boldsymbol{U}_I^{(n,s)} \in \Re^{I_n \times (R^{(s)}-L_n)}$, representing the same meanings with LCPTD model. $\underline{\boldsymbol{G}}^{(s)} \in \Re^{R^{(s)} \times R^{(s)} \times \cdots R^{(s)}}$ denotes the sth core tensor.

Figure 1 illustrates the conceptual model of dual-linked tensor decomposition based on CP model (all tensors are linked together by the common parts $\boldsymbol{U}_C^{(1)}$ and $\boldsymbol{U}_C^{(2)}$).

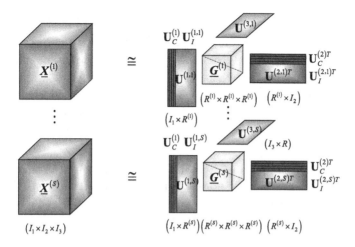

Fig. 1. Conceptual illustration of GLCPTD model with dual-linked parts [11]

3   Realization of GLCPTD Model

In this section, we aim to provide a solution of how to solve the above-mentioned GLCPTD model through HALS strategy [7]. The optimization criterion of squared Euclidean distance minimization is utilized to minimize the error between the original and estimated tensors. Therefore, the cost function can be expressed as:

$$min \sum_{s=1}^{S} \left\| \underline{X}^{(s)} - \sum_{r=1}^{R^{(s)}} \lambda_r^{(s)} u_r^{(1,s)} \circ u_r^{(2,s)} \circ \cdots \circ u_r^{(N,s)} \right\|_F^2 \qquad (3)$$

$$s.t. u_r^{(n,1)} = \cdots = u_r^{(n,S)}, \; r \le L_n,$$

$$\left\| u_r^{(n,s)} \right\| = 1, n = 1 \cdots N, r = 1 \cdots R^{(s)}, s = 1 \cdots S.$$

The above minimized optimization problem can be transformed into $max(R^{(s)})$ sub-problems via HALS strategy, which can be solved sequentially and iteratively as follows:

$$D_F^{(r)}(\lambda_r^{(s)}, u_r^{(n,s)}) = \sum_{s=1, r \le R^{(s)}}^{S} \left\| \underline{Y}_r^{(s)} - \lambda_r^{(s)} u_r^{(1,s)} \circ u_r^{(2,s)} \circ \cdots \circ u_r^{(N,s)} \right\|_F^2, \qquad (4)$$

where $\underline{Y}_r^{(s)} \doteq \underline{X}^{(s)} - \sum_{k \ne r}^{R^{(s)}} \lambda_k^{(s)} u_k^{(1,s)} \circ u_k^{(2,s)} \circ \cdots \circ u_k^{(N,s)}$. For the solution of $u_r^{(n,s)}$, we only set the derivative of $D_F^{(r)}(\lambda_r^{(s)}, u_r^{(n,s)})$ with respect to $u_r^{(n,s)}$ to zero. The learning rule of $u_r^{(n,s)}$ can be formulated as:

$$u_r^{(n,s)} = \begin{cases} \sum_s \left(\underline{Y}_{r,(n)}^{(s)} \lambda_r^{(s)} \{u_r^{(s)}\}^{\odot -n} \right) / \sum_s \left(\lambda_r^{(s)T} \lambda_r^{(s)} \right), & r \le L_n, \\ \underline{Y}_{r,(n)}^{(s)} \{u_r^{(s)}\}^{\odot -n} / \lambda_r^{(s)T}, & r > L_n, \end{cases} \qquad (5)$$

where $\underline{Y}_{r,(n)}^{(s)}$ is the mode-n matricization of $\underline{Y}_r^{(s)}$. $\{u_r^{(s)}\}^{\odot -n} = u_r^{(N,s)} \odot \cdots \odot$ $u_r^{(n+1,s)} \odot u_r^{(n-1,s)} \odot \cdots \odot u_r^{(1,s)}$ and '$\odot$' denotes the Khatri-Rao product. If $r \leq$ L_n, $u_r^{(n,s)}$ will be calculated by combining all tensor information and assigned to each s. Otherwise, it needs to be calculated separately. $u_r^{(n,s)}$ needs to be normalized to unit variance by $u_r^{(n,s)} \leftarrow u_r^{(n,s)}/\|u_r^{(n,s)}\|_2$ in each iteration. After N iterations of $u_r^{(n,s)}$, the $(r,r,...,r)$th element $\lambda_r^{(s)}$ of core tensors is updated as follows:

$$\lambda_r^{(s)} \leftarrow \underline{Y}_r^{(s)} \times_1 u_r^{(1,s)} \times_2 u_r^{(2,s)} \cdots \times_N u_r^{(N,s)}. \tag{6}$$

These $max(R^{(s)})$ stages are alternatively updated one after another until convergence. In order to impose non-negativity, a simple "half-rectifying" nonlinear projection is applied as $u_r^{(n,s)} \leftarrow \|u_r^{(n,s)}\|_+$ or $\lambda_r^{(s)} \leftarrow \|\lambda_r^{(s)}\|_+$ after (5) and (6). We summarize the extended GLCPTD-HALS algorithm in Algorithm 1.

Algorithm 1: GLCPTD-HALS algorithm

Input: $\underline{X}^{(s)}$, L_n and $R^{(s)}$, $n = 1, \cdots, N$, $s = 1, \cdots, S$
Initialization: $\underline{G}^{(s)}$, $U^{(n,s)}$, $u_r^{(n,s)} \leftarrow u_r^{(n,s)}/\|u_r^{(n,s)}\|_2$
$\underline{E}^{(s)} = \underline{X}^{(s)} - \sum_r^{R^{(s)}} \lambda_r^{(s)} u_r^{(1,s)} \circ u_r^{(2,s)} \circ \cdots \circ u_r^{(N,s)}$
while *not convergence* **do**
    for $r = 1, 2, \cdots, max(R^{(s)})$ **do**
        $\underline{Y}^{(s)} = \underline{E}^{(s)} + \lambda_r^{(s)} u_r^{(1,s)} \circ u_r^{(2,s)} \circ \cdots \circ u_r^{(N,s)}$, $r \leq R^{(s)}$, $s = 1, 2, \cdots, S$
        for $n = 1, 2, \cdots, N$ **do**
            update $u_r^{(n,s)}$, $r \leq R^{(s)}$, $s = 1, 2, \cdots, S$ via equation (5)
        end
        update $\lambda_r^{(s)}$, $s = 1, 2, \cdots, S$ via equation (6)
        $\underline{E}^{(s)} = \underline{Y}^{(s)} - \lambda_r^{(s)} u_r^{(1,s)} \circ u_r^{(2,s)} \circ \cdots \circ u_r^{(N,s)}$, $r \leq R^{(s)}$, $s = 1, 2, \cdots, S$
    end
end
Output: $\underline{G}^{(s)}$, $U^{(n,s)}$, $n = 1, ..., N$, $s = 1, ..., S$

---

4   Simulation Results

4.1   Synthetic EEG Data Analysis

In this part, we synthetically generate three types of factor matrices based on brain activities, respectively presenting topography, waveform and power spectrum, as shown in the Fig. 2(a)–(c). Through the back projection of factor matrices, four tensor blocks representing four subjects are constructed with the SNR of 10 dB, as shown in Fig. 2(d). SNR refers to the signal-to-noise ratio, which is defined as SNR $= 10\log_{10}(\sigma_s/\sigma_n)$. σ_s and σ_n denote the levels of signal and noise, respectively. To prove the usefulness of GLCPTD model, we set the number of components for four tensors as $\{4, 4, 3, 3\}$. Furthermore, factor matrices of topography and power spectrum consist of two common bases and one or two individual bases ($L_1 = L_2 = 2$), while the components of waveform are completely individual ($L_3 = 0$). The common bases represent that the occipital region in the mid-line and left-hemisphere of four subjects are activated with the alpha oscillations (8~13 Hz).

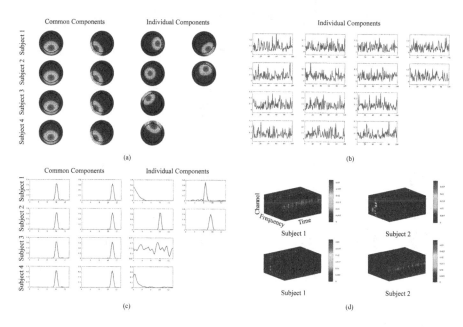

Fig. 2. Illustration of factor matrices of (a) topography, (b) waveform and (c) power spectrum and (d) tensors (frequency × time × channel) for four subjects. Factor matrices of topography and power spectrum for each subject consist of two common components and one/two individual components, while temporal components are individual for each subject.

We apply LCPTD-HALS [14], GLCPTD-HALS, and NTF-HALS [7] algorithms with nonnegative constraint to analyze the four tensor blocks. Solutions of topography learned by these algorithms are shown in Fig. 3(a)–(d). We can see that, GLCPTD-HALS and NTF-HALS algorithms can successfully extract the common components as well as individual components. The difference is that the components learned by NTF-HALS algorithm are disordered. Clustering and other post-ordering methods need to be applied to obtain the common bases. Although LCPTD-HALS algorithm can extract all the common components, only 3 components are extracted from subject 1 or subject 2 shown in Fig. 3(c) and 4 components are recovered from subject 3 or subject 4 shown in Fig. 3(d). The former causes potential components to be omitted (subject 1) or merged (subject 2). The latter depends on the magnitude of the particular component being redistributed in a certain way driven by algorithm (e.g. the 1st component of subject 3 in Fig. 2(a) is recovered to the 1st and 4th components in Fig. 3(d) corresponding to that its magnitude is divided into two parts from predefined 1), which makes group analysis more complicated especially when the number of components increases.

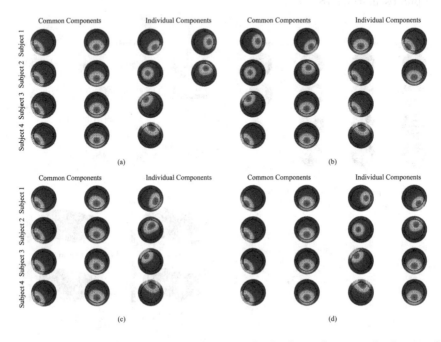

Fig. 3. Illustration of factor matrices of topography for four subjects under four conditions. (a)–(d) show the components learned by GLCPTD-HALS, NTF-HALS, LCPTD-HALS ($R = 3$) and LCPTD-HALS ($R = 4$) algorithms, respectively.

4.2   Image Reconstruction and Denoising

In this part, to examine and demonstrate the performance of the proposed algorithm, we apply the LCPTD and GLCPTD models to image reconstruction and denoising. There are 165 gray-scale images from 15 individuals in the Yale face database. Each individual has 11 images of different face expressions ('centerlight', 'glasses', 'happy', 'leftlight', 'noglasses', 'normal', 'rightlight', 'sad', 'sleepy', 'surprised', 'wink'), and the size of each image is 215×171 pixels. We construct the multi-block tensors by stacking corresponding face images under two conditions: (1) Face images from the same subject with different expressions, $I_1 = 215$, $I_2 = 171$, $I_3 = 11$, $S = 15$; (2) Face images from different subjects with the same expression, $I_1 = 215$, $I_2 = 171$, $I_3 = 15$, $S = 11$. Furthermore, 5% salt-and-pepper noises are added to all face images. We use the peak-signal-to-noise ratio (PSNR) to measure the quality of reconstructed images.

In terms of the number of components in each tensor, we set R to 40 in the LCPTD model, which is consistent with the original parameter in [14]. Differently, in the GLCPTD model, we use the following method to calculate it: we concatenate each tensor along the first mode to generate a matrix, and perform principle component analysis (PCA) on the matrices successively; when the percentage of the total variance explained by each principle component is greater than 99.6%, the number of corresponding principle components is chosen as the number of components. In this experiment, we assume that the coupling information exists in two modes of images so that we set the number of coupled components to $L_1 = L_2$, $L_3 = 0$, and the values of $L_{1,2}$ are changed in $\{10, 20, 30\}$.

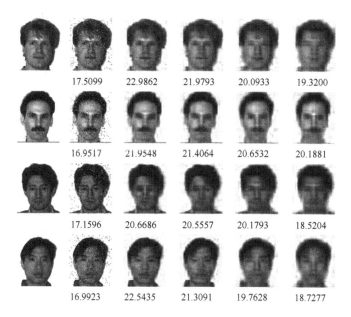

| | 17.5099 | 22.9862 | 21.9793 | 20.0933 | 19.3200 |

| | 16.9517 | 21.9548 | 21.4064 | 20.6532 | 20.1881 |

| | 17.1596 | 20.6686 | 20.5557 | 20.1793 | 18.5204 |

| | 16.9923 | 22.5435 | 21.3091 | 19.7628 | 18.7277 |

Fig. 4. Original, noisy and reconstructed face images of 'centerlight' from four subjects with PSNRs (dB). 1st column: original images, 2nd column: noisy images, 3rd column: GLCPTD model of condition I, 4th column: LCPTD model of condition I, 5th column: GLCPTD model of condition II, 6th column: LCPTD model of condition II.

By performing the LCPTD-HALS [14] and GLCPTD-HALS algorithms with nonnegative constraint on the above two models, we can compute the PSNRs of reconstructed images. Figure 4 depict the original, noisy and reconstructed face images from subject 1–4 with the same expression of 'centerlight' ($I_{1,2} = 10$). We can see that the images reconstructed by LPCTD model/condition II are more fuzzier or distorted than those from GLCPTD model/condition I. Table 1 shows the averaged PSNRs of the reconstructed images under the two conditions. The PSNRs obtained by GLCPTD model are higher than those obtained by the LCPTD model in both conditions. It can be considered that the proposed

GLCPTD model matches the real-world data more closely, which may make it more practical in real-world data analysis. The PSNRs obtained under condition I are higher than those under condition II, which means that it is more reliable to stack face images from the same subject with different expressions together. It seems that if the number of common components becomes larger, the PSNRs become smaller. The excessive number of common components may affect the fitness of the estimated tensors. However, the selection of parameter L_n is still an open issue in the current study, which will be one of our future works.

Table 1. Averaged PSNRs (dB) of reconstructed images

	Condition I			Condition II		
	$L_1, L_2 = 10$	20	30	$L_1, L_2 = 10$	20	30
LCPTD	21.3651	20.7517	19.9021	19.0809	18.8694	18.5321
GLPCTD	22.0421	21.5476	20.8134	19.9649	19.6537	19.4444

5   Conclusion

The main objective of this paper is to develop a generalized and flexible model of linked tensor decomposition which is more suitable for group analysis. We proposed the generalized LCPTD model as well as its realization, in which the common components, individual components and core tensors can be extracted simultaneously. Experiments of synthetic EEG data analysis and image reconstruction and denoising were conducted to compare the performance of proposed algorithm with LCPTD-HALS and NTF-HALS algorithms. The results illustrated the superior performance of the newly generalized model and its realization.

Acknowledgments. This work was supported by the National Natural Science Foundation of China (Grant No. 81471742), the Fundamental Research Funds for the Central Universities [DUT16JJ(G)03] in Dalian University of Technology in China, and the scholarships from China scholarship Council (No. 201706060262).

References

1. Zhou, G.-X., Zhao, Q.-B., Zhang, Y., et al.: Linked component analysis from matrices to high-order tensors: applications to biomedical data. Proc. IEEE. **104**(2), 310–331 (2016). https://doi.org/10.1109/JPROC.2015.2474704
2. Sorensen, M., De Lathauwer, L.: Multidimensional harmonic retrieval via coupled canonical polyadic decomposition – part II: algorithm and multirate sampling. IEEE Trans. Signal Process. **65**(2), 528–539 (2017). https://doi.org/10.1109/TSP.2016.2614797

3. Gong, X.-F., Lin, Q.-H., Cong, F.-Y., De Lathauwer, L.: Double coupled canonical polyadic decomposition for joint blind source separation. IEEE Trans. Signal Process. **66**(13), 3475–3490 (2016). https://doi.org/10.1109/TSP.2018.2830317
4. Acar, E., Bro, R., Smilde, A.-K.: Data fusion in metabolomics using coupled matrix and tensor factorizations. Proc. IEEE. **103**(9), 1602–1620 (2015). https://doi.org/10.1109/JPROC.2015.2438719
5. Zhou, G.-X., Cichocki, A., Xie, S.-L.: Fast nonnegative matrix/tensor factorization based on low-rank approximation. IEEE Trans. Signal Process. **60**(6), 2928–2940 (2012). https://doi.org/10.1109/TSP.2012.2190410
6. Cong, F.-Y., Zhou, G.-X., Cichocki, A., et al.: Low-rank approximation based nonnegative multi-way array decomposition on event-related potentials. Int. J. Neural Syst. **24**(8), 1440005 (2014). https://doi.org/10.1142/S012906571440005X
7. Cichocki, A., Zdunek, R., Amari, S.: Hierarchical ALS algorithms for nonnegative matrix and 3D tensor factorization. In: Davies, M.E., James, C.J., Abdallah, S.A., Plumbley, M.D. (eds.) ICA 2007. LNCS, vol. 4666, pp. 169–176. Springer, Heidelberg (2007). https://doi.org/10.1007/978-3-540-74494-8_22
8. Cong, F.-Y., Phan, A.-H., Zhao, Q.-B., et al.: Analysis of ongoing EEG elicited by natural music stimuli using nonnegative tensor factorization. In: 20th European Signal Processing Conference, pp. 494–498. Elsevier, Bucharest (2012)
9. Calhoun, V.-D., Liu, J., Adali, T.: A review of group ICA for fMRI data and ICA for joint inference of imaging, genetic, and ERP data. Neuroimage **45**(1), 163–172 (2009). https://doi.org/10.1016/j.neuroimage.2008.10.057
10. Gong, X. F., Wang, X.-L., Lin, Q.-H.: Generalized non-orthogonal joint diagonalization with LU decomposition and successive rotations. IEEE Trans. Signal Process. **63**(5), 1322–1334 (2015). https://doi.org/10.1109/TSP.2015.2391074
11. Cichocki, A.: Tensor decompositions: a new concept in brain data analysis. arXiv Prepr. arXiv1305.0395 (2013)
12. Mørup, M.: Applications of tensor (multiway array) factorizations and decompositions in data mining. Wiley Interdisc. Rev.: Data Min. Knowl. Disc. **1**(1), 24–40 (2011) https://doi.org/10.1002/widm.1
13. Cong, F.-Y., Lin, Q.-H., Kuang, L.-D., et al.: Tensor decomposition of EEG signals: a brief review. J. Neurosci. Methods **248**, 59–69 (2015). https://doi.org/10.1016/j.jneumeth.2015.03.018
14. Yokota, T., Cichocki, A., Yamashita, Y.: Linked PARAFAC/CP tensor decomposition and its fast implementation for multi-block tensor analysis. In: Huang, T., Zeng, Z., Li, C., Leung, C.S. (eds.) ICONIP 2012. LNCS, vol. 7665, pp. 84–91. Springer, Heidelberg (2012). https://doi.org/10.1007/978-3-642-34487-9_11
15. Hitchcock, F.-L.: The expression of a tensor or a polyadic as a sum of products. J. Math. Phys. **6**(1–4), 164–189 (1927). https://doi.org/10.1002/sapm192761164
16. Harshman, R.-A.: Foundations of the PARAFAC procedure: models and conditions for an 'explanatory' multimodal factor analysis. UCLA Work. Pap. Phonetics. **16**, 1–84 (1970)
17. Carroll, J.-D., Chang, J.-J.: Analysis of individual differences in multidimensional scaling via an n-way generalization of 'Eckart-Young' decomposition. Psychometrika **35**(3), 283–319 (1970). https://doi.org/10.1007/BF02310791
18. Kolda, T.-G., Bader, B.-W.: Tensor decompositions and applications. SIAM Rev. **51**(3), 455–500 (2008). https://doi.org/10.1137/07070111X

Intelligent and Distributed Data Warehouse for Student's Academic Performance Analysis

Jesús Silva[1]($\boxtimes$), Lissette Hernández[2], Noel Varela[2],
Omar Bonerge Pineda Lezama[3], Jorge Tafur Cabrera[4],
Bellanith Ruth Lucena León Castro[4], Osman Redondo Bilbao[4],
and Leidy Pérez Coronel[4]

[1] Universidad Peruana de Ciencias Aplicadas, Lima, Peru
`jesussilvaUPC@gmail.com`
[2] Universidad de la Costa, Street 58 #66, Barranquilla, Atlántico, Colombia
`{lhernand31,nvarela2}@cuc.edu.co`
[3] Universidad Tecnológica Centroamericana (UNITEC),
San Pedro Sula, Honduras
`omarpineda@unitec.edu`
[4] Corporación Universitaria Latinoamericana, Barranquilla, Colombia
`{jtafur,brleonc,oredondo}@ul.edu.co,`
`leidyperezcoronel89@gmail.com`

Abstract. In the academic world, a large amount of data is handled each day, ranging from student's assessments to their socio-economic data. In order to analyze this historical information, an interesting alternative is to implement a Data Warehouse. However, Data Warehouses are not able to perform predictive analysis by themselves, so machine intelligence techniques can be used for sorting, grouping, and predicting based on historical information to improve the analysis quality. This work describes a Data Warehouse architecture to carry out an academic performance analysis of students.

Keywords: Intelligent data retrieval · Data Warehouse ·
Unique Identification Number · Academic performance

1 Introduction

One of the most commonly used actions in educational institutions to give value to information and to support decision-making processes is the design of reports. The report designing is an exploratory action where certain crosses of data are made and, depending on the results, other criteria are analyzed until reaching a point in which the results are enough to make decisions about the organization. Support for the decision-making process can be provided by specially designed systems such as [1] DSS (Decision Support Systems), which can generate configurable reports on a regular, quick, and easy basis, as expressed in [2].

On the other hand, Data Warehouses (DW) are electronic data repositories specially designed for generating reports and data analyses [3, 4]. The distinctive features of DW about systems described above are the following: (i) they are flexible, (ii) integrate all

© Springer Nature Switzerland AG 2019
H. Lu et al. (Eds.): ISNN 2019, LNCS 11555, pp. 190–199, 2019.
https://doi.org/10.1007/978-3-030-22808-8_20

points of interest about the organization, (iii) can efficiently handle large amounts of data, and (iv) allow the creation and calculation of management indicators. In addition, the DW are designed with the aim to be efficient in the analysis requirements for strategic levels in organizations, directly considering the organizational strategic objectives [5]. In the same way, DW let efficiently analyze historical information and allow to visualize trends in the behavior of management indicators over time. However, even though the historical information can provide an indication of the historical trend followed by an indicator, it is not enough to certainly predict any particular indicator. A DW can certainly provide a solid basis for analysis and initial performance in the Machine Intelligence techniques [6] that allow learn the patterns of these indicators to predict future patterns. For the latter, the Artificial Neural Networks (ANN) are algorithms that can associate or classify patterns, compress data, control processes and approximate nonlinear functions [7, 8].

In this work, the DW has been designed for the behavior analysis of approval and advance in a curriculum with real data of the curricula vitae of students from different universities in Colombia that offer the Industrial Engineering career. The DW is not focused only on the analysis of historical behaviors of students, but also has been thought of as a base architecture for the prediction of future trends through ANN techniques. This research proposes the approach of data retrieval from an Intelligent Distributed Data Warehouse (IDDW), which is a hierarchical distributed data store of N levels.

2   Theoretical Review

2.1   Artificial Neural Networks

Artificial Neural Networks (ANN) can learn from data and can be used to construct reasonable input-output mapping, with no prior assumptions on the statistical model of the input data (Haykin, 2009) [9]. ANN have non-linear modeling capability with a data-driven approach so that the model is adaptively formed based on the features presented from the data (Zhang 2003) [10]. An introduction to ANN model specifications and implementation and their approximation properties has been provided from an econometric perspective (Kuan 2008) [11]. Several studies show that ANN can solve a variety of challenging computational problems, such as pattern classification, clustering or categorization, function approximation, prediction or forecasting, optimization (traveling salesman problem), retrieval by content, and control (Jain, Mao, and Mohiuddin, 1996) [12].

Some studies of ANN application related to financial early warning models have been conducted by Sevim et al. (2014) [13], as well as Sekmen and Kurkcu (2014) [14] who used ANN as a classifier with a categorical output. Other authors used ANN as financial forecasting models with continuous value. Some of them are Singhal and Swarup (2011) [15], as well as Mombeini and Yazdani-Chamzini (2015) [16] who implemented ANN with a single-step prediction output. A previous study on ANN forecasting model was also proposed by Kulkarni and Haidar (2009) [17] for a multi-step prediction with a direct strategy, so the number of models is equal to the number of the prediction horizon. In the context of basic commodity prices, the need for prediction is not limited to one-step forward but could be extended to include multi-step ahead predictions. Three strategies to tackle the multi-step forecasting problem can be considered, namely recursive, direct, and multiple output strategies (Bontempi, Ben Taieb, and Borgne 2013) [18]. The Multiple Input Multiple Output (MIMO) techniques train a single prediction model f that produces vector outputs of future prediction values. The study proposes to Multi-Layer Perceptron with Multiple Input and Multiple Output (MLP-MIMO) as an agricultural product price prediction model coupled with the variation coefficient from the Colombian state price reference to the criteria of warning level.

2.2 Data Retrieval

Initially, the data retrieval techniques were restricted only to the centralized processing, as discussed by Duan L. et al. (2009) [19]. But, according to Abhay et al. (2015) [20], the data retrieval from the distributed data warehouse refers to the implementation of the classic procedure for retrieving data in a distributed computing environment that seeks to maximize the use of available resources (communication networks, computers, and databases). Some algorithms and systems used for the distributed retrieval of databases are the following: the partition algorithm of Savasere et al. (1995) [21]; Multiagent system based on JAVA JAM by Stolfo et al. (1997) [22] and Prodromidis et al. (2000) [23]; Parthasarathy et al. (2000) [24] in D-DOALL uses the primitive distributed do-all to easily program the task of independent retrieval in a workstations network; Grossman et al. (1999) [25] proposed the Papyrus, a JAVA-based system which aims to wide-area distributed data on clusters and meta-clusters; and the system based on Java for distributed enterprises by Chattratichat et al. (1999) [26].

The data retrieval in a highly parallel environment on multiple processors was explained by Wang et al. (2013) [27]. There are two commonly used parallel programming models: Subprocesses (POSIX subprocesses by Butenhof (1997) [28]) and message passing (OpenMP by Duan et al. (2009)) [19]. Modern programming languages are also structured to efficiently use innovative architectures. There are parallel programming paradigms focusing on parallelizing the algorithms on multiprocessor systems and networks. OPENMP and MPI are used to achieve the parallelization of shared and distributed memory. CUDA is a programming language that is designed for parallel programming used by Garciarena et al. (2015) [29]. In CUDA, the threads access different memories of the GPU. CUDA offers a model of data parallel programming which is incomplete without discussing the more recent approach called MapReduce that can process large amounts of data in a highly parallel way, as shown by Bhaduri et al. (2008) [30]. Several data recovery algorithms have been modified for parallel processing architectures as discussed in Parthasarathy et al. (2000) [24].

3 Materials and Methods

3.1 Data

The databases were obtained from the Ministry of Higher Education in Colombia, the Colombian Institute for the Promotion of Higher Education (ICFES - Instituto Colombiano para el Fomento de la Educación Superior) [31] and four (4) private universities of this country. Such data consisted of the reports described in Table 1.

Table 1. Database of the study sample.

Database	Description
Student	Personal data of students and their status
Subject	Data of the subjects taught, and entry conditions of universities under study
Region	Regions and cities where students come from
Opportunity	Data on possible opportunities to study the subjects
Advance time	Permanence time of a student in the career, based on semesters
Geographical area	Geographical area where the student is located
Cohort	Cohort to which students belong

3.2 Methods

3.2.1 Implementation of the Data Warehouse

A DW system can be implemented under Molap approach (MultidimensionalOlap), Rolap (RelacionalOlap) or by using the hybrid Holap (allows both Molap and Rolap) [32]. In this study, Rolap approach was used. Independently from the approach, the main processes carried out in the development of a DW are as follows.

The process of conceptual modeling: The conceptual model is independent from technology and is essential for specifying the analysis requirements and information availability. When talking about DW conceptual models, there is no consensus in the scientific community about a standard model type for the representation of a DW. However, there are various proposals presented in [33–35]. During the process of conceptual modeling, a DW conceptual scheme is generated. In this study, the MCMD conceptual model was used [3] due to its notation simplicity and because its objective is precisely the conceptual specification of a DW.

Logical modeling process and physical implementation: The logical model formally specifies the multidimensional scheme, its restrictions, and capabilities. In the same way, the logical scheme is implemented directly in a database engine, becoming physical tables. In the case of DW schemes with logical design, they are the star scheme and snowflake scheme [32]. At the stage of physical implementation, dimension tables and fact tables are created depending on the type of scheme, whether star or snowflake.

ETL data load process: The ETL (Extraction, Transformation, Load) process is responsible for extracting, transforming, and loading the data from the original databases into the DW. The data retrieval approach is proposed from the Intelligent Distributed Data Warehouse (IDDW), which is a hierarchical distributed data store of N levels. Based on Abhay et al. (2017) [36], the data retrieval approach begins when the user enters the UIN (Unique Identification Number) corresponding to the data store located in IDDW. Once the data store is located, the desired data are retrieved. A flowchart of the IDDW data retrieval approach is shown in Fig. 1.

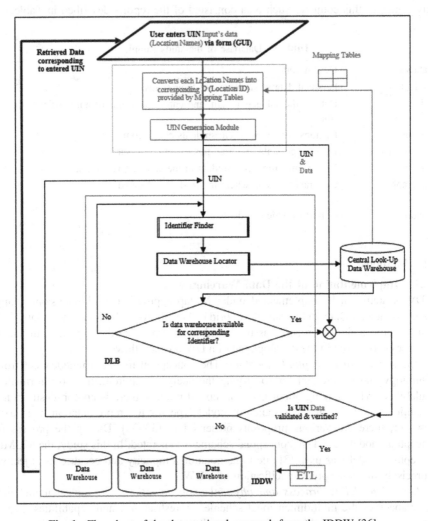

Fig. 1. Flowchart of the data retrieval approach from the IDDW [36].

The ETL process in Fig. 1 consists of extracting data from the system database of the university student's curriculum information, which is not supported by a relational engine and works through files (legacy systems). This system is accessible only through a user interface over the network via a console application inherited from the

COBOL language. To remove this information, the manual process of extraction was simulated by means of an application specially designed for this purpose, after which, the curriculum of each student was extracted in text format. These text files were transformed using a custom software and loaded to a relational database. Then, the files were transformed again by another application for loading them to the DW.

3.2.2   Implementation of the ANN Architecture

At this stage, the ANN architecture was created to be fed with some data obtained by means of the DW. After uploading the DW, an ANN architecture was designed for predicting student's performance using MATLAB algorithms. In this case, the ANN was used to estimate the behavior of a student in next semester. The neural network was trained with backpropagation algorithm and the sigmoid logarithmic function was used on both layers of the network [7].

The obtained results were validated using performance measures that indicate the generalization degree of the used model. Among the indices used are [8]: The Mean Square Error (MSE), the Residual Standard Error (RSE) and the Index of Adequacy (IA), shown in Eqs. (1), (2), and (3) respectively, where oi and pi are observed and predicted values respectively, in the time i, and N is the total number of data. In addition, pi' = pi − om and oi' = oi − om, om representing the average value of the observations.

The IA indicates the adjustment degree that the estimated values present with the actual values of a variable. A value close to 1 indicates a good estimate. On the other hand, MSE and RSE close to zero indicate a good adjustment quality [8].

$$RMS = \sqrt{\frac{\sum_{i=1}^{n}(o_i - p_i)^2}{\sum_{i=1}^{n} o_i^2}} \tag{1}$$

$$RSD = \sqrt{\frac{\sum_{i=1}^{n}(o_i - p_i)^2}{N}} \tag{2}$$

$$IA = 1 - \frac{\sum_{i=1}^{n}(o_i - p_i)^2}{\sum_{i=1}^{n}(|o_i'| + |p_i'|)^2} \tag{3}$$

4   Analysis and Results

This section analyzes the behavior of certain indicators over time through the DW architecture implemented and the prediction of any of these indicators through an ANN.

4.1 DW Analysis

In order to validate the IDDW operation when integrating the generated profiles, 1.500 queries were carried out, with a limit of 860 records to be retrieved for each by executing the data retrieval processes mentioned in Fig. 1. The effectiveness of the IDDW was evaluated in three aspects: (a) the storage of the links to the profiles, (b) the retrieval of the entity data, and (c) the registry of the relationships between the entities retrieved from the same document [36]. The results obtained can be seen in Table 2.

Table 2. Validation results

Metrics	Value
Number of profiles to generate	750
Effectiveness of persistence	90%
Effectiveness of retrieval	76%
Effectiveness of the relationship generation between entities	95%

Initially, the UIN "13302010410520017" is entered through the developed form (Abhay et al., 2017) [36]. The first identifier calculated by the identifier search engine for this UIN is "1330201041052001". The data store locator searches for the address of the machine, corresponding to this identifier in the Central Look-Up data store tables. For levels of hierarchy see Table 3.

Table 3. The percentage of correctly retrieved data from the common table of the data store located at different levels of hierarchy.

Level in hierarchy	Percentage (%)
1	90
2	91
3	93
4	94
5	95
6	98
7	99

Table 3 shows the correctly retrieved data (in percentage) from the data warehouse located in various levels of hierarchy. The correct data are the data that must be retrieved for the entered UIN. From the values in Table 2, it may be seen that as the data warehouse is placed at lower levels of hierarchy, the percentage of correct data retrieved increases. It is because the number of times the Identifier is calculated are less, and chances of error are less too.

4.2   Results of the Prediction Using ANN

The number of subjects enrolled by a student was estimated, analyzing the 760 recovered data, taking only one semester toward the future (number of courses approved). With respect to the foregoing, Table 4 shows the values of the indices obtained for estimating both variables.

Table 4. Indices of adequacy and errors in test data estimation.

Indices	Estimation of quantity of enrolled subjects	Estimation of quantity of approved subjects
IA	0.8714	0.8124
RMS	0.3001	0.3492
RSD	**0.0899**	**0.1199**

The results confirm that the prediction is adjusted to the DW historical trend. So, the complement between DW and ANN is a powerful tool to predict the future behavior of a management indicator.

5   Conclusions

The implementation of a Data Warehouse and the Artificial Neural Network architecture has been carried out for the analysis and prediction of academic performance in students of Industrial Engineering at a group of Colombian private universities. The main advantage of using a DW lies in the possibility of crossing different analysis dimensions in a simple and fast way to perform an exploratory analysis of data for the creation of reports. It can be noted that the process of ETL (Extraction, Transformation, and Loading) is the one that more time and resources demanded, mainly since the information should be cross-posted from different sources. Additionally, operational systems are not designed to analyze data, and the heterogeneity of the platforms where the information is located adds a greater difficulty that requires the creation of specific applications and systems to draw on historical data. The use of a multidimensional conceptual model to generate the IDDW conceptual scheme with UIN becomes a great tool that, independently from the platforms, allows to narrow down the analysis and give clarity to the ETL subsequent process.

To obtain summaries and reports using DW as a product of the historical analysis of data, a solid database can be created for the ANN architecture and the prediction of future behavior. Based on the above, the use of DW combined with the use of estimation or prediction techniques (in our case, the ANN), provides a complement to substantiate more extensive analyzes because, as shown in this study, it is possible to predict the management indicators obtained from the DW. This allows the institution to take steps to analyze, modify, and validate the management indicators or, perhaps, to generate new strategies to improve and/or optimize the management process, since knowledge is extracted from the same databases, thus giving value to the management information that is logged but that is not always considered.

References

1. Vasquez, C., Torres, M., Viloria, A.: Public policies in science and technology in Latin American countries with universities in the top 100 of web ranking. J. Eng. Appl. Sci. 12 (11), 2963–2965 (2017)
2. Aguado-López, E., Rogel-Salazar, R., Becerril-García, A., Baca-Zapata, G.: Presencia de universidades en la Red: La brecha digital entre Estados Unidos y el resto del mundo. Revista de Universidad y Sociedad del Conocimiento 6(1), 1–17 (2009)
3. Torres-Samuel, M., Vásquez, C., Viloria, A., Lis-Gutiérrez, J.P., Borrero, T.C., Varela, N.: Web visibility profiles of top100 Latin American universities. In: Tan, Y., Shi, Y., Tang, Q. (eds.) DMBD 2018. LNCS, vol. 10943, pp. 1–12. Springer, Cham (2018). https://doi.org/10. 1007/978-3-319-93803-5_24
4. Viloria, A., Lis-Gutiérrez, J.P., Gaitán-Angulo, M., Godoy, A.R.M., Moreno, G.C., Kamatkar, S.J.: Methodology for the design of a student pattern recognition tool to facilitate the teaching – learning process through knowledge data discovery (big data). In: Tan, Y., Shi, Y., Tang, Q. (eds.) DMBD 2018. LNCS, vol. 10943, pp. 1–12. Springer, Cham (2018). https://doi.org/10.1007/978-3-319-93803-5_63
5. Caicedo, E.J.C., Guerrero, S., López, D.: Propuesta para la construcción de un índice socioeconómico para los estudiantes que presentan las pruebas Saber Pro. Comunicaciones en Estadística 9(1), 93–106 (2016)
6. Mazón, J.N., Trujillo, J., Serrano, M., Piattini, M.: Designing data warehouses: from business requirement analysis to multidimensional modeling. In Proceedings of the 1st International Workshop on Requirements Engineering for Business Need and IT Alignment, Paris, France (2005)
7. Vásquez, C., et al.: Cluster of the Latin American universities top100 according to webometrics 2017. In: Tan, Y., Shi, Y., Tang, Q. (eds.) DMBD 2018. LNCS, vol. 10943, pp. 1–12. Springer, Cham (2018). https://doi.org/10.1007/978-3-319-93803-5_26
8. Haykin, S.: Neural Networks a Comprehensive Foundation, 2nd edn. Macmillan College Publishing Inc., USA (1999). ISBN 9780023527616
9. Isasi, P., Galván, I.: Redes de Neuronas Artificiales. Un enfoque Práctico. Pearson, London (2004). ISBN 8420540250
10. Haykin, S.: Neural Networks and Learning Machines. Prentice Hall International, New Jersey (2009)
11. Zhang, G.P.: Time series forecasting using a hybrid ARIMA and neural network model. Neurocomputing 50(1), 159–175 (2003)
12. Kuan, C.M.: Artificial neural networks. In: Durlauf, S.N., Blume, L.E. (eds.) The New Palgrave Dictionary of Economics. Palgrave Macmillan, Basingstoke (2008)
13. Jain, A.K., Mao, J., Mohiuddin, K.M.: Artificial neural networks: a tutorial. IEEE Comput. 29(3), 1–32 (1996)
14. Sevim, C., Oztekin, A., Bali, O., Gumus, S., Guresen, E.: Developing an early warning system to predict currency crises. Eur. J. Oper. Res. 237(1), 1095–1104 (2014)
15. Sekmen, F., Kurkcu, M.: An early warning system for Turkey: the forecasting of economic crisis by using the artificial neural networks. Asian Econ. Finan. Rev. 4(1), 529–543 (2014)
16. Singhal, D., Swarup, K.S.: Electricity price forecasting using artificial neural networks. IJEPE 33(1), 550–555 (2011)
17. Mombeini, H., Yazdani-Chamzini, A.: Modelling gold price via artificial neural network. J. Econ. Bus. Manag. 3(7), 699–703 (2015)
18. Kulkarni, S., Haidar, I.: Forecasting model for crude oil price using artificial neural networks and commodity future prices. Int. J. Comput. Sci. Inf. Secur. 2(1), 81–89 (2009)

19. Bontempi, G., Ben Taieb, S., Le Borgne, Y.-A.: Machine learning strategies for time series forecasting. In: Aufaure, M.-A., Zimányi, E. (eds.) eBISS 2012. LNBIP, vol. 138, pp. 62–77. Springer, Heidelberg (2013). https://doi.org/10.1007/978-3-642-36318-4_3
20. Duan, L., Xu, L., Liu, Y., Lee, J.: Cluster-based outlier detection. Ann. Oper. Res. **168**(1), 151–168 (2009)
21. Abhay, K.A., Badal, N.A.: Novel approach for intelligent distribution of data warehouses. Egypt. Inf. J. **17**(1), 147–159 (2015)
22. Savasere, A., Omiecinski, E., Navathe, S.: An efficient algorithm for data mining association rules in large databases. In: Proceedings of 21st Very Large Data Base Conference, vol. 5, no. 1, pp. 432–444 (1995)
23. Stolfo, S., Prodromidis, A.L., Tselepis, S., Lee, W., Fan, D.W.: Java agents for meta learning over distributed databases. In: Proceedings of 3rd International Conference on Knowledge Discovery and Data Mining, vol. 5, no. 2, pp. 74–81 (1997)
24. Prodromidis, A., Chan, P.K., Stolfo, S.J.: Meta learning in distributed data mining systems: issues and approaches. In: Kargupta, H., Chan, P. (eds.) Book on Advances in Distributed and Parallel Knowledge Discovery. AAAI/MIT Press, Cambridge (2000)
25. Parthasarathy, S., Zaki, M.J., Ogihara, M.: Parallel data mining for association rules on shared-memory systems. Knowl. Inf. Syst. Int. J. **3**(1), 1–29 (2001)
26. Grossman, R.L., Bailey, S.M., Sivakumar, H., Turinsky, A.L.: Papyrus: a system for data mining over local and wide area clusters and super-clusters. In: Proceedings of ACM/IEEE Conference on Supercomputing, vol. 63, pp. 1–14 (1999)
27. Chattratichat, J., Darlington, J., Guo, Y., Hedvall, S., Köhler, M., Syed, J.: An architecture for distributed enterprise data mining. In: Sloot, P., Bubak, M., Hoekstra, A., Hertzberger, B. (eds.) HPCN-Europe 1999. LNCS, vol. 1593, pp. 573–582. Springer, Heidelberg (1999). https://doi.org/10.1007/BFb0100618
28. Wang, L., Tao, J., Ranjan, R., Marten, H., Streit, A., Chen, J., Chen, D.: G-Hadoop: MapReduce across distributed data centers for data-intensive computing. Future Gener. Comput. Syst. **29**(3), 739–750 (2013)
29. Butenhof, D.R.: Programming with POSIX Threads. Addison-Wesley Longman Publishing Company, Boston (1997)
30. Bhaduri, K., Wolf, R., Giannella, C., Kargupta, H.: Distributed decision-tree induction in peer-to-peer systems. Stat. Anal. Data Min. **1**(2), 85–103 (2008)
31. Instituto colombiano para la Evaluación de la Educación - ICFES. Informe nacional de resultados Saber Pro 2015–2018. ICFES, Bogotá (2018)
32. Rafailidis, D., Kefalas, P., Manolopoulos, Y.: Preference dynamics with multimodal user-item interactions in social media recommendation. Expert Syst. Appl. **74**(1), 11–18 (2017)
33. Zheng, C., Haihong, E., Song, M., Song, J.: CMPTF: contextual modeling probabilistic tensor factorization for recommender systems. Neurocomputing **205**(1), 141–151 (2016)
34. Hidasi, B., Tikk, D.: Fast ALS-based tensor factorization for context-aware recommendation from implicit feedback. In: Flach, P.A., De Bie, T., Cristianini, N. (eds.) ECML PKDD 2012. LNCS (LNAI), vol. 7524, pp. 67–82. Springer, Heidelberg (2012). https://doi.org/10.1007/978-3-642-33486-3_5
35. Lee, J., Lee, D., Lee, Y.C., Hwang, W.S., Kim, S.W.: Improving the accuracy of top-n recommendation using a preference model. Inf. Sci. **348**(1), 290–304 (2016)
36. Abhay, K.A., Neelendra, B.: Data storing in intelligent and distributed data warehouse using unique identification number. Int. J. Grid Distrib. Comput. **10**(9), 13–32 (2017)

Multi-dimension Tensor Factorization Collaborative Filtering Recommendation for Academic Profiles

Jesús Silva[1](✉), Noel Varela[2], Omar Bonerge Pineda Lezama[3],
Hugo Hernández-P[4], Jairo Martínez Ventura[4], Boris de la Hoz[4],
and Leidy Pérez Coronel[4]

[1] Universidad Peruana de Ciencias Aplicadas, Lima, Perú
jesussilvaUPC@gmail.com
[2] Universidad de la Costa, Street 58 #66, Barranquilla, Atlántico, Colombia
nvarela2@cuc.edu.co
[3] Universidad Tecnológica Centroamericana (UNITEC),
San Pedro Sula, Honduras
omarpineda@unitec.edu
[4] Corporación Universitaria Latinoamericana, Barranquilla, Colombia
hugohernandezpalma@gmail.com,
leidyperezcoronel89@gmail.com,
{academico, bdelahoz}@ul.edu.co

Abstract. The choice of academic itineraries and/or optional subjects to attend is not usually an easy decision since, in most cases, students lack the information, maturity, and knowledge required to make right decisions. This paper evaluates the support of Collaborative Systems for helping and guiding students in this decision-making process, considering the behavior and impact of these systems on the use of data different from the formal information the students usually use. For this purpose, the research applied the clustering based Multi-dimension Tensor Factorization approach to build a recommendation system and confirm that the increment in tensors improves the recommendation accuracy. As a result, this approach permits the user to take advantage of the contextual information to reduce the sparsity issue and increase the recommendation accuracy.

Keywords: Collaborative filtering · Context aware recommendation system · Contextual Modeling · Item recommendations · Multi-dimensionality · Tensor Factorization

1 Introduction

People are continually making important decisions, sometimes facing many alternatives to consider. There are three main elements that play a fundamental role in the decision-making process: (i) the maturity degree of the individual, (ii) the level of knowledge, and (iii) the information available to make the best decision [1, 2]. Sometimes, inexperienced individuals in a specific field of education may not reach the

© Springer Nature Switzerland AG 2019
H. Lu et al. (Eds.): ISNN 2019, LNCS 11555, pp. 200–209, 2019.
https://doi.org/10.1007/978-3-030-22808-8_21

desirable level of knowledge for making the best choices, so it is important to provide tools to assist them either by providing relevant information or by defining the different options to get orientation for a better decision making.

During the educational training stage of every individual, there are moments when the student must make certain decisions regarding the future. Some questions arise: what kind of training suits me? what area to choose? what academic itinerary to follow? which subjects to choose...? This fact is inevitable and happens in most educational stages, starting with Secondary Education in which the degree of responsibility, maturity, and knowledge of the students when making these important decisions is questionable. Is there any way to help students in these proposed tasks either by defining the spectrum of possibilities or by orienting towards an educational itinerary? [3–5]. This research intends to answer to these questions by proposing a Recommendation System based on Collaborative Filtering algorithms (hereinafter CF).

A generic multi-dimensional framework based on Tensor Factorization is presented to address context aware recommendations with MD-TFCF (Multi-dimension Tensor Factorization Collaborative Filtering). Tensor Factorization is used as it can handle any number of contextual variables. Tensor Factorization allows flexible assimilation of contextual information by modeling the context associated with user and products. The contextual information is related to additional dimensions that are represent in the form of tensors. The factorization of this tensor helps in building a unified model of data which provides context aware recommendations. The proposed approach allows integrating more than one context at a time and helps predicting the missing ratings.

The contribution of the research are the following: (1) an efficient 5-mode Tensor Factorization approach is proposed to factorize the tensors, (2) uses Tensor Factorization for the explicit generation of recommendations in which model based clustering and Tensor Factorization learning method is combined to predict missing ratings, (3) Comparative analysis of higher order tensors with lower order tensors is done and confirms that the proposed approach, so the MD-TFCF, leads to more promising results when more contextual dimensions are considered. The results confirm that as the number of contextual dimensions increases, more accurate the recommendations are.

2   Theoretical Review

Various recommendation systems are used on the basis of content based collaborative filtering or hybrid-based approach. Most of the work on CF has been done on traditional 2D-matrix, i.e. user-item rating matrix, but recently, context has become an important factor to be integrated into the recommendation generation algorithms as context plays an important role in real applications such as temporal effect while doing online shopping or selecting places [6]. So, the relevant work of the study in this domain focuses in this point.

Recommendation Systems have been initially devised to improve the decision strategy of users under complex information environments [7, 8]. Recommendation Systems reduce the problem of information overload by recommending the users most relevant information. Recommendation Systems use content based [9, 10], collaborative filtering [11], and hybrid filtering [12] techniques for efficient recommendations.

The collaborative filtering approach is the most prevailing approach which is further divided into implicit feedback and explicit feedback [13] and [14] methods. In the implicit feedback method, the user's interaction is analyzed in clicks, time spent, and other indicators, and in explicit feedback about the ratings assigned to specific items, questionnaires filled by the user, and others are considered. Then, based on these factors, recommendations are given [15]. CF approach can also be broadly categorized in two types: memory-based, and model-based [16]. In the memory-based method, user or product rating vectors are used to compute analogy among users or products which further operate on a neighborhood-based method. But the major challenge faced in memory-based collaborative filtering approach is the sparsity of the user-item rating matrix, i.e. several entries in the rating matrix might be NULL as there are many non-rated products available in the data pool. This sparsity problem can be reduced by using the model-based approach. In this approach, the generalized model is built to discover latent factors or use the contextual information of users or items for capturing user's preferences. The most common model-based approach is the Matrix Factorization technique as it considers latent factors that reduce the sparsity of the matrix and gives better results than the User-based Collaborative Filtering approach which simply uses neighborhood approach to find similar users [8]. But the Matrix Factorization technique [16] cannot integrate the contextual information in a straightforward way, so this concept has been extended to multi-dimensional matrices known as Tensor Factorization [15]. that, in this contextual information, can be integrated in more easy ways to give more accurate results than the Matrix Factorization.

The more related work in this domain is elaborated like there are various Tensor Factorization models available which can be used to incorporate contextual information which increases the flexibility and quality of the recommendation systems [17]. Tensor factorization models are applicable in almost every domain due to the increase of computational complexity and the need of a dynamic environment. [18] issued a thorough survey on tensor models, their application domains and the available software. The authors [18] propose various tensor decomposition models such as PARAFAC, DEDICOM, PARATUCK2. Other successful recommendation approaches are the Context-Aware filtering techniques which are broadly categorized as Contextual Pre-Filtering, Contextual Post-Filtering, and Contextual Modeling [15]. The comparative analysis of the three approaches is done by [19] to determine which approach is better and under what situation in relation to accuracy and diversity. The factors considered for evaluating the performance are the dataset type, type of recommendation, and context granularity.

Similarly, [20] presented the Tensor Factorization and Tag Clustering Model (TCM) for recommendations in social tagging systems in which content information is processed to find tags among comparable items, then the tag clusters are formed and finally, association among users, items, and topics are discovered by working upon the Tensor Factorization technique, i.e. Higher Order Singular Value Decomposition (HOSVD). But this work is limited to just three dimensions whereas the proposed approach extends to 5 dimensions and confirms that higher dimensions gives better results. In the same way, [21] proposed a new model Multiverse Recommendation in which contextual information has been integrated with the traditional user-item rating matrix which is not as easy for integrating the contextual information in other model-based approaches like Matrix Factorization. This contextual information represents additional dimensions to original user-item rating matrix as tensor. This approach outperforms other traditional methods which do not involve contextual information in terms of Mean Absolute Error up to 30% whereas the proposed work implements up to 5 dimensions while the performance of proposed recommendation system is assessed against various evaluation metrics.

Recently, [22] introduced the Contextual Modeling Probabilistic Tensor Factorization (CMPTF) model which is basically abstraction of the Probabilistic Tensor Factorization (PTF). In PTF model, the entire information like ratings, item content, context, and social relationship is integrated into a single model which was not possible in earlier approaches. CMPTF further integrates topic modeling information which improves the quality of recommendation systems, and experimental results prove that this approach is superior than traditional approaches. [23] proposed other generic context-aware implicit feedback recommendation algorithms and employ a fast, ALS-based tensor factorization learning method that linearly scales with the number of non-zero elements in the tensor while maintaining the computational efficiency.

Thus, considering the mentioned confrontations by various researchers, the proposed MD- TFCF approach integrates the contextual information as higher order tensors and results support that increment in tensors improves the recommendation performance.

3 Data and Methods

The formal teaching that allows some degree of choice present the following structural patterns: (i) there are students who are enrolled in subjects and obtain certain qualifications; (ii) the subjects are associated to a course, level, or degree, and can be of different types depending on whether they are mandatory, optional, referring to a specific modality or profile, with groupings of subjects that form profiles or educational itineraries in the case of attending to all or a group of them. An academic record can be defined as a set of grades obtained by a student in a series of subjects taken over a certain time period.

The main objective of this contribution is to answer to the following question: is it possible to use people's academic records to offer suggestions when choosing their future? Initially, the answer is not entirely clear since subjective, psychological, and aptitude factors come into play.

Since qualifications provide reliable information about the skills of a student, the areas where people perform best, and even their preferences, a Collaborative Recommendation System is evaluated, estimating the possible qualification that a student would obtain in a subject in case of studying it, to observe if it provides relevant information which, properly linked to future information, could help individuals to make decisions about their future. To this purpose, a series of experiments was conducted to obtain a reliable output to this issue.

3.1 Data

The used data set consists of a total of 7315 anonymous students from primary, secondary, and university levels from several private education institutions in Colombia, considering up to 100 subjects and a total of 155,022 qualifications, which involve values from 0 and 5.

3.2 The Proposed MD-TFCF Mechanism

This section presents the framework of the Multi-dimension Tensor Factorization Collaborative Filtering (MD-TFCF) approach. The work flow of the proposed framework is shown in Fig. 1 [24], which illustrates that the process starts from the data processing and continues to predictions according to the wishes of the users.

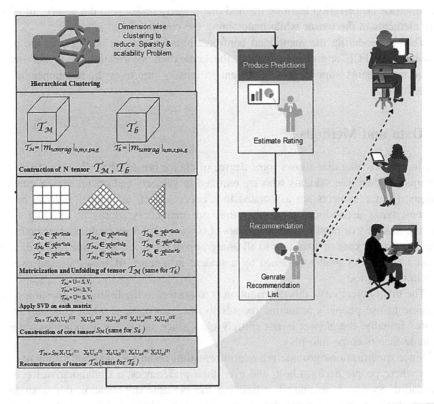

Fig. 1. Proposed Multi-dimension Tensor Factorization Collaborative Filtering (MD-TFCF) Framework, based in Lee et al. [24]

3.2.1 Hierarchical Clustering Approach

Hierarchical Clustering is one of the coherent clustering techniques [12] in which hierarchies of clusters are formed and every formed cluster is part of another cluster.

This research applies the agglomerative hierarchical clustering-based approach in which the clustering process starts with one initial cluster and then a pair of clusters are merged up together. So, clusters based on age are formed first including users grouped by age, as shown in Table 1 [4].

3.2.2   Decomposition of the Singular Value of Higher Order

In the consulted literature, several tensor decomposition models are available [18] such as PARAFAC, Tucker, Canonical, HOSVD, etc. In the study, the Higher Order Singular Value Decomposition (HOSVD) Model is used to factor the tensors in matrices obtained from the qualification matrix. The main benefit of using HOSVD is to address the high dimensionality of the data in an effective way [14, 20], which helps to discover the relationship between users, the qualifications, and other contextual dimensions such as age, gender, and academic term.

Table 1. Categorization according to age

Age Group	Group Name
0–12	Kids
13–17	Teenager
18–25	Youth
26–50	Middle
51–73	Aged

The Higher Order Singular Value Decomposition (HOSVD) Model is constituted by the following stages, for more details see Lee et al. [24]:

- Initial Construction of Tensor
- Matricization of Tensor (T_i)
- Apply SVD on each matrix (TM_i)
- Construction of Core Tensor (SM)
- Reconstruction of Tensor (TM')
- Recommendation List.

3.3   Experimental Setup

The Pareto Principle which is also known as 80/20 rule is used for the verification of the predicted rating allotted through the projected MD-TFCF approach. According to the Pareto Principle the dataset is divided and evenly distributed into training and test set in the ratio of 80% and 20% respectively. The data is evenly distributed in 80–20 ratio so that the entire dimensions data are distributed conceptually. The approach is experimented and assessed on cluster sets formed through the hierarchical clustering approach, for dataset each experiment is run 26 times. Henceforth, the prediction error is minimized using Pareto Principle as it arbitrates in evaluating the efficiency of the proposed MD-TFCF approach.

3.3.1 Evaluation Metrics

The peculiarity of a recommendation algorithm can be assessed using different forms of metrics. The suitability of the metrics used reckons on the recommendation approach, dataset, and what the recommender system will perform. Moreover, Mean Absolute Error (MAE), precision, and recall [13, 17] are statistical measures to assess the accuracy and peculiarity of the recommendation system.

Mean Absolute Error (MAE): the MAE is the most popular and simplest form of metrics [15] for measuring the accuracy. The MAE basically measures the average absolute difference between the predicted and the actual rating. It is simply, as the name suggests, the mean of the absolute error. It is a measure of deviation of the recommendation or absolute error between the predicted value and the user specific rating value. It is formally calculated using Eq. (1) as:

$$MAE = \frac{1}{N} \sum_{u,i \in N} |p_{u,i} - r_{u,i}| \qquad (1)$$

Where $p_{u,i}$ is the predicted rating for user u on subject i, $r_{u,i}$ is the actual rating, and N is the total number of ratings. The lower is the value of MAE, the more accurate the recommendation system is for predicting ratings of users. It tells how big an error can be expected from the approach. Other metric measures used for evaluation are classic measure-precision and the recall.

Precision: The Precision is basically the measurement of the probability that the retrieved record is a relevant record [15]. The precision rate is the fraction of successful rating prediction that is predicted by users. The precision is computed using the Eq. (2) as:

$$precision = \frac{Correctly\,Predicted\,Rating}{Total\,No.\,of\,Correctly + Incorrectly\,Predicted\,Rating} \qquad (2)$$

Therefore, the precision identifies the ratio of the number of the correctly predicted rating retrieved to the total number of incorrectly and correctly predicted ratings.

Recall: It is defined as fraction of relevant prediction retrieved to the total number of the user prediction in the dataset. The recall is computed using the Eq. (3) as:

$$Recall = \frac{Correctly\,Predicted\,Rating}{Total\,number\,of\,User\,Assigned\,Prediction} \qquad (3)$$

4   Results and Discussions

The proposed MD-TFCF approach is different from existing approaches as an integrated framework is developed in the proposed approach to unanimously represent the five dimensions. Figure 2 shows that there are remarkable improvements in results in form of precision, recall, and mean absolute error for the datasets.

Figure 2 infers that precision varies from 0.54 to 0.96; recall varies from 0.30 to 0.80, and the mean absolute error decreases from 2.2 to 0.38 for dataset, while similarly, precision varies from 0.753 to 0.916, recall varies from 0.50 to 0.73, and the mean absolute error decreases from 2.2 to 0.38 showing that the MD-TFCF approach

Measuring Metrics →	Recall			Precision			MAE		
Approaches ↓	Max	Min	Avg	Max	Min	Average	Max	Min	Avg
User-Item Based Neighborhood Collaborative Filtering									
Normal U-I	0.5	0.1	0.23	0.5	0.05	0.38	3	1.2	2.4
Multi-Dimensional Tensor Factorization Collaborative Filtering									
4-order Tensor	0.6	0.30	0.50	0.77	0.50	0.65	1.86	1.02	1.46
5- order Tensor	0.80	0.30	0.65	0.96	0.54	0.81	2.2	0.38	1.06

Fig. 2. Comparative analysis of higher order tensor with lower order tensor results

achieves more promising results than the traditional user-item based collaborative filtering approach. In the same way, on adding even one dimension, i.e. 5-tensor approach, is better than 4-tensor as accuracy in results has been improved as precision varies from 0.50 to 0.77, recall varies from 0.30 to 0.60, and the mean absolute error decreases from 1.86 to 1.02. Thus, a new technique is concurrently proposed to deal with 5 dimensions and used for comparative analysis with traditional user-item based approach and with lower dimensional spaces.

It is empirically validated that MD-TFCF approach gains about 49% accuracy in form of precision, 20% in form of recall and 32% in terms of mean absolute error for the studied dataset. Thus, the proposed approach is achieving more desirable results whenever more contextual parameters are considered. Figure 3 shows results of conventional user-item based neighborhood CF process and MD-TFCF (higher order tensors with lower order tensors) approach in comparison to each other in form of graph. As shown in Fig. 3, conventional algorithm's precision and recall varies from 5%–50% and 1%–5% respectively, for dataset.

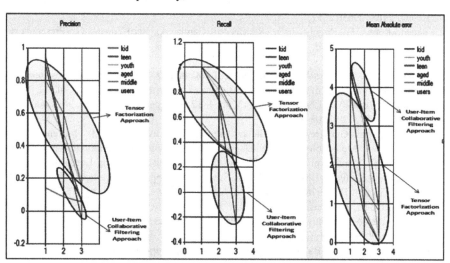

Fig. 3. Comparative analysis results for the set of data studied

The precision and recall values show improvement in dataset because of the large data size. The following graphs (x-axis represent number of folds and y-axis represents recall, precision and mean absolute error respectively) validate that tensor factorization approach provides more accuracy in results in form of precision, recall, and mean absolute error as evaluation metrics than traditional.

5 Conclusions

In this research, a novel Multi-dimension Tensor Factorization Collaborative Filtering (MD-TFCF) approach is introduced to mitigate the sparsity problem as this is the major challenge of the Collaborative Filtering approach. In traditional user-item based Collaborative Filtering approach, the user-item matrix is formed by considering only ratings accredited by users to different products, but several entries in rating matrix are NULL because there are diverse set of items that are generally not rated by users. So, to overcome this problem, User-Item based approach is extended to Model based approach MD-TFCF and mainly comparative analysis of MD-TFCF with user-item based collaborative filtering and lower order dimensional spaces is done.

After analyzing the recommendation systems based on proposed collaborative filtering, it has been proved that their use can be useful for making personalized recommendations to students about educational itineraries [25] when choosing optional subjects and foreseeing which common subjects will present greater learning difficulties or specific needs of reinforcement in the student.

References

1. Vásquez, C., Torres, M., Viloria, A.: Public policies in science and technology in Latin American countries with universities in the top 100 of web ranking. J. Eng. Appl. Sci. **12**(11), 2963–2965 (2017)
2. Aguado-López, E., Rogel-Salazar, R., Becerril-García, A., Baca-Zapata, G.: Presencia de universidades en la Red: La brecha digital entre Estados Unidos y el resto del mundo. Revista de Universidad y Sociedad del Conocimento **6**(1), 1–17 (2009)
3. Torres-Samuel, M., Vásquez, C., Viloria, A., Lis-Gutiérrez, J.P., Borrero, T.C., Varela, N.: Web Visibility Profiles of Top100 Latin American Universities. In: Tan, Y., Shi, Y., Tang, Q. (eds.) DMBD 2018. LNCS, vol. 10943, pp. 1–12. Springer, Cham (2018). https://doi.org/10.1007/978-3-319-93803-5_24
4. Viloria, A., Lis-Gutiérrez, J.P., Gaitán-Angulo, M., Godoy, A.R.M., Moreno, G.C., Kamatkar, S.J.: Methodology for the design of a student pattern recognition tool to facilitate the teaching – learning process through knowledge data discovery (Big Data). In: Tan, Y., Shi, Y., Tang, Q. (eds.) DMBD 2018. LNCS, vol. 10943, pp. 1–12. Springer, Cham (2018). https://doi.org/10.1007/978-3-319-93803-5_63
5. Caicedo, E.J.C., Guerrero, S., López, D.: Propuesta para la construcción de un índice socioeconómico para los estudiantes que presentan las pruebas Saber Pro. Comun. Estadística **9**(1), 93–106 (2016)
6. Reihanian, A., Minaei-Bidgoli, B., Alizadeh, H.: Topic-oriented community detection of rating-based social networks. J. King Saud Univ. Comput. Inf. Sci. **28**(3), 303–310 (2016)

7. Rashid, A.M.: Getting to know you: learning new user preferences in recommender systems. In: Proceedings of the 7th International Conference on Intelligent User Interfaces, pp. 127–134. ACM (2002)
8. Gómez, S., Zervas, P., Sampson, D.G., Fabregat, R.: Context-aware adaptive and personalized mobile learning delivery supported by UoLmP. J. King Saud Univ. Comput. Inf. Sci. **26**(1), 47–61 (2014)
9. Lops, P., de Gemmis, M., Semeraro, G.: Content-based recommender systems: state of the art and trends. In: Ricci, F., Rokach, L., Shapira, B., Kantor, P.B. (eds.) Recommender Systems Handbook, pp. 73–105. Springer, Boston (2011). https://doi.org/10.1007/978-0-387-85820-3_3
10. Malik, F., Baharudin, B.: Analysis of distance metrics in content-based image retrieval using statistical quantized histogram texture features in the DCT domain. J. King Saud Univ. Comput. Inf. Sci. **25**(2), 207–218 (2013)
11. Su, X., Khoshgoftaar, T.M.: A survey of collaborative filtering techniques. Adv. Artif. Intell. **4**(1), 1–12 (2009)
12. De Campos, L.M., Fernández-Luna, J.M., Huete, J.F., Rueda-Morales, M.A.: Combining content-based and collaborative recommendations: a hybrid approach based on Bayesian networks. Int. J. Approx. Reason. **51**(7), 785–799 (2010)
13. Isinkaye, F.O., Folajimi, Y.O., Ojokoh, B.A.: Recommendation systems: principles, methods and evaluation. Egypt. Inform. J. **16**(3), 261–273 (2015)
14. Buder, J., Schwind, C.: Learning with personalized recommender systems: a psychological view. Comput. Hum. Behav. **28**(1), 207–216 (2012)
15. Taneja, A., Arora, A.: Cross domain recommendation using multidimensional tensor factorization. Expert Syst. Appl. **92**(1), 304–316 (2018)
16. Frolov, E., Oseledets, I.: Tensor methods and recommender systems. Wiley Interdisc. Rev. Data Min. Knowl. Discov. **7**(3), 1–12 (2017)
17. Kolda, T.G., Bader, B.W.: Tensor decompositions and applications. SIAM Rev. **51**(3), 455–500 (2009)
18. Panniello, U., Tuzhilin, A., Gorgoglione, M.: Comparing context-aware recommender systems in terms of accuracy and diversity. User Model. User-Adap. Inter. **24**(2), 35–65 (2014)
19. Rafailidis, D., Kefalas, P., Manolopoulos, Y.: Preference dynamics with multimodal user-item interactions in social media recommendation. Expert Syst. Appl. **74**(1), 11–18 (2017)
20. Karatzoglou, A., Amatriain, X., Baltrunas, L., Oliver, N.: Multiverse recommendation: n-dimensional tensor factorization for context-aware collaborative filtering. In: Proceedings of the Fourth ACM Conference on Recommender Systems, pp. 79–86. ACM (2010)
21. Zheng, C., Haihong, E., Song, M., Song, J.: CMPTF: contextual modeling probabilistic tensor factorization for recommender systems. Neurocomputing **205**(1), 141–151 (2016)
22. Hidasi, B., Tikk, D.: Fast ALS-based tensor factorization for context-aware recommendation from implicit feedback. In: Flach, P.A., De Bie, T., Cristianini, N. (eds.) ECML PKDD 2012. LNCS (LNAI), vol. 7524, pp. 67–82. Springer, Heidelberg (2012). https://doi.org/10.1007/978-3-642-33486-3_5
23. Lee, J., Lee, D., Lee, Y.C., Hwang, W.S., Kim, S.W.: Improving the accuracy of top-n recommendation using a preference model. Inf. Sci. **348**(1), 290–304 (2016)
24. Vásquez, C., et al.: Cluster of the latin american universities top100 according to webometrics 2017. In: Tan, Y., Shi, Y., Tang, Q. (eds.) DMBD 2018. LNCS, vol. 10943, pp. 1–12. Springer, Cham (2018). https://doi.org/10.1007/978-3-319-93803-5_26
25. Torres-Samuel, M., Vásquez, C.L., Viloria, A., Varela, N., Hernández-Fernandez, L., Portillo-Medina, R.: Analysis of patterns in the university world rankings webometrics, Shanghai, QS and SIR-SCimago: case Latin America. In: Tan, Y., Shi, Y., Tang, Q. (eds.) DMBD 2018. LNCS, vol. 10943, pp. 188–199. Springer, Cham (2018). https://doi.org/10.1007/978-3-319-93803-5_18

The Goal Programming as a Tool for Measuring the Sustainability of Agricultural Production Chains of Rice

Jesús Silva[1]([⊠]), Noel Varela[2], Omar Bonerge Pineda Lezama[3],
David Martínez Sierra[4], Jainer Enrique Molina Romero[5],
John Anderson Virviescas Peña[6], and Rubén Dario Munera Ramirez[7]

[1] Universidad Peruana de Ciencias Aplicadas, Lima, Peru
jesussilvaUPC@gmail.com
[2] Universidad de la Costa, Street 58 #66, Barranquilla, Atlántico, Colombia
nvarela2@cuc.edu.co
[3] Universidad Tecnológica Centroamericana (UNITEC), San Pedro Sula,
Honduras
omarpineda@unitec.edu
[4] Facultad de Ingeniería, Universidad Simón Bolívar, Barranquilla, Colombia
dmartinez@unisimonbolivar.edu.co
[5] Universidad Libre Facultad Ciencia de Salud, Barranquilla, Colombia
jainer.molina@unilibre.edu.co
[6] Corporación Universitaria Minuto de Dios - UNIMINUTO, Bello,
Antioquia, Colombia
john.virviescas@uniminuto.edu
[7] Corporación Universitaria Lasallista, Caldas, Antioquia, Colombia
rumunera@lasallista.edu.co

Abstract. Agricultural activity is characterized by an intensive use of capital and a considerable dependence on external financing. Access to credit is often limited by the scarcity of resources and lack of guarantees, seriously affecting the productivity and economic performance of agricultural exploitations. The objective of this paper is to assess the sustainability of agricultural production chain of rice in Latin America using multi-criteria analysis tools to facilitate decision-making through a benchmarking process to contribute to their economic sustainability. The implementation of the model in an exploitation typy depending on financing sources (conservative, intermediate, and innovative) has revealed the conflict between the goals, being the intermediate exploitation, which gets the best results. The conclusions show that the flexibilization of financing options positively affects the economic performance.

Keywords: Agricultural financing · Multicriteria programming ·
Sustainability rice farming · Decision making · Goals programming

1 Introduction

The rice in Latin America and the Caribbean (LAC) is a crop of great social and economic importance. Rice consumption in LAC has significantly increased during the last few years, presenting a current average of 30 kg per person per year. The

© Springer Nature Switzerland AG 2019
H. Lu et al. (Eds.): ISNN 2019, LNCS 11555, pp. 210–221, 2019.
https://doi.org/10.1007/978-3-030-22808-8_22

particularity of the rice produced in LAC lies in its high grain quality, and a production that is most often performed under mechanized systems with direct seeding. In LAC, rice is produced under irrigation and rainfed areas in different eco-regions (temperate, tropical moist, tropical dry). Food security and climate change constitute a challenge for rice market in LAC, facing the need to increase production, but stabilizing the yields and the grain quality [1].

The objective of this research is to evaluate the economic sustainability of the agricultural production chain of rice in Latin America through a multicriteria programming model that represents the rice exploitations in Latin America, considering the short-term financing and production. This model will facilitate the formulation of agricultural policies and will serve to reorient the services now provided by institutions. At the same time, the study will allow to deepen in the understanding of real goals of farmworkers and the importance they attach to the different criteria for selecting sources.

2 Literature Review

Mathematical programming has been widely used for analyzing financing alternatives in enterprises through financial planning models in the short-term under both certainty [2], and uncertainty conditions [3, 4]. These models consider alternatives such as long-term loans, credit lines, loans postponement, trade credits, and pledge.

In the field of agriculture, for incorporating credit in mathematical programming models, the year is divided into time periods and the circulating capital requirements are added in each of them [5]. If credits are acquired, the interest payments must be added as a cost in the objective function. Alternative financing sources such as traditional lenders, credit unions, or banks can be added as separate activities and it is possible to integrate restrictions on credit limitations by source type [6, 7].

When several objectives must be integrated, including qualitative data, the multicriteria programming can be used. This technique assumes that economic agents seek to find a balance or compromise between a set of objectives, usually in conflict [8, 9]. The commonly used methodologies for solving this issue are the restrictions method, the weighting method, and the goals and commitments programming method [10].

The financial management of agricultural exploitations has been presented by models that use the multicriteria analysis and its different resolution methods, integrating objectives related to agricultural planning [11], which consider the objective of indebtedness minimization. The following researches are the basis for the development of this study [12, 13, 14].

3 Method

The research was developed in three phases, according to [15, 16, 17]. The first of them consisted of a documentary monographic research on the statistical information about the risk sources in rice production, mainly about the interest rate, inflation, and the financial problems, among others.

The second one consisted of a study of the reality of a sample of the rice farmers in 5 Latin American countries, among them: Brazil, Cuba, Colombia, Venezuela, and Uruguay. A population of 1648 producers was considered, from which a sample was extracted by means of stratified random sampling. For this purpose, the sample was divided into 6 strata (from 1 to 50 hectares, from 51 to 100 hectares, from 101 to 150 hectares, from 151 to 200 hectares, from 201 to 500 hectares, and greater than 500 hectares).

Finally, the sample was constituted by 160 producers. An optimal allocation was made, which is the distribution method for a given sample size in n units since it produces more accurate results [10, 11], thus constituting the strata. The producers were randomly selected by the method of random numbers [6].

In order to characterize these producers, a technical survey was applied and 18 of the qualified entities of the Latin American financial institutions were interviewed, applying both instruments at the end of 2018.

Additionally, interviews were conducted with experts in the field of rice financing and cultivation. Given the limitation to apply the model to each exploitation, a typology was made of the exploitations under study, based on the characteristics of an innovative producer in relation with its financial decisions [15]. The variables considered were diversification of the number of sources, dominance of economic criteria over personal criteria, lower total cost per hectare, lower financial cost, lower percentage of external financing, desire to explore other sources, and higher performance [16].

These variables were entered into the Statgraphics plus 5 software to establish the cluster, using the maximum distance method (Furthest Neighbor) and the Squared Euclidean distance measure [18].

The third phase centered on the design of a multicriteria programming model that represents the economic operation of the exploitations in the typology, using the average of the explotations as data from each group. The model is mathematically presented in Gams language and developed through non-linear programming, adding some scenarios to see the behavior of the model against some changes in its variables [11, 12].

4   Results and Analysis

4.1   Presentation of the Area Under Study and Survey Results

The results of the surveys applied to rice producers in the Latin American sample show that the most important general problems were the limitations of small producers to access private credits, high dependence on production with external financing, and scarcity of medium and long-term credits. The benefits of the crops are highly influenced by inflation, the exchange rate, the State price fixing, and the political conditions at the time of the survey.

In rice production, there is a horizontal integration between producers and between financing sources, as well as vertical integration between the producer and the sources through the harvest; scarce problems of asymmetric information, differences in credit conditions in relation to interest rates, granted amounts, time of granting, closeness to the client, technical assistance, and responsibility, among other features. The most used criteria for choosing the source by the part of the producers were: the opportunity (referred to the granting of credit in the required time), interest rate, amount, client, trust, input availability, technical assistance, deadline, organization, and proximity.

The most used financing institutions were the associations, followed by the commercial houses, banks, public institutions, and the agroindustries.

4.2   Farmers Typology

Three groups were formed: innovative, conservative, and intermediate exploitation depending from their attitude toward financing [19]. The results are shown in Table 1.

Table 1. Results of the cluster and centroids

Cluster	No of producers	%	NS	OPP	CLI	TC	FC	FFX	CA	RY
1	87	54.38	1.85	0.92	0.71	1.2342	0.1023	61	0.62	**5.102**
2	43	26.88	1.95	0.832	0.12	1.5821	0.115	60.2	0.93	5.839
3	**30**	**18.75**	**0.795**	**0.42**	**0**	1.1025	**0.0452**	32	**0.7**	4.93

NS: number of sources, OPP: opportunity; CLI: client; TC: total cost; FC: financing cost; FFX: foreign financing percentage; CA: desire to explore other forms of financing; RY: rice yield (kg/ha).

4.2.1   Intermediate Exploitation

This type presents average values with respect to the other groups in variables such as yields, income, financial cost for interest payments, profit, depreciation, and total cost. About the criteria for choosing the financing source, they give more value to trust, being a client of the financial institution, opportunity, and interest. By having these first two values higher than the rest of the types, it can be said that they give an important weight to personal criteria and are not interested in exploring new sources. On the other hand, this group includes those farmworkers who sow in a staggered way and diversify the sources. It can be concluded that, from a technical point of view, they are innovative or have no limitations to carry out a greater number of sowing but are more reluctant to explore new sources.

4.2.2   Innovative Exploitation

In this group, the income is greater but the variable costs, financing costs, benefit, and the depreciation are also higher. For this type, the most important criteria are opportunity, amount, technical assistance, and interest, prevailing the economic criterion. They also value the shared risk by preferring technical assistance. In addition, the data suggest an open disposition to innovations.

4.2.3   Conservative Exploitation

This type presents low returns, income, total cost, benefit, and average depreciation. Interests are more important than the opportunity and show low values in relation to the other criteria. In relation to the ratios, this group shows the greatest relationships, so it can be said that they are efficient from an economic point of view.

4.3 Design of the Optimization Model

The fundamental features of the model can be summarized as follows: descriptive, with qualitative and quantitative variables; economic-productive, integrating linear and non-linear functions; in the short term, with random elements and multiple objectives which include the criteria for choosing farmworkers with respect to their financing source.

For the model, the following assumptions are assumed [15]: the alternative activities are the financing sources; the introduction of other crops or technologies is not proposed; only rice production with two crops per year is considered; the same variety of rice is used, and with the same price. The producer is solvent and has the necessary guarantees to take loans, no agricultural insurance is taken, and pays the established interest; there are no additional charges for technical assistance, insurance, or contingency funds. The credit restrictions are taken according to the information provided by the financial sources; credits are requested at the beginning of the sowing and are paid at the end.

Following is the programming model used to optimize the sample through algorithms 1 and 2, following the model of [21]:

Algorithm 1. Optimize objectives and agronomic constraints
(1) Objective - profit maximization in thousands of dollars:

$$Z = (\sum_{p}^{12} (IT_p - CVT_P - Cfij_p)) - CFT \tag{1}$$

(2) Objective - satisfaction of producer's preferences regarding the financing source:

$$W = \sum_{f=1}^{12} \sum_{s=1}^{12} (Ef_f * X_{fs}) \tag{2}$$

(3) Objective - risk minimization:

$$V = \frac{1}{n} \sum_{n=1}^{7} (ZN_n - Z^2) \tag{3}$$

Subject to:
Restrictions on surface occupation per period:

$$\sum_{s=1}^{12} usosu_{sp} * NHA_s \leq d \text{ sup} \tag{4}$$

Restrictions on occupation of the total area per planting:

$$\sum_{s=1}^{12} NHA_s = 2 * d \text{ sup} \tag{5}$$

Equation of Random Benefit Calculation:

$$ZN_n = (\sum_{s=1}^{12} \sum_{p=1}^{12} (ren_s * pr_{pn}) * NHA_s) - (\sum_{p=1}^{12} (CVT_p + Cfij_p))$$
$$- (\sum_{f=1}^{5} \sum_{s=1}^{12} \sum_{p=1}^{12} (Vs_{ps} * X_{fs} * ti_{fn} * vu_s)) \tag{6}$$

Where:

Z: total exploitation benefit; ITp: total income; CVTp: total variable cost; Cfijp: fixed cost per period; CFT: total financial cost and p: period; W: measure that establishes the order of the financing sources according to the producer's preferences; Eff: distance L1 from the weight of the score provided by the producers to each criterion by rating of each standardized source; Xfs: amount of credit requested by source; V: benefit variance; ZNn: benefit in each state of nature (n); usosusp: period of time in which each planting of rice occupies the soil surface; NHAs: surface to be sown; dsup: maximum surface availability in hectares; rens: rice yields in kg/ha per planting; prpn: sale price of rice production for each state of nature (thousands US$/kg); Vsps: month of the production's sale; vus: credit useful life corresponding to months (periods); and tifn: monthly interest rate for each state of nature.

Algorithm 2. Financial restrictions

Restrictions on the total amount of credit to be requested by planting:

$$\sum_{f=1}^{12} X_{fs} < CVT_p \tag{7}$$

Restrictions of maximum amount granted by financing source:

$$X_{fs} \leq mf_f * d \text{ sup} \tag{8}$$

Financial cost per period:

$$CFp_p = (\sum_{f=1}\sum_{s=1} Vs_{sp} * X_{fs}) * (tr_f * Vu_s) \tag{9}$$

Credit payment per period:

$$AMOR_p = (\sum_{f=1}^{12}\sum_{s=1}^{12} Vs_{sp} * X_{fs}) * (1 + (tr_f * Vu_s)) \tag{10}$$

Total financial cost:

$$CFT = \sum_{p=1}^{12} CFp_p \tag{11}$$

General Balance per period:

$$(\sum_{f=1}^{12}\sum_{s=1}^{12} X_{fs} * is_{sp}) + liq0_p + LIQ_{(p-1)} + IT_p = CVT_p + CONS_p + LIQ_p + AMOR_p + Cfij_p \tag{12}$$

Short-term financial balance equation per period:

$$(\sum_{f=1}^{5}\sum_{s=1}^{12} X_{fs} * is_{sp}) + liq0_p + LIQ_{(p-1)} \geq CVT_p \tag{13}$$

Restrictions on the amounts covered by financing sources, planting, and period:
In the case of financing by agrocommerce.

$$X_{agrocomercio,s} \leq CVL_{s,cmprod,p} + CVL_{s,cpprod,p} + CVL_{s,fprod,p} + CVL_{s,rprod,p} \tag{14}$$

In the case of financing by agroindustry.

$$X_{agroindustria,s} \leq CVL_{s,semilla,p} + CVL_{s,cmprod,p} + CVL_{s,cpprod,p} + CVL_{s,fprod,p} + CVL_{s,rprod,p} \tag{15}$$

Where:
X_{fs}: amount of credit that is requested by source and by seeding; CVT_p: total variable cost; mff: amount to be financed by the source (thousands of US\$/ha); Vssp: month of the production sale in each sowing; trf: interest rate per period for each source monthly percentage; Vus: useful life of the credit which corresponds to the months (periods) between the granting and its payment; CFp_p: financial cost period, liq0p: initial liquidity per planting; $LIQp-1$: liquidity of the previous period; IT_p: total income; CVT_p: total variable cost; $CONS_p$: monthly family consumption; LIQ_p: liquidity of the period; $AMOR_p$: payment of the credit; $Cfij_p$: fixed cost, $X_{agrocomercio,s}$: amount of credit that is requested to the agrocommerce; Cvlslp: variable cost per task; cmprod: cost per products for weed control; cpprod: cost per products for pest control; fprod: cost per products for fertilization; rprod: cost per products for the reboot; $X_{agroindustria,s}$: amount of credit requested by the agroindustry; and, $CVL_{s,semilla,p}$: variable costs by seed purchase for planting rice.

As the exclusive means of external financing used by producers is credit, the alternatives presented correspond to the existing sources of credit. Five types of institutions are established (banking, association, public, agrocommerce, and agroindustry), while the data assumed for the model is calculated using the average or the mode of the characteristics in the eighteen surveyed sources.

The objectives are [21]: maximize the benefit (max Z), satisfaction of the producer's preferences with respect to the financing source (min W) and minimize the risk (min V). These objectives contribute equal weight. The objective of min W is used as an ordinal condition (a fictitious objective) to establish a measure that reflects the order of funding sources, according to the producer's preferences. For this purpose, the amounts of credits to be requested are used, multiplied by the weight attributed by the producer to each chosen source according to their criteria. Therefore, the smaller its value (minimization), the closer it is to the observed producer's preferences with respect to its financing source.

The procedure for converting the qualitative selection criteria into quantitative variables was as follows [1]: identification of opinions related to financing sources, quantification of the attribute according to the number of producers who vote for each of them, introduction of the distance between the assigned values and the ideal value. Each choice or non-dominated alternative is considered an efficient endpoint. Calculation of the distances between each alternative or efficient endpoint with respect to the ideal point, for the L1 metric.

The model incorporates the risk in the cultivation of rice, which is affected by some variables, including climatology (rainfall), productivity (attack of weeds and pests), economy by the change in price of the product, and financial variables by the variation of the credit interest rate. In the model, the risk is measured by means of the variance in benefit, integrating the variation of the interest rate (as a financial risk) and the price of the rice (as business risk) using its historical series of the last seven years. The producer can improve its financing if its management and productive processes are improved. Therefore, several scenarios and sensitivity analyses were considered, as follows [14, 21]:

Scenario E1: Base Model. The multi-objective model of exploitation described above is considered.

Scenario E2: Credits per Periods. A credit modality is introduced in the base model in which the producer can have the credit in the month it needs it and pays interests according to the months elapsed from its granting to the payment date. The difference of this type of credit, given by items, is that the interest rate is the same as at the beginning of the sowing when it was requested. However, in this case, the period for calculating the loan payment and interests will be that of the months that elapsed from the withdrawn until the payment date.

Scenario E3: Stepped Planting. It integrates the possibility of selecting up to twelve plantings of rice per year, one for each month. The stepped planting is recommended by technicians of the zon, because it generates a periodicity in the income that will favor the producer and diminishes the foreign financing requirements. It is achieved by making time periods in a circular way where the planting number thirteen (s13) becomes the first planting (s1) again.

Sensitivity Analysis: new prices are introduced so that two possibilities are appraised, namely: a 10% increase in prices, which should improve the results in the base model and a price decrease by 10%. In this way, the sensitivity of the base model to this change was identified.

4.4   Results of the Model

For the resolution of the base model, it is assumed that the exploitation is cultivated in its entirety, for all eploitations, in both sowings (summer and winter). This is introduced to achieve that the analysis of the selection behavior about the amounts and financing sources are comparable in the three objectives [13]. The model calculates total revenues and costs according to the financial cost generated by the interest accrued by the credits requested from the financing source, considered as optimal by the model (Table 2).

Table 2. Economic calculations of the base model for profit maximization

Exploitation	Income	Costs	Costs financial	Benefit
Intermediate	3.068952346	2.196523741	0.092	0.957
Innovative	3.41256327	2.423058941	0.132	0.899
Conservative	2.895742385	2.0005211	0.093	0.735

The payment matrix of the base model objectives is presented in Table 3. In this matrix, the conflict that represents the variation between the ideal (in bold) and the anti-ideal (underlined) between the objectives when max Z, for the INTERMEDIA exploitation is 31.7% and min W represents 120,15%; however, in minV, the conflict is 18.41%. In the case of INNOVATIVE exploitation, max Z presents a conflict of 33.23%. On the other hand, min W of 170.01%, and the risk of 18.52%. However, the CONSERVATIVE exploitation presents a conflict of 40.03% in profits, higher than the rest of the holdings, while for the case of min W, it presents a lower conflict than the rest of 64.29%, and 18.03% for min V.

Table 3. Payments matrix of the base model

Exploitation	FO	Z (thousands of $ US)	W	V
Intermediate	Max Z	**72.542**	5,754.32	**2,682.10**
	Min W	59.98	**2,512.42**	2,410.15
	Min V	55.242	4,341.23	2,299.42
Innovative	Max Z	**45.524**	5,252.13	**1,615.12**
	Min W	34.899	**1,852.10**	1,442.12
	Min V	34.312	3,754.44	1,412.25
Conservative	Max Z	**79.015**	5,098.91	**3,852.12**
	Min W	78.998	**4,785.14**	3,851.41
	Min V	**56.725**	7,995.73	3,751.10

Despite the little conflict manifested in the minimization of risk, its integration into the model introduces changes in the source selection since risk is a fundamental element in the granting of a loan. The low conflict is since the variation of the interest rate affects the financial cost, which is a part of the total costs. Therefore, its magnitude is small in relation to the random benefits. Starting from the payment matrix, the extreme points closest to the ideal were obtained through the programming goals. The goal set was calculated, which defines the restricted efficient set to obtain efficient solutions of max Z, min W, and min V together. In relation to the benefit, the target solution per hectare shown in Table 4 is between 0.795 to 0.802 thousand US$/ha for the INTERMEDIATE operation, at 0.707 US$/ha for the INNOVATIVE, and between 0.761 and 0.710 US$/ha for the CONSERVATIVE; therefore, it is higher in the INTERMEDIA exploitation.

Table 4. Results for L1 and L? in the base model

Metric and exploitation	Z (thousands of US$)	W	V
Metric L1			
Intermediate	60,042	2,595.46	**2,481.22**
Innovative	36,508	1,966.61	**1,405.12**
Conservative	76,568	4,891.05	**3,814.29**
Metric L?			
Intermediate	60,591	2,668.39	**2,464.87**
Innovative	36,508	1,966.61	**1,405.12**
Conservative	**71.38**	**5,373.19**	**3,702.58**

L1: distance one. L?: infinite distance

Figure 1 compares the choice of financing source between the producers and the Base Model for the different objectives.

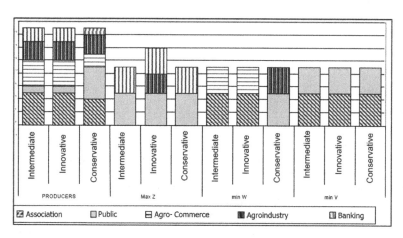

Fig. 1. Comparison of the choice of financing source between producers and the base model for the different objectives.

When observing the financing sources chosen in the base model by objective (Fig. 1) where the thickest color strip is the first choice and the last one is the thinnest, it means that for the max Z, the INTERMEDIA exploitation first chooses the public source and then banking. For INNOVADORA, the order will be public, banking, and agroindustry; for the CONSERVATIVE, it will be public and banking, for min W, the INTERMEDIA and INNOVATIVE exploitations use the association and agro-commerce source, while the CONSERVATIVE chooses the public sources and agroindustry.

In the objective min V, they use association and public. The model reproduces the producer's behavior in the second objective when it establishes the satisfaction of the producer's preferences with respect to the financing source, except in the case of the CONSERVATIVE, which uses agrocommerce in the second place instead of the association, as happens with the producers.

5   Conclusions

The studied producers give great value to their criteria for choosing the financing source, including the opportunity (credit granting time) and shared risk (evidenced through the criteria of being a client, technical assistance, and trust with the source). Not always the most innovative producers obtain greater benefits with respect to financing, probably because they are not looking for better financial opportunities. The resulting alternatives privilege the public source and the banks in the maximization of benefits. The same occurs with the association and agrocommerce source in the satisfaction of preferences, and with the association and public source in the risk minimizing.

The model shows results that positively respond to an increase in profits when the price increase is introduced, the stepped planting, the credit per period and, negatively, to the decrease in prices. The benefits are highly affected by price variations. As the analyzed scenarios and variations can be used together, it is expected that a producer who obtains periodic credits and staggers in a price increase, reaches the maximum benefit. When facing a price reduction, producers of the conservative type have a lower impact on the reduction of benefits.

References

1. Rebolledo, M.C., et al.: Modelación del arroz en Latinoamérica: Estado del arte y base de datos para parametrización. Publications Office of the European Union, Luxembourg (2018)
2. Carrijo, D.R., Lundy, M.E., Linquist, B.A.: Rice yields and water use under alternate wetting and drying irrigation: a meta-analysis. Field Crops Res. 203(1), 173–180 (2017)
3. Pérez, M.P., Cortiza, M.A.P.: Los rendimientos arroceros en cuba: propuesta de un sistema de acciones. Revista Economía y Desarrollo (Impresa) 152(2), 138–154 (2016)
4. Lezama, O.B.P., Izquierdo, N.V., Fernández, D.P., Dorta, R.L.G., Viloria, A., Marín, L.R.: Models of multivariate regression for labor accidents in different production sectors: comparative study. In: Tan, Y., Shi, Y., Tang, Q. (eds.) DMBD 2018. LNCS, vol. 10943, pp. 43–52. Springer, Cham (2018). https://doi.org/10.1007/978-3-319-93803-5_5

5. Suárez, J.A., Beatón, P.A., Escalona, R.F., Montero, O.P.: Energy, environment and development in Cuba. Renew. Sustain. Energy Rev. **16**(5), 2724–2731 (2012)
6. Sala, S., Ciuffo, B., Nijkamp, P.: A systemic framework for sustainability assessment. Ecol. Econ. **119**(1), 314–325 (2015)
7. Singh, R.K., Murty, H.R., Gupta, S.K., Dikshit, A.K.: An overview of sustainability assessment methodologies. Ecol. Ind. **9**(2), 189–212 (2009)
8. Varela, N., Fernandez, D., Pineda, O., Viloria, A.: Selection of the best regression model to explain the variables that influence labor accident case electrical company. J. Eng. Appl. Sci. **12**(1), 2956–2962 (2017)
9. Yao, Z., Zheng, X., Liu, C., Lin, S., Zuo, Q., Butterbach-Bahl, K.: Improving rice production sustainability by reducing water demand and greenhouse gas emissions with biodegradable films. Sci. Rep. **7**(1), 1–12 (2017)
10. Suárez, D.F.P., Román, R.M.S.: Consumo de água em arroz irrigado por inundação em sistema de multiplas entradas. IRRIGA **1**(1), 78–95 (2016)
11. Stuart, A.M., et al.: The application of best management practices increases the profitability and sustainability of rice farming in the central plains of Thailand. Field Crops Res. **220**(1), 78–87 (2018)
12. Aprianti, E., Shafigh, P., Bahri, S., Farahani, J.N.: Supplementary cementitious materials origin from agricultural wastes–a review. Constr. Build. Mater. **74**(1), 176–187 (2015)
13. Gomes, A.D.S., Scivittaro, W.B., Petrini, J.A., Ferreira, L.H.G.: A água: distribuição, regulamentação e uso na agricultura, com enfase ao arroz irrigado. EMBRAPA Clima Temperado-Documentos (INFOTECA-E) (2018)
14. Donkor, E., Owusu, V.: Effects of land tenure systems on resource-use productivity and efficiency in Ghana's rice industry. Afr. J. Agric. Resour. Econ. **9**(4), 286–299 (2014)
15. Baloch, M.A., Thapa, G.B.: The effect of agricultural extension services: date farmers' case in Balochistan, Pakistan. J. Saudi Soc. Agric. Sci. **17**(3), 282–289 (2018)
16. Izquierdo, N.V., Lezama, O.B.P., Dorta, R.G., Viloria, A., Deras, I., Hernández-Fernández, L.: Fuzzy logic applied to the performance evaluation. honduran coffee sector case. In: Tan, Y., Shi, Y., Tang, Q. (eds.) ICSI 2018. LNCS, vol. 10942, pp. 164–173. Springer, Cham (2018). https://doi.org/10.1007/978-3-319-93818-9_16
17. Bezerra, B.G., Da Silva, B.B., Bezerra, J.R.C., Brandão, Z.N.: Evapotranspiração real obtida através da relação entre o coeficiente dual de cultura da FAO-56 e o NDVI. Revista Brasileira de Meteorologia **25**(3), 404–414 (2010)
18. Diaz-Balteiro, L., González-Pachón, J., Romero, C.: Forest management with multiple criteria and multiple stakeholders: An application to two public forests in Spain. Scand. J. For. Res. **24**(1), 87–93 (2009)
19. Hák, T., Janoušková, S., Moldan, B.: Sustainable development goals: a need for relevant indicators. Ecol. Ind. **60**(1), 565–573 (2016)
20. Lampayan, R.M., Rejesus, R.M., Singleton, G.R., Bouman, B.A.: Adoption and economics of alternate wetting and drying water management for irrigated lowland rice. Field Crops Res. **170**(1), 95–108 (2015)
21. Delgado, A., Blanco, F.M.: Modelo Multicriterio Para El Análisis De Alternativas De Financiamiento De Productores De Arroz En El Estado Portuguesa. Venezuela. Agroalimentaria **28**(1), 35–48 (2009)

Application of Machine Learning Based Technique for High Impedance Fault Detection in Power Distribution Network

Katleho Moloi[✉], Jaco Jordaan[✉], and Yskandar Hamam[✉]

Department of Electrical Engineering, Tshwane University of Technology,
Pretoria, South Africa
Moloikt023@gmail.com, {JordaanJA, HamamA}@tut.ac.za

Abstract. High-impedance faults (HIFs) detection with high reliability has been a prominent challenge for protection engineers over the years. This is mainly because of the nature and characteristics this type of fault has. Although HIFs do not directly pose danger to the power system equipment, they pose a serious threat to the public and agricultural environment. In this paper, a technique which comprises of a signal decomposition technique, feature extraction, feature selection and fault classification is proposed. A practical experiment was conducted to validate the proposed method. The scheme is implemented in MATLAB and tested on the machine intelligence platform WEKA. The scheme was tested on different classifiers and showed impressive results for both simulations and practical cases.

Keywords: Fault classification · Fault detection · Feature extraction ·
High impedance fault · Power system · Fault detection

1 Introduction

The ever-growing demand and expansion of electricity has brought complexity within the power utilities. Nonetheless, power system protection philosophies are still required to reliably detect any faults which may arise in the system. This requirement has been well-achieved by conventional protection for dynamic fault detection, which results in a drastic increase in fault current at the location of the fault. However, conventional protection has proven to have limitations in detecting high impedance faults (HIFs) [1]. HIFs usually occur when an energized overhead conductor makes contact with a high impedance object, either by falling on the ground or making contact with the object. In both cases, the fault current is limited below the threshold value detectable by over-current protection devices. In either case, HIFs can ignite fires, which may negatively impact the community [2]. It is for this reason that reliable detection of HIFs is required. A comprehensive literature review of methods proposed for HIF detection from classical to heuristically has been presented in [1]. The authors provide a summary of about 255 HIF proposed methods between the years 1960 to 2008. This may form part of a useful guide in tracking the progress made in an attempt to find a more suitable technique for HIFs detection. Fractal theorem (FT) [3], discrete wavelet

H. Lu et al. (Eds.): ISNN 2019, LNCS 11555, pp. 222–231, 2019.
https://doi.org/10.1007/978-3-030-22808-8_23

transforms (DWT) [4, 5], neural network (NN) [6, 7] and Fuzzy logic (FL) [8] were amongst her methods proposed to detect HIFs.

In later years, there were more developments in techniques proposed to detect HIFs. An inter-harmonic content algorithm was developed to detect HIFs [9]. A real-time based method using transients induced by HIFs was incorporated using with wavelet transform (WT) to calculate the energy content was proposed to detect HIFs [10]. Another method based on using sum approach to detect HIFs was presented in [11]. A technique based on smart meters for HIF detection and location in distribution network was proposed in [12] this technique uses the voltage unbalanced based approach. The method uses Mathematical Morphology (MM) and Spectral Analysis (SA) to diagnose HIFs. In [13] the method based on neuro-fuzzy learning is proposed. The method uses WPT for feature extraction combined with a variation of evolving neuro-fuzzy network. The lengthy discussion of the attempts made to develop a more suitable technique for HIFs detection reveals that, although much work has been done there is still a need to enhance the development of a more practical solution. In this paper, we introduce the concept of data window reduction to improve the window shift method. In Sect. 2, the signal processing tool is introduced, Sect. 3 discusses the feature extraction, selection and fault classification schemes. The modelling of the power system under study is outlined in Sect. 4. The proposed method is discussed in Sect. 5. The results of the proposed method based on simulations and experimental work is also outlined in Sect. 5. Lastly, the conclusion and future work are discussed in Sect. 6.

2 Signal Processing Technique

2.1 Discrete Wavelet Transform (DWT)

Discrete wavelet transform (DWT) has over the years emerged as a powerful signal processing tool. This is because of DWT has the ability to give both frequency and time information simultaneously. This has proven to be effective for transient signal analysis and showed good results in power system applications [14]. The discrete wavelet can be mathematically expressed as:

$$x(t) = \sum_{k} c_j(k)\phi(t - k) + \sum_{k}\sum_{j-1} d_j(k)\varphi(2^{-j}t - k) \tag{1}$$

where c_j and d_j are the approximation and detail coefficient respectively. ϕ is the scaling function and φ is the wavelet function. The approximation and detail coefficients represent the low and high pass filters outputs respectively.

2.2 Multiresolution Analysis (MRA)

The MRA technique is used to decompose the signal into its c_j and d_j, this is done by passing the signal through a series of high and low filters. The scaling and the wavelet coefficients of a signal $x(n)$ at different levels can be determined by the following equations:

$$c_j(n) = \sum_K x(n).h(2n - k),$$ (2)

$$d_j(n) = \sum_K x(n).g(2n - k).$$ (3)

3 Feature Extraction and Selection

Feature extraction forms the basis of most pattern recognition and classification schemes. In order to effectively classify different types of faults and incidents in a power system, the set of features should be compatible in size and have minimum computational burden [15]. In this work statistical features are extracted from the decomposed signal and subsequently used as inputs to classifiers. Four cases are considered in this work namely the HIFs, capacitor switching (CS), normal operation (NO) and load switching (LS). Feature selection involves the critical selection of an appropriate feature which may be used in classification problems. The viability of the extracted features using DWT is studied using three different machine learning techniques: (i) information gain, (ii) gain ratio and (ii) support vector machine. In this work i_{HIF}, i_{CS}, i_{NC}, and i_{LS} represent the HIF current, CS current, NO current and LS current features respectively. Similarly, v_{HIF}, v_{CS}, v_{NC}, and v_{LS} represent the HIF voltage, CS Voltage, NO voltage and LS voltage features.

Table 1. The feature set ranking matrix.

Feature	IG	GR	SVM
i_{HIF}	0.35	0.91	3
I_{IC}	0.40	0.92	6
I_{NO}	0.35	0.90	2
I_{LS}	0.51	1.23	9
v_{HIF}	0.23	0.33	1
V_{IC}	0.23	0.42	5
v_{NC}	0.23	0.44	11
v_{LS}	0.40	0.42	14

3.1 Fault Classification

In order to effectively distinguish HIFs from other power system operations, a classification scheme is required. In this work, three classifiers are studied namely the SVM, J48 and neural network (NN). In this work, delay samples are used.

4   Modelling of the Radial Distribution System – A Case Study

The distribution system under study in this work is presented in Fig. 1. The network is a representation of an Eskom power system. Eskom is the largest power utility in Africa. The system comprises of the three feeders, (i) Middleburg, (ii) Witbank and (iii) Trans-alloys, with the length of the lines of 55 km, 28.3 km and 16.9 km at 22 kV supply voltage. The positive resistance and reactance sequences parameters of the conductor represented by the power systems are 0.195 (Ω/km) and 0.183 (Ω/km) respectively. The zero phase resistance and reactance sequences parameters are 0.453 (Ω/km) and 0.453 (Ω/km) respectively. The source parameters of the system are presented in Table 1. The loading on the Middleburg, Witbank and Trans-alloys are 12.8 MVA, 9.2 MVA and 7.6 MVA respectively. In this work, the Emmanuel arc model for HIF is used in this work [16] (Table 2).

Table 2. Source parameters.

Source	Short circuit power (MVA)	Short circuit current (KA)	X/R ratio	X0/X1 ratio	R0/R1 ratio
Hillside	61.9	16.3	133.1	5.52	12.12

5   The Proposed Method for HIF Detection

The proposed method for HIF detection follows a sequence of various schemes to adequately address the problem. The first stage is the current measuring, subsequent to that, the current is decomposed to level 5 using DWT. Features are then extracted from the decomposed signal and the best feature scheme is employed to select the best features to minimize the data sample window. The features are then used to train different classifiers in order to detect and distinguish HIFs from other power system operation. The schematic diagram of the proposed method is presented in Fig. 2.

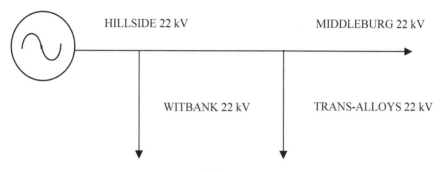

Fig. 1. Eskom's distribution system.

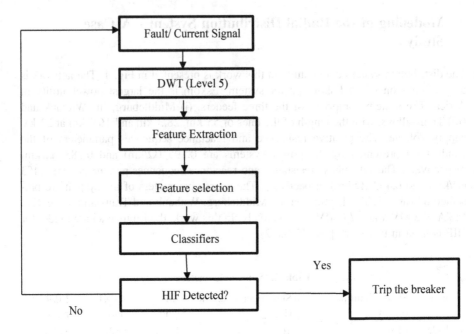

Fig. 2. Proposed method schematic diagram.

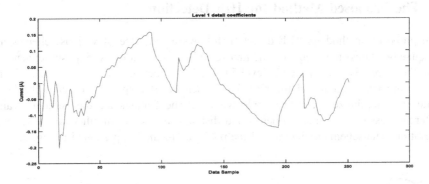

Fig. 3. Level 1 detail coefficient of HIF using sym4.

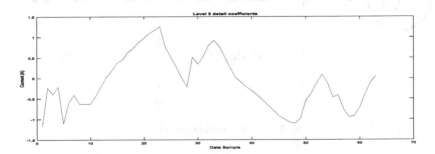

Fig. 4. Level 2 detail coefficient of HIF using sym4.

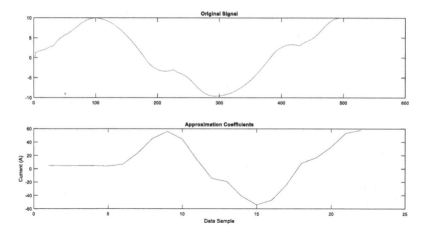

Fig. 5. HIF original and approximated signal using sym4.

5.1   Simulations and Discussions

The simulations performed in this work were carried out using MATLAB/Simulink platform and the proposed method was tested using WEKA. When using DWT it is important to select carefully the mother wavelet. In this work symlet4 (sym4) mother wavelet was used as it has been shown that it performs better for transient signals [14]. The decomposed signal at level 5 (level 1 and 2) for HIF is presented from Figs. 3, 4 and 5. The performance of the classifiers (SVM, J48 and NN) based on the proposed method for different time delays are presented in Tables 3, 4 and 5.

Table 3. SVM classifier performance for different time delays.

SVM classifier parameters	4 samples delay	¼ cycle delay	½ cycle delay
Time taken to build the model	12.5	11.3	15.8
Number of cases	16 250	25 025	32 580
Correct classified	2 028	24300	31 625
Incorrectly classified	14 222	725	1 225
Correctly classified %	87.5	97.1	96.24
Incorrectly classified %	12.5	2.9	3.76
Mean absolute error	0.0158	0.0098	0.0168
Root mean	0.0113	0.0	0.0781
Kappa statistic	0.883	0.938	0.965
Coverage cases (0.95 level)%	95.231	95.661	96.281

Table 4. J48 classifier performance for different time delays.

J48 classifier parameters	4 samples delay	¼ cycle delay	½ cycle delay
Time taken to build the model	12.5	11.3	15.8
Number of cases	16 250	25 025	32 580
Correct classified	14 350	23 980	31 435
Incorrectly classified	1 900	1 045	1 145
Correctly classified %	88.3	95.8	96.5
Incorrectly classified %	11.7	4.2	3.5
Mean absolute error	0.0009	0.0098	0.0397
Root mean	0.0185	0.0985	0.0892
Kappa statistic	0.889	0.959	0.966
Coverage cases (0.95 level)%	95.101	96.261	97.181

Table 5. NN classifier performance for different time delays.

NN classifier parameters	4 samples delay	¼ cycle delay	½ cycle delay
Time taken to build the model	12.5	11.3	15.8
Number of cases	16 250	25 025	32 580
Correct classified	15 890	24 110	30 080
Incorrectly classified	360	915	2500
Correctly classified %	97.8	96.3	92.4
Incorrectly classified %	2.2	3.7	7.6
Mean absolute error	0.0085	0.0089	0.00667
Root mean	0.0867	0.0785	0.0858
Kappa statistic	0.976	0.938	0.921
Coverage cases (0.95 level)%	96.335	95.661	95.152

5.2   Experimental Work and Discussions

In order to practically validate the proposed method, an experiment was conducted. The experiment was performed at the high voltage laboratory, at the Tshwane University of

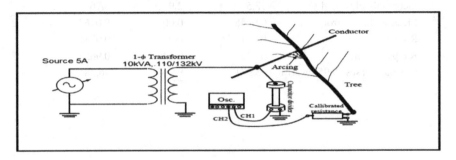

Fig. 6. Experimental configuration.

Technology, eMalahleni campus. The experimental configuration and setup are presented in Figs. 6 and 7. respectively. When the conductor is energized, an arc is created. The instrument used to perform the experiment is the ICM8 power analyzer to detect the presence of the arc (HIF). The experimental parameters are presented in Table 6.

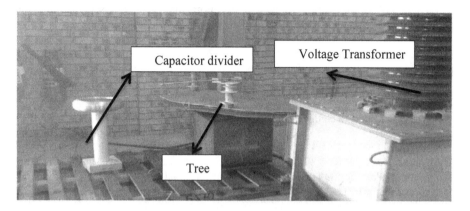

Fig. 7. Laboratory setup.

Table 6. Experimental set up parameters.

Source	Transformer	Mean absolute deviation	Atmosphere conditions
5A, 50 Hz, 2.5% source impedance	10kVA, 11/132 kV, 4.5%	HV 100 pF, 132 kV, LV 100nF	T = 31 °C

Table 7. Experimental Results.

Classifier	No. of tested cases	Correctly classified	Incorrectly classified	Accuracy (%)
SVM	100	92	8	92
J48	100	87	13	87
NN	100	96	4	96
Total	300	275	25	92

The experiment was conducted to determine the arc current and voltage, which are associated with HIF. A dry tree was used to bridge the gap between the electrodes to create a high impedance path and subsequently a HIF. The proposed method is subsequently tested using experimental data; the results are presented in Table 7, showing that NN performed better than the others.

6 Conclusion

In this paper, a technique for HIF is proposed. The method uses a signal decimation tool (DWT), a feature extraction and selection scheme (LG, GR and SVM) to select best possible features to be used as inputs to test and train classifiers (SVM, J48 and NN). The method showed good classification results both on simulated and experimental results. Future work will include location and prediction schemes and the application of this method to multi-grounded power distribution systems.

References

1. Sedighizadeh, M., Rezazadeh, A., Elkalashy, N.I.: Approaches in high impedance fault detection detection - a chronological review. Adv. Electr. Comput. Eng. **10**, 114–118 (2010)
2. Laaksonen, H., Hovila, P.: Straightforward detection method for high-impedance faults. Int. Rev. Electr. Eng. (IREE). **12**(2) (2017)
3. Mamishev, A.V., Russell, B.D., Benner, C.L.: Analysis of high impedance faults using fractal techniques. IEEE Trans. Power Sys. **11**(1), 435–440 (1996)
4. Lai, T.M., Snider, L.A., Lo, E.: High impedance fault detection using discrete wavelet transform and frequency range and RMS conversion. IEEE Trans. Power Deliver **20**(1), 397–407 (2003)
5. Michalik, M., Rebizant, W., Lukowicz, M., Lee, S.J., Kang, S.H.: High-Impedance fault detection in distribution networks with use of wavelet-based algorithm. IEEE Trans. Power Deliver. **21**(4), 1793–1802 (2006)
6. Ebron, S.: A neural network processing strategy for the detection of high impedance fault. Masters's Thesis, Electrical and Computer Engineering Department, NCSU (1988)
7. Ebron, S., Lubkeman, D.L., White, M.: A neural network approach to the detection of incipient fault on power distribution feeders. In: IEEE/PES Transmission and Distribution Conference, New Orleans, LA (1989)
8. Jota, F.G., Jota, P.S.: High-impedance fault identification using a fuzzy reasoning system. IEEE Prot., Gener., Transm., Distrib. **45**(6), 656–661 (1998)
9. Macedo, J.R., Resende, J.W., Bissochi, C.A.: Proposition of an Interharmonic-based methodology for high-impedance fault detection in distribution systems. IET Gener. Transm. Distrib. **9**(16), 2293–2601 (2015)
10. Costa, F.B., Souza, B.A., Brito, N.S.D., Silver, J.A.C.B., Santos, W.C.: Real-time detection of transients induced by high-impedance faults based on the boundary wavelet transform. IEEE Trans. Ind. Appl. **51**(6), 5312–5323 (2015)
11. Sarwagya, K., De, S., Nayak, P.K.: High-impedance fault detection in electrical power distribution systems using moving sum approach. IET The Institution of Engineering and Technology. 1–8 (2017)
12. Vieira, F.L., Filho, J.M.C., Silveira, P.M., Guerrero, C.A.V., Leite, M.P.: High impedance fault detection and location in distribution networks using smart meters. In: 18th International Conference on Harmonics and Quality of Power (ICHQP), Ljubljana, Slovenia (2018)
13. Silver, S., Costa, P., Santana, M., Leite, D.: Evolving neuro-fuzzy network for real-time high impedance fault detection and classification. Neural Comput. Appl. **40**(175), 1–14 (2018)

14. Magagula, X.G., Hamam, Y., Jordaan, J.A., Yusuff, A.A.: Fault Detection and classification method using DWT and SVM in a power distribution network. In: IEEE PES-/IAS, Ghana, Accra (2017)
15. Yusuff, A.A., Jimoh, A.A., Munda, J.L.: Determinant-based feature extraction for fault detection and classification for power transmission lines. IET Gener. Transm. Distrib. 5(12), 1259–1267 (2011)
16. Sekar, K., Mohanty, N.K.: Combined mathematics morphology and data mining based high impedance fault detection. In: 1st International Conference on Power Engineering, Computing and Control. VIT University, Chennai Campus (2017)

Artificial Neural Network Surrogate Modeling of Oil Reservoir: A Case Study

Oleg Sudakov[1], Dmitri Koroteev[1], Boris Belozerov[2], and Evgeny Burnaev[1(✉)]

[1] Skolkovo Institute of Science and Technology, Moscow, Russia
e.burnaev@skoltech.ru
[2] Gazprom Neft, Science and Technology Center, St. Petersburg, Russia

Abstract. We develop a data-driven model, introducing recent advances in machine learning to reservoir simulation. We use a conventional reservoir modeling tool to generate training set and a special ensemble of artificial neural networks (ANNs) to build a predictive model. The ANN-based model allows to reproduce the time dependence of fluids and pressure distribution within the computational cells of the reservoir model. We compare the performance of the ANN-based model with conventional reservoir modeling and illustrate that ANN-based model (1) is able to capture all the output parameters of the conventional model with very high accuracy and (2) demonstrate much higher computational performance. We finally elaborate on further options for research and developments within the area of reservoir modeling.

Keywords: Reservoir modeling · Machine learning ·
Surrogate modeling · Artificial neural networks

1 Introduction and Justification

Hydrodynamical reservoir modeling based on variations of Darcy's law has been an acknowledged foundation for field development for several decades [11,21,30]. This modeling is typically based on numerical integration of diffusion-type equations within a computational domain represented by several hundreds to several tens of billions of cells. The number of cells depends on the complexity of reservoir geology, field development system and placement of the wells. Some of the modern reservoir simulation software packages support multi-CPU and sometimes multi-GPU environments, which allow to speed up the computations. But even highly parallelized software releases can execute a single field development scenario for several days and even weeks for complex reservoir systems.

We strive to build models for fast approximation, or the so-called surrogate models, to estimate selected properties based on the results of physical modeling (for details, see example of surrogate model application in [5,7]). Surrogate models are a well-known way of solving various industrial engineering problems [4,6,12,15,17,27–29].

© Springer Nature Switzerland AG 2019
H. Lu et al. (Eds.): ISNN 2019, LNCS 11555, pp. 232–241, 2019.
https://doi.org/10.1007/978-3-030-22808-8_24

Previous efforts on surrogate modeling for reservoir development could be found in [14, 16, 18–20, 26]. These approaches were able to predict either well-by-well or a field scale production. Our approach differs significantly from the ones presented in [14, 16, 18–20, 26]. We aim to predict the evolution of distribution of pressures and reservoir fluids within all the computational cells.

Therefore, in this paper we consider utilization of recent advances in deep Artificial Neural Network based models (ANNs) to speed up the simulations by substituting specific time consuming portions of physics-driven simulation by relatively fast data-driven forecast.

In Sect. 2 we describe the initial physical simulator, used ANN architecture and its training. In Sect. 3 we provide results of computational experiments. In Sect. 4 we draw conclusions and discuss possible future research directions.

2   Methods

2.1   Description of Features

The information about the state of a hydrocarbon reservoir is typically stored for numerous time steps. It might contain either time series of actual records of fluid production from the wells or simulated data with a fixed time resolution. The latter contains distribution of pressure and saturation of reservoir fluids over 3D cells, which represent the reservoir model. The pressure and saturation values are considered to be constant within a cell for a given time step. These values are obtained using conventional mathematical modeling software, such as Eclipse, TNavigator or CMG software products [1–3].

Depending on several factors, a cell might be either active or inactive. An inactive cell does not play significant role in petrophysical processes in the reservoir and is disregarded both in conventional reservoir modelling and in surrogate modelling procedure. The same cell can be active (contain a numerical value) for some of the physical characteristics, and be inactive for others. For example, a given cell might simultaneously have pressure value (be active for pressure) and lack value for another physical characteristic (be inactive for it).

In this study we use the reservoir model represented by a parallelogram computational domain containing $70 \times 145 \times 114$ cubic cells. The number of active cells for the most of physical characteristics is 715112. This amount makes straightforward application of conventional machine learning techniques, such as linear regression, or decision trees [13] relatively infeasible even for two sequential timesteps, due to the size of the sample vector and of complete dataset. 3D convolutions, which were explicitly designed to work with similar data [22], are impractical for the same reason when using conventional workstations. The total number of timesteps exceeds 100, which also complicates the usage of conventional algorithms.

The information about oil reservoir for a given timestep is not limited to 3D cubic structure. For each timestep, we have a set of directives for every oil well. For our purposes, we considered three kinds of directives: oil extraction, water

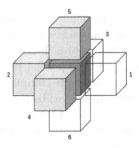

Fig. 1. Target cell with its local set

extraction and water injection. Depending on the timestep, these can be historical values or the ones, acquired with petrophysical mathematical modelling software. Generally, extraction of the coordinates of each oil well and conversion of them to a set of indices of 3D cubic structure, mentioned above, might be straightforward as well.

Therefore, for each timestep we have a parallelogram cubic structure, containing petrophysical information about the reservoir, and a set of directives for every oil well in the reservoir. Our goal, given information about the current timestep and all the previous ones, is to predict physical values for all cells for the next timestep. As we have indicated above, straightforward application of conventional machine learning techniques without any preprocessing is infeasible for regular workstations.

To overcome these difficulties, we use a general approach for grid-based Surrogate Reservoir Modeling: to predict physical values for a given cell, we use its state, the state of local neighbor cells, the set of directives for all oil wells for a considered timestep, and geometrical information about the given cell. We will denote examined cell as a target cell. The reasoning behind such approach is that the cell state mostly depends on the state of neighboring cells and the nearest oil wells. The influence of distant cells or oil wells is insignificant if we consider a single timestep. Therefore, such information can be safely disregarded. The cells, which have a common face with considered cell, were chosen as a local set, forming a kind of cross shape. The local set consists of seven cells.

In Fig. 1 we introduced an enumeration for neighboring cells. Green cell in the center corresponds to the target cell, gray cells are active, and transparent cells are inactive.

To introduce geometrical information to the algorithm, coordinates of a target cell in parallelogram computational domain and distances to all wells from the target cell were introduced as additional features, as well as running sums of oil extraction, water extraction and injection for all wells. To sum up, a single sample vector contains physical values for a target cell, neighboring cells, oil wells directives, running sums, and geometrical information for the target cell.

As all these values differ in absolute size and have different units of measurement, it is reasonable to normalize them before applying any machine learning algorithm.

Permeability for all 3 axes and porosity were chosen as static physical characteristics for each cell. Oil saturation and pressure were chosen as dynamic physical characteristics, and will have to be predicted by the resulting algorithm.

2.2 Surrogate Modeling Based on Artificial Neural Networks

As we are going to predict values for oil saturation and pressure, which, at least explicitly, are governed by different physical laws, we will use a separate ANN for each kind of target variable. Their architecture is almost similar, and the same input values will be used for both networks for each target variable. For initialization and training of ANNs we used algorithms from [5,8,10].

As target cells might have different patterns of activity of neighbors ($2^6 - 1$ neighbors to be exact, excluding the one with all cells in the local set being inactive), one network will be used for each possible pattern. Simply replacing input values corresponding to a given inactive cell with zeros will be not sufficient, as such approximation does not have any implicit physical meaning. This approach might also greatly obscure the calculations. It is convenient to use a dictionary for storing these networks, as each pattern can be represented with 6-digit binary code, where i'th position is 0, if cell i is inactive and 1 if it is active. These codes are used as dictionary keys to retrieve neural networks during training.

As we are working with simple numerical data and 3D convolutional neural networks might be impractical for conventional workstations due to the size of the input, standard feedforward architecture will be used for all the networks. In addition, complex high-dimensional structure of the input data, which reflects terrain features of the oil deposit, prevents from using recurrent neural network models.

To sum up, there is a total of $n = 2(2^6 - 1)$ ANNs, used in learning and prediction processes. Each ANN is designed to process active and neighbouring cells with specific pattern of activity as input and predict one type of dynamic physical characteristic. The number of ANNs is relatively high, but such quantity provides the flexibility, needed for the group of networks to correctly process more peculiarities of the structure of oil deposit for all time steps. For each timestep we iteratively acquire predictions of physical characteristics for each cell using ensemble of ANNs. Prediction for an arbitrary number of timesteps is acquired by sequentially processing individual ones. The outputed dynamic physical characteristics are used as input for the next timestep. Since $n \gg 1$, we used rather simple network architecture to speed up learning process. Each network contains five hidden layers with varying numbers of hidden units. This number of layers was chosen to facilitate fast training and inference even on machines without GPUs without significantly sacrificing representational power. All layers use rectified linear unit as activation function, except for the final layer of saturation prediction model. As saturation values belong to $[0, 1]$ interval, sigmoid activation function allows to effectively restrict output to provide only acceptable saturation values.

2.3 Model Training and Prediction Routines

The sheer amount of target cells and neighbors configurations, paired with corresponding geometrical and well information, effectively prevents storing all training set directly. Another procedure needs to be implemented in order to perform training and prediction using described group of ANNs.

For convenience, each parallelogram computational domain for a given timestep is padded with inactive cells. For each previously active cell on a given timestep we determine the pattern of neighboring active cells, their physical values and geometrical information, then flatten and concatenate them. Well directives for a given timestep are also added. After that, the resulting vector is stored in a dictionary with a key, corresponding to the observed pattern. State of the target cell on the next step is also stored to serve as an output value for training. As the number of vectors in the dictionary for a given pattern reaches predetermined batch size, two corresponding networks are trained on the assembled dataset. After that, values, corresponding to the used batch, are removed. When the final cell for the timestep has been processed, training is performed on all datasets, still remaining in the dictionary before their removal. The algorithm then proceeds to the next timestep.

We have found out, that normalization of input and target cell values, so that they all have zero mean and unit variance among all timesteps, greatly improves prediction quality. Normalizing values only for a given timestep is not enough, as the same normalized values will correspond to different values on the previous timesteps. This preprocessing takes $O(n)$ time, requires constant memory to store means and variances of values, and can easily be reversed. If input or target cell value already belongs to $[0, 1]$ interval, such as saturation for example, no normalization is required.

Both models are trained using the Adam optimizer with default parameters. All the remaining parameters were set to their default values. Pressure prediction model uses L_2-loss function, and saturation prediction model is trained with binary cross entropy loss. Using binary cross entropy loss is counterintuitive, but provides better results in practice.

Prediction is done in a similar fashion: for each cell, activity pattern of its neighbors is examined, their values are flattened, concatenated and passed as an input to trained network. By processing all currently active cells we estimate the state of the reservoir on the next step, and by consequently repeating the procedure we recover the state of the reservoir on all examined timesteps.

3 Results

We used data from currently developed oil reservoir with size of $70 \times 145 \times 114$ in order to evaluate models' performance. 20 timesteps, starting from step 0, were used to train the models and 50 timesteps, starting from the same step, were used to evaluate its performance.

Below are histograms of error distributions for pressure and saturation, see Figs. 2 and 3. Predictions for timesteps were obtained sequentially: starting from timestep 0, we use its values as an input to predict values for timestep 1, then we use timestep 1 values to predict timestep 2 values and so on. Note, that the histograms are normalized.

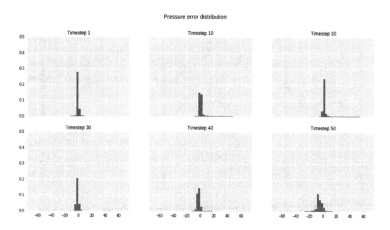

Fig. 2. Histograms of errors in pressure prediction

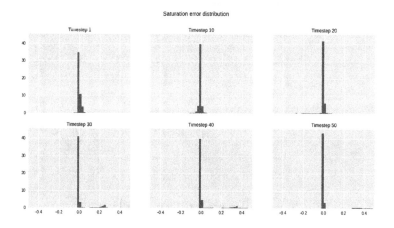

Fig. 3. Histograms of errors in saturation prediction

Not all observations are included in the plot. Insignificant number of strong outliers was excluded as abscissa axis was limited to interval $[-75, 75]$, where the majority of observations are located. Error for the most of cells on timesteps from 1 to 20, on which the models were trained, stays close to 0. And, the models were able to provide reasonable estimations for a significant amount of cell configurations on timesteps it had not seen during training. This leads to conclusion that chosen approach generalizes well and could be used for other applications.

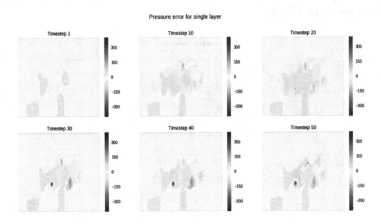

Fig. 4. Pressure prediction error for layer 33

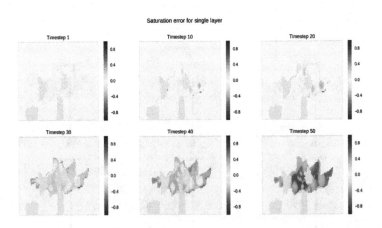

Fig. 5. Saturation prediction error for layer 33

The following set of error heatmaps for the single layer for different timesteps illustrates the same idea: difference between prediction and true value increases in time, see Figs. 4 and 5. For all layers, error clusters in a subset of cubes, corresponding to geologically heterogeneous regions, which might have peculiarities not covered by a chosen subset of physical characteristics.

Prediction was done on conventional workstation using NVIDIA GTX 970 GPU. Inference for a single timestep, containing 715105 active cells, requires approximately 104.87344 s, or 0.00015 s per cell.

4   Conclusions and Discussion

This paper illustrates that surrogate modeling is a powerful tool for acceleration of routine simulations representing a critical element of field development planning workflow. The area of applicability of the presented approach requires re-training of the data-driven model for each new reservoir which might be considered as a limitation. However, the tool is of practical use for a very fast scenario modeling of reservoirs, for which there exist a history of conventional models usage. The computational performance enables very efficient optimization of field development schemes aimed at minimizing financial risks of cost-intensive decisions on reservoir development and redevelopment planning.

One potential direction of algorithm improvement includes adding more types of values to the model. Besides simply increasing the amount of value types predicted, one can also improve the quality of inference by employing data augmentation. Data augmentation is a method, mostly used in Computer Vision. It is used to increase the size of the dataset by making use of possible invariances derived from the nature of the dataset samples. For example, if our goal is to recognize, whether the image contains a dog, we can train the model not just on the initial image, but on the rotated, mirrored, and slightly noised images also.

Finally, one could consider to apply adaptive design of experiments, devised for industrial engineering problems, to simultaneously increase efficiency of sensitivity analysis and to improve utilization of computational budget of generating a training sample [9, 23–25].

As both training and prediction are done for all cubes independently, one may argue that using a distributed programming model, such as MapReduce can be a viable choice. Implementation of both map and reduce phases both for training and prediction is straightforward, but its description goes outside of the scope of this article.

Acknowledgements. The work was supported by the Russian Science Foundation under Grant 19-41-04109.

References

1. Computer modelling group ltd. https://www.cmgl.ca/software
2. Eclipse reservoir simulator. https://www.software.slb.com
3. TNavigator - rock flow dynamics. http://rfdyn.com/tnavigator/
4. Alestra, S., et al.: Surrogate models for spacecraft aerodynamic problems. In: Proceedings of WCCM - ECCM - ECFD 2014 Congress, Barcelona, Spain, 20–25 July 2014
5. Belyaev, M., Burnaev, E.: Approximation of a multidimensional dependency based on a linear expansion in a dictionary of parametric functions. Inform. Appl. **7**(3), 114–125 (2013)

6. Belyaev, M., et al.: Building data fusion surrogate models for spacecraft aerodynamic problems with incomplete factorial design of experiments. Adv. Mater. Res. **1016**, 405–412 (2014)
7. Belyaev, M., et al.: Gtapprox: surrogate modeling for industrial design. Adv. Eng. Softw. **102**, 29–39 (2016)
8. Burnaev, E., Erofeev, P.: The influence of parameter initialization on the training time and accuracy of a nonlinear regression model. J. Comm. Tech. and Electr. **61**(6), 646–660 (2016)
9. Burnaev, E., Panov, M.: Adaptive design of experiments based on gaussian processes. In: Gammerman, A., Vovk, V., Papadopoulos, H. (eds.) SLDS 2015. LNCS (LNAI), vol. 9047, pp. 116–125. Springer, Cham (2015). https://doi.org/10.1007/978-3-319-17091-6_7
10. Burnaev, E., Prikhodko, P.: On a method for constructing ensembles of regression models. Autom. Remote Control **74**(10), 1630–1644 (2013)
11. Chavent, G., Jaffré, J.: Mathematical Models and Finite Elements for Reservoir Simulation: Single Phase, Multiphase and Multicomponent Flows Through Porous Media, vol. 17. Elsevier, Amsterdam (1986)
12. Grihon, S., Burnaev, E., Belyaev, M., Prikhodko, P.: Surrogate modeling of stability constraints for optimization of composite structures. In: Koziel, S., Leifsson, L. (eds.) Surrogate-Based Modeling and Optimization, pp. 359–391. Springer, New York (2013). https://doi.org/10.1007/978-1-4614-7551-4_15
13. Hastie, T., Tibshirani, R., Friedman, J.: The Elements of Statistical Learning. SSS, 2nd edn. Springer, New York (2009). https://doi.org/10.1007/978-0-387-84858-7
14. Kalantari-Dahaghi, A., Mohaghegh, S., Esmaili, S.: Coupling numerical simulation and machine learning to model shale gas production at different time resolutions. J. Nat. Gas Sci. Eng. **25**, 380–392 (2015)
15. Klyuchnikov, N., et al.: Data-driven model for the identification of the rock type at a drilling bit. J. Petrol. Sci. Eng. (2019)
16. Lasdon, L., Shirzadi, S., Ziegel, E.: Implementing crm models for improved oil recovery in large oil fields. Opt. Engin. **18**(1), 87–103 (2017)
17. Makhotin, I., Burnaev, E., Koroteev, D.: Gradient boosting to boost the efficiency of hydraulic fracturing. J. Petrol. Explor. Prod. Tech. (2019)
18. Mohaghegh, S., Popa, A., Ameri, S.: Design optimum frac jobs using virtual intelligence techniques. Comput. Geosci. **26**(8), 927–939 (2000)
19. Mohaghegh, S.D.: Converting detail reservoir simulation models into effective reservoir management tools using SRMs; case study-three green fields in saudi arabia. Int. J. Oil Gas Coal Technol. **7**(2), 115–131 (2014)
20. Mohaghegh, S.D., Amini, S., Gholami, V., Gaskari, R., Bromhal, G.S., et al.: Grid-based surrogate reservoir modeling (SRM) for fast track analysis of numerical reservoir simulation models at the gridblock level. In: SPE Western Regional Meeting. Society of Petroleum Engineers (2012)
21. Nguyen, A.P.: Capacitance resistance modeling for primary recovery, waterflood and water-CO flood. Ph.D. thesis (2012)
22. Notchenko, A., Kapushev, Y., Burnaev, E.: Large-scale shape retrieval with sparse 3D convolutional neural networks. In: van der Aalst, W.M.P., Ignatov, D.I., Khachay, M., Kuznetsov, S.O., Lempitsky, V., Lomazova, I.A., Loukachevitch, N., Napoli, A., Panchenko, A., Pardalos, P.M., Savchenko, A.V., Wasserman, S. (eds.) AIST 2017. LNCS, vol. 10716, pp. 245–254. Springer, Cham (2018). https://doi.org/10.1007/978-3-319-73013-4_23

23. Burnaev, E., Panin, I.: Adaptive design of experiments for sobol indices estimation based on quadratic metamodel. In: Gammerman, A., Vovk, V., Papadopoulos, H. (eds.) SLDS 2015. LNCS (LNAI), vol. 9047, pp. 86–95. Springer, Cham (2015). https://doi.org/10.1007/978-3-319-17091-6_4

24. Burnaev, E., Panin, I., Sudret, B.: Effective design for sobol indices estimation based on polynomial chaos expansions. In: Gammerman, A., Luo, Z., Vega, J., Vovk, V. (eds.) COPA 2016. LNCS (LNAI), vol. 9653, pp. 165–184. Springer, Cham (2016). https://doi.org/10.1007/978-3-319-33395-3_12

25. Panin, I., Sudret, B., Burnaev, E.: Efficient design of experiments for sensitivity analysis based on polynomial chaos expansions. Ann. Math. Artif. Intell. **81**(1–2), 187–207 (2017)

26. Pyrcz, M.J., Deutsch, C.V.: Geostatistical Reservoir Modeling. Oxford University Press, Oxford (2014)

27. Sterling, G., Prikhodko, P., Burnaev, E., Belyaev, M., Grihon, S.: On approximation of reserve factors dependency on loads for composite stiffened panels. Adv. Mater. Res. **1016**, 85–89 (2014)

28. Sudakov, O., Burnaev, E., Koroteev, D.: Driving digital rock towards machine learning: predicting permeability with gradient boosting and deep neural networks. Comput. Geosci. **127**, 91–98 (2019)

29. Temirchev, P., et al.: Deep neural networks predicting oil movement in a development unit. J. Petrol Sci. Eng. (2019)

30. Tyson, S.: An Introduction to Reservoir Modelling. Piper's Ash Limited, United Kingdom (2007)

Efficient Reservoir Encoding Method for Near-Sensor Classification with Rate-Coding Based Spiking Convolutional Neural Networks

Xu Yang[1,2] , Shuangming Yu[1,2], Liyuan Liu[1,2(✉)], Jian Liu[1,2], and Nanjian Wu[1,2,3(✉)]

[1] State Key Laboratory of Superlattices and Microstructures, Institute of Semiconductors, Chinese Academy of Sciences, Beijing 100083, People's Republic of China
liuly@semi.ac.cn, nanjian@red.semi.ac.cn
[2] Center of Materials Science and Optoelectronics Engineering, University of Chinese Academy of Sciences, Beijing 100049, People's Republic of China
[3] Center for Excellence in Brain Science and Intelligence Technology, Chinese Academy of Sciences, Beijing 100083, People's Republic of China

Abstract. This paper proposes a general and efficient reservoir encoding method to encode information captured by spike-based and analog-based sensors into spike trains, which helps to realize near-sensor classification with rate-coding based spiking neural networks in real applications. The concept of reservoir is proposed to realize long-term residual information storage while encoding. This method has two configurable parameters, integration time and threshold, and they are determined optimal based on our analysis about encoding requirements. Trough different setting we proposed, reservoir encoding method can be configured as compression mode to compress sparse spike trains obtained from spike-based sensors, or conversion mode to convert pixel values captured by analog-based sensor into spike trains respectively. Verified on MNIST and SVHN dataset, the mapping relationship of information before and after encoding are linear, and the experimental results prove that rate-coding based spiking neural networks with our reservoir encoding method can realize high-accuracy and low-latency classification in two modes.

Keywords: Rate coding · Reservoir encoding ·
Near-sensor classification · Spiking neural networks

1 Introduction

Object classification is one of the most important problems in computer vision. Recently, deep-learning based neural networks, particularly the convolutional

This work was supported by National Natural Science Foundation of China (Grant No. 61704167, 61434004), Beijing Municipal Science and Technology Project (Z181100008918009), Youth Innovation Promotion Association Program, Chinese Academy of Sciences (No.2016107), the Strategic Priority Research Program of Chinese Academy of Science, Grant No.XDB32050200.

H. Lu et al. (Eds.): ISNN 2019, LNCS 11555, pp. 242–251, 2019.
https://doi.org/10.1007/978-3-030-22808-8_25

neural networks (CNNs), achieve state-of-the-art classification accuracy but always require high-performance data center that consume large energy. As a candidate for solving this problem, spiking neural networks (SNNs) are considered as the power efficient solution with the help of emerging ultra-low power neuromorphic hardware platforms. Comparing with some brain-like learning rules, converted SNNs perform higher accuracy, especially the converted spiking convolutional neural networks (Spiking CNNs) [2,3,8]. They inherit high-accuracy and noise-robustness from CNN while maintaining the spike-based computing paradigm of SNN. In the whole network, information is represented as the spike density of neuron's output spike train, which is a typical rate coding method. Also, the input of converted SNNs must be in the form of spike train with rate coding. As the result, abundant non-spiking information cannot be used directly and a spike generator or encoder is necessary.

In general, sensor's output information is in one of the following forms: sequence of spikes (events) and real-value. The sensors with that form of output are called as spike-based sensor and analog-based sensor respectively in following for convenience. For spike-based sensors, pre-processing circuits are included in each pixel to generate and encode events into rate-based AER information encoding [7]. Besides, due to the property of some photodiodes themselves, their output is rate coded spike train naturally without any pre-processing nor encoding, like single photon avalanche photodiode [1]. These sensors fit for the converted SNNs naturally, but there are also some shortcomings for direct use. Both the low photo detection efficiency of sensors and low illumination in scenes can cause sparse spike out [1], which increase the accuracy convergence time and deteriorate the power efficiency when SNNs are implemented in hardware. However, these problems have not been researched yet, and we still lack of efficient compression encoding method to increase the information density of spike-based sensors' sparse output.

For analog-based sensors, their outputs are in the form of real-values within a dynamic range [6]. Rueckauer et al. proposed that analog input can be fed into converted SNNs directly [5]. The analog input is convoluted by the weight matrix of first hidden layer and the result is accumulated in the membrane potential of spiking neuron for firing spike to following layer. So extra multiplications are needed and power efficiency is worsen. Therefore, an encoding method for spiking conversion is needed before feeding into SNNs. Recently, some spike generator and encoding method have been researched and verified for improving the latency and accuracy of SNNs [2,9,10]. But, logarithm function or random number generator in uniform distribution are used while encoding. Both of them are difficult to be implemented in hardware precisely and efficiently. So it is still a challenge that encodes analog-based into rate coding spike train through a simple and hardware-friendly method for near-sensor processing.

To realize near-sensor classification with rate-coding based SNNs, we proposed a general and efficient method to encode imaging information captured by various sensors into rate-coding spike trains with high information density.

2 Encoding Requirements and Issues

The converted Spiking CNN consists of convolution layers with convolutional kernels mapped from a well-trained CNN, pooling layers, a fully connected linear classifier and a spike counter. The classification result is the label with maximum number of spikes emitted during a period of time Tdt counting by the spike counter lied in the end of whole network. Integrate-and-fire (IF) neuron is one of spiking neurons commonly used in converted SNNs to realize rate coding. Based on the conversion theory proposed by Rueckauer et al. [5], the firing rate of IF neuron in SNNs approximate the real-value of its corresponding neuron in CNNs linearly, which is the key to minimize the accuracy drop when mapping CNNs into Spiking CNNs. Therefore, to prevent information distortion while encoding, the mapping relationship of information after and before encoding should be linear too.

For converted SNNs, the time step dt is period of time for once spiking computation, which is various for different hardware platforms. Also, the imaging information need to be fed into network in same step. Therefore, we use the spike density of pixel's output train to describe whether information is sparse or not, that is:

$$d = \frac{N}{T}, \tag{1}$$

where T is the maximum number of spikes that pixel can generate during Tdt and N represents the total number of spikes actually generate during Tdt. So spike density is a normalized index ranging from 0 to 1 without notion of time. For realizing low-latency classification, the density of spike trains fed into network should as high as possible, which means that the maximum spike density after encoding should equal or near 1 at best.

The encoding issues are different for two types of sensors because their different encoding goals. For encoding sparse spike-based output, the goal is compressing spike trains to increase their spike density. But the times that spikes (events) appear are random and difficult to predict, which increase the complexity of encoding. For encoding analog-based output, the goal is converting the information representation field, which is also the main difficulty. In summary, to realize near-sensor classification with rate-coding based Spiking CNNs efficiently, there are some requirements for encoding:

- 1 the encoding method must be linear;
- 2 the maximum spike density within pixel array after encoding should equal or near 1;
- 3 the encoding method is general for two types of sensors through simple parameters configuration.

3 Reservoir Encoding Method

To solve the encoding issues mentioned above, we present a reservoir encoding method. Encoding-timestep is the basic encoding unit that reservoir obtain and process sensor's original information then judge whether an encoded spike output in current timestep. The information stored in reservoir is updated during each encoding-timestep by the following equation:

$$R_{ij}(n) = X_{ij}(n) + R_{ij}(n-1), \tag{2}$$

where $X_{ij}(n)$ is the processed original information obtained from pixel (i, j) during n^{th} encoding-timestep. In the end of every encoding-timestep, if $R_{ij}(n) \geq \tau$, a spike is generated in encoded train and reset $R_{ij}(n) = R_{ij}(n) - \tau$, where τ is the encoding threshold. During encoding, reservoir play a role like buffer, storing the residual information that have not been encoded. Therefore, the residual information left in reservoir is the total information obtained from sensor except for the part released as spikes in encoded train, that is:

$$R_{ij}(n) = \sum_{n} X_{ij}(n) - \tau N_{ij}^e(n), \tag{3}$$

where $N_{ij}^e(n)$ indicates the total number of spikes in encoded train during n encoding-timesteps. Through different parameters' settings, reservoir encoding method can be configured as different modes to encode two types of sensors' output respectively.

3.1 Reservoir Encoding in Compression Mode

Reservoir encoding method works in compression mode to encode the sparse output of spike-based sensors. The original output of pixel (i, j) is a spike train which can be described as $\sum_{s \in S_{ij}} \delta_{ij}(t - s)$ with $S_{ij} = \{s_{ij}^1, s_{ij}^2, ...\}$ indicated the spike time. The time resolution of spike train is decided by various circuits design, which is abstracted as "timestep" without any notion of time. To increase the spike density, we compress several successive original output in the length of T_i timesteps into one encoded output (spike or not) in one encoding-timestep. Therefore, every encoding-timestep contains T_i timesteps, and $X_{ij}(n)$ is the total number of spikes generated during n^{th} encoding-timestep by pixel (i, j), that is:

$$X_{ij}(n) = \sum_{t=(n-1)T_i+1}^{nT_i} \sum_{s \in S_{ij}} \delta_{ij}(t - s). \tag{4}$$

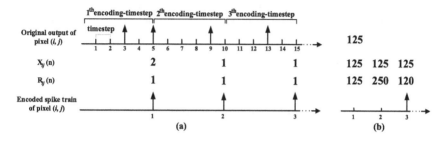

Fig. 1. Reservoir encoding. (a) compression mode. (b) conversion mode.

Therefore, the residual information in reservoir can be obtained by putting Eq. (4) into Eq. (3):

$$R_{ij}(n) = \sum_{t=1}^{nT_i} \sum_{s \in S_{ij}} \delta_{ij}(t-s) - \tau N_{ij}^e(n) = N_{ij}^o(nT_i) - \tau N_{ij}^e(n). \tag{5}$$

According to Eq. (5) and (1), the spike density of encoded train can be written as:

$$d_{ij}^e(n) = \frac{T_i}{\tau} \cdot d_{ij}^o(nT_i) - \frac{R_{ij}(n)}{n\tau}, \; R_{ij}(n) \in [0, \tau), \tag{6}$$

which are mapped from the spike density d_{ij}^o of original output linearly. Though an additional encoding error exists, it can be ignored when n becomes larger. Within pixel array, the highest spike density of original output is D_M. According to Eq. (6), the highest spike density before encoding are mapped into:

$$d_M^e(n) = \frac{T_i}{\tau} \cdot D_M - \frac{R_{ij}(n)}{n\tau}, \; R_{ij}(n) \in [0, \tau). \tag{7}$$

To satisfy the second requirement listed in Sect. 2, the maximum spike density after encoding $d_M^e(n) \approx 1$, so $\frac{T_i}{\tau} \approx 1/D_M$. In compression mode, τ is the threshold of reservoir where stores the number of spikes, so τ is an integer starting from 1. Therefore, to minimize the integrate time T_i and realize fast encoding, we configure $T_i = 1/D_M$ and $\tau = 1$.

Figure 1(a) shows the encoding process of pixel (i, j). For example $T_i = 5$ and $\tau = 1$, in every encoding-timestep, each reservoir $R_{ij}(n)$ accumulates $X_{ij}(n)$, which is the number of spikes generated from pixel (i, j) during integrate time T_i, with the value left from previous encoding-timestep $R_{ij}(n-1)$. In the end of current encoding-timestep, if $R_{ij}(n) \geq 1$, a spike output in encoded train and reset $R_{ij}(n) = R_{ij}(n) - 1$. Repeating several encoding-timesteps, the encoded train is generated. In compression mode, reservoir acts like a buffer to eliminate the encoding issues caused by the spike time randomness of original output.

3.2   Reservoir Encoding in Conversion Mode

Reservoir encoding works in conversion mode to encode the output of analog-based sensors. Their original output of pixel (i, j) is a real-value within sensor's dynamic range, indicated as I_{ij}, like typical RGB/gray image with pixel value ranging from 0 to 255. The original output of pixel (i, j) without any notion of time, so reservoir receives same information in every encoding-timestep:

$$X_{ij}(n) = I_{ij}. \tag{8}$$

According Eq. (3), residual information in conversion mode is:

$$R_{ij}(n) = nI_{ij} - \tau N_{ij}^e(n). \tag{9}$$

Based on Eq. (9) and (1), the spike density after encoding is:

$$d_{ij}^e(n) = \frac{N_{ij}^e(n)}{n} = \frac{I_{ij}}{\tau} - \frac{R_{ij}(n)}{n\tau}, \; R_{ij}(n) \in [0, I_M]. \tag{10}$$

Same as the mapping relationship shown in Eq. (6), the spike density $d_{ij}^e(n)$ of encoded train approximates the analog input I_{ij} linearly with an encoding error that can be ignored when n is large enough. After encoding, the maximum value within sensor's dynamic range I_M is converted into a spike train with spike density equals to:

$$d_M(n) = \frac{I_M}{\tau} - \frac{R_{ij}(n)}{n\tau}. \tag{11}$$

Therefore, the threshold are set as $\tau = I_M$ to map the maximum pixel value into the spike train with density nearing 1. For example, the pixel value $I_{ij} = 125$ and $\tau = I_M = 255$. The reservoir encoding method works in conversion mode are shown in Fig. 1(b). Sensor captures an image first, and the image is converted into the form of spike trains through several encoding-timesteps. In every encoding-timestep, each reservoir $R_{ij}(n)$ accumulates the analog input $X_{ij} = I_{ij} = 125$ with value left from previous encoding-timestep $R_{ij}(n-1)$. In the end of current encoding-timestep, if $R_{ij}(n) \geq 255$, a spike generate in encoded train and reset $R_{ij}(n) = R_{ij}(n) - 255$.

Through above analysis, we derive the information mapping relationship when reservoir encoding method works in compression and conversion mode respectively, Eqs. (7) and (10), which prove that the mapping relationship between the spike density of encoded train and their original information are linear for both modes. And the optimal configurable parameter setting ensure that the maximum spike density after encoding near 1. Besides, in conversion mode, $X_{ij}(n)$ can be written as: $X_{ij}(n) = \sum_{t=1}^{T_i} I_{ij}(t)$, where $I_{ij}(t)$ remains constant as pixel value I_{ij} and $T_i = 1$, and the result is same as Eq. (8). Therefore, through different configuration of threshold τ and integration time T_i shown in Table 1, reservoir encoding method becomes general for two types of sensors' output. In compression mode, reservoir encoding method encodes the sparse spike trains from spike-based sensor with integration time set as the inverse of maximum spike density within all pixels and threshold set as 1. In conversion mode, reservoir encoding method encodes the pixel value from analog-based sensor with integration time set as 1 and threshold set as the maximum value within sensor's dynamic range. So far, all requirements for encoding listed in Sect. 2 are satisfied.

4   Experimental Results and Discussions

4.1   Experimental Setup

To test conversion performance of reservoir encoding method, we use images in MNIST and SVHN datasets directly, which contains gray handwritten digit images and color street view house number images respectively in 10 classes.

Table 1. Parameters configuration for different modes of reservoir encoding method.

Mode	Reservoir	$X_{ij}(n)$	T_i	τ
Compression	$R_{ij}(n) = X_{ij}(n) + R_{ij}(n-1)$	$\sum_{t=(n-1)T_i+1}^{nT_i} \sum_{s \in S_{ij}} \delta_{ij}(t-s)$	$\frac{1}{D_M}$	1
Conversion		$\sum_{t=1}^{T_i} I_{ij}(t)$	1	I_M

However, there lack of dataset that contains enough samples captured by spike-based sensor in rate-coding. So, we generate the spike-based output based on MNIST and SVHN datasets. At timestep t, pixel (i,j) in k^{th} channel produces a spike if: $rand() < cI_{ijk}$, where $rand()$ is a random number generator with uniform distribution on $(0,1)$ and all images are normalized $I_{ijk} \in [0,1]$, $k = 1,2,3$ for RGB images, and $k = 1$ for gray images. To test the compression performance, we set $c = 0.05, 0.1, 0.2, 0.3, 0.4, 0.5$ to generate spike-based output with different ideal maximum spike density $D_M = c$. While in real-time encoding, maximum spike density D_M is an estimated value that obtained by counting the total number of spikes generated by all pixels within a period of time and selecting the maximum one. For MNIST dataset classification, the network architecture of Spiking CNN is same as [3] that $28 \times 28 - 12c5 - 2s - 64c5 - 2s - 10o$. The parameters of convolutional layer are denoted as "<numbers of channels>c<receptive field size>", and the stride of non-overlapping average pooling is shown as "<stride>s". "<number of classes>o" indicates a fully connected layer with the output class number. For SVHN dataset, the network architecture of Spiking CNN is $32 \times 32 - 32c5 - 2s - 96c5 - 2s - 64c3 - 3s - 10o$. ReLU activation after convolution are used in both networks.

4.2 Encoding Efficiency and Linearity

The encoding results are shown in Fig. 2 from pixel and image level both when the maximum spike density of original spike-based output within pixel array is $D_M = 0.1$. As shown in Fig. 2(a), a bright and a dark pixel are selected to compare the encoding result. For compression mode, two pixels' original spike train are in the length of 1000 timesteps, and their encoded spike train are compressed into 100 timesteps after encoding. The spike density of encoded train increase dramatically, and the original output with higher density are encoded into trains with higher spike density. For conversion mode, the pixel value are 245 and 19 respectively. They are encoded into two spike train in the length of 100 timesteps both, and the spike density of their encoded spike train are proportional to their pixel value. The encoding results in image level are shown in Fig. 2(b). The images in last three columns are reconstructed with all pixels' spike train in same length (100 timesteps). Obviously, the reconstructed images with original sparse spike-based output are blurred or almost informationless. While for both MNIST and SVHN dataset, our method shows well performance in compression and conversion mode, encoding sensor's original information into high spike density that can be reconstructed with less timesteps. Figure 3 shows

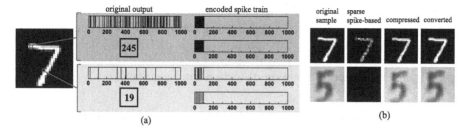

Fig. 2. Encoding results. (a) comparison of original output and encoded spike train in pixel level. The encoding results of two pixels lied in a sample selected from MNIST dataset are shown with blue and green background respectively. (b) comparison of original output and encoded output in image level. The encoding results of samples selected from MNIST and SVHN dataset are shown in the first and second row respectively. (Color figure online)

the curves describing the information relationship before and after encoding. The mapping curves of conversion mode shown in Fig. 3(c) are almost perfect linear for two datasets. And the mapping relations are also linear in compression mode for both datasets with all different levels of maximum spike density D_M. For two modes, the spike density after encoding are range from 0 to 1, covering the whole representation range. Thus, reservoir encoding method are satisfied the requirements listed in Sect. 2, and verified general for all types of outputs from analog-based to spike-based sensors.

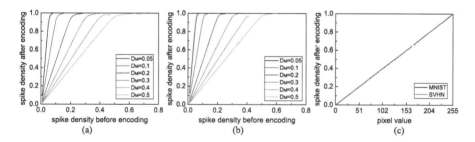

Fig. 3. Linearity. (a) linearity of compression mode on MNIST dataset. (b) linearity of compression mode on SVHN dataset. (c) linearity of conversion mode on MNIST and SVHN dataset.

4.3 Accuracy and Convergence Time

Though encoding errors always exist, we assess the information distortion severity by using the classification accuracy directly. The accuracy of converted Spiking CNNs with our reservoir encoding method are listed in Table 2, comparing with other methods. For compression mode, the result listed in Table 2 is an average accuracy over six different $D_M = 0.05, 0.1, 0.2, 0.3, 0.4, 0.5$. Our method realizes 99.09% and 99.10% classification accuracy on MNIST dataset, 96.20%

and 96.32% accuracy on SVHN dataset for compression and conversion mode respectively. For both modes, our encoding method achieves similar accuracy with [3] on MNIST and state-of-the-art accuracy on SVHN with converted Spiking CNN. Therefore, it proved that the impact on classification accuracy caused by encoding errors and information distortion can be ignored.

Table 2. Comparison of classification accuracy for different methods.

Dataset	Method	Accuracy
MNIST	converted Spiking CNN+Weight Normalization [3]	99.11%
	Spiking CNN+Perceptron-Inception [8]	88.00%
	converted Spiking CNN (compression)	99.09%
	converted Spiking CNN (conversion)	99.10%
SVHN	converted Spiking CNN+Threshold rescaling [9]	93.66%
	Spiking CNN+Synaptic filter [4]	93.92%
	converted Spiking CNN (compression)	96.20%
	converted Spiking CNN (conversion)	96.32%

Table 3. Classification accuracy and times of latency improvement.

	D_M	0.05	0.1	0.2	0.3	0.4	0.5
MNIST	Accuracy	99.13%	99.07%	99.07%	99.08%	99.08%	99.11%
	Speed up	×20	×7.7	×4	×3.3	×1.6	×1.5
SVHN	Accuracy	96.22%	96.19%	96.18%	96.26%	96.18%	96.14%
	Speed up	×19	×5.6	×2.5	×2.3	×1.5	×1.2

Classification latency is various for different sparsity of input spike trains. We found that latency of networks without compressed input becomes worse with maximum spike density D_M within pixel array going down on both two datasets. However, after compression, the convergence time for all different D_M are near 100 and 300 on MNSIT and SVHN dataset respectively. The precise times of latency improvement on MNIST and SVHN dataset with different D_M are listed in Table 3 with accuracy. Overall, the results show that the latency of converted Spiking CNN are shortened on two dataset obviously without much accuracy drop. And we test the conversion mode, the latency on two datasets are also near those achieved by reservoir encoding method in compression mode. Therefore, both compression and conversion mode can realize efficient spike encoding to ensure low classification latency with converted Spiking CNN.

5 Conclusion

To realize near-sensor classification with rate-coding based SNNs, we proposed a general and efficient encoding method. Through different setting about two configurable parameters, integration time and threshold, our reservoir encoding method can be configured as conversion and compression modes to encode output information captured by analog-based and spike-based sensor into rate-coding spike trains respectively. Verified on MNIST and CIFAR-10 dataset, the experimental results prove that Spiking CNNs achieve high-accuracy and low-latency classification with the help of our reservoir encoding method in both modes. In future work, we plan to test our encoding method with raw data obtained from spike-based sensor rather than simulation. And our reservoir encoding method can be improved by introducing the concept of temporal coding to fit for more types of SNNs.

References

1. Al Abbas, T., Dutton, N., Almer, O., Pellegrini, S., Henrion, Y., Henderson, R.: Backside illuminated SPAD image sensor with 7.83 μm pitch in 3D-stacked CMOS technology. In: 2016 IEEE International Electron Devices Meeting (IEDM), pp. 1–8. IEEE (2016)
2. Cao, Y., Chen, Y., Khosla, D.: Spiking deep convolutional neural networks for energy-efficient object recognition. Int. J. Comput. Vis. **113**(1), 54–66 (2015)
3. Diehl, P.U., Neil, D., Binas, J., Cook, M., Liu, S.C., Pfeiffer, M.: Fast-classifying, high-accuracy spiking deep networks through weight and threshold balancing. In: 2015 International Joint Conference on Neural Networks (IJCNN), pp. 1–8. IEEE (2015)
4. Hunsberger, E., Eliasmith, C.: Training spiking deep networks for neuromorphic hardware. arXiv preprint arXiv:1611.05141 (2016)
5. Rueckauer, B., Hu, Y., Lungu, I.A., Pfeiffer, M., Liu, S.C.: Conversion of continuous-valued deep networks to efficient event-driven networks for image classification. Front. Neurosci. **11**, 682 (2017)
6. Shi, C., et al.: A 1000 fps vision chip based on a dynamically reconfigurable hybrid architecture comprising a pe array processor and self-organizing map neural network. IEEE J. Solid-St. Circ. **49**(9), 2067–2082 (2014)
7. Wu, N.: Neuromorphic vision chips. Sci. China Inf. Sci. **61**(6), 060421 (2018). https://doi.org/10.1007/s11432-017-9303-0
8. Xu, Q., Qi, Y., Yu, H., Shen, J., Tang, H., Pan, G.: CSNN: an augmented spiking based framework with perceptron-inception. In: IJCAI, pp. 1646–1652 (2018)
9. Xu, Y., Tang, H., Xing, J., Li, H.: Spike trains encoding and threshold rescaling method for deep spiking neural networks. In: 2017 IEEE Symposium Series on Computational Intelligence (SSCI), pp. 1–6. IEEE (2017)
10. Yu, Q., Tang, H., Tan, K.C., Li, H.: Rapid feedforward computation by temporal encoding and learning with spiking neurons. IEEE Trans. Neural Netw. Learn. Syst. **24**(10), 1539–1552 (2013)

Fault Diagnosis of Gas Turbine Fuel Systems Based on Improved SOM Neural Network

Zhe Chen, Yiyao Zhang, Hailei Gong, Xinyi Le, and Yu Zheng[✉]

Institute of Intelligent Manufacturing and Information Engineering,
School of Mechanical Engineering, Shanghai Jiao Tong University,
Shanghai, China
{chenzhecz96, sjtuzyy89, chenliang1994, lexinyi,
yuzheng}@sjtu.edu.cn

Abstract. Considering the difficulties in constructing physical models to simulate fuel systems of gas turbines, we introduce an improved approach based on SOM neural network for the fault diagnosis of fuel systems. In this model the competitive layer structure is decided by an objective function, the parameter functions are selected with genetic algorithm, and the weight vectors are initialized with a pre-segmentation method. Meanwhile, before the data is inputted, PCA dimensional reduction is used to decrease the training consumption. Eventually, practical dataset verification suggests that this improved SOM neural network performs better in recognition rate than other classification algorithms and original SOM network in fault diagnosis of gas turbine fuel system.

Keywords: Fuel system · Fault diagnosis · SOM (Self-Organized Map)

1 Introduction

Gas turbine is one of the most advanced and complicated industrial equipment, which has been widely used in aviation, nautical industry, power generation and other industrial fields. Due to the sophisticated mechanical structure and its severe working environment of high temperature, high pressure and high rotate speed, various components of gas turbines are likely to malfunction and gradually age, decreasing the ability and efficiency of gas turbines.

The fuel system, a vital part of a gas turbine that directly affects the combustion process and determines the performance of the gas turbine, consists of the fuel injection pump, delivery valve, needle valve, high pressure fuel pipe and fuel injector. Data shows that fuel system failure accounts for 27% of the causes of gas turbine shutdown. However, for the complex multi-layer system, it is difficult for researchers to construct an accurate physical model to effectively describe it. And owing to the diversity and strong specificity of different gas turbines, establishing a universal physical model for the fuel system of various gas turbines is almost impossible. Factually, most physical models are only equal to the lab simulation for a single component.

Consequently, models based on data are now one of the most efficient methods for fault diagnosis of fuel system, by which the fault information can be extracted and

© Springer Nature Switzerland AG 2019
H. Lu et al. (Eds.): ISNN 2019, LNCS 11555, pp. 252–265, 2019.
https://doi.org/10.1007/978-3-030-22808-8_26

analyzed, and the eventual judgement can be made according to artificial neural networks or other algorithms. Concretely, in engineering applications, the most popular way to extract working information of the fuel system is to collect the pressure data of the fuel in high-pressure fuel pipe provided by pressure sensors. When a fault occurs in a certain component, the waveform of the fuel supply state will change, so the measured waveform information will also change. A large number of experiments have shown that the pressure information of high-pressure fuel pipe can adequately reflect the working state of a fuel system and be used to diagnose whether faults have occurred or not. However, with a severe working environment of high temperature, high pressure and strong vibration, many helpful characteristic signals are interfered by noises, which leads to the difficulty for traditional statistic models to detect fault information contained in pressure data. Therefore, a more effective and general data-driven method should be proposed to support the fault diagnosis for fuel systems of gas turbines.

2   Related Works

In the past 40 years, fault diagnosis can be roughly divided into two categories: methods based on physical models and methods based on data [1]. In the early stage, physical models are usually constructed to accurately represent real mechanical systems. But as systems becoming more and more complicated and comprehensive, the value of traditional physical models is becoming more and more limited. Intelligent fault diagnosis, a data-driven method utilizing digital signal processing, modeling, artificial intelligence, have now shown obvious superiorities compared with physical models for large scale equipment fault diagnosis [2–4]. Specifically, some machine learning algorithms, such as logistic regression, decision tree or random forest, support vector machine (SVM), and various artificial neural networks, have been widely used and improved to diagnose faults for complicated systems and shown excellent performance [5–8].

Jianmin, Yupeng et al. proposed a fault monitoring method by analyzing the cylinder head variations and proved in the paper that peak characteristics in combustion stage signal of cylinder vibration under different work conditions can accurately describe injection information and can be regarded as input data for fault diagnosis [9]. In Albarbar, Gu and Ball's work, an ICA (Independent Component Analysis) based scheme was developed to decompose air-borne acoustic signals. Then the fuel injection process characteristics were processed and monitored by WVD (Wigner-Ville Distribution) technique in time-frequency domain, which was testified to be valid by signal simulations and empirical measurements [10]. Zhiling, Bin et al. presented an expert system to diagnose gear box in wind turbine timely and accurately, which was based on fault tree analysis and developed by C# on the .NET platform [11]. Amirat, Choqueuse et al. provided a method on the basis of the generator stator current data collection and the Hilbert transformation and validly detected failures in DFIG (Doubly-Fed Induction Generator) based wind turbine for both stationary and nonstationary conditions [12].

More concretely, considering the application of neural networks in fault diagnosis, Fengming proposed the design and implementation of the fault diagnosis of diesel fuel system based on wavelet analysis, of which the overall approach is to preprocess the dataset by wavelet transform, extract some features of faulty data, and then make them the input samples of the neural network. One of the advantages of this approach is that both the value of dataset and physical significance of model are taken into account [13]. Cheng presented a new fault identification method based on H-H wavelet transform and SOM (Self Organized Map) neural network, whose main innovations lay on effectively separating the frequency family of the gear vibration signals by EMD (Empirical Mode Decomposition), obtaining Hilbert spectrum and Hilbert edge spectrum according to the Hilbert transformation of IMFs (Intrinsic Mode Functions), and then using the energy percentage of the first six IMFs as the input vector of SOM neural network [14]. Based on SOM network, Jafari-Marandi proposed a SOED (Self-Organizing and Error-Driven) artificial neural network clustering method and showed that SOED is a reliable technique by testing several datasets [15]. Delgado, Higuera et al. developed an original computational approach based on SOM prototype for cluster analysis, where topology-preserving and connectivity functions were used in the process of clustering [16]. Additionally, the improved method of preprocessing original data with principle components analysis proposed by Wei was testified to be capable to enhance the accuracy and speed of training process [17].

3   Establishment of Improved SOM Network for Fuel System

3.1   Basic Introduction of SOM Network

SOM, a kind of visual unsupervised learning algorithm, is short for Self-Organized Map. A typical 2-dimensional SOM neural network structure is as Fig. 1 shows, which usually contain only two layers of neurons. The same as other artificial neural networks, update of the weight matrix is the key to SOM algorithm.

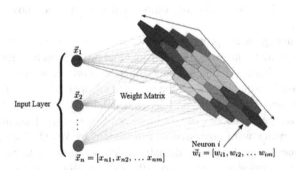

Fig. 1. A 2-dimensional SOM neural network

The learning algorithm of SOM network consists of two steps: competitive learning and self-stabilizing learning. Competitive learning means that if the distance between an input vector and a weight vector is shorter, the adjustment of the weight vector will be greater, so that a series of competitive layer neurons that resemble one specific winning neuron will gather. The most classical SOM algorithm is Kohonen training algorithm, which can be presented as Fig. 2 below.

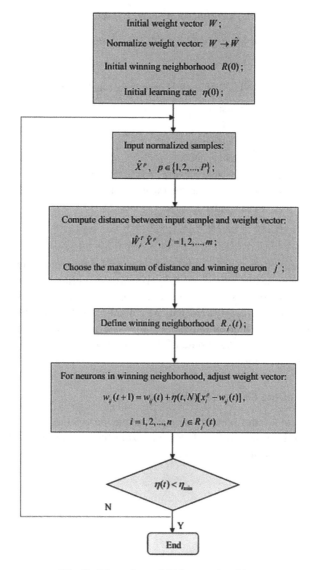

Fig. 2. Flow chart of Kohonen algorithm

What calls for special attention is that SOM network is more suitable to those problems with features that samples in the same category are concentrated in output layer plane and that samples in different categories are relatively far away from each other. Besides, the input training data should have obvious characteristics for classification.

3.2 Description of Fault Phenomena of Fuel System

There are eight main fault phenomena for the fuel system of gas turbines, which are shown in Table 1.

Table 1. Main fault phenomena of fuel system

a.	Normal state
b.	75% fuel supply
c.	25% fuel supply
d.	Idle fuel supply
e.	Needle valve sticking (small fuel volume)
f.	Needle valve sticking (calibrated fuel volume)
g.	Needle valve leakage
h.	Delivery valve failure

These eight fault phenomena can be attributed to the failure of three components: oil gauge in fuel pipe, needle valve, and delivery valve. Experimental facts indicate that once the physical state of these three components change, the pressure signal in high-pressure fuel pipe of the fuel system will be directly affected. Meanwhile, effective information extraction of high-pressure fuel pipe and proper data analysis can help researchers determine the working state of these three components and distinguish the eight fault phenomena validly.

3.3 Extraction of the Characteristic Data

A typical waveform of the pressure in fuel system describing the variation of fuel pressure over time is shown in Fig. 1. By analyzing the working state of fuel system, one can divide the pressure signal curve into several stages, which are marked below the time axis in Fig. 3.

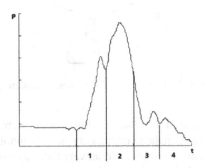

Fig. 3. Waveform of the pressure in fuel pipe

The first stage starts at the moment delivery valve opens and ends at the moment the needle valve of fuel injector opens. In this stage, the pressure curve rises from the initial state into a higher level, during which a gap will always appear in the middle of this period. Some specific pressure data, reflecting the state information of the needle, can be extracted from the gap curve.

The second stage is the period while fuel keeps moving forward after the opening of needle valve of fuel injector. There is only one component that is engaged in this stage, so the waveform in the second stage contains the working state information of the needle valve.

The third stage is the free expansion process, in which the working state of delivery valve changes.

The fourth stage refers to the free attenuation process of the residual pressure. Components who affect this period mainly include delivery valve and fuel circuit, whose sealing may matter.

Besides, for a long time period, the pressure waveform is regular and shows a certain periodicity. In consequence, it is convenient and useful for us to monitor the pressure sensors of fuel pipe with a specific sample frequency and obtain a clear waveform and pressure data.

Based on a large number of experiments, eight characteristic parameters, including starting pressure, maximum pressure, and so on, can give a complete and accurate description of the fuel system, which are shown in Fig. 4 in more detail.

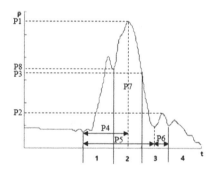

Fig. 4. Diagram of characteristic extraction

The meaning of characteristic parameters that marked in the diagram above are shown in Table 2.

Table 2. Characteristic parameters of pressure waveform

P1	Max pressure
P2	Secondary pressure
P3	Wave range
P4	Rising edge width
P5	Wave width
P6	Max repercussion width
P7	Wave integral area
P8	Starting pressure

3.4   The Structure of SOM Network

The structure of input layer of SOM network is usually a simple 1-dimensional array (a vector) that are directly decided by the number of attributes of original data. Therefore, in this case, as Table 2 shows, input layer can be represented by an 8-dimensional vector $(P1, P2, P3, P4, P5, P6, P7, P8)$.

As for the competitive layer (output layer), the structure is related to the number of clusters of samples and their topological structure [18]. In this paper, there are eight fault phenomena that can be converted into eight modes as Table 1 shows, which can be regarded as priori knowledge. Considering that the characteristic parameters of these eight modes are not separated but correlated with each other and full of continuity in some dimensions, a 2-dimensional plane competitive layer is more suitable for the fault diagnosis of fuel system.

To ensure both a good fitting rate of the training set and a high predictive accuracy for the test set, a variable indicating the error of the fitting process can be used to design the structure of competitive layer.

$$err = \sum_{i=1}^{D} (x_i - w_{ij})^2 \tag{1}$$

For training set, err_{train} expresses the training error, and for test set, err_{test} expresses the generalization error. So, an objective function is defined as Eq. (2).

$$obj : \min err_{train} \times err_{test} \tag{2}$$

Under this objective, the structure of competitive layer can be determined based on the priori knowledge and an iteration test. In terms of competitive layer, for $x = 7, 8, 9$ and $y = 7, 8, 9$, nine kinds of 2-dimensional plane are tested with each 1600 times iterations, from which the conclusion that the optimal structure is a 2-dimensional 8×8 plane is drawn (Fig. 5).

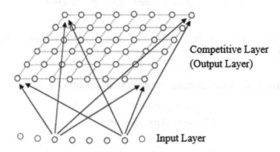

Fig. 5. Structure of SOM network of fuel system

3.5   Parameters of SOM Network

After the topological structure of SOM network decided, some structure parameters related to the performance of network need to be chosen properly: learning rate, weight adjustment function, and winning neighborhood [19]. Here for every parameter, several alternatives are considered and the most suitable one is finally chosen by using genetic algorithm.

Learning Rate. At the beginning of training, learning rate should be relatively large, which helps the algorithm quickly obtain a rough structure of weight vector. And as number of training increases, to make the model more accurate, learning rate should be designed to gradually decline and eventually reach zero. Therefore, a linear attenuation function and an exponential attenuation function are optional as Eqs. (3) and (4).

$$\eta(t) = C\left(1 - \frac{t}{T}\right) \tag{3}$$

$$\eta(t) = Ce^{-Bt/T} \tag{4}$$

Weight Adjustment Function. The most common and useful alternative weight adjustment functions are shown in Fig. 6, in which (a) is called "Mexican straw hat function", (b) is called "Top hat function", and (c) is called "Chef cap function". The interval $(-R, R)$ on horizontal axis is the winning neighborhood.

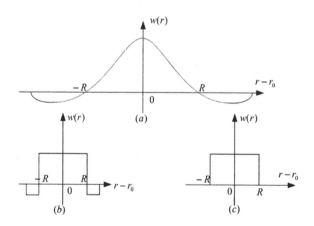

Fig. 6. Response neighborhood functions

Because of a relatively simple competitive layer structure and a small number of fault modes, the "Top hat function" and "Chef cab function" are more appropriate with a low time consumption and an adequate training accuracy.

Winning Neighborhood. The basic design criterion of winning neighborhood is that range of neighborhood shrinks as number of training increases, which guarantees the existence of both differences and similarities between the weight vectors corresponding to adjacent neurons on the competitive layer plane. In result, similar to the design method of learning rate, two winning neighborhood attenuation functions are designed to be chosen as Eqs. (3) and (4).

Elements in alternate function set are discrete, so genetic algorithm can be applied to optimize the selection of these parameter functions, of which the detailed steps are as follows.

(1) Encode each function that can be selected for each structure parameter.
(2) Randomly generate the initial population structure, keeping the objective function still min $err_{train} \times err_{test}$.
(3) Determine the fitness function.
(4) Generate new individuals according to the fitness criterion: individuals with high fitness have high genetic probability, while individuals with low fitness have low genetic probability.
(5) Generate a new generation of population according to a certain cross-genetic method.
(6) Generate some variants according to a certain degree of variability.
(7) Evaluate the new generation of population.
(8) Repeat steps 2–7 until the optimal solution is selected.

Finally, after experiments for each parameter, the result of genetic algorithm indicates that learning rate function and winning neighborhood function should be both linear attenuation function, and the weight adjustment function should be the "Chef cab function".

4 Data Implementation of Fault Diagnosis for Fuel System

4.1 Data Pre-processing

Experimental data is collected in the actual operation process of gas turbine, which has been divided into training set (40 samples) and test set (8 samples). Each sample contains an 8-dimensional vector expressing the eight characteristic parameters mentioned in 3.3 and a tag in range of 1–8 expressing the fault modes.

Normalization. Dimensional variables in different ranges can be transformed into dimensionless variables by normalization. By simple calculation, values in different dimensions are limited in the interval $[-1, 1]$, in case that values in different dimensions affect the training process differently. Equation (5) is the normalization formula.

$$y_i = \frac{2(x_i - x_{min})}{x_{max} - x_{min}} - 1 \qquad (5)$$

Part of normalized data is as Table 3 shows.

Table 3. Part of normalized data

Fault modes	Number	P1	P2	P3	P4	P5	P6	P7	P8
Normal state	1	1.000	0.802	0.027	0.549	0.556	0.400	0.706	0.631
	2	1.000	0.837	0.092	0.901	0.556	0.400	0.882	0.620
	3	0.920	0.744	0.114	0.930	1.000	0.600	0.824	0.464
	4	0.952	0.674	0.481	0.479	0.556	0.400	0.765	0.473
	5	0.936	0.605	0.341	0.747	1.000	0.600	0.924	0.486
...	...	...	...	...	...	...	...	...	...

PCA and Data Dimensional Reduction. PCA (Principle Component Analysis) algorithm is used to transform the original normalized data into a series new vector and meanwhile speed up the training process, with most of the information in the original data remaining [20]. It has been calculated that the largest four eigenvalues of the covariance matrix account for more than 95% of the sum of eight eigenvalues, so four principle components are chosen to approximate the whole information of original data. Eventually, the dimensionality of normalized dataset declines from eight to four.

4.2 Initialization of the Network Neurons

Now there are mainly three initialization method of SOM network [21]. First, the network can be directly initialized by random normalized constant vectors. The second method is to use part of data from training set as initial weight vectors, which saves much training time compared with the first method. The third method is to initialize the network at the center of competitive layer plane first, and then add some small random values to the other neurons as the distance increases.

However, in this case, on account of the fact that the data we have obtained is already tagged, a pre-segmentation makes it more convenient and appropriate to initialize the neurons in each segment. The specific operation steps are as follows.

(1) Create a 2-dimensional 8 × 8 SOM neural network.
(2) Append two types of unknown fault modes to eight fault modes ever known before and notate them with number 1–10.
(3) Transform the discrete competitive layer plane into continuous plane and divide it in to 10 pieces evenly.
(4) Choose eight proper pieces corresponding to the fault mode 1–8 and initialize each piece of network with the second initialization method.
(5) Initialize the other two pieces of network with random weight vectors.

By this means, the network has a certain topological structure at the beginning of training process, which guarantees the convergence of algorithm and a relatively low time consumption.

4.3 Verification of the Algorithm

Directly Validation. After initialization, the improved network is trained with pre-processed training set data for 1600 iterations and the weight vectors are up to date.

Then the test set data is inputted into the input layer of SOM network, in order to verify the algorithm. The result is as Fig. 7 shows, in which the color of small rectangles reflects the distance between the neuron and other neurons.

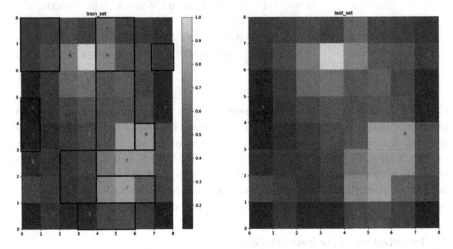

Fig. 7. Result of SOM network

From the result picture of training set, it is clear that the eight fault modes are identified in clusters in the competitive layer plane of SOM network, whose topological structure has been presented. According to the result of test set, the recognition rate of the network is almost 100%. Additionally, when it is observed that a cluster of winning neurons are distant from all the other winning neurons, the cluster can be regarded as a brand-new fault mode we have never known before.

Cross Validation. For the available dataset is extremely small, the verification result lacks convincingness. So as Fig. 8 shows that the training set is divided into many subsets (6 subsets in this case) and for each iteration, a subset acts as the test fold and the other act as training folds. Consequently, the original dataset is adequately utilized, and the training accuracy can be improved. Furthermore, over-fitting problem can also be avoided by cross validation.

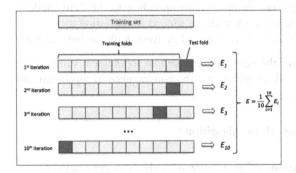

Fig. 8. Schematic diagram of cross validation

After cross validation, the precision rate of this improved SOM network algorithm eventually lies at 89.6%.

4.4   Analysis of the Algorithm

Comparison Between Original SOM Network. The time consumptions of both original SOM network and the improved algorithm are firstly calculated, where the training time is an average of 100 times training, as Table 4 shows.

Table 4. Comparison of time consumption

Methods	Original SOM	Improved SOM
Training time (s)	1.71	1.17

It can be seen that the improvement of algorithm helps reduce the training time by about 30%. On analysis, it is the PCA process that play the dominant role in reducing training time. Therefore, SOM network combined with PCA algorithm is superior in time consumption, especially when dealing with a big and high-dimensional dataset.

Then in consideration of the convergence result, as Fig. 9 shows, the convergence speed and training error of improved SOM network are both better than original one. The reason why improved SOM is worse than the original one at the beginning may lies in that random initialization of original one happened to reach a higher accuracy.

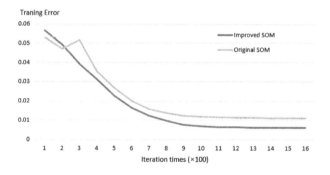

Fig. 9. Comparison of convergence curve

Comparison Between Other Algorithms. To verify the classification accuracy of this improved network, some other algorithms are also used to be compared with, such as original SOM neural network, NB (Naïve Bayes Classification) and KMeans clustering algorithm (When a clustering algorithm are applied to a classification problem, it is essential for experts to judge which categories the clusters belong to by professional experience after the clustering process. Then the number of correctly classified samples can be counted, and the precision rate can be calculated). The test accuracies of these algorithms are as Table 5 shows.

Table 5. Comparison of precision rate

Algorithms	Improved SOM	Original SOM	Naïve Bayes	Kmeans
Recognition rate	89.6%	80.1%	79.2%	85.4%

5 Conclusion

In consideration of the difficulties in establishing physical models for fuel system of gas turbines, a data-driven method is proposed and applied in fault diagnosis of fuel system. Concretely, in this paper, an improved SOM neural network is designed and constructed. In this process, the structure of competitive layer plane is designed based on a quantitative test, the parameters of network are chosen using genetic algorithm, and the weight vectors are initialized with a specific pre-segmentation method. Besides, PCA algorithm is utilized before the samples are inputted, improving both the convergence and training speed of the network. Finally, the network is verified by a dataset and proved to reach a satisfying performance for fault diagnosis compared with other existing algorithms.

From the verification result of this fault diagnosis algorithm we can draw the conclusion that this improved SOM neural network has the advantages of data-driven characteristic, self-organized characteristic and unsupervised characteristic, which are of vital importance on condition that it is almost impossible to create physical models for complicated fuel systems of various gas turbines.

However, limited by the size of dataset, this paper does not testify the performance of network with a big size of training data and hence lacks the analysis of space consumption of the algorithm in more detail.

References

1. Isermann, R.: Fault-Diagnosis Systems: An Introduction from Fault Detection to Fault Tolerance. Springer Science & Business, Berlin (2006)
2. Yazici, B., Kliman, G.B.: An adaptive statistical time-frequency method for detection of broken bars and bearing faults in motors using stator current. IEEE Trans. Ind. Appl. **35**(2), 442–452 (1999)
3. Yen, G.G., Lin, K.C.: Wavelet packet feature extraction for vibration monitoring. IEEE Trans. Ind. Electron. **47**(3), 650–667 (2000)
4. Liu, B., Ling, S.F., Meng, Q.: Machinery diagnosis based on wavelet packets. J. Vib. Control **3**(1), 5–17 (1997)
5. Muralidharan, V., Sugumaran, V.: Feature extraction using wavelets and classification through decision tree algorithm for fault diagnosis of mono-block centrifugal pump. Measurement **46**(1), 353–359 (2013)
6. Hui, K.H., Ooi, C.S., Lim, M.H., Leong, M.S.: A hybrid artificial neural network with Dempster-Shafer theory for automated bearing fault diagnosis. J. Vibroeng. **18**(7), 4409–4418 (2016)
7. FernáNdez-Francos, D., MartíNez-Rego, D., Fontenla-Romero, O., Alonso-Betanzos, A.: Automatic bearing fault diagnosis based on one-class ν-SVM. Comput. Ind. Eng. **64**(1), 357–365 (2013)

8. Cai, B., Liu, Y., Fan, Q., Zhang, Y., Liu, Z., Yu, S., Ji, R.: Multi-source information fusion based fault diagnosis of ground-source heat pump using bayesian network. Appl. Energy **114**, 1–9 (2014)
9. Jianmin, L., Yupeng, S., Xiaoming, Z., Shiyong, X., Lijun, D.: Fuel injection system fault diagnosis based on cylinder head vibration signal. Procedia Eng. **16**, 218–223 (2011)
10. Albarbar, A., Gu, F., Ball, A.D.: Diesel engine fuel injection monitoring using acoustic measurements and independent component analysis. Measurement **43**(10), 1376–1386 (2010)
11. Zhi-Ling, Y., Bin, W., Xing-Hui, D., Hao, L.I.U.: Expert system of fault diagnosis for gear box in wind turbine. Syst. Eng. Procedia **4**, 189–195 (2012)
12. Amirat, Y., Choqueuse, V., Benbouzid, M.H.: Wind turbines condition monitoring and fault diagnosis using generator current amplitude demodulation. In: 2010 IEEE International Energy Conference, pp. 310–315 (2010)
13. Fengming, L.: Design and Implementation of Fault Diagnosis of Diesel Engine Fuel System Based on Wavelet Neural Network (in Chinese). Master's thesis. Shandong University (2010)
14. Cheng, G., Cheng, Y.L., Shen, L.H., Qiu, J.B., Zhang, S.: Gear fault identification based on Hilbert-Huang transform and SOM neural network. Measurement **46**(3), 1137–1146 (2013)
15. Jafari-Marandi, R., Khanzadeh, M., Smith, B.K., Bian, L.: Self-Organizing and Error Driven (SOED) artificial neural network for smarter classifications. J. Comput. Design Eng. **4**(4), 282–304 (2017)
16. Delgado, S., Higuera, C., Calle-Espinosa, J., Morán, F., Montero, F.: A SOM prototype-based cluster analysis methodology. Expert Syst. Appl. **88**, 14–28 (2017)
17. Li, W., Peng, M., Wang, Q.: False alarm reducing in PCA method for sensor fault detection in a nuclear power plant. Ann. Nucl. Energy **118**, 131–139 (2018)
18. Jiang, X., Liu, K., Yan, J., Chen, W.: Application of improved SOM neural network in anomaly detection. Phys. Procedia **33**, 1093–1099 (2012)
19. Guanglan, L., Tielin, S., Nan, J., Shiyuan, L.: Research on feature selection technology based on som network (in Chinese). J. Mech. Eng. **41**(2), 46–50 (2005)
20. Burgas Nadal, L., Meléndez Frigola, J., Colomer Llinàs, J., Massana Raurich, J., Pous Sabadí, C.: N-dimensional extension of unfold-PCA for granular systems monitoring. Eng. Appl. Artif. Intell. **71**, 113–124 (2018)
21. Yan, Y.: Research on Clustering and Data Analysis Method Based on One-Dimensional SOM Neural Network (in Chinese). Tianjin University, Diss (2010)

Power System Fault Detection
and Classification Using Wavelet Transform
and Artificial Neural Networks

Paul Malla, Will Coburn, Kevin Keegan, and Xiao-Hua Yu[✉]

Department of Electrical Engineering, California Polytechnic State University,
San Luis Obispo, CA 93407, USA
xhyu@calpoly.edu

Abstract. Power system fault detection has been an import area of study for power distribution networks. The power transmission systems often operate in the kV range with significant current flowing through the lines. A single fault, even lasting for a fraction of a second, can cause huge losses and manufacturing downtime for industrial applications. In this research, we develop an approach to detect, classify, and localize different types of phase-to-ground and phase-to-phase faults in three-phase power transmission systems based on discrete wavelet transform (DWT) and artificial neural networks (ANN). The multi-resolution property of wavelet transform provides a suitable tool to analyze the irregular transient changes in voltage or current signals in the network when fault occurs. An artificial neural network is employed to discriminate the types of fault based on features extracted by DWT. Computer simulation results show that this method can effectively identify various faults in a typical three-phase transmission line in power grid.

Keywords: Power systems · Fault detection · Fault classification ·
Wavelet transform · Artificial neural networks

1 Introduction

In today's society, electricity is a necessity for our daily lives. From large industrial companies to small households, energy is consumed and always needed to be readily available. A major issue that power companies face is the power transmission discontinuity due to various faults along transmission lines. It is known that the power transmission systems often operate at high voltage (in kV range) for lesser resistive losses over long distance; thus when fault occurs, excessively high current flows through the power network which may cause severe damages to equipment and devices ([1–3]). A single fault, even when lasting only for a fraction of a second, may affect potentially millions of customers on the grid and result in huge losses and manufacturing downtime in industry. These power quality events (PQEs) can be caused by natural disasters, equipment failures, or human errors. For example, a line-to-ground fault may be caused by a fallen tree limb that makes contact with one transmission phase line and the ground. If an object, such as a bird or other animal, makes a contact with two transmission phase lines may result in a short current of these two phases called a line-to-line fault.

© Springer Nature Switzerland AG 2019
H. Lu et al. (Eds.): ISNN 2019, LNCS 11555, pp. 266–272, 2019.
https://doi.org/10.1007/978-3-030-22808-8_27

The conventional approach to discover and identify faults in power networks is to manually analyze the system. However, this method is usually time-consuming and not very efficient. In recent years, there have been some developments in the applications of computational intelligent models and algorithms for fault detection and diagnosis. In [4], support vector machine (SVM) is employed to detect and classify four different types of faults in a distributed power network. The output of SVM is binary; thus four SVMs are used in the system and each SVM is trained to detect and classify fault for a particular phase. In [5], transformer disturbances in power networks are discussed. Two artificial neural networks (ANN) are connected in cascade form; one for fault detection and one for classification. Once the disturbance is detected by the first neural network, the algorithm enables the second ANN to discriminate different types of faults appropriately. Reference [6] considers the application of a probabilistic neural network (PNN) with discrete wavelet transform (DWT). The details of dataset and simulation results are not given. In [7], DWT and neural networks are combined to detect three different faults for a typical three-phase inverter used in power systems. The inputs to neural network are the normalized approximate coefficients of level 1, 2, and 3 from wavelet transform. The performance of ANN is tested on a limited dataset (12 tests total), with satisfactory results.

This paper focuses on the development of a hybrid approach to detect, classify, and localize different types of phase-to-ground and phase-to-phase faults in three-phase power transmission systems based on discrete wavelet transform and artificial neural networks. The multi-resolution property of wavelet transform provides a suitable tool to extract and analyze the transient changes in voltage or current signals when a network fault occurs. Note this "irregular" change in time domain also results in the change of signal power distribution in frequency domain. In this research, instead of using DWT coefficients directly as proposed in literature, the power of the subband signal (decomposed by DWT) is used as the feature vector. An artificial neural network is then employed to discriminate various types of faults in the network. Seven different cases are considered, namely, no fault, phase A line-to-ground fault, phase B line-to-ground fault, phase C line-to-ground fault, phase A and B line-to-line fault, phase B and C line-to-line fault, and phase A and C line-to-line fault. Computer simulation results show that this method can effectively identify various faults in a typical three-phase transmission line in power grid.

This paper is organized as follows. Section 2 provides the background information on discrete wavelet transform (DWT), artificial neural networks (ANN), as well as the hybrid approach based on DWT and ANN. Section 3 discusses the computer simulation results. Section 4 concludes the paper and gives direction for future work.

2 The Hybrid Approach for Power System Fault Identification

In this section, the background information on discrete wavelet transform and artificial neural networks is introduced first; then the hybrid approach for power grid fault detection based on DWT and ANN is discussed.

For a three-phase power grid, there are two different types of faults, i.e. the symmetric fault and the asymmetric fault. About 5% of power transmission line faults are symmetric (or balanced) faults, which affect each of the three phases equally ([1–3]). Typical symmetrical faults include line to line to line (L-L-L) and line to line to line to ground (L-L-L-G). Asymmetric faults (or unbalanced faults), which do not affect each of the three phases equally, are more common in power systems. Asymmetric faults include line-to-line, line-to-ground, as well as double line-to-ground faults. In this research, we consider six different types of asymmetric faults, i.e., phase A line-to-ground fault, phase B line-to-ground fault, phase C line-to-ground fault, phase A and B line-to-line fault, phase B and C line-to-line fault, and phase A and C line-to-line fault.

Wavelet transform is a powerful mathematical tool for signal processing. It decomposes signals into multiple frequency bands with different resolutions, and is especially suitable to analyze non-stationary signals or to detect irregular transient changes in signals. The continuous wavelet transform of a signal $f(t)$ can be written as:

$$T(\tau, s) = \frac{1}{\sqrt{s}} \int_{-\infty}^{\infty} f(t)\psi^*\left(\frac{t-\tau}{s}\right) dt \tag{1}$$

where $\psi(\cdot)$ is called the "mother wavelet" and $\psi^*(\cdot)$ represents its complex conjugate; s is the scaling factor and τ is the shifting factor. In computer simulations, discrete wavelet transform is performed by selecting

$$s = 2^a, \quad \tau = 2^b \tag{2}$$

where a and b are positive integers.

The wavelet transform can be considered as passing a signal through a set of low-pass (LP) and high-pass filters (HP). Through this process, a signal can be decomposed into various levels. At each level, it contains a set of detail coefficients (D) and approximation coefficients (A). In this research, we use DWT to extract features from the voltage or current signals in the power network. Instead of using DWT coefficients directly as proposed in literature, the power of the subband signal (decomposed by DWT) is used as the feature vector. Daubechies (db4) wavelet is employed for decomposition to level 4; then the power of the subband signal is calculated using detail coefficients of level 4:

$$E_s = \sum_{i=1}^{L_m} (D_i)^2 \tag{3}$$

where D_i is the ith detail coefficient of the decomposition level m which contains totally L_m detail coefficients.

After feature extraction, a neural network model is proposed to classify different types of faults, with the subband signal power obtained from wavelet transform as its inputs. The weights of the neural network are initialized randomly, and then updated with the Levenberg-Marquardt algorithm [8]:

$$W(k + 1) = W(k) + \eta(J_a^T J_a + \mu I)^{-1} J_a^T e \qquad (4)$$

where J_a is the first order derivative of the error function with respect to the neural network weight (also called the Jacobian matrix); e is the output error (i.e., the difference between the neural network outputs and the desired outputs. In this application, it represents the classification error); μ and η are learning parameters; and k is the index of iterations.

A typical multi-layer feedforward neural network has an input layer, an output layer, and one or more hidden layer(s). The wavelet transform is performed on three voltage or line current signals, one for each phase (i.e., phase A, B, and C); thus the neural network classifier has three inputs. In this research, by trial and error, we choose the neural network with one hidden layer and ten hidden neurons for the simulation in Sect. 3. The output of neural network represents the type of each fault, or no fault. Therefore, the neural network classifier can have either a single output, or three outputs with the fault type ID binary encoded for seven different cases. Initial training and test results show that the single output neural network classifier performs slightly better than the binary encoded outputs. As a result, in Sect. 3, we choose the neural network with a single output. The overall system diagram is shown in Fig. 1.

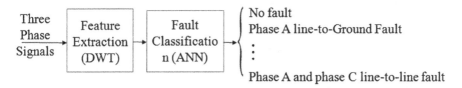

Fig. 1. The power system fault detection and classification using DWT and ANN

3   Simulation Results

A typical configuration of a power grid consists of three major modules, i.e. generation, transmission, and distribution. Figure 2 illustrates the Matlab Simulink model of such a network. In this section, the performance of the proposed approach based on DWT and ANN is tested by computer simulations.

In Fig. 2, the three-phase transmission line between the 50 Hz power generator site and the load is divided into two series-connected portions by the fault location in network, where the first portion shows the connection between the source and fault location; the second portion illustrates the connection between the fault location and the load. The resistance, capacitance, and inductance per unit length for the positive- and zero-sequence of the transmission lines are summarized in Table 1. These values of parameters are chosen to match with the parameters of the power source. Note the transmission line is continually transposed; i.e., the positive and negative sequences are equal. Also, a small non-zero ground resistor should be included. In this simulation, it is chosen to be 0.001 Ω. All of these values of parameters can be varied upon one's choice.

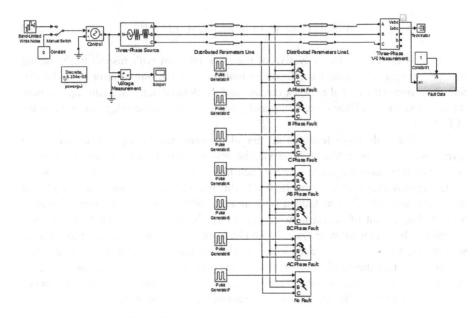

Fig. 2. The three-phase power transmission line

Table 1. Parameters of transmission lines

Parameter	Positive sequence	Zero sequence
Resistance per unit length (Ohms/km)	0.01273	0.3864
Capacitance per unit length (H/km)	12.74×10^{-9}	7.751×10^{-9}
Inductance per unit length (F/km)	0.9337×10^{-3}	4.1264×10^{-3}

The computer simulation results are shown in Tables 2, 3, 4 and 5. In Tables 2 and 3, "N" represents "no fault"; "A" represents phase A line-to-ground fault; "B" represents phase B line-to-ground fault; and "C" represents phase C line-to-ground fault. Similarly, "AB" represents phase A to B line-to-line fault; "BC" represents phase B to C line-to-line fault; and "AC" represents phase A to C line-to-line fault. The percentages on main diagonal positions are the percentage of correct classification; while all the other percentages on the off-diagonal positions are the percentage of misclassification. For example, the first column shows that for the true "no fault" case, the algorithm yields a 99.95% correct classification rate while 0.05% of the true "no fault" cases are misclassified as phase A to B line-to-line fault. The "no-fault" accuracy is obtained based on 3800 input/output data pairs and for each fault case, the accuracy of detection is obtained based on 300 input/output data pairs. Table 2 shows the confusion matrix if the data used are line current measurements; Table 3 shows similar results but with data taken on voltage signals. The average accuracy for the seven different cases is 83.33%.

Table 2. Fault identification confusion matrix based on current signals

	N	A	B	C	AB	BC	AC
N	99.95%	16.67%	8.33%	8.33%	0	0	0
A	0	83.33%	8.33%	16.33%	8.33%	0	8.33%
B	0	0	75.00%	0.33%	0	8.33%	8.33%
C	0	0	8.33%	75.00%	8.33%	0	0
AB	0.05%	0	0	0	83.33%	0	0
BC	0	0	0	0	0	91.67%	8.33%
AC	0	0	0	0	0	0	75.00%

Table 3. Fault identification confusion matrix based on voltage signals

	N	A	B	C	AB	BC	AC
N	99.95%	16.67%	8.33%	8.33%	0	0	0
A	0	83.33%	8.33%	16.67%	8.33%	0	8.33%
B	0	0	75.00%	0	0	8.33%	8.33%
C	0.05%	0	8.33%	75.00%	8.33%	0	0
AB	0	0	0	0	83.33%	0	0
BC	0	0	0	0	0	91.67%	8.00%
AC	0	0	0	0	0	0	75.33%

Table 4. Fault location determination confusion matrix based on current signals

	0 km	50 km	100 km	150 km	200 km
0 km	99.95%	22.22%	11.11%	5.56%	0
50 km	0	66.67%	16.67%	33.33%	20.22%
100 km	0	0	55.56%	0	18.67%
150 km	0	11.11%	5.56%	55.56%	5.56%
200 km	0.05%	0	11.11%	5.56%	55.56%

Tables 4 and 5 shows the simulation results on fault localization (for simplicity, only phase A line-to-ground fault is considered) using the current or voltage signals, respectively. The distances in the tables indicate the fault location. For example, "0 km" indicates the fault occurs right at the end of the power lines where measurements are taken; "50 km" indicates the fault occurs 50 km away from the endpoint (or 150 km to the power source); etc. For simplicity, only four locations are considered in this proof-of-concept study. For example, the neural network correctly identifies the fault location with an accuracy of 66.67% if the fault occurs 50 km away from the measurement data acquisition location; while the neural network misclassifies 11.11% of the fault locations as 150 km away from the point where measurement data are taken. For each distance, 450 sample pairs are generated in this simulation. In general, the fault detection accuracy decreases as the distance between fault location and measurement point increases, except at the midpoint in Table 5 (100 km away from both source and load) which needs further analysis.

Table 5. Fault location determination confusion matrix based on voltage signals

	0 km	50 km	100 km	150 km	200 km
0 km	100%	22.22%	11.11%	5.56%	0
50 km	0	66.67%	16.67%	33.33%	16.67%
100 km	0	0	55.56%	0	22.22%
150 km	0	11.11%	5.56%	61.11%	0
200 km	0	0	11.11%	0	61.11%

4   Conclusions

In this paper, we develop an approach to detect, classify, and localize different types of phase-to-ground and phase-to-phase faults in three-phase power transmission systems based on discrete wavelet transform and artificial neural networks. Satisfactory computer simulation results are obtained and presented. For future work, we plan to consider the situation when measurement data contain noise and/or outliers. Pre-processing noisy data using adaptive filtering and/or outlier detection may speed up neural network learning and improve the neural network generalization ability. More tests will be conducted to further investigate the performance of this hybrid approach.

References

1. Anderson, P.M.: Analysis of Faulted Power Systems. IEEE Press Series on Power Engineering. Wiley-IEEE Press, Piscataway (1995)
2. Glover, J.D., Sarma, M.S., Overbye, T.J.: Power System Analysis and Design, 5th edn. Cengage Learning, Boston (2012)
3. Kothari, D.P., Nagrath, I.J.: Modern Power System Analysis, 4th edn. McGraw-Hill Education, New York City (2011)
4. Magagula, X.G., Hamam, Y., Jordaan, J.A., Yusuff, A.A.: Fault detection and classification method using DWT and SVM in a power distribution network. In: 2017 IEEE PES Power Africa, Accra, pp. 1–6 (2017)
5. Fernandes, J.F., Costa, F.B., de Medeiros, R.P.: Power transformer disturbance classification based on the wavelet transform and artificial neural networks. In: 2016 International Joint Conference on Neural Networks (IJCNN), Vancouver, BC, pp. 640–646 (2016)
6. Ramaswamy, S., Kiran, B.V., Kashyap, K.H., Shenoy, U.J.: Classification of power system transients using wavelet transforms and probabilistic neural networks. In: Conference on Convergent Technologies for Asia-Pacific Region (IEEE TENCON 2003), vol. IV, pp. 1272–1276 (2003)
7. Charfi, F., Sellami, F., Al-Haddad, K.: Fault diagnostic in power system using wavelet transforms and neural networks. In: 2006 IEEE International Symposium on Industrial Electronics, Montreal, Quebec, pp. 1143–1148 (2006)
8. Haykin, S.: Neural Networks and Learning Machines. Prentice Hall/Pearson, New York (2009)

A New Diffusion Kalman Algorithm Dealing with Missing Data

Shuangyi Xiao[✉], Nankun Mu, and Feng Chen

College of Electronic and Information Engineering, Southwest University,
Chongqing, People's Republic of China
xsy2016@email.swu.edu.cn, nankun.mu@qq.com,
Fengchen.uestc@gmail.com

Abstract. In this paper, we propose a novel modified distributed Kalman algorithm, which is a diffusion strategy that the state estimation is more precise while the system model is time-varying. Our focus is on the missing data gathered by a set of sensor nodes that may obtain incomplete information because of the harsh environment. Simulation results evaluate the performance of the proposed distributed Kalman filtering algorithm.

Keywords: Diffusion estimation · Kalman filtering · Missing data

1 Introduction

The sensor network is deployed in a large number of micro sensor nodes distributed spatially independent in the monitoring area. Due to the limited energy, sensing range, communication and computing ability of the node, distributed algorithms exhibit superior performance over traditional centralized algorithms, especially the diffusion strategy of LMS distributed algorithm [1].

However, LMS algorithm is usually used for estimating constant parameters instead of dynamic model such as target tracking [2] which is also an important research area of sensor networks no matter military or civilian. In time-varying model, Kalman filter algorithm is one of the most popular recursion algorithm since it was proposed in 1960s [3]. Cattivelli and Sayed [4] studied DKF using a diffusion strategy, whose technical challenge is how to migrate mature central (or traditional) Kalman filtering methods to complex large dynamic systems and distribute the measurements across a large geographic area.

Most of Previous distributed algorithms in sensor networks assume that the information sensed by sensor nodes are lossless. As the size of the deployed sensor network grows, raw data typically has significant data loss because data collection is heavily influenced by hardware and wireless conditions [5]. Another advantage of Kalman filter algorithm is that estimating the states of dynamic system from an incomplete and noisy measurement [6]. But Kalman filter algorithm cannot be directly implemented for improving estimations with missing data because of the unique data loss patterns of sensor network [7].

© Springer Nature Switzerland AG 2019
H. Lu et al. (Eds.): ISNN 2019, LNCS 11555, pp. 273–281, 2019.
https://doi.org/10.1007/978-3-030-22808-8_28

In this paper, a sensor network is considered which is developed to track targets. Our work is to derive a diffusion algorithm based on KF, then obtain accurate values even the measurements are missing.

The reminder of the paper is arranged as follows: In the next section, the system model is presented and the problem is formulated. Section 3 derives the novel diffusion Kalman estimation algorithm with some data missing. Simulations results of a target tracking example are presented and discussed in Sect. 4 as well as with other existing strategy, showing improvement in performance. Finally, we sum up the article in Sect. 5.

2   The System Model and Problem Formulation

2.1   System Model

We consider the connection network composed of N nodes with limited communication ranges which are spatially distributed over a remote and inaccessible region, as shown in Fig. 1. Nodes that can share information with each other are connected by edges [8]. The neighborhood of node k is represented by N_k, which consists of all nodes connected to k by edges (including node k itself). Assume an undirected graph. If node k is the neighbor of node l, then node l is also the neighbor of node k. That is to say, the information between two adjacent nodes is bidirectional flow.

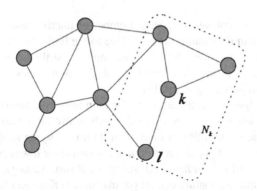

Fig. 1. Sensor network

Consider a state-space model associated with the environment for a target system described by

$$x_{i+1} = F_i x_i + n_i \tag{1}$$

$$y_i = H_i x_i + v_i \tag{2}$$

where $i = 1, 2, 3, \ldots$ is a the time index, $x_i \in R^M$ is the state (signal) vector with initial state x_0 distributed as a zero mean Gaussian vector with covariance q_0, $y_i \in R^{r_i}$ is the (possibly) time-varying observation measurement vector by any node k at time i. $F_i \in R^{M \times M}$ and $H_i \in R^{M_y \times M}$ denote the sparse localized transition matrix and the local observation matrix of the system, respectively. The signal $n_i \in R^M$ is the state noise with covariance $Q_i \geq 0$, and v_i is the measurement noise with covariance $R_i(t) > 0$. Both of them are assumed to be zero-mean i.i.d. Gaussian noises denoted by

$$E \begin{bmatrix} n_i \\ v_i \end{bmatrix} \begin{bmatrix} n_j \\ v_j \end{bmatrix}^* = \begin{bmatrix} Q_i & 0 \\ 0 & R_i \end{bmatrix} \delta_{ij} \tag{3}$$

where the operator $*$ denotes complex conjugate transposition and δ_{ij} is the Kronecker delta. Also, the noise at different nodes are independent of each other,

To describe this problem well, Eq. (2) is rewritten as

$$y_k^i = C^i x_k + v_k^i \tag{4}$$

2.2 Missing Data Problem

We assume that the data is missing at random (MAR) and the missing does not relay on the data [9]. Let $p_{i,k}$ denote the probability of missing data by sensor k at time i in the sensor network and the missing is independent of time and space. O_{ik} is binary value, which indicate if the data values gathered by node K at time i are missing or not. O_{ik} is defined as

$$O_{ik} = \begin{cases} 0 & \text{if } y_{k,i} \text{ is missing} \\ 1 & \text{otherwise} \end{cases} \tag{5}$$

When any node k lose the signal, it can receive nothing but noise which lead to deviations in estimation results.

3 Diffusion Kalman Algorithm with Missing Data

Kalman filtering (KF) algorithm was proposed in dynamic system model like target tracking for decades. The iterations are shown in (6) and (7)

Measurement update:

$$\begin{aligned} K_i &= P_{i|i-1} H_i^* \left(R_i + H_i P_{i|i-1} H_i^* \right)^{-1} \\ \widehat{x}_{i|i} &= \widehat{x}_{i|i-1} + K_i \left(y_i - H_i \widehat{x}_{i|i-1} \right) \\ P_{i|i} &= P_{i|i-1} - K_i H_i P_{i|i-1} \end{aligned} \tag{6}$$

Time update:

$$\widehat{x}_{i+1|i} = F_i x_{i|i} + u_i$$
$$P_{i+1|i} = F_i P_{i|i-1} F_i^* \tag{7}$$

where $\widehat{x}_{i|j}$ denote the linear minimum mean-square error estimate of x_i and $P_{i|j}$ denote the covariance matrix of the estimation error $\widetilde{x}_{i|j} \triangleq x_i - \widehat{x}_{i|j}$. K_i is the Kalman filtering gain.

On the basis of Kalman filtering which refer to the local measurement information, nodes obtained the current estimates of the target state by communicating with their neighbors in the sensor network [10]. Therefore, we could add the intermediate estimate information from its neighbors to node k to update the Eq. (5). Formally, Eq. (5) is rewritten as

$$P_{k,i|i}^{-1} = P_{k,i|i-1}^{-1} + \sum_{l \in N_k} H_{l,i}^* R_{l,i}^{-1} H_{l,i}$$
$$\widehat{x}_{k,i|i} = \widehat{x}_{k,i|i-1} + P_{k,i|i} \sum_{l \in N_k} H_{l,i}^* R_{l,i}^{-1} \left(y_{l,i} - H_{l,i} \widehat{x}_{k,i|i-1} \right) \tag{8}$$
$$\widehat{x}_{k,i+1|i} = F_i \widehat{x}_{k,i|i}$$
$$P_{k,i+1|i} = F_i P_{k,i|i} F_i^* + G_i Q_i G_i^*$$

In this paper, we investigate how to get state estimate as accurately as possible in a dynamic system when the missing sensor data exist. For deal with missing data problem, we introduce an imputation strategy into the step 1 of the diffusion Kalman filter [11] to reconstructing missing information.

Before designing the imputation, we device a simple detection system to detect the nodes with missing data. Let's set a threshold γ_{ik} which refers to environment noise. The threshold is used to size up the missing data, i.e., the indicate O_{ik} is determined by γ_{ik} as the following:

$$O_{ik} = \begin{cases} 0, & y_{k,i} < \gamma_{ik} \\ 1, & y_{k,i} > \gamma_{ik} \end{cases} \tag{9}$$

For the node detected by the detection system that the data was lost, its measure message $y_{k,i}$ is replaced by the estimate values of Kalman algorithm at time $i - 1$, otherwise remain unchanged:

$$y_{k,i} = \begin{cases} \widehat{x}_{k,i+1|i}, & O_{ik} = 0 \\ y_{k,i}, & O_{ik} = 1 \end{cases} \tag{10}$$

Let $\varphi_{k,i}$ denote the intermediate estimate for the node k at time i. Then $\varphi_{l,i}$ means the intermediate estimate for the node l at time i as well. We rewrite the second of Eq. (8) using all neighbors' intermediate estimate as following:

$$\varphi_{k,i/i} = \varphi_{k,i/i-1} + P_{k,i/i}(\sum_{l \in N_k} H_{l,i}^T R_{l,i}^{-1} y_{l,i} - \sum_{l \in N_k} H_{l,i}^T R_{l,i}^{-1} H_{l,i} \varphi_{k,i/i-1}) \tag{11}$$

$P_{k,i/i}^{-1}$ can be rewritten as the same.

Algorithm 1: Imputation diffusion Kalman algorithm with missing data

---

Initialization: For each node $k \in N_i$,

Start with $\hat{x}_{k,0|-1} = x_0$, $P_{k,0|-1}^{-1} = P_0$ 和 $y_{k,i} = y_0$;

for $t = 1:T$:

  for each node k :

  1）Detect: $O_{ik} = \begin{cases} 0, & y_{k,i} < \gamma_{ik} \\ 1, & y_{k,i} > \gamma_{ik} \end{cases}$

  2）Adaptation：

    Imputation: $y_{k,i} = \begin{cases} \hat{x}_{k,i+1|i}, & O_{ik} = 0 \\ y_{k,i}, & O_{ik} = 1 \end{cases}$

$$\varphi_{k,i/i} = \varphi_{k,i/i-1} + P_{k,i/i}(\sum_{l \in N_k} H_{l,i}^T R_{l,i}^{-1} y_{l,i} - \sum_{l \in N_k} H_{l,i}^T R_{l,i}^{-1} H_{l,i} \varphi_{k,i/i-1})$$

$$P_{k,i/i}^{-1} = P_{k,i/i-1}^{-1} + \sum_{l \in N_k} H_{l,i}^T R_{l,i}^{-1} H_{l,i}$$

  3）Combination：

$$x_{k,i/i} = \sum_{l \in N_k} c_{k,l} \varphi_{l,i/i}$$

$$\varphi_{k,i+1/i} = F_i x_{k,i/i}$$

$$P_{k,i+1/i} = F_i P_{k,i/i-1} F_i^T$$

---

4 Simulation

In this section, the performance of the proposed algorithm is demonstrated by numerical simulation. We present the MSD performance of the non-cooperate Kalman filter (KF) with missing and non-missing data. In order to illustrate the performance of the IDKF, we also make a comparison of the proposed strategy and the usual strategy to the step 1 of the diffusion Kalman filter. To achieve this objective, we build a time-varying random system.

4.1 Simulation Environment

For simplicity, we consider the problem of tracking the trajectory of a Parabolic object observed by a network consists of 20 nodes with the topology shown in Fig. 2, where nodes obtain incomplete measurements (noise or missing) of the object's position independently and communicate with its neighbors.

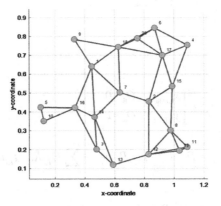

Fig. 2. Topology of sensor network

In our simulation, the system model is given by

$$F = \begin{bmatrix} 1 & 0 & T & 0 \\ 0 & 1 & 0 & T \\ 0 & 0 & 1 & 0 \\ 0 & 0 & 0 & 1 \end{bmatrix},$$

$$H = \begin{bmatrix} 1 & 0 & 0 & 0 \\ 0 & 1 & 0 & 0 \end{bmatrix},$$

where $T = 0.1$, $Q = 5I$, $R_k = 5$.

The missing probability for all nodes are set up to 0.1.

For the object, the initial position state $x_0 = 1$ $y_0 = 31$ and the initial speed $v = 15$, the gravity $g = 10$, angel $\theta = \pi/3$

Figure 3(a) shows noisy measurements at node 5 without missing, indicating that the noise can cause deviations in measurement but is not damaged. Compared with (a), Fig. 3(b) demonstrate that missing data can lead to bad influence to the measurements even if the probability is 0.1.

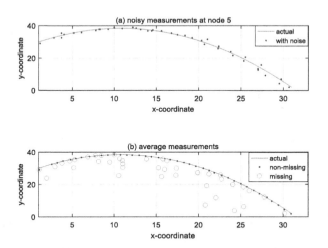

Fig. 3. Noisy measurements at node 5 (a), average measurements among 3 patterns (b)

4.2 Simulation Results

In the following simulations, we use the mean-square deviation (MSD) to evaluate the influence of the performance of missing as shown in Fig. 4. The MSD is defined as

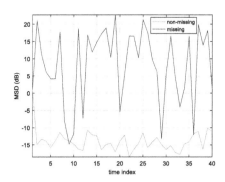

Fig. 4. MSD of non-imputation DKF with and without missing

$$MSD_i = \frac{1}{N} \sum_{k=1}^{N} E(x - \widehat{x}_{k,i|i}) \tag{12}$$

For a more intuitive explanation, error is used directly to testing the performance of the imputation strategy compared to the normal strategy and the measurements. Figure 5 demonstrate the proposed IDKF perform better than the DKF and can dealing with missing data problem effectively.

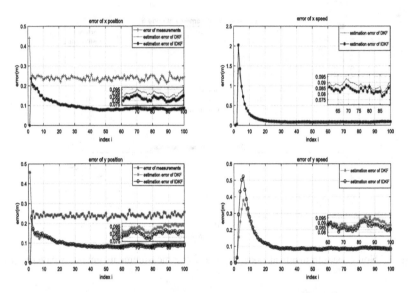

Fig. 5. Error of position and speed of x-coordinate and y-coordinate

5   Conclusions

In this paper, we provided a new imputation diffusion Kalman filter (IDKF) algorithm to track the object with missing data at random. We reconstruct the missing data by using last estimation result of Kalman filter replacement. Our simulation examples have demonstrated that the proposed algorithm can deal with the missing data problem and improve the accuracy of the sensor network. The modifications of detection and imputation can recognize the nodes who lose data when the noise is Gaussian white noise and perform better than usual strategy in the first step of DKF.

References

1. Cattivelli, F.S., Sayed, A.H.: Diffusion LMS strategies for distributed estimation. IEEE Trans. Sign. Process. **58**(3), 1035–1048 (2010)
2. Bhuiyan, M.Z., Wang, G., Vasilakos, A.V.: Local area prediction-based mobile target tracking in wireless sensor networks. IEEE Trans. Comput. **64**(7), 1968–1982 (2015)
3. Huo, Y., Cai, Z., Gong, W., Liu, Q.: A new adaptive Kalman filter by combining evolutionary algorithm and fuzzy inference system. In: Proceedings of the IEEE Congress on Evolutionary Computation, pp. 2893–2900 (2014)

 4. Balazinska, M., et al.: Data management in the worldwide sensor web. IEEE Pervasive Comput. **6**(2), 30–40 (2007)
 5. Cattivelli, F.S., Lopes, C.G., Sayed, A.H.: Diffusion strategies for distributed Kalman filtering: formulation and performance analysis. In: Proceedings Cognitive Information Processing. Santorini, Greece (2008)
 6. Spanos, D.P., Olfati-Saber, R., Murray, R.M.: Approximate distributed Kalman filtering in sensor networks with quantifiable performance. In: IPSN, pp. 133–139. Los Angeles, CA (2005)
 7. Kulakov, A., Davcev, D.: Tracking of unusual events in wireless sensor networks based on artificial neural-networks algorithms. Inf. Technol. Coding Comput. (ITCC) **2**, 534–539 (2005)
 8. Estrin, D., Girod, L., Pottie, G., Srivastava, M.: Instrumenting the world with wireless sensor networks. In: ICASSP, vol. 4, pp. 2033–2036. Salt Lake City (2001)
 9. Candes, E., Tao, T.: Near optimal signal recovery from random projections: universal encoding strategies. IEEE Trans. Inf. Theor. **52**(12), 5406–5425 (2006)
10. Cattivelli, F.S., Sayed, A.H.: Diffusion mechanisms for fixedpoint distributed Kalman smoothing. In: European Signal Processing Conference. Lausanne, Switzerland (2008)
11. Cattivelli, F.S., Sayed, A.H.: Diffusion strategies for distributed Kalman filtering and smoothing. IEEE Trans. Autom. Control **55**(9), 2069–2084 (2010)

Separability Method for Homogeneous Leaves Using Spectroscopic Imagery and Machine Learning Algorithms

Bolanle Tolulope Abe$^{(\boxtimes)}$ and Jaco Jordaan

Department of Electrical Engineering, Tshwane University of Technology,
eMalahleni Campus, South Africa
abe_tolulope@yahoo.com, abebt@tut.ac.za,
jordaan.jaco@gmail.com

Abstract. Agathosma Buchu plants are a holistic healing system used as alternative medicine in dealing with numerous diseases and also for cooking oil production. There are two types namely, Betulina and Crenulata. The plants are difficult to separate if mixed up after harvest. Furthermore, the high rate of the plants' cultivation poses challenges in separating them for specific functions. Hence, other identification methods are crucial. This paper presents an implementation of machine learning algorithms based on spectroscopic imagery properties for automatic recognition of the plants' species. Image Local Polynomial Approximation method is used for the image processing to reduce classification error and dimensionality of classification challenges. To demonstrate the efficacies of the processed dataset, K-Nearest Neighbour, Naïve Bayes, Decision Tree, and Neural Network classifiers were used for the classification procedures in different data mining tools. The classifiers' performances are valuable for decision-makers to consider tradeoffs in method accuracy versus method complexity.

Keywords: Image · Agathosma · Separability · Classification

1 Introduction

Agathosma (Buchu) grown in the Southern part of Africa and known as genus plant in the family of Rutaceae (citrus) has over 120 species flowering plants. Among the various species is Agathosma (A.) Betulina, a fragrant shrub with round leaves, which grows to a height of 2 m and the Agathosma (A.) Crenulata plant, a pungently aromatic, woody, single- stemmed shrub that reaches a height of 2.5 m. As an alternative medicine, traditional healers use the plants as an antispasmodic, antipyretic cough remedy and diuretic for health care purposes [1]. Furthermore, the plants are used to treat urinary tract infectivity, stomach pains, for treating wounds, kidney problems, and symptomatic treatment for rheumatism [2, 3]. In addition, the plants are also used for oil production. A. Betulina produces good quality oil and available for sales while A. Crenulata oil is of a less quality [4]. The plants are well-known plants used for medicinal purposes in South Africa [5]. The leaves are normally recognized based on their morphology. A. Betulina leaves are round in shape while A. Crenulata leaves are oval in shape. Figures 1 and 2 show the pictures of the leaves.

© Springer Nature Switzerland AG 2019
H. Lu et al. (Eds.): ISNN 2019, LNCS 11555, pp. 282–291, 2019.
https://doi.org/10.1007/978-3-030-22808-8_29

Fig. 1. Agathosma Betulina

Fig. 2. Agathosma Crenulata

However, the morphological method of recognition cannot be used for the plants' identification due to the extensive cultivation of the plants. New crossbreeds of the plants now exist which cannot be clearly separated. A. betulina and A. crenulata have been studied extensively by investigating the volatile oil consistents [5], while few studies have been done on the identification of the leaves using hyperspectral data processing and classification methods for the leave separability. This research aims at exploring the Local Polynomial approximation (LPA) method of hyperspectral image processing and thereafter classifies the processed data using machine-learning algorithms as tools for separating the species.

Hyperspectral images are made up of a large number of spectral bands, which are about 200–250 imageries with identical views. The spectral bands found applications in various fields for analyzing objects and materials within the electromagnetic spectrum [6–9]. Research studies have shown that leaf image classification is an acceptable technique for leaf identification and classification [9, 10]. The process involves taking images of different leaves under question and subjects them for processing in a computer so as to extricate necessary information from the images. This is followed by the identification of the leaves' patterns through the application of machine learning tools.

The study is carried out with the application of the first and second order Local Polynomial Approximation techniques for data processing. The paper is structured as follows: discussions on the Local Polynomial Approximation (LPA) method is presented in section two, section three presents the classification algorithm, description of the study site is presented in section four, we discussed the experimental procedures and results in section five, while section six concludes the work.

2 Local Polynomial Approximation

Consider a set of data points $\{x_k\}$, where the points are evenly spaced. Piecewise polynomials are used in many places to smooth data and to calculate the derivatives of a set of data. Savitzky and Golay made this popular by using a least-square technique [11, 12] for designing a polynomial filter. The idea of local polynomial approximation is to take the raw data within a certain interval (this interval, also called a window, is moving across the whole data set, one data point at a time) and fit a polynomial of a certain degree to the data inside the interval. The length of the interval, also referred to as the window length, is either a fixed value chosen based on prior knowledge of the data, or based on some adaptive window length selection method, like that used in [13]. The use of a least-squares technique enables fitting the polynomial and finding the LPA filter coefficients, which define the impulse response of the filter [12, 14].

There are 3 window types, namely right, left and center. The *right-sided* window implies that all the samples in that particular window are to the right of the sample to be estimated (or smoothed). Likewise, for the *central* and *left-sided* windows the data point to be estimated is in the middle and far-right of the window respectively. For the purpose of this research, we only use the central window.

Consider only the data samples in the window of interest. To derive the LPA model, let the sample to be estimated have an index $k = 0$. A continuous-time polynomial function of order p is given by:

$$f(t) = c_0 + c_1 t + c_2 t^2 + c_3 t^3 + \cdots + c_p t^p, \tag{1}$$

where the polynomial coefficients are c_i. Sampling the polynomial with period T, the time becomes $t = kT$. The sampled Eq. (1) becomes

$$f(k) = c_0 + c_1 kT + c_2 (kT)^2 + \cdots + c_p (kT)^p. \tag{2}$$

To solve c_i (central window) using least squares, the following objective function is obtained:

$$J = \sum_{k=-w_n}^{w_n} (x_k - f(k))^2, \tag{3}$$

where the window length is $2w_n + 1$, with x_k the $k - th$ point in the current window and $f(k)$ the function value based on the model in (1). Setting the gradient $\frac{\partial J}{\partial c_i} = 0$ enables solving the set of equations for coefficients c_i. For this technique [12], there is only interest in the point $k = 0$. In this case the $s - th$ derivative requires an expression for only c_s. Considering a single polynomial term $f_i(k) = c_i(kT)^i$, the $s - th$ order derivative becomes:

$$f_i^s(k) = (i - s + 1)(i - s + 2) \cdots (i - 1)ic_i(kT)^{i-s} = \frac{i!}{(i - 1)!}c_i(kT)^{i-s}. \qquad (4)$$

Using Eq. (4), and then taking the $s - th$ order derivative of (2), one gets

$$f^{(s)}(k) = \sum_{i=s}^{p} f_i^{(s)}(kT). \qquad (5)$$

See [15] for more details regarding LPA. Local Polynomial Approximation is used to generate the derivative signal of the original data. The derivative gives information about the shape of the original data, and this could be a useful feature in classification techniques.

3   Classification Algorithms

The data classification algorithms used in the research are the Naive Bayes (NB), Decision tree (DT), Neural Network (NN), and K-Nearest Neighbor (KNN). These tools are implemented in Waikato Environment for Knowledge Analysis (WEKA) and Orange. The process involves constructing the classifier model by learning the training set with the class labels. This is followed by applying the test dataset in the classifiers to predict the accuracy of the classification algorithms.

K-Nearest Neighbor: This algorithm classifies new objects using the Euclidean equation [16, 17]. The classification is based on the similarity measure. The algorithm is easy and normally good for handling noise and uses large computerization.

Naive Bayes: This is known to be a fast learning and testing algorithm. This is a simplified version of the Bayesian classifier and operates on two assumption that is, (i) the latent attribute has no effect on the prediction process; (ii) for the class label, attributes are conditionally independent [18].

Neural Networks: This tool can handle problems with lots of parameters and can classify objects, even when the distribution of object in n-dimensional parameter space is very complex. They are nonlinear models and this makes them flexible in modeling real-world complex relationships. The method of classification is easy but also very popular because as easy as it is, it is very efficient [19].

Decision Tree: This algorithm is efficiently used in data mining because of its ability to solve difficult problems through the generating simple computer-readable graphical illustrations, which can easily be understood by experts and non-experts. The algorithm has been successfully used in proffering solutions to real-world problems [20].

3.1   Classification Tools Description

All the experimental tools used are open-source machine learning software and are freely available online for download.

WEKA: This machine learning tool was firstly developed by the University of Waikato, New Zealand. The tool has been numerously used by researchers for the application of machine learning and data mining. Weka software written in Java contains a huge collection machine learning with visualization tools for data analysis and predictive modeling.

Orange: This tool contains the machine learning component and has visual programming or Python scripting use for data analysis and visualization. In addition, the tool has add-ons for bioinformatics, text mining and also has various features for data analysis. The components in the tool are refer to as widgets and ranges from simple data visualization, subset selection and preprocessing up to the experimental assessment of classifiers and predictive modeling. Orange was developed and maintained by the Bioinformatics Laboratory, Faculty of Computer and Information Science, University of Ljubljana, Slovenia.

4   Study Sites

The experimental site is the Buchu Moon farm, Wellington, Western Cape, South Africa. The farm is divided into various blocks in which different type of Agathosma (Buchu) are planted. There are more than 130 species of Buchu, but only two are commercially viable. These are the Betulina and crenulata spiecies. Seven samples of the different Buchu types were collected and used for the research. They are named as follows:

- Big tank – Betulina (BTB)
- Block 1 – Crenulata (CR)
- Block 2 – Betulina, fast grower, low diosphenol Chemotype (BFG)
- Block 3 – Betulina, mid-range diosphenol Chemotype (BMR)
- Block 5 – Betulina, low diosphenol Chemotype (BL)
- Block 7 – Betulina, high diosphenol Chemotype (BH)
- Mother Block – Betulina, mid to high diosphenol Chemotype (MB)

The different types of Buchu leaves produce different quality of Buchu oils, and if the plants get mixed up after harvesting, it is difficult to distinguish which type is which. This research aims to help with classification of the leaves by means of multi-spectral analysis.

5   Experimental Procedures

5.1   Dataset

The total number of instances and attributes used for the experiment is 4200 and 257 respectively. The data is further divided into two so that the train dataset is 60% (2520 instances) and the remaining 40% (1680 instances) is for test data. The dataset is subjected for classification procedures using various machine learning tools as mentioned earlier.

5.2   Classification Algorithm Evaluation Using WEKA

The classifiers' performance is evaluated using a Confusion Matrix. The following tables present the correlation of the predicted instances values. Tables 1, 2, 3 and 4 present the confusion matrices for K-NN, NB, DT and NN.

Table 1. Performance evaluation using K-NN

TRUE:	BTB	CR	BFG	BMR	BL	BH	MB	Σ
BTB:	239	0	0	0	0	0	0	239
CR:	0	226	6	0	0	1	0	233
BFG:	0	0	232	0	0	0	3	235
BMR:	0	0	0	232	0	0	0	232
BL:	0	0	0	0	241	0	0	241
BH:	0	0	1	0	0	262	0	263
MB:	0	0	8	0	0	0	229	229
Σ	239	226	247	232	241	263	232	**1680**

Table 2. Performance evaluation using decision tree

TRUE:	BTB	CR	BFG	BMR	BL	BH	MB	Σ
BTB:	231	0	0	6	2	0	0	239
CR:	0	225	6	0	0	2	0	233
BFG:	0	8	214	1	0	1	11	235
BMR:	0	0	0	232	0	0	0	232
BL:	2	0	0	0	239	0	0	241
BH:	0	6	2	0	0	255	0	263
MB:	0	1	17	1	0	1	217	237
Σ	233	240	239	240	241	259	228	**1680**

Table 3. Performance evaluation using neural network

TRUE:	BTB	CR	BFG	BMR	BL	BH	MB	Σ
BTB:	231	0	0	6	2	0	0	239
CR:	0	225	6	0	0	2	0	233
BFG:	0	8	214	1	0	1	11	235
BMR:	0	0	0	232	0	0	0	232
BL:	2	0	0	0	239	0	0	241
BH·	0	6	2	0	0	255	0	263
MB:	0	1	17	1	0	1	217	237
Σ	233	240	239	240	241	259	228	**1680**

Table 4. Performance evaluation using Naïve Bayes

TRUE:	BTB	CR	BFG	BMR	BL	BH	MB	Σ
BTB:	225	0	0	0	14	0	0	239
CR:	0	199	6	0	0	28	0	233
BFG:	0	15	162	0	0	17	41	235
BMR:	0	0	0	232	0	0	0	232
BL:	10	0	0	0	231	0	0	241
BH:	0	18	1	0	0	244	0	263
MB:	0	0	32	0	0	0	205	237
Σ	235	232	201	232	245	289	246	1680

5.3 Classification Algorithm Evaluation Using Orange

Evaluation of the classifiers' performance is also evaluated using the Confusion Matrix. We present the correlation of the predicted instances in Tables 5, 6, 7 and 8. The K-NN, NB, DT and NN are used in Orange Machine Learning tool.

Table 5. Performance evaluation using K-NN

TRUE:	BTB	CR	BFG	BMR	BL	BH	MB	Σ
BTB:	240	0	0	0	0	0	14	254
CR:	0	223	8	0	0	9	0	240
BFG:	0	5	218	0	0	3	0	226
BMR:	0	0	0	240	0	0	0	240
BL:	0	0	0	0	240	0	0	240
BH:	0	1	2	0	0	239	0	240
MB:	0	0	16	0	0	0	224	240
Σ	240	229	242	240	240	251	238	**1680**

Table 6. Performance evaluation using decision tree

TRUE:	BTB	CR	BFG	BMR	BL	BH	MB	Σ
BTB:	239	0	0	1	0	0	0	240
CR:	0	228	8	0	0	44	0	240
BFG:	0	10	212	0	0	1	17	240
BMR:	0	0	1	238	0	0	1	240
BL:	0	0	0	0	240	0	0	240
BH:	0	6	5	0	0	229	0	239
MB:	0	1	13	3	0	0	223	240
Σ	239	245	239	242	240	234	241	**1680**

Table 7. Performance evaluation using neural network

TRUE:	BTB	CR	BFG	BMR	BL	BH	MB	Σ
BTB:	240	0	0	0	0	0	0	240
CR:	0	240	0	0	0	0	0	240
BFG:	0	0	240	0	0	0	0	240
BMR:	0	0	0	240	0	0	0	240
BL:	0	0	0	0	240	0	0	240
BH:	0	0	0	0	0	240	0	240
MB:	0	0	0	0	0	0	240	240
Σ	240	240	240	240	240	240	240	**1680**

Table 8. Performance evaluation using Naïve Bayes

TRUE.	BTB	CR	BFG	BMR	BL	BH	MB	Σ
BTB:	193	0	0	0	47	0	0	240
CR:	0	205	25	0	0	10	0	240
BFG:	0	18	181	1	0	13	27	240
BMR:	0	0	0	240	0	0	0	240
BL:	10	0	0	0	230	0	0	240
BH:	0	27	26	0	0	187	0	240
MB:	0	0	78	33	0	0	129	240
Σ	203	250	310	274	277	210	156	**1680**

6   Conclusion

In this research, we used Algathosma (Buchus) datasets from the Buchu moon farm, Wellington, Western Cape, South Africa. The dataset has 7 independent variables, 4200 instances and 256 attributes are used for the experiments. The Local Polynomial Approximation method is used for the data processing and the classification procedures are carried out using various machine learning algorithms. The implementations are carried out in WEKA and Orange data mining tools. The performance of the classification models based on the accuracy is compared as presented in Table 9:

Table 9. Summary of accuracy measure of classification algorithm

	WEKA (%)	Orange (%)
K-Nearest neighbor	98.87	96.60
Decision tree	96.01	95.70
Neural network	99.88	100
Naïve Bayes	89.17	81.30

From the table, it is evident that the WEKA tool generally performed better when compared with the Orange tool on the processed datasets, although the Orange estimate on neural network gives 100%, which is not significant as compared to WEKA's performance. Naïve Bayes has the lowest performance on the two machine tools. Further work will be to apply other machine learning tools in addition to the tools used in this research on the raw dataset and then validate the experimental results. The outcome will give room for better analysis and assessment of the of results which will assist in taking adequate decisions on the choice of plants.

Acknowledgement. Our appreciation goes to Allen Harris the owner of the Buchu moon farm, near Cape Town, South Africa, for his support, taking us through his farm and giving us Buchu specimens used for this research. This work is based on the research supported wholly by the National Research Foundation of South Africa (Grant specific unique reference number (UID) 85745). The Grant holder acknowledges that opinions, findings and conclusions or recommendations expressed in any publication generated by the NRF supported research are that of the author(s), and that the NRF accepts no liability whatsoever in this regard.

References

1. Mavimbela, T., Viljoen, A., Vermaak, I.: Differentiating between Agathosma betulina and Agathosma crenulata–a quality control perspective. J. Appl. Res. Med. Aromat. Plants 1(1), e8–e14 (2014)
2. Van Wyk, B.E., Oudtshoorn, B.V., Gericke, N.: Medicinal Plants of South Africa. Briza Publications, Pretoria (1997)
3. Thring, T.S.A., Weitz, F.M.: Medicinal plant use in the Bredasdorp/Elim region of the southern overberg in the western cape province of South Africa. J. Ethnopharmacol. 103, 261–275 (2006)
4. Sandasi, M., et al.: Hyperspectral imaging and chemometric modeling of Echinacea-a novel approach in the quality control of herbal medicines. Molecules 19(9), 13104–13121 (2014)
5. Moolla, A., Viljoen, A.M.: Buchu – Agathosma betulina and Agathosma crenulata (Rutaceae): a review. Elsevier J. Ethnopharmacol. 119, 413–419 (2008)
6. Abe, B.T., Olugbara, O.O., Marwala, T.: Hyperspectral image classification using random forest and neural network. Lecture Notes in Engineering and Computer Science. In: Proceedings of the World Congress on Engineering and Computer Science, WCECS 2012, pp. 522–527, San Francisco, 24–26 October 2012
7. Van der Meer, F.D., van der Werff, H.M.A., van Ruitenbeek, F.J.A., Hecker, C.A., Bakker, W.H., Noomen, M.F., et al.: Multi- and hyperspectral geologic remote sensing: a review. Int. J. Appl. Earth Obs. Geoinf. 14, 112–128 (2012)

8. Landmann, T., et al.: Application of hyperspectral remote sensing for flower mapping in African savannas. Remote Sens. Environ. **166**, 50–60 (2015)
9. Abe, B.T., Jordaan, J.A.: Identifying agathosma leaves using hyperspectral imagery and classification techniques. Lecture Notes in Engineering and Computer Science. In: Proceedings of the World Congress on Engineering and Computer Science, WCECS 2016, pp. 476–479, San Francisco, 19–21 October 2016
10. Sardogan, M., Tuncer, A., Ozen, Y.: Plant leaf disease detection and classification based on CNN with LVQ algorithm. In: 3rd International Conference on Computer Science and Engineering (UBMK), pp. 382–385, Sarajevo (2018)
11. Savltzky, A., Golay, M.J.E.: Smoothing and differentiation of data by simplified least squares procedures. Anal. Chem. **1084**(36), 1627–1639 (1964)
12. Gorry, P.: General least-squares smoothing and differentiation by the convolution (Savitzky-Golay) method. Anal. Chem. **62**(6), 570–573 (1990)
13. Katkovnik, V.: A new method for varying adaptive bandwidth selection. IEEE Trans. Signal Process. **47**(9), 2567–2571 (1999)
14. Bialkowski, S.: Generalized digital smoothing filters made easy by matrix calculations. Anal. Chem. **61**(11), 1308–1310 (1989)
15. Jordaan, J.A.: Fast and accurate spectral estimation algorithms for power system applications. Doctoral thesis, Tshwane University of Technology, South Africa (2006)
16. Larose, D.T.: Discovering Knowledge in Data an Introduction to Data Mining, pp. 90–106. Wiley Interscience, Hoboken (2005)
17. Okfalisa, I., Gazalba, M., Reza, N.G.I.: Comparative analysis of k-nearest neighbor and modified k-nearest neighbor algorithm for data classification. In: 2nd International Conferences on Information Technology. Information Systems and Electrical Engineering (ICITISEE), Yogyakarta, pp. 294–298 (2017)
18. Jahromi, A.H., Taheri, M.: A non-parametric mixture of Gaussian Naive Bayes classifiers based on local independent features. In: 2017 Artificial Intelligence and Signal Processing Conference (AISP), Shiraz, pp. 209–212 (2017)
19. Gui, J., Liu, T., Tao, D., Sun, Z., Tan, T.: Representative vector machines: a unified framework for classical classifiers. IEEE Trans. Cybern. **46**(8), 1877–1888 (2016)
20. Trabelsi, A., Elouedi, Z., Lefevre, E.: Decision tree classifiers for evidential attribute values and class labels. Fuzzy Sets Syst. ISSN 0165-0114 (2018)

Learning Ensembles of Anomaly Detectors on Synthetic Data

Dmitry Smolyakov[1], Nadezda Sviridenko[1], Vladislav Ishimtsev[1], Evgeny Burikov[2], and Evgeny Burnaev[1(✉)]

[1] Skolkovo Institute of Science and Technology, Moscow Region, Russia
{Dmitrii.Smoliakov,Nadezda.Sviridenko}@skolkovotech.ru,
{V.Ishimtsev,E.Burnaev}@skoltech.ru
[2] PO–AO "Minimaks-94", Moscow, Russia
Burikov@mm94.ru

Abstract. The main aim of this work is to develop and implement an automatic anomaly detection algorithm for meteorological time-series. To achieve this goal we develop an approach to constructing an ensemble of anomaly detectors in combination with adaptive threshold selection based on artificially generated anomalies. We demonstrate the efficiency of the proposed method by integrating the corresponding implementation into "Minimax-94" road weather information system.

Keywords: Anomaly detection · Predictive maintenance · RWIS

1 Introduction

Effective operation of federal highways in Russia during the winter period is a very complicated and important task which reduces the number of road accidents and incidents significantly. Sleet, frost, low visibility are several examples of meteorological conditions on the road that increase the car crash chances. Preventing such conditions and reducing their consequences is a complicated problem to deal with for two main reasons. First, they require immediate measures like involving snow removal machinery, reducing the speed limits or even closing certain parts of the road. Secondly, dangerous conditions are difficult to recognize using global weather reports as they depend strongly on local road conditions like the presence of water bodies, forests, traffic intensity, etc. Very strong locality is a real challenge in this case.

An urgent need of constant monitoring road surface meteorological conditions caused development and implementation of what is called *Road Weather Information Systems* (RWIS) [7,21,30]. A standard RWIS consists of three main components: monitoring, forecasting, and decision supporting.

As in any multi-component system, there are a lot of sources of errors. In addition to the entirely autonomic and automatic mode of meteorological stations, they work in an aggressive environment with sharp temperature changes,

The research was partially supported by the Russian Foundation for Basic Research grants 16-29-09649 ofi m.

H. Lu et al. (Eds.): ISNN 2019, LNCS 11555, pp. 292–306, 2019.
https://doi.org/10.1007/978-3-030-22808-8_30

heavy precipitation and a wide range of mechanical impacts. Breakage of sensors, external impacts, server connection errors result in providing unreliable data and creating incorrect forecasts. The consequences of errors can be different from road accidents caused by improper road service to environmental problems caused by an excessive amount of road salts poured onto the surface.

Clearly, in the case of the road information system, the mentioned problem of malfunctioning equipment and incorrect data delivery detection can be formulated as a problem of outliers/anomalies detection in the sensor data, collected by RWIS (in this work we use words "outlier" and "anomaly" interchangeably).

In this work, we propose the ensemble-based approaches for creating an outlier detector to be used for different kinds of outliers and artificial outliers generation for the optimal threshold selection. Ensemble techniques have achieved significant success in many data mining problems in recent years [15,25,27]. Although the ensemble approach for outlier detection is not as widely studied as for classification and clustering tasks, several algorithms showed considerable improvement and allowed to achieve significant advances [1,2,32]. The proposed method allows coping with heterogeneous outliers caused by different factors due to its ensemble structure, as different individual algorithms in the ensemble compensate errors of each other.

Moreover, the proposed method allows selecting a threshold for the given consensus function which is very close to the threshold, which is optimal w.r.t. F_1-score metric.

We conducted experiments on the data provided by "Minimax-94" to verify the proposed approach. We used records for five years collected with the frequency of two measurements per hour from 59 meteorological stations: we utilized data from 50 stations to train the anomaly detector and data from nine stations to test the model. To prove the general applicability of the method for multidimensional datasets without time series structure we also conducted experiments on the so-called Shuttle dataset.

The paper is organized as follows. In Sect. 2 we discuss related work. In Sect. 3 we describe the proposed methods. We discuss our experimental setup in Sect. 4 and the obtained experimental results in Sect. 5. In Sect. 6 we draw conclusions.

2   Related Work

In this section, we introduce a survey of existing outlier detection techniques, some of which are used in this paper as base learners to build an ensemble of anomaly detectors. We describe their motivations, comparative advantages, disadvantages and underlying assumptions.

Local Outlier Factor (LOF) [6] is a typical representative of density-based methods. It is based on computing a local density of each point and comparing it with the density of its neighborhood. Thus, the anomaly score of LOF is the ratio of the local density of this point and the local density of its nearest neighbors. The points with density lower than that of their k neighbors are considered to be outliers. There are also several extensions and modifications of

Local Outlier Factor algorithm. For example, Local Correlation Integral (LOCI) [20], which uses Multi-Granularity Deviation Factor (MDEF) as an abnormal measure. MDEF expresses how the number of data points in the neighborhood of a particular point compares with that of the points in its neighborhood.

Outlier detection methods based on statistical parametric approaches work under the assumption that data is generated from a known distribution with unknown parameters. Thereby, the way for detecting outliers with such models includes two steps: training and testing. Train step implies the estimation of the distribution parameters. Test step involves checking whether a new instance was generated from the probabilistic model with parameters determined on the previous step or came from another distribution. Usually, a class of distributions to model normal data is based on some prior assumptions and a user personal experience. For example, Elliptic Envelope uses multivariate Gaussian distribution assumption and robust covariance matrix estimation by MinCovarianceDet estimator [23].

Outlier detection methods based on regression analysis have been extensively used for time series data. The procedure of regression-based outlier detection is generally the following: a regression model is fitted to the data during the training phase, during the test phase the fitted model is applied to the new data instances and the abnormality of each instance is evaluated according to the obtained residuals. If the instance is located far from the regression line, we mark it as an outlier. We can use regression conformal confidence measures to estimate an abnormality threshold [10,12,18]. Note that we can efficiently utilize the same methodology for conformal measures construction to define non-parametric anomaly detectors in time-series data [16,24,31].

One-class Support Vector Machine (OCSVM) [26] is an outlier detection modification of Support Vector Machine (SVM). While in supervised tasks SVM aims to find a maximal margin hyperplane separating two classes, OCSVM aims to separate training data instances from the origin. In [9] they consider approaches to OCSVM model selection. Generalization of OCSVM taking into account privileged information is provided in [11,29].

Isolation Forest (IForest) [19] is a successful modification of the standard decision tree-based algorithm for outlier detection. It randomly selects a feature and then randomly selects a split value from the interval between the minimum and the maximum values of the chosen parameter. The process continues until each leaf of the tree is assigned to only one observation from the dataset. The primary assumption of the algorithm is that anomalous observations can be separated during the first steps of this process because the algorithm requires fewer conditions to isolate the anomalies from normal instances. Normal observations, on the contrary, require more conditions to be separated from each other. The resulting anomalous score equals the length of the path from the root to the leaf.

If at least partial labeling of anomalies is known, we can use approaches to imbalanced classification for the anomaly detector construction, see [8,28].

Algorithm 1. Threshold Selection Using Artificial Anomalies

Input:
D – given dataset; E – set of base ensemble models;
G – distribution to generate anomalies from; N – number of outliers;
Output:
$\tilde{E}$ – trained ensemble base methods; T – threshold

1: Divide D into D_{train} for training and D_{thresh} for selecting threshold
2: **for** i in range $(1, N)$ **do**
3:     Generate an outlier $o \sim G$
4:     $D_{thresh} = D_{thresh} \bigcup\{o\}$
5: **end for**
6: $S_{thresh} = \emptyset$ – scores obtained on D_{thresh}
7: **for** algorithm e in E **do**
8:     Train e on D_{train}
9:     Obtain s_e – output score of e on D_{thresh}
10:     $S_{thresh} = S_{thresh} \bigcup\{s_e\}$
11: **end for**
12: Combine scores S_{thresh} and get s_{final}
13: Select threshold T for s_{final}, which optimizes the quality metric

---

3   Proposed Method

The meteorological time series dataset explored in this work is notable for the diverse nature of occurring outliers. Anomalies in the given dataset are caused by many different factors like sensor malfunction, some external events like a bird sitting on the station, server connection errors, etc. An individual algorithm can hardly cope with the whole range of outliers, due to strong underlying assumptions. Hence, exploiting ensemble analysis seems to be an appropriate choice. In this section, we propose an ensemble outlier detection method with artificial anomalies generation for selecting an optimal threshold. We affirm that the proposed approach has several advantages over the existing methods, especially in the case of outlier detection in meteorological time series.

Most algorithms require either defining an exact threshold for outlier scores or percentage of outliers in the given data to convert the scores into binary labels. The proposed method is aimed at solving the mentioned problem of selecting an optimal threshold. The first step of the algorithm includes dividing the data into two samples: for training ensemble base models and for selecting the threshold. The next and essential step is generating some amount of artificial anomalies and adding them to the second part of the data. The pseudo code for the proposed method is given in Algorithm 1.

4   Experiment Setup

4.1   Dataset Description

Company "Minimax-94" provided the primary dataset used in this work. "Minimax-94" is the research and production company working in the road industry and specializing in the creation of intelligent transport meteorological control systems. Archived data received from RWIS and provided for the research contains records for the period from 2012 to 2017. The stations aggregate data and send it to the data storage with a frequency of approximately two measurements per hour. For the train part, we selected 50 stations covering all main Russian climate zones. Then we divided them into two subsets.

The first subset of stations consists of 35 stations ($\sim 2 \cdot 10^6$ time ticks). There are two different versions of data from these stations: initial with noise, and clean data. The cleaning process was applied only to road surface temperature; other components remain in their original form. We removed each anomaly point with a rather big neighborhood of length equal to several days.

The second part of the training sample consists of only clean records from the rest 15 meteorological stations ($\sim 1 \cdot 10^6$ time ticks). It is used for creating artificial outliers and the threshold selection.

The test contains records from nine stations with labeled data; we marked each anomaly point with its 2–3 h local neighborhood, i.e., we labeled all data points from anomaly's vicinity as outliers. In total test data contains about $8 \cdot 10^5$ records, and $\sim 1\%$ of which we marked as outliers.

We can roughly categorize typical outliers behavior into two categories: short-term anomalies usually caused by some external impact, like exhaust pipe directed at the sensor (Fig. 1) and long-term anomalies caused by serious station malfunction (Fig. 2).

To prepare data we conducted the following steps. If the time gap between two nearest records from RWIS is longer than two hours, then the time series is divided into two parts. We remove patterns with duration less than 12 h from the sample. Finally, we linearly interpolate the data.

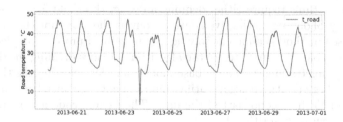

Fig. 1. Example of a single outlier

Also, we consider another dataset to test the proposed approach—the Shuttle dataset. It has been generated to classify which type of control to use during

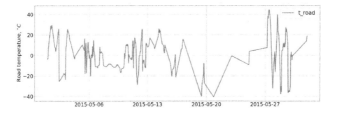

Fig. 2. Example of long-term malfunction

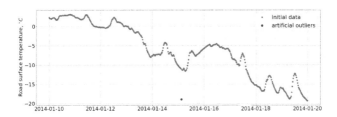

Fig. 3. Example of artificially generated single outlier

the landing of the spacecraft depending on the external conditions. The dataset contains 58000 instances, described by nine integer attributes and divided into seven classes: 1: Rad Flow, 2: Fpv Close, 3: Fpv Open, 4: High, 5: Bypass, 6: Bpv Close, 7: Bpv Open. We attribute classes 1 and 4 to normal instances while assigning the rest five categories to outliers. Two normal classes constitute about 94% of the sample, so the number of outliers is 6%.

For the test part, we selected 25% of the dataset. Remaining 75% were split into two parts: 2/3 is used to train and validate outlier detection algorithms and consists of both normal and abnormal classes, the other 1/3 is used to select an optimal threshold for an ensemble of algorithms using artificially generated outliers and includes only instances of the normal classes.

4.2   Anomaly Generation

Since real outliers probably come from different distributions and have different nature, we model the most frequent examples of outliers occurring in the given time series. They are single outliers (see Algorithm 2 and Fig. 3), short-term outliers (see Algorithm 3 and Fig. 4) and long-term sensor malfunctions (see Algorithm 4 and Fig. 5). For each station we generate 30 single outliers, 20 short-term and 3 long-term anomalies. Artificial labels are generated based on real-life data which was labeled by experts. During the labeling, the experts eliminated the sudden weather changes and so these events were not marked as sensors malfunctions.

The algorithm for generating artificial outliers for the Shuttle dataset consists of three steps. Transform the train data using Principal Component Analysis

Algorithm 2. Single Outliers Generation

Input:

S – list of meteorological stations;

t_{road} – road surface temperature time series;

$N = 30$ – number of outliers per each station in S;

$a_{low} = 2$, $a_{up} = 5$ – lower and upper boundaries of the uniform distribution

Output: $\tilde{t}_{road}$ – road surface temperature time series along with artificially generated anomalies

1: $\tilde{t}_{road} = t_{road}$
2: **for** each station s in S **do**
3:     Randomly select N time stamps from $\tilde{t}_{road}[s,:]$
4:     **for** each selected time stamp t **do**
5:         Generate perturbation $p \sim U(a_{low}, a_{up})$
6:         Randomly select $sign$ from $\{-1, 1\}$
7:         $\tilde{t}_{road}[s, t] = t_{road}[s, t] + sign * p$
8:     **end for**
9: **end for**

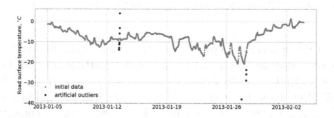

Fig. 4. Example of artificially generated short-term anomalous series

(PCA) decomposition, consider only 2 components corresponding to the largest singular values. Randomly generate 450 instances with the first component from $Uniform(-10000, 10000)$ and the second component equal to zero. Randomly generate 450 instances with the first component equal to zero and the second component from $Uniform(-5000, 5000)$. Conduct inverse PCA transformation.

4.3   Evaluation

The meteorological test sample is labeled as follows: for each anomaly point we mark its 2–3 h neighborhood as anomalous.

The developed quality metric is an extension of simple F_1-score to the considered case. The standard F_1-the score is defined as

$$F_1 = 2 \cdot \frac{recall \times precision}{recall + precision},\tag{1}$$

where recall is the fraction of true positive instances over the number of all real positive instances, while precision is the fraction of true positive points among all

Algorithm 3. Short-Term Anomaly Generation

Input:
t_{road} – road surface temperature time series; S – list of meteorological stations;
$N = 20$ – number of anomalies per each station in S;
$\lambda = 2$ – rate parameter of exponential distribution;
$d_{low} = 3$, $d_{up} = 12$ – lower and upper boundaries of anomalous duration
Output: $\tilde{t}_{road}$ – time series along with artificially generated anomalies

1: $\tilde{t}_{road} = t_{road}$
2: **for** each station s in S **do**
3:    Randomly select N time stamps from $\tilde{t}_{road}[s,:]$
4:    **for** each selected time stamp t **do**
5:      $d \sim RandInteger(d_{low}, d_{up})$
6:      Randomly select $sign$ from $\{-1, 1\}$
7:      Perturbation array $p = zeros(d)$
8:      **for** i in range(2, d) **do**
9:        $p[i] \sim Exp(\lambda)$
10:        $p[i] = p[i] + p[i-1]$
11:      **end for**
12:      $\tilde{t}_{road}[s, t : (t+d)] = t_{road}[s, t : (t+d)] + sign * p$
13;   **end for**
14: **end for**

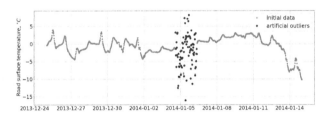

Fig. 5. Example of artificially generated long-term sensor malfunction

the instances labeled as positive. The modified version of the quality metric is the
following. Recall is the fraction of instances labeled as outliers (by the algorithm)
in the vicinity of which there is at least one instance marked as an outlier (real
label) to the total amount of positive instances (real labels). Precision is the
fraction of instances labeled as outliers (real label) in the vicinity of which there
is at least one instance marked as an outlier (by the algorithm) to the total
amount of positive instances (by the algorithm).

4.4   Base Learners for Anomaly Detection

We consider several outlier detection approaches to test the effectiveness of the
proposed method for threshold selection: single regression models, model averaging, feature bagging ensemble. All of them except for Multi Layer Perceptron

---
Algorithm 4. Long-Term Anomaly Generation

Input:
t_{road} – road surface temperature time series;
S – list of meteorological stations;
$N = 3$ – number of anomalies per each station in S;
$a_{low} = 30,\ a_{up} = 200$ – lower and upper boundaries of multiplier;
$d_{low} = 30,\ d_{up} = 200$ – lower and upper boundaries of anomalous duration;
$\mu = 0,\ \sigma = 5$ – mean and standard deviation of perturbation
Output: $\tilde{t}_{road}$ – time series along with artificially generated anomalies

1: $\tilde{t}_{road} = t_{road}$
2: **for** each station s in S **do**
3:     Randomly select N time stamps from $\tilde{t}_{road}[s,:]$
4:     **for** each selected time stamp t **do**
5:         $d \sim RandInteger(d_{low}, d_{up})$
6:         Randomly select multiplier $mult$ from $Uniform(m_l, m_u)$
7:         Perturbation array $p = zeros(d)$
8:         **for** i in range(1, d) **do**
9:             $p[i] \sim N(\mu, \sigma)$
10:         **end for**
11:         $\tilde{t}_{road}[s, t : t + d] = mult * t_{road}[s, t : (t + d)] + p$
12:     **end for**
13: **end for**

(MLP) use 35 clean stations as a training set. Forecasting based anomaly detection approach predicts target value 30 min ahead.

As features, we used air, road surface and subsurface temperatures, pressure, and humidity for several previous hours; besides that we used differences between the first six lags, solar azimuth, and altitude angles, road id, longitude and latitude of the station, sine and cosine of the hour, day of the year and month.

We tested several algorithms: XGBoost Regression [14], MLP (2 hidden layers with 64 and 16 neurons on the first and the second layers relatively, rectified linear unit (ReLU) activation function), XGBoost Air, One-class SVM (OCSVM), Elliptic Envelope (Ell. Env.), Local Outlier Factor (LOF), Isolation Forest (IForest).

In case of the Shuttle dataset model averaging includes the following algorithms: Local Outlier Factor, Isolation Forest, Elliptic Envelope, One-class SVM, Ridge (we use the first feature as a target variable, and others as independent variables).

We created two feature bagging based ensembles on the base of Ridge Regression, Elliptic Envelope for the meteorological dataset. We implemented feature bagging in combination with Ridge Regression, Elliptic Envelope, Local Outlier Factor and One-Class SVM for the Shuttle dataset.

Ensembles built for the meteorological dataset include some minor modifications concerning the number of selected features, i.e., the amount of features is generated from an interval that is different from the initial $[\lfloor d/2 \rfloor, d - 1]$ interval,

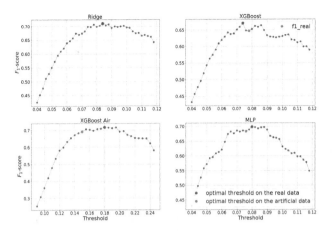

Fig. 6. Threshold selection using artificial anomalies for the meteorological dataset

where d is the complete number of features: (1) Ridge Regression—20 models in the ensemble, number of features is a random integer from $\lfloor d/6 \rfloor$ to $\lfloor d/2 \rfloor$; (2) Elliptic Envelope—10 models in the ensemble, number of features for each model is a random integer from 2 to $\lfloor (2 \cdot d)/3 \rfloor$.

Ensembles built for the Shuttle dataset completely repeat the described method. Shuttle dataset ensembles consist of 20 models each, and the number of features is a random integer from $\lfloor d/2 \rfloor$ to $d - 1$ interval.

4.5 Combination of Scores

To build a combination of predictions, we normalize predictions to $[0, 1]$ and try three approaches: simple averaging, averaging with weights equal to Pearson correlation between scores on the artificial data sample and the vector with labels, and finally training logistic regression with model scores as an input feature vector and artificial data as labels.

5 Experimental Results

The proposed threshold selection method shows excellent performance in case of applying individual regression algorithms to the meteorological dataset. The main advantage of the method is a rather precise estimation of the F_1-optimal threshold obtained on the artificial data. As can be seen in Fig. 6 the threshold selected using the generated data is very close to the threshold optimizing real F_1-score on the actual data.

As described in Sect. 4, we examined three different ensembles for the meteorological dataset: model-centered model averaging and two data-centered ensembles based on feature bagging with Ridge and Elliptic Envelope base algorithms (denoted as "Ridge FB", "Ell.Env. FB"). For each algorithm different combination functions have been used: an average of the scores linearly transformed

Table 1. Results: meteorological dataset

Algorithm	Ridge	XGBoost	MLP	XGBoost air	LOF	Ell.Env	OCSVM	IForest	
Recall	0.668	0.640	0.637	0.662	0.513	0.719	0.904	0.395	
Precision	0.758	0.655	0.738	0.722	0.502	0.482	0.130	0.252	
F_1-score	0.710	0.647	0.684	0.691	0.507	0.577	0.227	0.308	
Ensemble	Model averaging			Ridge FB			Ell.Env FB		
Combination	LT	WLT	LogReg	LT		WLT	LogReg	LT	WLT	LogReg
Recall	0.701	0.668	0.753	0.671		0.085	0.690	0.691	0.696	0.76
Precision	0.750	0.802	0.797	0.789		0.731	0.829	0.739	0.708	0.630
F_1-score	0.725	0.729	0.774	0.725		0.707	0.753	0.714	0.701	0.670

into $[0, 1]$ interval ("LT"), a weighted average of the linearly transformed scores ("WLT") and a Logistic Regression ("LogReg").

We compared the ensemble approach with individual outlier detectors on the meteorological dataset. In this experiment, all individual algorithm results serve as a baseline for the ensemble approach. We provided the results of the comparison in Table 1. As can be observed from the table, almost all ensemble techniques outperformed individual algorithms and, what is more critical, outperformed all included base ensemble models. In case of the meteorological data Logistic Regression consensus function turned out to show the best performance, although it is a kind of "risky" choice as it tends to overfit and as a result is somewhat unstable, which we can observe when using Elliptic Envelope. The similar situation happens with the Shuttle dataset.

Table 2. Results: shuttle dataset

Algorithm	Ridge	LOF	Ell.Env	OCSVM	IForest		
Recall	0.955	1.00	0.704	0.461	0.799		
Precision	0.767	0.512	0.680	0.193	0.791		
F_1-score	0.851	0.677	0.692	0.272	0.795		
Ensemble	Model Averaging			Ridge FB			Ell.Env FB		
Combination	LT	WLT	LogReg	LT	WLT	LogReg	LT	WLT	LogReg
Recall	0.957	0.924	0.756	0.956	0.953	0.681	0.625	0.567	0.476
Precision	0.979	0.986	0.815	0.871	0.938	0.801	0.937	0.930	0.915
F_1-score	0.968	0.953	0.784	0.911	0.946	0.736	0.750	0.704	0.626
Ensemble	LOF FB			OCSVM FB					
Combination	LT	WLT	LogReg	LT	WLT	LogReg			
Recall	0.587	0.577	0.311	0.981	0.999	0.999			
Precision	0.891	0.850	0.882	0.992	0.998	0.998			
F_1-score	0.704	0.688	0.460	987	0.999	0.999			

Table 2 shows that ensembling techniques improve the results of the individual algorithms significantly almost in all cases for the Shuttle Dataset. The most considerable improvement of F_1-score has been achieved by One-class SVM, which increased it from 0.272 to 0.999. As a result, feature bagged OCSVM achieved the best result for the Shuttle dataset.

6   Conclusions

In this paper, we proposed a new method for outlier detection in the RWIS data. The method allows to detect anomalies of heterogeneous nature efficiently, makes the selection of the decision rule less subjective and also turns out to be applicable not only to the RWIS data but also to more general cases of multidimensional datasets.

The results of the experiments show us that ensembles tend to outperform individual algorithms for both datasets. Ensembles tested on the meteorological data show better performance than all compared individual methods, which proves the assumption about benefits of ensemble-based methods for detecting anomalies in the RWIS data. There is no such strong tendency for the Shuttle data, for example, single Ridge Regression or Isolation Forest obtain higher F_1-scores than LOF feature bagging ensemble. However, even for the Shuttle dataset, the combination of different ensembles received higher F_1-score than any individual algorithm included in the combination.

The comparison of different combination functions showed that although linear transformation seems to perform better than weighted linear transformation, there is no significant difference between these two methods and choosing the best one largely depends on the specific setting of the problem, ensemble technique and included individual algorithms. Ensembles achieve the best results on both datasets with Logistic Regression. However, this combination method can easily overfit especially in case of the Shuttle dataset and is considered to be the most unstable, as providing the best and the worst results depending on the ensemble technique.

The proposed threshold selection technique showed significant performance for both ensemble approaches and individual algorithms. We can observe that the thresholds optimizing F_1-score for the artificial data are very close to the ones for the real data.

An ensemble with the best results, i.e., a model averaging (Sect. 4.5) with Logistic Regression consensus function, has been chosen as the principal method for outlier detection in "Minimax-94" road information system. It was implemented in Python as a separate module and integrated into the company system in the test mode. According to the experiments conducted on the archive records the selected method allows to detect 75% of occurring anomalies with 20% false alarm rate, which is sufficient for "Minimax-94". The future work assumes aggregation of statistics obtained in online mode during the test period and incorporation of the module into the working cycle of the "Minimax-94" system. To perform online aggregation, we are going to use long-term aggregation strategies [17] along with approaches to model quasi-periodic data [5] and extraction of trends in the presence of non-stationary noise with long tails [3,4]. Approaches to multidimensional time-series prediction [22] and multichannel anomaly detection [13] will allow detecting complex anomalies related to change of dependencies between time-series components.

We also would like to note that the resulting approach is highly dependent on the algorithm of synthetic data generation. We selected this technique based on certain domain knowledge. In order to adapt the technique for any new problem we have to find out what are the most common patterns of anomalous data appearing due to malfunctioning sensors. If due to some reason behavior of the sensors changes and malfunctioning sensors start to produce other patterns of anomalous data or some drastic weather change happens, then there is a chance that our algorithm fails to detect anomalies. However, this is a common problem for all machine learning algorithms.

References

1. Aggarwal, C.C., Sathe, S.: Outlier Ensembles: An Introduction. Springer, Cham (2017). https://doi.org/10.1007/978-3-319-54765-7
2. Artemov, A., Burnaev, E.: Ensembles of detectors for online detection of transient changes. In: Eighth International Conference on Machine Vision (ICMV 2015), 98751Z, 8 December 2015, Proceedings SPIE, vol. 9875 (2015)
3. Artemov, A., Burnaev, E.: Detecting performance degradation of software-intensive systems in the presence of trends and long-range dependence. In: 2016 IEEE 16th International Conference on Data Mining Workshops (ICDMW), pp. 29–36 (2016). https://doi.org/10.1109/ICDMW.2016.0013
4. Artemov, A., Burnaev, E.: Optimal estimation of a signal perturbed by a fractional brownian noise. Theor. Probab. Appl. **60**(1), 126–134 (2016)
5. Artemov, A., Burnaev, E., Lokot, A.: Nonparametric decomposition of quasi-periodic time series for change-point detection. In: Eighth International Conference on Machine Vision (ICMV 2015), 987520, 8 December 2015, Proceedings SPIE, vol. 9875 (2015)
6. Breunig, M.M., Kriegel, H.P., Ng, R.T., Sander, J.: LOF: identifying density-based local outliers. SIGMOD Rec. **29**(2), 93–104 (2000). https://doi.org/10.1145/335191.335388. http://doi.acm.org/10.1145/335191.335388
7. Buchanan, F., Gwartz, S.: Road weather information systems at the ministry of transportation, Ontario. In: 2005 Annual Conference of the Transportation Association of Canada (2005)
8. Burnaev, E., Erofeev, P., Papanov, A.: Influence of resampling on accuracy of imbalanced classification. In: Eighth International Conference on Machine Vision (ICMV 2015), 987521, 8 December 2015, Proceedings SPIE, vol. 9875 (2015)
9. Burnaev, E., Erofeev, P., Smolyakov, D.: Model selection for anomaly detection. In: Proceedings SPIE. vol. 9875, pp. 9875–9876 (2015). https://doi.org/10.1117/12.2228794
10. Burnaev, E., Nazarov, I.: Conformalized kernel ridge regression. In: 2016 15th IEEE International Conference on Machine Learning and Applications (ICMLA), pp. 45–52. December 2016. https://doi.org/10.1109/ICMLA.2016.0017

11. Burnaev, E., Smolyakov, D.: One-class SVM with privileged information and its application to malware detection. In: 2016 IEEE 16th International Conference on Data Mining Workshops (ICDMW), pp. 273–280. December 2016. https://doi.org/10.1109/ICDMW.2016.0046

12. Burnaev, E., Vovk, V.: Efficiency of conformalized ridge regression. In: Balcan, M.F., Feldman, V., Szepesvári, C. (eds.) Proceedings of The 27th Conference on Learning Theory. Proceedings of Machine Learning Research, vol. 35, pp. 605–622. PMLR, Barcelona, Spain, 13–15 June 2014. http://proceedings.mlr.press/v35/burnaev14.html

13. Burnaev, E.V., Golubev, G.K.: On one problem in multichannel signal detection. Problems of Information Transmission **53**(4), 368–380 (2017). https://doi.org/10.1134/S0032946017040056

14. Chen, T., Guestrin, C.: XGBoost: a scalable tree boosting system. In: Proceedings of the 22nd ACM SIGKDD International Conference on Knowledge Discovery and Data Mining, pp. 785–794. KDD 2016, ACM, New York, USA (2016). https://doi.org/10.1145/2939672.2939785

15. Da Silva, N.F., Hruschka, E.R., Hruschka Jr., E.R.: Tweet sentiment analysis with classifier ensembles. Decis. Support Syst. **66**, 170–179 (2014)

16. Ishimtsev, V., Bernstein, A., Burnaev, E., Nazarov, I.: Conformal k-NN anomaly detector for univariate data streams. In: Gammerman, A., Vovk, V., Luo, Z., Papadopoulos, H. (eds.) Proceedings of the Sixth Workshop on Conformal and Probabilistic Prediction and Applications. Proceedings of Machine Learning Research, vol. 60, pp. 213–227. PMLR, Stockholm, Sweden, 13–16 June 2017. http://proceedings.mlr.press/v60/ishimtsev17a.html

17. Korotin, A., V'yugin, V., Burnaev, E.: Aggregating strategies for long-term forecasting. In: Gammerman, A., Vovk, V., Luo, Z., Smirnov, E., Peeters, R. (eds.) Proceedings of the Seventh Workshop on Conformal and Probabilistic Prediction and Applications. Proceedings of Machine Learning Research, vol. 91, pp. 63–82. PMLR, 11–13 June 2018. http://proceedings.mlr.press/v91/korotin18a.html

18. Kuleshov, A., Bernstein, A., Burnaev, E.: Conformal prediction in manifold learning. In: Gammerman, A., Vovk, V., Luo, Z., Smirnov, E., Peeters, R. (eds.) Proceedings of the Seventh Workshop on Conformal and Probabilistic Prediction and Applications. Proceedings of Machine Learning Research, vol. 91, pp. 234–253. PMLR, 11–13 June 2018. http://proceedings.mlr.press/v91/kuleshov18a.html

19. Liu, F.T., Ting, K.M., Zhou, Z.H.: Isolation forest. In: Eighth IEEE International Conference on Data Mining (ICDM), 2008, pp. 413–422. IEEE (2008)

20. Papadimitriou, S., Kitagawa, H., Gibbons, P.B., Faloutsos, C.: LOCI: fast outlier detection using the local correlation integral. In: Proceedings of 19th International Conference on Data Engineering, 2003, pp. 315–326. IEEE (2003)

21. Pinet, M., Lo, A.: Development of a road weather information system (RWIS) network for Alberta's national highway system. In: Intelligent Transportation Systems (2003)

22. Rivera, R., Nazarov, I., Burnaev, E.: Towards forecast techniques for business analysts of large commercial data sets using matrix factorization methods. J. Phys.: Conf. Series **1117**(1), 012010 (2018). http://stacks.iop.org/1742-6596/1117/i=1/a=012010

23. Rousseeuw, P.J., Driessen, K.V.: A fast algorithm for the minimum covariance determinant estimator. Technometrics **41**(3), 212–223 (1999)

24. Safin, A., Burnaev, E.: Conformal kernel expected similarity for anomaly detection in time-series data. Adv. Syst. Sci. Appl. **17**(3), 22–33 (2017). https://doi.org/10.25728/assa.2017.17.3.497

25. Salehi, M., Zhang, X., Bezdek, J.C., Leckie, C.: Smart sampling: a novel unsupervised boosting approach for outlier detection. In: Kang, B.H., Bai, Q. (eds.) AI 2016. LNCS (LNAI), vol. 9992, pp. 469–481. Springer, Cham (2016). https://doi.org/10.1007/978-3-319-50127-7_40

26. Schölkopf, B., Williamson, R.C., Smola, A.J., Shawe-Taylor, J., Platt, J.C.: Support vector method for novelty detection. In: Advances in Neural Information Processing Systems, pp. 582–588 (2000)

27. Seni, G., Elder, J.F.: Ensemble methods in data mining: improving accuracy through combining predictions. Synth. Lect. Data Min. Knowl. Disc. 2(1), 1–126 (2010)

28. Smolyakov, D., Korotin, A., Erofeev, P., Papanov, A., Burnaev, E.: Meta-learning for resampling recommendation systems. In: Eleventh International Conference on Machine Vision (ICMV 2018); 110411S (2019). Proceedings SPIE, vol. 11041 (2019)

29. Smolyakov, D., Sviridenko, N., Burikov, E., Burnaev, E.: Anomaly pattern recognition with privileged information for sensor fault detection. In: Pancioni, L., Schwenker, F., Trentin, E. (eds.) ANNPR 2018. LNCS, vol. 11081, pp. 320–332. Springer, Cham (2018). https://doi.org/10.1007/978-3-319-99978-4_25

30. Toivonen, K., Kantonen, J.: Road weather information system in finland. Transp. Res. Rec.: J. Transp. Res. Board **1741**, 21–25 (2001)

31. Volkhonskiy, D., Burnaev, E., Nouretdinov, I., Gammerman, A., Vovk, V.: Inductive conformal martingales for change-point detection. In: Gammerman, A., Vovk, V., Luo, Z., Papadopoulos, H. (eds.) Proceedings of the Sixth Workshop on Conformal and Probabilistic Prediction and Applications. Proceedings of Machine Learning Research, vol. 60, pp. 132–153. PMLR, Stockholm, Sweden, 13–16 June 2017. http://proceedings.mlr.press/v60/volkhonskiy17a.html

32. Zimek, A., Gaudet, M., Campello, R.J., Sander, J.: Subsampling for efficient and effective unsupervised outlier detection ensembles. In: Proceedings of the 19th ACM SIGKDD International Conference on Knowledge Discovery and Data Mining, pp. 428–436. ACM (2013)

Blind Noise Reduction for Speech Enhancement by Simulated Auditory Nerve Representations

Anton Yakovenko[1(✉)], Aleksandr Antropov[1], and Galina Malykhina[1,2]

[1] Peter the Great St. Petersburg Polytechnic University, St. Petersburg, Russia
yakovenko_aa@spbstu.ru, malykhina@ftk.spbstu.ru
[2] Russian State Scientific Center for Robotics and Technical Cybernetics,
St. Petersburg, Russia

Abstract. Background and environmental noises negatively affect the quality of verbal communication between humans as well as in human-computer interaction. However, this problem is efficiently solved by a healthy auditory system. Hence, the knowledge about the physiology of auditory perception can be used along with noise reduction algorithms to enhance speech intelligibility. The paper suggests an approach to noise reduction at the level of the auditory periphery. The approach involves an adaptive neural network algorithm of independent component analysis for blind source separation using simulated auditory nerve firing probability patterns. The approach has been applied to several categories of colored noise models and real-world acoustic scenes. The suggested technique has significantly increased the signal-to-noise ratio for the auditory nerve representations of complex sounds due to the variability in spatial positioning of sound sources and a flexible number of sensors.

Keywords: Speech enhancement · Noise reduction ·
Blind source separation · Independent component analysis ·
Machine hearing · Auditory periphery model ·
Auditory nerve responses · FastICA

1 Introduction

Background noises given by single or multiple sound sources are always present in the environment. In engineering practice, they have a significant impact on acoustic signal processing. However, human or mammalian auditory systems are less responsive to noise than computational systems, as they process sensory signals of the auditory periphery using high-level neuronal structures that form biological neural networks.

Humans can concentrate attention on a certain acoustic source, e.g. the speaker's voice, despite the variability of the sound environment. This allows for verbal communication in noisy environments, including conditions of multi-talker babble noise. This phenomenon of auditory perception is widely known as the "cocktail party effect", accentuated by E.C. Cherry back in 1953. It represents a unique hearing ability that enables extracting the necessary acoustic

© Springer Nature Switzerland AG 2019
H. Lu et al. (Eds.): ISNN 2019, LNCS 11555, pp. 307–316, 2019.
https://doi.org/10.1007/978-3-030-22808-8_31

signal source in the presence of varied background noises. In psychoacoustics, this feature is associated with auditory scene analysis (ASA) [1]. ASA is related to the problem of acoustic scene, event, or source recognition through the perceptual mechanisms of the auditory system. The principles of ASA underlie various biologically relevant computational studies, which examined the systems of practical acoustic signal processing and used one or two microphone recordings in the experimental setup. These studies are known as computational auditory scene analysis (CASA) [2].

However, there are certain conditions where technical systems of automatic speech signal processing and acoustic scene analysis may have an advantage over biological auditory systems. This advantage is due to the fact that microphone sensors used in the setup can be arbitrarily allocated and optimally positioned in space. Besides, unlike monaural or binaural hearing, a technical system can have multiple microphone sensors and channels, which allows improving the quality of results through information redundancy and appropriate signal processing [3–5]. Thus, if there are no restrictions on the number and relative position of microphones, the limitations of CASA can be circumvented.

To create machine hearing and audition systems, it is advisable to combine the advantages of auditory signal processing with technical capabilities. Auditory peripheral coding of an input acoustic signal in the form of neural responses provides a robust representation against background noises [6] due to neural phase-locking [7]. Furthermore, the representation and parametrization of speech signals based on the responses of auditory nerve (AN) fibers provides noise-robust features for automatic speech recognition, outperforming common mel-frequency cepstral coefficients under certain noise conditions [8–11].

Our study aims to develop a signal processing algorithm for noisy vowel phoneme representations in the form of simulated AN responses with the purpose of noise reduction. The approach described in the present paper imitates some features of biological neural processing. It employs a computational model of the auditory periphery and an artificial neural network for blind separation of AN responses. Three different spontaneous rates for signal and noise mixture were considered in the study.

2 Simulation of the AN Responses

A physiologically-motivated computer model of the auditory periphery by R. Meddis [9] was used to obtain neural responses of auditory nerve fibers. This model simulates the temporal fine structure of AN firing for three types of fibers corresponding to the input speech signal: low spontaneous rate (LSR)—less than 0.5 spikes/s, middle spontaneous rate (MSR)—0.5–18 spikes/s, and high spontaneous rate (HSR)—18–250 spikes/s [12]. In the present study, the model was set to generate a probabilistic firing rate pattern.

The model requires a digitalized speech signal in the WAV format, sampled at 44.1 kHz. The sound pressure level of the input signal was adjusted to 60 dB, as it must correspond to the preferred listening level for conversational speech. Further, the signal passes through a processing cascade that simulates the functions of the outer, middle, and inner ear. The nonlinear mechanical behavior of the basilar membrane is modeled by a dual-resonance nonlinear filterbank (DRNL) [13]. Each segment of the basilar membrane provides the most pronounced response for a specific frequency of the acoustic stimulus, which is defined as best frequency (BF) for sounds near threshold. Thus, DRNL decomposes the signal into 41 frequency bands, logarithmically spaced from 250 to 8,000 Hz, corresponding to BFs in the most significant range for speech. The subsequent processing stages simulate stereocilia movement, inner hair cells transduction, synaptic exocytosis, and AN firing.

At the output, the auditory periphery model generates a signal encoded by the average firing rate of the auditory nerve fibers. The present study compares the results for three types of AN fibers as mentioned above. Figure 1 demonstrates a sequence of five English long vowels – clean (first column) and with additive white Gaussian noise at 0 dB SNR (second column). The duration of each vowel sound is 300 ms. The figure illustrates AN responses for LSR, MSR, and HSR fibers correspondingly. Every BF channel of the obtained AN firing probability pattern provides responses, which are then smoothed using a 20 ms Hann window and a 10 ms frame shift to extract feature data. Thus, each input signal is represented by its own multivariate data matrix consisting of 41 spectral features and an equal number of samples.

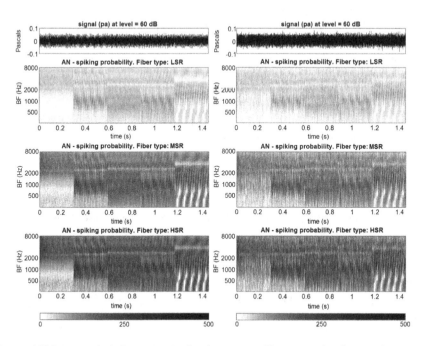

Fig. 1. AN firing probability patterns for three nerve fiber types for the vowel sequence: first column – clear speech signal, second column – signal corrupted by AWGN with 0 dB SNR.

3   Blind Signal Processing for AN Responses

Let us suppose that the sources of background noise are localized in environment about the target signal source. In that case, all sound signals are received by all sensors, but with different intensity, thus forming a linear mixture. Using the information about signal intensity difference on various sensors, we can solve the noise reduction problem for the target signal as a problem of blind source separation (BSS) [14]. Let us assume two mutually independent sound sources. The first source corresponds to the speech signal represented by a sequence of vowel phonemes. The second source is localized background environmental noise. In this case, the blind noise reduction problem [15] comes down to the task of independent component analysis (ICA).

The paper suggests using blind signal processing to solve the problem of noise reduction for a speech signal at the level of the auditory periphery. However, a technical system allows for arbitrary placement of multiple sensors in space, as distinct from the mammalian auditory system. Therefore, the intensity of signals may vary significantly, depending on the positions of sources in relation to sensors. Every sensor is represented by an auditory periphery model that encodes information in the form of stationary AN firing probability patterns. In this case, the mixing model of the speech signal and the background noise remains indeterminate, and source separation is based only on the AN responses on different sensors.

Let us consider a case where the two aforementioned sound sources are separated with the use of two biologically relevant sensors. The mixing model is a transformation of two AN output signals by a non-singular mixing matrix $\mathbf{H}$, the dimension of which depends on the number of mixed sound sources. If the sources of mixing are significantly different in amplitude or if the location of the sensors is chosen poorly, the mixing matrix is ill-conditioned. For a stable operation of the separation algorithm, it is advisable to perform a decorrelated transformation of the signal mixture in advance. The use of a decorrelation matrix makes it possible to present mixed signals in such a way that their correlation matrix is identity: $\mathbf{R}_{x_1 x_1} = E\left\{\mathbf{x}_1 \mathbf{x}_1^T\right\} = \mathbf{I}$. The mixing matrix will take the form $\mathbf{A} = \mathbf{QH}$, where $\mathbf{H}$ is the original unknown mixing matrix.

At the output of the auditory periphery model built-in each sensor, a mixture of sources is formed. Some of the signals represent the neural responses to the target signal, and others represent the responses to the noise caused by the sound environment: $\mathbf{s}(t) = [s_1(t), s_2(t)]^T$. The speech signal and the noise are mixed, and the additional mixture can affect the formed mixture to represent the intrinsic noise $\mathbf{n}(t) = [n_1(t), n_2(t)]^T$ of the system elements. The result of the conversion is the observed and measured signal $\mathbf{x}(t) = \mathbf{As}(t) + \mathbf{v}(t)$, where $\mathbf{v}(t) = [v_1(t), v_2(t)]^T$. The task is reduced to the search for the separation matrix $\mathbf{W}$ of the observed signal vector $\mathbf{x}(t)$ by means of an artificial neural network. The matrix $\mathbf{W}$ should be such that the estimate $\mathbf{y}(t)$ of the unknown signal vector $\mathbf{s}(t)$ would be the result of applying the separation matrix to the measured signal: $\mathbf{y}(t) = \mathbf{Wx}(t)$. In other words, the BSS task for the AN firing rate pattern is reduced to estimating the original signal by searching for the inverse mixing operator.

We separate the stationary AN response patterns into components attributable to signal or noise, assuming the independence of neural responses to these two factors. The independence condition is determined by the minimum of information that the neural responses to signal and noise have in common. The transformation of two-dimensional signals of the AN output $\mathbf{X}_1(t), \mathbf{X}_2(t)$ into vectors is performed according to the equation: $x_{n(i-1) \times j} = x_{i,j}$. Therefore, the goal of source separation is to minimize the Kullback-Leibler divergence between the two distributions – the probability density function (PDF) $f(\mathbf{y}, \mathbf{W})$, which depends on the coefficients of the matrix $\mathbf{W}$ and the factorial distribution:

$$f_1(\mathbf{y}, \mathbf{W}) = \prod_{i=1}^{m} f_{1,i}(y_i, \mathbf{W}) . \tag{1}$$

$$D_{f \| f_1}(\mathbf{W}) = -h(\mathbf{y}) + \sum_{i=1}^{m} h_1(y_i) . \tag{2}$$

where $h(\mathbf{y})$ is the entropy at the output of the separator, $h_1(y_i)$ is the entropy of the i-th element of the vector. For BSS, we used an approximation of probability density $f_{1,i}(y_i)$ by truncating the Gram-Charlier decomposition:

$$f_{1,i}(y_i) \approx N(y_i) \left[1 + \frac{\kappa_{i,3}}{3!} H_3(y_i) + \frac{\kappa_{i,2}}{4!} H_4(y_i) + \frac{\kappa_{i,6} + 10\kappa_{i,3}^2}{6!} H_6(y_i) \right] . \tag{3}$$

where $\kappa_{i,k}$ is the cumulant k-order of the variable y_i; $H_3(y_i) = y_i^3 - 3y_i, H_4(y_i) = y_i^4 - 6y_i^2 + 3, H_6(y_i) = y_i^6 - 15y_i^4 + 45y_i^2 - 15$ are Hermite polynomials; $N(y_i) = \frac{1}{\sqrt{2\pi}} exp\left(\frac{-y_i^2}{2}\right)$ is a PDF of a random quantity. The rule of weights correction when adapting a shared matrix is:

$$\mathbf{W}(n+1) = \mathbf{W}(n) + \mu(n) \left[\mathbf{I} - \varphi(\mathbf{y}(n)) \mathbf{y}^T(n) \right] \mathbf{W}^{-T}(n) . \tag{4}$$

where $\mu(n)$ is the convergence rate parameter, $\varphi(\mathbf{y}(n)) = [\varphi(y_1(n)), \varphi(y_2(n))]^T$ is a vector consisting of activation functions, the form of which changes in the course of adaptation. The activation functions change in the learning process, since their magnitude depends on the observed values $y_i(n)$.

4 Results and Discussion

4.1 Experimental Setup

This study addresses the problem of blind noise reduction. A series of computational experiments was conducted in which noise with different spectral power distributions was removed from the signal. The study aimed to investigate the impact of noise on the stationary AN firing probability pattern distortion and included different kinds of colored noises on the first stage: white, pink, red, blue, and violet. White Gaussian noise is a widespread noise model in robustness studies. In application areas, it is also important to remove pink (or flicker) noise, whose power spectral density is inversely proportional to frequency. The spectral density of red noise decreases in proportion to the square of the frequency. The spectral density of blue noise is specular with respect to pink noise, i.e. it increases with increasing frequency. Blue noise was synthesized using spectrum inversion. The spectral density of violet noise is inverted with respect to the red noise frequency spectrum.

On the second stage of blind noise reduction computational experiments, mixtures with real-world environmental noise were considered, including eight categories of urban acoustic scenes: airport, travelling by a bus, travelling by an underground metro, travelling by a tram, street with medium level of traffic, public square, metro station and indoor shopping mall. To obtain such categories of the environmental noises a TUT Urban Acoustic Scenes 2018 dataset [16] from DCASE Challenge was used.

Here is a summary of the experimental setup of our study. In accordance with the problem statement, we used two sensors and two sound sources. The first source was a clean speech signal. The second source was interference represented by one of the aforementioned noise types. On the first stage of computational experiments with colored noise interferences, the speech signal was a sequence of English long vowels represented by a multi-frequency complex tone that was synthesized as a sum of the first five formant frequencies – a speech-related model sound. Sound mixture had a duration of 1.5 s. On the second stage, the vowel sequence was pronounced several times by a male speaker. Sound mixture with real-world noise interferences had a duration of 10 s.

Each sensor received a sound mixture of two sources, with different mixing parameters specified in the mixing matrix. In this way, a certain spatial location of each sound source was simulated. Then, auditory peripheral representation was modelled for the sound mixtures in the form of AN average firing rate probability pattern. The output data matrices served as inputs for the FastICA algorithm of blind source separation [17]. The resulting data matrices describe the unmixed patterns of the corresponding sound sources. Finally, the impact on the blind noise reduction quality by the increase in the number of sensors from 2 to 8 has been considered.

4.2   Blind Noise Reduction Evaluation

The noise reduction performance for simulated AN fibers response patterns was evaluated through the signal-to-noise ratio (SNR) and noise intensity measurements. SNR allows estimating the ratio of target signal power to the power of background noise. For denoised response pattern Y of AN fibers, SNR is defined as follows:

$$SNR_y = 10log_{10} \left(\frac{\|X\|^2}{\|Y - X\|^2} \right) . \tag{5}$$

where X is the response pattern for the clean vowel sequence. Also, SNR was calculated for the response patterns for initial signal mixtures S_1 and S_2. The tables below demonstrate SNR estimation results corresponding to different spontaneous rate types of AN fibers and colored noise interferences. Table 1 presents the initial SNR for sound mixtures on two sensors provided by the mixing matrix, averaged for the vowel sequence. Table 3 presents the SNR values for the unmixed AN response pattern for the vowel sequence – a result of blind noise reduction.

Table 1. Initial SNR/dB for a mixture on two sensors by spontaneous rate

Noise	White		Pink		Red		Blue		Violet	
Sensor	1st	2nd	1st	2nd	1st	2nd	1st	2nd	1st	2nd
LSR	10.8	6.9	9.5	2.6	11.5	6.3	11.3	5.4	11.5	8.2
MSR	9.6	5.3	8.2	2.2	9.7	5.8	9.3	5.4	9.5	8.5
HSR	9.1	4.4	7.6	1.9	8.5	5.2	8.2	5.1	8.4	8.1

Table 2. Initial noise intensity for a mixture on two sensors by spontaneous rate

Noise	White		Pink		Red		Blue		Violet	
Sensor	1st	2nd	1st	2nd	1st	2nd	1st	2nd	1st	2nd
LSR	26.7	20.5	27.5	24.1	25.7	15.6	25.6	15.7	25.4	13.9
MSR	106.4	85.6	109.8	100.7	65.9	43.3	101.2	62.6	100.1	54.8
HSR	154.1	128.6	159.2	150.8	93.3	63.4	145.4	91.3	143.7	79.2

The noise intensity for each of 41 BF bands of the AN firing probability pattern can be approximated by the standard deviation. For the resultant response pattern Y, it can be defined as follows:

$$\sigma_y = \sqrt{\frac{1}{T}\sum_{t=1}^{T}\left(y(t) - \left[\frac{1}{T}\sum_{t=1}^{T}y(t)\right]\right)^2}. \tag{6}$$

where T represents the total number of samples. Table 2 shows initial noise intensity values for the AN response pattern of the two-sensor sound mixture. Table 4 lists the resulting noise intensity values for the output AN response pattern obtained through blind noise reduction algorithm.

As can be seen, the results of noise reduction are most representative for HSR AN fibers. As for MSR and LSR fibers, the performance of blind noise reduction was poorer for colored noises. While the AN response pattern turned out to be less sensitive to red noise, the most distortion was made by violet noise. Let us consider the second stage of computational experiments. Table 5 summarizes the

Table 3. Resultant SNR/dB for mixture with colored noise

Noise	White	Pink	Red	Blue	Violet
LSR	10.9	10.1	15.8	12.7	12.6
MSR	11.4	11.4	16.2	11.9	11.7
HSR	11.5	12.1	16.2	11.4	11.3

Table 4. Resultant noise intensity for mixture with colored noise

Noise	White	Pink	Red	Blue	Violet
LSR	44.3	43.8	45.8	45.1	45.1
MSR	45.3	45.3	35.6	45.6	45.4
HSR	45.9	46.1	36.1	45.8	45.6

Table 5. Results of two-sensor blind noise reduction (SNR/dB) for HSR AN fibers: mixture with real-world environmental noise

Noise	Airport		Bus		Metro		Tram		Traffic		Square		Station		Mall	
Sensor	1st	2nd	1st	2nd	1st	2nd	1st	2nd	1st	2nd	1st	2nd	1st	2nd	1st	2nd
Input	6.3	−4.4	6.2	−4.4	5.5	−6.5	6.1	−5.1	4.6	−7.4	4.7	−7.3	7.4	−4.6	4.5	−7.6
Average	0.9		0.9		−0.5		0.5		−1.4		−1.3		1.4		−1.5	
Output	7.6		7.5		6.6		6.9		7.2		7.1		6.2		7.2	

blind noise reduction results in terms of SNR for a vowel sequence mixed with real-world environmental noises represented by eight categories of urban acoustic scenes. These noises largely overlap with the speech range, so their removal is the challenging task. The location of sound sources with respect to sensors was set by the mixing matrix so that the average SNR value for the sound mixtures was approximately 0 dB. As can be seen from the obtained results, the approach allowed us to improve the average value of SNR by 7 dB. This is a good result for an initial study.

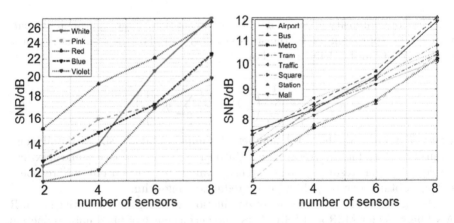

Fig. 2. Blind noise reduction performance for vowel sequence represented by HSR AN fibers, depending on the number of sensors: left panel – mixture with colored noises, right panel – mixture with real-world environmental noises.

As mentioned in the introduction, technical systems of speech signal processing allow the use of multiple microphone sensors. Therefore, we also evaluated blind noise reduction performance with increasing number of sensors for an HSR AN fiber response pattern. As seen from Fig. 2, performance was improved for the considered types of noise interference with increasing number of sensors, both for colored noise models and for real-world environmental noises.

5   Conclusions

The paper has suggested an approach to enhancement of noisy speech intelligibility by means of processing the signals of the auditory periphery. We have considered the task of designing a blind noise reduction system, which uses the information about the sound sources that is received by biologically relevant sensors distributed in space. The sensors simulates the processes of encoding information at the AN level of the auditory periphery. The speech signal, represented by a sequence of English long vowels, was separated from noise by means of independent component analysis of stationary AN firing probability patterns.

Two stages of computational studies were carried out – the first stage involved colored noise models, and the second dealt with background noises of real-world acoustic scenes. The quality of noise reduction largely depends on the mutual position of sound sources and sensors. In our case, arbitrary positions were chosen, modelled by a well-conditioned mixing matrix. The suggested approach has improved the SNR of the stationary AN firing activity pattern for colored and real-world noises. Besides, an increased number of sensors has demonstrated an improved quality of blind noise reduction.

An increase in SNR values can also be achieved through the optimization of quantity and relative placement of sensors in a given acoustic environment. Further elaboration of the approach will involve methods of blind signal extraction and real-time processing of dynamic AN firing activity patterns. The developed methodology can be used at the stage of pre-processing in machine hearing and biologically-inspired speech signal classification systems, such as [6,11,18]. Ultimately, it can become part of the new generation of neurocomputer interfaces and find use in cochlear implants [19].

Acknowledgments. The reported study was funded by the Russian Foundation for Basic Research according to the research project № 18-31-00304.

References

1. Bergman, A.S.: Auditory Scene Analysis: The Perceptual Organization of Sound. MIT Press, Cambridge (1994)
2. Wang, D.L., Brown, G.J.: Computational Auditory Scene Analysis: Principles, Algorithms, and Applications. Wiley-IEEE Press, Hoboken (2006)
3. Nugraha, A.A., Liutkus, A., Vincent, E.: Deep neural network based multichannel audio source separation. In: Makino, S. (ed.) Audio Source Separation. SCT, pp. 157–185. Springer, Cham (2018). https://doi.org/10.1007/978-3-319-73031-8_7

4. Schwartz, O., David, A., Shahen-Tov, O., Gannot, S.: Multi-microphone voice activity and single-talk detectors based on steered-response power output entropy. In: 2018 IEEE International Conference on the Science of Electrical Engineering in Israel (ICSEE), pp. 1–4 (2018)
5. Bu, S., Zhao, Y., Hwang, M.Y., Sun, S.: A robust nonlinear microphone array postfilter for noise reduction. In: 2018 16th International Workshop on Acoustic Signal Enhancement (IWAENC), pp. 206–210 (2018)
6. Alam, M.S., Jassim, W.A., Zilany, M.S.A.: Neural response based phoneme classification under noisy condition. In: Proceedings of International Symposium on Intelligent Signal Processing and Communication Systems, pp. 175–179 (2014)
7. Miller, R.L., Schilling, J.R., Franck, K.R., Young, E.D.: Effects of acoustic trauma on the representation of the vowel "eh" in cat auditory nerve fibers. J. Acoust. Soc. Am. **101**(6), 3602–3616 (1997)
8. Kim, D.-S., Lee, S.-Y., Kil, R.M.: Auditory processing of speech signals for robust speech recognition in real-world noisy environments. IEEE Trans. Speech Audio Process. **7**(1), 55–69 (1999)
9. Brown, G.J., Ferry, R.T., Meddis, R.: A computer model of auditory efferent suppression: implications for the recognition of speech in noise. J. Acoust. Soc. Am. **127**(2), 943–954 (2010)
10. Jurgens, T., Brand, T., Clark, N.R., Meddis, R., Brown, G.J.: The robustness of speech representations obtained from simulated auditory nerve fibers under different noise conditions. J. Acoust. Soc. Am. **134**(3), 282–288 (2013)
11. Yakovenko, A., Sidorenko, E., Malykhina, G.: Semi-supervised classifying of modelled auditory nerve patterns for vowel stimuli with additive noise. In: Kryzhanovsky, B., Dunin-Barkowski, W., Redko, V., Tiumentsev, Y. (eds.) NEUROINFORMATICS 2018. SCI, vol. 799, pp. 234–240. Springer, Cham (2019). https://doi.org/10.1007/978-3-030-01328-8_28
12. Liberman, M.C.: Auditory nerve response from cats raised in a low noise chamber. J. Acoust. Soc. Am. **63**(2), 442–455 (1978)
13. Lopez-Poveda, E., Meddis, R.: A human nonlinear cochlear filterbank. J. Acoust. Soc. Am. **110**, 3107–3118 (2001)
14. Houda, A., Otman, C.: Blind audio source separation: state-of-art. Int. J. Comput. Appl. **130**(4), 1–6 (2015)
15. Vorobyov, S., Cichocki, A.: Blind noise reduction for multisensory signals using ICA and subspace filtering, with application to EEG analysis. Biol. Cybern. **86**(4), 293–303 (2002)
16. Heittola, T., Mesaros, A., Virtanen, T.: TUT Urban Acoustic Scenes 2018, Development dataset [Data set]. Zenodo. https://doi.org/10.5281/zenodo.1228142
17. Miettinen, J., Nordhausen, K., Taskinen, S.: fICA: FastICA algorithms and their improved variants. R J. **10**(2), 148–158 (2018)
18. Yakovenko, A.A., Malykhina, G.F.: Bio-inspired approach for automatic speaker clustering using auditory modeling and self-organizing maps. Procedia Comput. Sci. **123**, 547–552 (2018)
19. Kokkinakis, K., Azimi, B., Hu, Y., Friedland, D.R.: Single and multiple microphone noise reduction strategies in cochlear implants. Trends Amplif. **16**(2), 102–116 (2012)

Automatically Generate Hymns Using Variational Attention Models

Han K. Cao, Duyen T. Ly, Duy M. Nguyen, and Binh T. Nguyen[✉]

Inspectorio Research Lab, University of Science, Ho Chi Minh City, Vietnam
hancao@inspectorio.com, ngtbinh@hcmus.edu.vn

Abstract. Building an intelligent system for automatically composing music like human beings has been actively investigated during the last decade. In this work, we propose a new approach for automatically creating hymns by training a variational attention model from a large collection of religious songs. We compare our method with two other techniques by using Seq2Seq and attention models and measure the corresponding performance by BLEU-N scores, the entropy, and the edit distance. Experimental results show that the proposed method can achieve a promising performance that is able to give an additional contribution to the current study of music formulation. Finally, we publish our dataset online for further research related to the problem.

Keywords: Seq2Seq · Variational attention model · Music generation

1 Introduction

With the recent rise in popularity of artificial neural networks especially from deep learning methods, many successes have been found in the various machine learning tasks covering classification, regression, prediction, and content generation. These techniques have achieved great results in many aspects of artificial intelligence including the generation of visual art [7] as well as language modelling problem in the field of natural language processing [21]. During the last decade, there is a new trend for applying machine learning techniques in music generation.

Music is an artistic and creative domain which is an important part of human life since it plays a significant role in many social and cultural activities, for example, singing, dancing, or playing instruments in an orchestra. Therefore, music is human in its nature. Automatic music generation is versatilely applicable. For instance, it can provide an access to a huge amount of copyright-free music for multiple purposes such as aiding the narration of videos and improving the moods of computer game scenes. The special thing of this is that it does not require users to have knowledge in music theory in order to produce a piece of music, thus simplifying the process of content creation. Automatic music composition has had a long history of researches since Hiller and Isaacson [8] proposed the use of Markov chains for this problem in 1956. As music can be interpreted

© Springer Nature Switzerland AG 2019
H. Lu et al. (Eds.): ISNN 2019, LNCS 11555, pp. 317–327, 2019.
https://doi.org/10.1007/978-3-030-22808-8_32

as a sequence of musical events, it can be modelled as conditional probabilities between these events. This can be observed in a typical example of chord progressions rules: some chords are more likely to occur after the given chords than others, and they are all dependent on the central key of the musical piece. By assuming that the current state only depends on the states in the past events, one can generate a sequence of musical events given a primer sequence as a seed.

One notable property of music, which is also one of the biggest challenges in the field of automatic music composition, is the non-locality. For example, musical patterns are often repeated with or without variations in different time in a score. Most computer-generated music sounds good in a short duration, however, lacks any larger structure such as sections or recurring and developing themes. Recurrent Neural Networks (RNNs) are suited for modelling this property by reason of their ability to represent long-term dependency. The idea of experimenting with RNNs started with the use of a simple version of RNNs, Jordan net [11], which was used to generate chord sequences [17]. A system called CONCERT [14] was proposed to generate melodies by training on 10 Bach pieces. Theoretically, RNN models can remember infinitely long sequences; however, said ability is limited in practice due to the vanishing gradient problem [9]. To solve this limit, Long Short-Term Memory Networks (LSTMs) [10] are used to retain information for much longer periods of time. Eck and co-workers proposed an LSTM system to learn 12-bar Blues chord progressions and melodies [6].

Musical events generation can mean a wide variety of tasks ranging from the composition of melody, harmony to writing lyrics [3,4,19,26]. Regarding harmony, musicians have written thousands of treatises to propose rules for building beautiful harmony from given melodies. One example is the treatise of Arnold Schoenberg [20], which is the most accomplished treatise of tonal music. As for lyrics, this is more likely to be a natural language processing problem. Consequently, RNNs [23] can be applied to the task of generating folk music [22] using a text-based representation of music. This method covers the generation of music titles, along with other metadata of the music piece. In this paper, we mainly focus on investigating a machine-learning method to automatically generate melodies, especially for hymns. Precisely, we consider the problem of composing a sequence of notes, which at most one note is played at the same time. To this aim, we present an approach by using variational attention models to create a list of melodies from a given sequence of notes as an initial seed. We measure the performance by comparing with other methods using Seq2Seq (sequence-to-sequence) models and attention models. Experimental results show that using variational attention models can surpass two other techniques and achieve a very good performance in BLEU scores, the entropy, and the edit distance. The results can give an additional contribution for the current research of music generation.

2   Methodology

In what follows, we formulate the original problem and present an approach for generating melodies by using variational attention models. After that, we compare the proposed method with a variety of algorithms by measuring a number of metrics in a specific dataset including thousands of hymns.

2.1 Problem Formulation

We study a machine-learning method such that given a sequence of notes (which can be initially composed by a musician), it can automatically predict the most appropriate melodies to play next. The predicted melodies are combined with the previous ones to act as a new input for doing another generation and so on. The corresponding flow of the problem can be illustrated in Fig. 1. It is important to note that the most popular format of symbolic music available on the Internet is MIDI (Musical Instrument Digital Interface) files. MIDI is a protocol to connect electronic instruments and digital musical tools - such as synthesizer, samplers, and computers - using a series of MIDI messages. These messages specify information such as note (pitch, velocity) and control signals for parameters (volume, vibrato, instruments, etc.). Especially, they instruct MIDI instruments to trigger sound. There are two most important kinds of messages that one can concern:

- *note-on* messages containing a MIDI note number (the note's pitch, an integer within the interval $[0, 127]$; for example, a MIDI number of 60 corresponds to the middle C), a velocity (how forcefully the note is played, an integer within the interval $[0, 127]$);
- *note-off* messages indicating that a note ends in a similar manner.

Using MIDI messages, one can convert a song into an array of MIDI notes. For example, an array $A = a_1 a_2 \ldots a_n$ satisfying $a_i \in [-2, 127]$ represents a melody of n time steps. A value a_i from 0 to 127 indicates a MIDI note played at the i-th time step; the value -1 stands for no action (the previous note keeps on being played); and the value -2 represents a rest (the playing note is released). In this work, one can ignore the notes' velocity at this point. As a result, it gives the possibility to transform the current problem to a new problem by producing the next sequence of melodies from a given list of MIDI notes. It turns out that the problem can be considered as an "NLP-like" problem and usually, one can leverage the current state-of-the-art NLP approaches for finding the most suitable solution. Figure 2 illustrates a sequence of melodies from the song "Twinkle Twinkle Little Star". This English lullaby can be translated into a NoteSequence of $(60, -1, 60, -1, 67, -1, 67, -1, 69, -1, 69, -1, 67, -1, -1, -1, 65, -1, 65, -1, 64, -1, 64, -1, 62, -1, 62, -1, 60, -1)$.

2.2 Problem Solution

There are several ways to construct an appropriate model for the problem. Usually, for a given input sequence of notes $x = (x_1, \ldots x_T)$, a recurrent neural network (RNN) can be trained to compute a sequence of hidden vectors

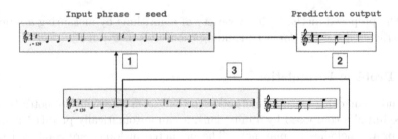

Fig. 1. In the formulated problem, one can use an initial seed (1) to generate a predicted sequence of melodies (2) by a trained model and then combine those sequences together as another input (3) for the next generation.

Fig. 2. A sequence of melodies extracted from the song "Twinkle Twinkle Little Star".

$h = (h, \ldots h_T)$ and an output sequence $y = (y_1, \ldots y_T)$ by iterating the following equations from $t = 1$ to T:

$$h_t = \mathcal{H}\left(W_{xh}x_t + W_{hh}h_{t-1} + b_h\right)$$
$$y_t = W_{hy}h_t + b_y \qquad (1)$$

where the $W_{(..)}$ terms denote weight matrices and the $b_{(.)}$ terms denote bias vectors and $\mathcal{H}_{(.)}$ is the hidden layer function. Here, $\mathcal{H}_{(.)}$ is often an element-wise application of a sigmoid function.

It is important to emphasize that recurrent neural network is a popular class of artificial neural networks that have shown great results in various tasks, especially ones dealing with sequential information such as speech, text, audio, and video. They were initially presented in the 1980's, but can only show their real potential recently, which is due to the development of computational power and an increase of the amount of available data. RNNs are called recurrent as they perform the same function for each element of one sequence, with the output being dependent on the previous computations. This is different from traditional neural networks as their outputs are independent of previous computations. RNNs can be considered an a special type of machine-learning models which is capable of "remembering" information calculated so far. Theoretically, RNNs can take into consideration the past information in arbitrarily long sequences, but in practice, they are limited to looking back only a few steps [25].

Normally, an RNN can be interpreted as a sequence of neural networks which are trained one after another with the back-propagation procedure. This training process is called Back Propagation Through Time (BPTT) and the corresponding errors can be back-propagated from the last step to the first-time step. The error corresponding to each step can be calculated and then weights of RNNs can be updated. When there is a large number of time steps, the computation becomes very intensive. Due to the fact that repetitive calculations could lead to over minimizing or amplifying effects, RNNs are prone to problems with the gradients: exploding gradients and vanishing gradients [9]. The former problem happens when the algorithm assigns an unreasonably high importance to the weights and can be solved by truncating or squashing the gradients. Whereas, the vanishing gradients problem is much harder to solve when the values of a gradient are too small causing the model to stop learning efficiently. To solve this, a new variation of RNNs was introduced by Hochereiter and Schimidhuber [10], which is called Long Short-Term Memory (LSTM) Networks.

In this work, we consider three different models leveraging architectures of LSTMs and bi-RNNs to investigate the music generation, especially for automatically composing hymns, including Sequence-to-Sequence (Seq2Seq) model, the original attention model, and the variational attention model. We briefly describe those techniques in the following paragraphs.

Long Short-Term Memory Networks - Seq2Seq. Seq2seq models can be trained from a given dataset and then take a sequence of input notes to predict a sequence of output notes as the expected results. Sutskever and co-workers propose an appropriate architecture for Seq2Seq [24] by including two LSTMs: the first LSTM (encoder) maps the input sequence to a fixed-length vector representation, and the second LSTM (decoder) is a recurrent neural network model which predicts the target sequence step by step. It has been ubiquitously applied in the field of natural language processing and demonstrated the state-of-the-art performance in many studies, especially in machine translation problems. Figure 3 describes an example of Seq2Seq architecture.

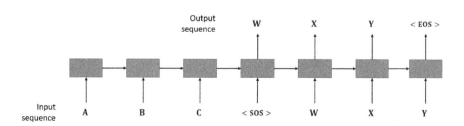

Fig. 3. An example for a Seq2Seq architecture. Here, an input sentence A B C is passed through an LSTM layer - the encoder. The final output is used to initiate the state of the decoder which regressively produces an output sentence from an $< SOS >$ (start of sentence) token until a $< EOS >$ (end of sentence) token is met.

Attention Models. Attention model has been one of the well-known techniques in Neural Machine Translation (NMT). Although there are a huge number of applications using the Seq2Seq model [24], this model seems not to have enough effect for very long sentences, which are quite common in the music generation

problem. As a consequence, the attention model [13] has been proposed to overcome that weakness. It can generate quite well for long-series inputs with an
encoder using Bi-RNN [27] instead of a single RNN cell and a decoder using an
attention layer. This technique is quite close to the visual attention mechanism
found in humans. The structure of the attention model can be found at [1].

Variational Attention Models. Variational attention models have been extensively investigated during the last few years. Variational auto-encoder (VAE)
model [2] uses a variational approach for the latent representation learning. It
includes an encoder, a decoder, and a loss function. The encoder maps the input
data into latent variables with a given probabilistic distribution and the decoder
then reconstructs the input data from the latent variables. During the training process, the model optimizes the loss function, which is usually chosen as a variational
lower bound of the log-likelihood of the data [2]. The VAE model learns to reproduce the original content. In attempt to solve the common issue of a sequence-to-
sequence model mentioned previously, it is extended to another framework called
variational encoder-decoder (VED), proposed by Zhou and Neubig [28], by assuming the output data is a function of the input data. Bahuleyan and colleagues
[2] present a variational attention mechanism for Variational Encoder-Decoder
(VED) by modelling attention vectors as Gaussian distributed random variables.
We leverage the power of variational attention models for resolving the hymn generation problem by using an architecture depicted at Fig. 4.

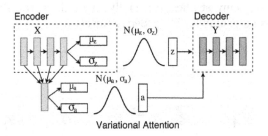

Fig. 4. A variational attention model [2] is applied for the hymn generation problem.

Evaluation Metrics. In order to compare the performance of different methods described above, we first use Bleu-N scores ($N = 1, 2, 3, 4$). It is important
to note that BLEU (BiLingual Evaluation Understudy) [18] scores are ubiquitously selected to measure the average N-gram precision on a set of generated
sentences in the machine translation. They count up to N-grams for comparing
the candidate translation and the reference translation. The value of each metric
has a range from 0 to 1 with a perfect match attaining a score of 1.0. Two other
metrics chosen for evaluation are the entropy [5] and the edit distance [15]. The
minimum edit distance measures how dissimilar two strings are to one another

by calculating the minimum number of insertions, deletions, and substitutions that are required to transform one string into the other one. These metrics are useful to analyze the performance of a music generation model.

3   Experiments and Discussion

In this section, we introduce all experiments related to three methods discussed in the previous section. We firstly describe the data collection process and how to set all experiments up. After that, we show experimental results and all possible discussion.

3.1   Data Collection

We create a dataset for training and testing various models by collecting from the website Hymnal.net. This website provides a large number of hymns which predate copyright laws and whose copyright has expired. Every hymn is available in a lead sheet, mp3, MIDI, and Tune (MIDI). We use the Tune format to extract all possible melodies as training data and a Python library "music21" to convert all MIDI files into NoteSequences. Since most of the hymns consist of simple melodies, training melodies are quantized at the granularity of an eighth note instead of the usual approach of a sixteenth note. This means each time step corresponds to one-eighth of a bar of music. Then, each melody can be transposed to the key of C. We filter out songs which do not have the time signature of 4/4, resulting in a total of 1471 songs and publish this dataset in [16].

3.2   Setup

All models are implemented in Keras by using TensorFlow as the backend. They are trained with the Adam optimizer [12] with a fixed learning rate $1e-5$ and the epoch size is 20 (samples). We construct all algorithms in one personal computer with 8 GB RAM and 4 cores (2.9 GHz Intel Core i7). For generating training, validating, and testing datasets, we use an 1-bar sliding window to extract all training sequences where the input and output lengths are 4 bars (32 time steps) and 1 bar (8 time steps) accordingly. As a result, there are 19846 samples in total. Among those, a model can be trained by using 15876 samples and validated on 3970 samples.

To train a Seq2Seq model, we use two Bi-LSTM layers with 256 units per cell as the encoder and two normal LSTM layers with the same amount of units (256) as the decoder with a teacher forcing ratio of 0.5. The decoder regressively produces each time step event, passing its output through a linear layer and SoftMax to create a distribution over 130 classes. Here, the total number of classes is exactly the number of values of MIDI notes which vary from -2 to

127. For the original attention model, the encoder is trained by using a Bi-LSTM layer with 64 hidden units whose length of the input sequence is 64. Meanwhile, the decoder is trained by using a LSTM layer with 64 hidden units in which the length of the input sequence is 8. We choose the same value 0.1 for hyper-parameters of both attention and latent KL terms. For measuring the performance of the variational attention model, we use a Bi-LSTM as the encoder with 40 hidden units whose the length of the input sequence is 64 and the decoder is constructed by a LSTM layer with 40 hidden units whose input length 8. In the experiments, we select the same value 1.0 for hyper-parameters of both attention and latent KL terms.

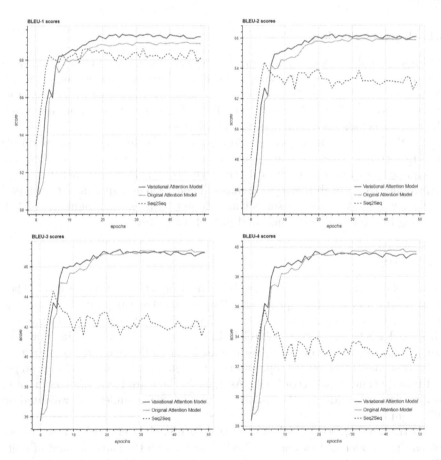

Fig. 5. The performance of three methods for different epochs by using BLEU-N scores ($N = 1, 2, 3, 4$).

Table 1. The performance of three methods with different metrics.

Method	BLEU-1	BLEU-2	BLEU-3	BLEU-4	Entropy	Edit distance
seq2seq	70.15%	56.43%	45.84%	37.12%	**1.49**	3.27
att	69.32%	57.22%	48.19%	41.00%	1.32	**2.99**
var-att	**70.51%**	**58.44%**	**49.59%**	**42.53%**	1.42	3.00

3.3   Results

Table 1 and Fig. 5 indicate the performance of these three models with different measurements: BLEU-N scores ($N = 1, 2, 3, 4$). One can easily see that all models have achieved good results on BLEU-1 scores while the Seq2Seq model has suffered from low scores in the rest. For the entropy, the computed results are very interesting in the sense of the ground-truth labels on the testing dataset have an entropy of 1.7860. Figure 6 also shows that the Seq2Seq model can achieve the best result which is the closest to the ground-truth entropy. Although the variational attention model and the original attention model may be good at keeping the structure of a music, they somehow lose the variability, in other words, the creativity of the pieces. For the edit distance, all models have quite close values together (nearly to 3.0) as depicted in Fig. 6. Experimental results show that the variation attention model can achieve a better performance in most of measurements and outperforms other techniques for all BLEU-N scores. These results are very encouraging and illustrate how the variational attention models can be well implemented to the hymn generation problem.

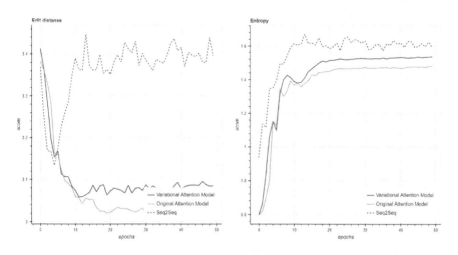

Fig. 6. The performance of three methods for different epochs by using the edit distance and the entropy.

4 Conclusion

We have investigated the problem of composing hymns given a sequence of melodies by using MIDI messages and deep learning approaches. We have completed a number of experiments by comparing three different methods: the Seq2Seq model, the attention model, and the variational attention model by using one dataset of thousands of hymns. Experimental results show that the variational attention model can achieve a promising performance and surpass other approaches in BLEU scores. We publish our dataset for additional contribution to the research community related to the musical generation problem. In the future, we will extend our research for different types of musical products and study other advance techniques for improving the performance of the proposed algorithm.

Acknowledgments. CKH and BTN would like to thank The National Foundation for Science and Technology Development (NAFOSTED), University of Science, and Inspectorio Research Lab in Viet Nam for supporting two authors throughout this paper.

References

1. Bahdanau, D., Cho, K., Bengio, Y.: Neural machine translation by jointly learning to align and translate. CoRR abs/1409.0473, September 2014. arXiv e-prints https://arxiv.org/abs/1409.0473
2. Bahuleyan, H., Mou, L., Vechtomova, O., Poupart, P.: Variational attention for sequence-to-sequence models. CoRR abs/1712.08207 (2017)
3. Brunner, G., Wattenhofer, R., Wiesendanger, J.: JamBot: music theory aware chord based generation of polyphonic with LSTMs. CoRR (2017)
4. Choi, K., Fazekas, G., Sandler, M.B.: Text-based LSTM networks for automatic music composition. CoRR (2016)
5. Coffman, D.: Measuring musical originality using information theory. Psychol. Music **20**(2), 154–161 (1992). https://doi.org/10.1177/0305735692202005
6. Eck, D., Schmidhuber, J.: A first look at music composition using LSTM recurrent neural networks. Technical report no. IDSIA-07-02. IDSIA/USI-SUPSI, Instituto Dalle Molle di studi sull' intelligenza artificiale, Manno, Switzerland (2002). http://www.iro.umontreal.ca/~eckdoug/blues/IDSIA-07-02.pdf
7. Gregor, K., Danihelka, I., Wierstra, D.: DRAW: a recurrent neural network for image generation. CoRR (2015)
8. Hiller, L.A., Isaacson, L.M.: Experimental Music; Composition with an Electronic Computer. Greenwood Publishing Group Inc., Westport (1979). ISBN: 0313221588. https://dl.acm.org/citation.cfm?id=578548
9. Hochreiter, S., Bengio, Y., Frasconi, P., Schmidhuber, J.: Gradient flow in recurrent nets: the difficulty of learning long-term dependencies (2001)
10. Hochreiter, S., Schmidhuber, J.: Long short-term memory. Neural Comput. **9**(8), 1735–1780 (1997)
11. Jordan, M.I.: Artificial neural networks. In: Jordan, M.I. (ed.) Attractor Dynamics and Parallelism in a Connectionist Sequential Machine, pp. 112–127. IEEE Press, Piscataway (1990)

12. Kingma, D.P., Ba, J.: Adam: a method for stochastic optimization. CoRR abs/1412.6980 (2014). http://arxiv.org/abs/1412.6980
13. Luong, T., Pham, H., Manning, C.D.: Effective approaches to attention-based neural machine translation. In: Proceedings of the 2015 Conference on Empirical Methods in Natural Language Processing, pp. 1412–1421 (2015)
14. Mozer, M.C.: Neural network music composition by prediction: exploring the benefits of psychoacoustic constraints and multi-scale processing. Connect. Sci. **6**, 247–280 (1994)
15. Navarro, G.: A guided tour to approximate string matching. ACM Comput. Surv. **33**(1), 31–88 (2001). https://doi.org/10.1145/375360.375365
16. Nguyen, B.: A public dataset for hymn generation (2019). https://sites.google.com/site/ntbinhpolytechnique/datasets
17. Lewis, J.P.: Algorithms for music composition by neural nets: improved CBR paradigms. In: Proceeding of the International Computer Music Conference (1989)
18. Papineni, K., Roukos, S., Ward, T., Zhu, W.J.: Bleu: a method for automatic evaluation of machine translation. In: Proceedings of the 40th Annual Meeting on Association for Computational Linguistics, ACL 2002, pp. 311–318 (2002)
19. Pudaruth, S., Amourdon, S., Anseline, J.: Automated generation of song lyrics using CFGS. In: The Seventh International Conference on Contemporary Computing (IC3) (2014)
20. Schoenberg, A.: Theory of Harmony. University of California Press, California (1983)
21. Shi, D.: A study on neural network language modeling. CoRR abs/1708.07252 (2017). http://arxiv.org/abs/1708.07252
22. Sturm, B.L., Santos, J.F., Ben-Tal, O., Korshunova, I.: Music transcription modelling and composition using deep learning. CoRR abs/1604.08723 (2016)
23. Sutskever, I., Martens, J., Hinton, G.: Generating text with recurrent neural networks. In: Proceedings of the 28th International Conference on International Conference on Machine Learning, pp. 1017–1024. Omnipress (2011)
24. Sutskever, I., Vinyals, O., Le, Q.V.: Sequence to sequence learning with neural networks. CoRR abs/1409.3215 (2014)
25. Tomczak, M.: BachBot. Master Thesis (2016)
26. Yi, L., Goldsmith, J.: Automatic generation of four-part harmony. In: Proceedings of the Fifth UAI Conference on Bayesian Modeling Applications Workshop, vol. 268 (2007)
27. Zeyer, A., Doetsch, P., Voigtlaender, P., Schluter, R., Ney, H.: A comprehensive study of deep bidirectional LSTM RNNS for acoustic modeling in speech recognition. In: ICASSP 2017, pp. 2462–2466 (2017)
28. Zhou, C., Neubig, G.: Morphological inflection generation with multi-space variational encoder-decoders. In: SIGMORPHON 2017, pp. 58–65 (2017)

Image Recognition, Scene Understanding, and Video Analysis

Mixed-Norm Projection-Based Iterative Algorithm for Face Recognition

Qingshan Liu[1(✉)], Jiang Xiong[2], and Shaofu Yang[3]

[1] School of Mathematics, Southeast University, Nanjing 210096, China
qsliu@seu.edu.cn
[2] Key Laboratory of Intelligent Information Processing and Control of Chongqing Municipal Institutions of Higher Education, Chongqing Three Gorges University, Chongqing 404100, Wanzhou, China
xjcq123@sohu.com
[3] School of Computer Science and Engineering, Southeast University, Nanjing 210096, China
sfyang@seu.edu.cn

Abstract. In this paper, the mixed-norm optimization is investigated for sparse signal reconstruction. Furthermore, an iterative optimization algorithm based on the projection method is presented for face recognition. From the theoretical point of view, the optimality and convergence of the proposed algorithm is strictly proved. And from the application point of view, the mixed norm combines the L_1 and L_2 norms to give a sparse and collaborative representation for pattern recognition, which has higher recognition rate than sparse representation algorithms. The algorithm is designed by combining the projection operator onto a box set with the projection matrix, which is effective to guarantee the feasibility of the optimal solution. Moreover, numerical experiments on randomly generated signals and three face image data sets are presented to show that the mixed-norm minimization is a combination of sparse representation and collaborative representation for pattern classification.

Keywords: Mixed norm · Projection method · Iterative algorithm · Face recognition

1 Introduction

Recently, sparse representation, especially L_1-norm-based sparse representation algorithms have been investigated and used for signal processing and pattern classification [1,2]. In general, we assume a vector $b \in \mathbb{R}^m$ to be a linear combination of column vectors of matrix $A \in \mathbb{R}^{m \times n}$ with full row-rank; i.e., $b = Ax$,

Q. Liu—This work was supported in part by the National Natural Science Foundation of China under Grant 61876036, the "333 Engineering" Foundation of Jiangsu Province of China under Grant BRA2018329, and the Fundamental Research Funds for the Central Universities.

H. Lu et al. (Eds.): ISNN 2019, LNCS 11555, pp. 331–340, 2019.
https://doi.org/10.1007/978-3-030-22808-8_33

where $x \in \mathbb{R}^n$. The aim of sparse representation is to recover or find the solution of vector x from the signals in b and A, which is solved by the L_0-min problem

$$
\begin{aligned}
\text{minimize} \quad & \|x\|_0, \\
\text{subject to} \quad & Ax = b,
\end{aligned}
\tag{1}
$$

in which $\| \cdot \|_0$ is the L_0-norm. Generally, if the L_0-min problem has sufficiently sparse solution, it can be converted into the following L_1-min problem [3]:

$$
\begin{aligned}
\text{minimize} \quad & \|x\|_1, \\
\text{subject to} \quad & Ax = b,
\end{aligned}
\tag{2}
$$

where $\| \cdot \|_1$ is the L_1-norm.

The L_0-min and L_1-min problems are both widely used to sparse signal reconstruction, including data clustering [4], face recognition [1], image restoration [5], and image classification [6]. At the same time, many sparse algorithms are presented to solve the L_0-min and L_1-min problems, such as orthogonal matching pursuit algorithm [7], gradient projection method [8], and primal-dual interior-point method [9]. In addition, the sparse reconstruction algorithms are widely used for face recognition [10,11].

In the literature, recurrent neural networks (RNNs) as one of the real-time optimization methods, described as continuous-time or discrete-time dynamic systems, have been investigated for engineering optimization with applications [12,13]. The Lyapunov method is developed for the analysis of the convergence of the systems, which is generally guaranteed to be globally convergent or initial free. In particular, based on projection method, the projection neural networks and their collective networks have been developed to solve linear and nonlinear optimization problems [14,15] and distributed optimization problems [16,17], which can be applied to face recognition [2], image processing [18], support vector machine learning [19], and robot motion planning [20].

In this paper, the L_1-norm minimization problem (2) is converted into a mixed-norm problem, which considers the influence of both sparse representation and collaborative representation [21]. To seek the optimal solution of mixed-norm problem, a projection-based algorithm is designed for face recognition. Compared with existing sparse representation algorithms, the proposed algorithm shows good performance in the examples.

2   Preliminaries

Definition 1. *A matrix P is said to be a projection matrix if P is symmetric and $P^2 = P$. For a general projection operator $g(x)$ onto a closed convex set $\mathcal{X}$, it can be defined as*

$$
g(x) = \arg \min_{\gamma \in \mathcal{X}} \|x - \gamma\|_2,
$$

where $\| \cdot \|_2$ is the L_2-norm.

Here the closed convex set $\mathcal{X}$ is assumed to be $\mathcal{X} = \{\mu \in \mathbb{R}^n : -1 \leq \mu_i \leq 1\}$, then the element of g is generally described as a piecewise-linear function

$$g(\mu_i) = \begin{cases} 1, & \mu_i > 1, \\ \mu_i, & -1 \leq \mu_i \leq 1, \\ -1, & \mu_i < -1. \end{cases}$$

Lemma 1. [22] *For the projection operator, we have*

$$(u - g(u))^T(g(u) - v) \geq 0, \quad \forall\, u \in \mathbb{R}^n, v \in \mathcal{X}.$$

Then the following lemma is got directly from the previous one.

Lemma 2. *For the projection operator, we have*

$$(u - v)^T(g(u) - g(v)) \geq \|g(u) - g(v)\|_2^2, \quad \forall\, u, v \in \mathbb{R}^n.$$

In this paper, the mixed-norm optimization is studied for solving the pattern classification problem. Combining the L_1 and L_2 norms, the following minimization problem is considered

$$\begin{aligned} \text{minimize} \quad & \|x\|_1 + \frac{1}{2}\|x\|_2^2, \\ \text{subject to} \quad & Ax = b. \end{aligned} \tag{3}$$

Since the objective function in (3) is strictly convex and not less than $\|x\|_2^2/2$, the optimal solution is always finite provided that the feasible set $\{x \in \mathbb{R}^n : Ax = b\}$ is nonempty.

Lemma 3. [2] $x^* \in \mathbb{R}^n$ *is an optimal solution to* (3) *if and only if there exists a point* $y^* \in \mathbb{R}^n$ *such that* $y^* - Pg(y^*) - q = 0$ *and* $x^* = y^* - g(y^*)$, *where* $P = A^T(AA^T)^{-1}A$ *and* $q = A^T(AA^T)^{-1}b$.

3   Algorithm Description and Convergence Analysis

According the result in Lemma 3, the projection-based iterative algorithm for solving (3) is formulated as

$$y_{k+1} = Pg(y_k) + q, \tag{4a}$$
$$x_k = y_k - g(y_k). \tag{4b}$$

Theorem 1. *For any initial value* $z_0 \in \mathbb{R}^n$, *the vector* x_k *in* (4) *is globally convergent to the optimal solution of* (3).

Proof. From Lemma 3, $\bar{x} = \bar{y} - g(\bar{y})$ is an optimal solution of (3), where $\bar{y}$ is a convergence point (equilibrium point) of (4a) satisfying

$$\bar{y} = Pg(\bar{y}) + q. \tag{5}$$

Substituting (5) into (4a) follows $y_{k+1} - \bar{y} = P(g(y_k) - g(\bar{y}))$. Then $\|y_{k+1} - \bar{y}\|_2^2 - \|y_k - \bar{y}\|_2^2 = \|P(g(y_k) - g(\bar{y}))\|_2^2 - \|y_k - \bar{y}\|_2^2$.

Let $J_1 = \|P(g(y_k) - g(\bar{y}))\|_2^2 - \|y_k - \bar{y}\|_2^2$ and $J_2 = \|y_k - \bar{y} - P(g(y_k) - g(\bar{y}))\|_2^2$. Then

$$\|y_{k+1} - \bar{y}\|_2^2 - \|y_k - \bar{y}\|_2^2 = J_1,$$

and

$$J_2 = \|y_k - \bar{y}\|_2^2 + \|P(g(y_k) - g(\bar{y}))\|_2^2 - 2(y_k - \bar{y})^T P(g(y_k) - g(\bar{y})).$$

It follows

$$J_1 + J_2 = 2\|P(g(y_k) - g(\bar{y}))\|_2^2 - 2(y_k - \bar{y})^T P(g(y_k) - g(\bar{y})).$$

From (4a), we have $Py_{k+1} = Pg(y_k) + q$ due to $P^2 = P$ and $Pq = q$, which follows $Py_{k+1} = y_{k+1}$ and $P\bar{y} = \bar{y}$. Then

$$J_1 + J_2 = 2\|P(g(y_k) - g(\bar{y}))\|_2^2 - 2(y_k - \bar{y})^T (g(y_k) - g(\bar{y})).$$

Combining with Lemma 2 results in

$$J_1 + J_2 \leq 2\|(g(y_k) - g(\bar{y}))\|_2^2 - 2(y_k - \bar{y})^T (g(y_k) - g(\bar{y})) \leq 0,$$

in which the first inequality holds due to $\|P\|_2 = 1$.

Based on above analysis, we have

$$\begin{aligned}
\|y_{k+1} - \bar{y}\|_2^2 - \|y_k - \bar{y}\|_2^2 &\leq -J_2 \\
&= -\|y_k - \bar{y} - P(g(y_k) - g(\bar{y}))\|_2^2 \\
&= -\|y_k - Pg(y_k) - q\|_2^2,
\end{aligned}$$

where the last equality holds due to (5).

The candidate Lyapunov function is considered as

$$V(y) = (y - \bar{y})^T (y - \bar{y}).$$

Then

$$V(y_{k+1}) - V(y_k) \leq -\|y_k - Pg(y_k) - q\|_2^2 \leq 0. \tag{6}$$

From the LaSalle invariance principle in [23], y_k is convergent to S, the largest invariant subset of $W = \{y_k \in \mathbb{R}^n : V(y_{k+1}) - V(y_k) = 0\}$.

According to (6), if $V(y_{k+1}) - V(y_k) = 0$, one gets $y_k - Pg(y_k) - q = 0$. Then combining with above analysis, y_k converges to the equilibrium set of (4a).

Furthermore, following (6), it results in

$$\|y_{k+1} - \bar{y}\|_2^2 \leq \|y_k - \bar{y}\|_2^2 \leq \cdots \leq \|y_0 - \bar{y}\|_2^2.$$

Then y_k is bounded. Furthermore, there exists an increasing subsequence $\{k_l\}_{l=1}^{\infty}$ with a limit point $\tilde{y}$ such that $\lim_{l \to \infty} y_{k_l} = \tilde{y}$, where $\tilde{y}$ is an convergence point.

On one hand, from above analysis, we have $\|y_{k+1} - \tilde{y}\|_2^2 \leq \|y_k - \tilde{y}\|_2^2$. On the other hand, for any k, there exists k_l such that $\|y_k - \tilde{y}\|_2^2 \leq \|y_{k_l} - \tilde{y}\|_2^2$. Let $l \to \infty$, then $k \to \infty$. It results in $\lim_{k \to \infty} \|y_k - \tilde{y}\|_2^2 = \lim_{l \to \infty} \|y_{k_l} - \tilde{y}\|_2^2 = 0$. That is, $\lim_{k \to \infty} y_k = \tilde{y}$. Furthermore, due to $V(y) = (y - \bar{y})^T (y - \bar{y})$ being radially unbounded, one gets that for any initial point $y_0 \in \mathbb{R}^n$, the global convergence of y_k in (4) is guaranteed. As $x_k = y_k - g(y_k)$, the convergence of x_k is derived directly from that of y_k. □

4   Application to Face Recognition

Next, the proposed mixed-norm algorithm is applied to face recognition. Generally, the frontal face images are expressed by a low dimensional linear subspace spanned by

$$A_i = \{a_{i1}, a_{i2}, \ldots, a_{in_i}\},$$

where a_{ij} $(j = 1, 2, \ldots, n_i)$ in vector form is training image in the ith subject. Assume c subjects for the recognition and let $A \in \mathbb{R}^{m \times n}$ be the matrix for collecting training images

$$A = (A_1, A_2, \ldots, A_c).$$

Our aim is to find the sparsest linear representation of the test image from all the training ones, which can be written as

$$b = Ax, \tag{7}$$

where $x = (x_1, x_2, \ldots, x_n)^T \in \mathbb{R}^n$. Generally, if b is valid from one of the c subjects, the above representation x in (7) can be expressed as a sparse vector $x = (0, \ldots, 0, x_{i1}, x_{i2}, \ldots, x_{in_i}, 0, \ldots, 0)^T \in \mathbb{R}^n$ with zero entries except those associated with the ith subject. Herein, we solve the mixed-norm problem (3) to get the sparse solution. To perform the sparse representation classification (SRC), similar to the algorithm in [2] for face recognition, the proposed algorithm in (4) is utilized and it is stated as the Algorithm 1.

5   Experiments

5.1   ORL Data Set

The ORL data set [24] contains 400 images from 40 individuals. Some example images of the faces are depicted in Fig. 1. In the experiment, The dimension of image is reduced to 50 by the Principal Component Analysis (PCA) method [25]. The first 7 images are selecting for training. The recognition rates of the MPZSC with different sparsity levels as well as the sparse representation classification (SRC) [1], zero-to-sparse classification (ZSC) [2] and collaborative representation classification (CRC) [21] are shown in Fig. 2. In Fig. 2, it shows that the recognition rates of MPZSC are higher than the others, such as SRC, ZSC and CRC, if the sparsity levels are set properly.

Fig. 1. Some image samples of the ORL data set.

Algorithm 1. Mixed-norm Projection-based Zero-to-Sparse Classification (MPZSC)

Initialization:
1: Initialize $k = 0$, $z_0 = x_0 = 0$, sparsity level K and reconstruction error ϵ. Input training images $A_1, A_2, \ldots, A_c$ and a test image b.

Iteration:
2: **while** $\|Ax - b\|_2 > \epsilon$ and $\|x_k\|_0 < K$ **do**
3:    $y_{k+1} \leftarrow Pg(y_k) + q$,
4:    $x_{k+1} \leftarrow y_{k+1} - g(y_{k+1})$.
5: **end while**
6: **return** $x \leftarrow x_k$.

Classification:
7: Calculate the residuals

$$e_i(b) = \|b - A_i x^i\|_2 / (\|x^i\|_0 \|x^i\|_2),$$

where $x^i = (x_{i1}, x_{i2}, \ldots, x_{in_i})^T$ is the sparsity representation vector with the ith subject.
8: The index of b is guaranteed by

$$\mathrm{index}(b) = \arg\min_i \{e_i\}.$$

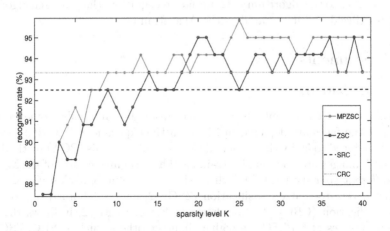

Fig. 2. Recognition rates on the Extended Yale B data set for different algorithms (MPZSC, ZSC, SRC and CRC).

5.2   Extended Yale B Data Set

The Extended Yale B (EYB) data set includes 2, 414 frontal face images from 38 individuals [26]. Figure 3 shows some example images of the data set. In the experiment, the images are cropped into the size of 192 × 168. The number of image samples for each class is set as 20 and the samples are selected randomly for training and testing. The PCA method is also used to reduce the dimension of images to 200. The recognition rates of the MPZSC under different sparsity levels compared with the other algorithms are given in Fig. 4. We can observe that the MPZSC algorithm displays better performance than the others if the sparsity levels are set properly.

Fig. 3. Some image samples of the EYB data set.

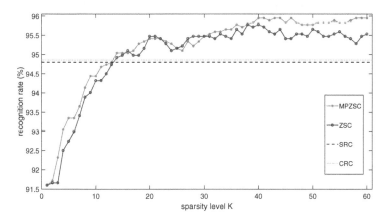

Fig. 4. Recognition rates on the Extended Yale B data set for different algorithms (MPZSC, ZSC, SRC and CRC).

5.3   CMU PIE Data Set

The CMU PIE data set consists of 41, 368 face images of 68 individuals under 13 different poses, four different expressions and 43 different illumination conditions [27]. Figure 5 depicts some example images of the data set. In the experiments, two

Fig. 5. Some image samples of the CMU PIE data set.

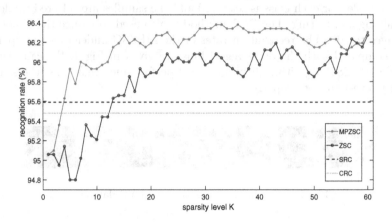

Fig. 6. Recognition rates on the CMU PIE data set (camera 05) for MPZSC, ZSC, SRC and CRC.

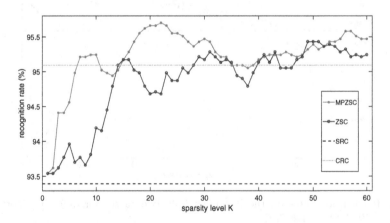

Fig. 7. Recognition rates on the CMU PIE data set (camera 27) for different algorithms (MPZSC, ZSC, SRC and CRC).

subsets of the data set are used, in which one is got near frontal (camera 27) and the other one is with horizontal rotation (cameras 05). In the experiment, we randomly select 10 images as training samples for each class, and the remainders as testing samples. The number of image dimension is compressed to 100. The experimental results on recognition rates of MPZSC, ZSC, SRC and CRC are shown in Figs. 6 and 7, which show the better performance of MPZSC than the others.

6   Conclusions

In this paper, a mixed-norm projection-based signal reconstruction algorithm has been proposed for face recognition with convergence analysis by using the Lyapunov method. To illustrate the algorithm's performance, experiment on random Gaussian signals is presented, which shows that the proposed algorithm is a combination of the sparse reconstruction and collaborative reconstruction. In addition, experiments on three publicity available data sets are provided to show the better performance of the proposed MPZSC algorithm for face recognition under proper sparsity level setting.

References

1. Wright, J., Yang, A.Y., Ganesh, A., Sastry, S.S., Ma, Y.: Robust face recognition via sparse representation. IEEE Trans. Pattern Anal. Mach. Intell. **31**, 210–227 (2009)
2. Liu, Q., Wang, J.: L_1-minimization algorithms for sparse signal reconstruction based on a projection neural network. IEEE Trans. Neural Netw. Learn. Syst. **27**, 698–707 (2016)
3. Chen, S.S., Donoho, D.L., Saunders, M.A.: Atomic decomposition by basis pursuit. SIAM Rev. **43**, 129–159 (2001)
4. Elhamifar, E., Vidal, R.: Sparse subspace clustering: algorithm, theory, and applications. IEEE Trans. Pattern Anal. Mach. Intell. **35**, 2765–2781 (2013)
5. Mairal, J., Elad, M., Sapiro, G.: Sparse representation for color image restoration. IEEE Trans. Image Proc. **17**, 53–69 (2008)
6. Wang, J., Yang, J., Yu, K., Lv, F., Huang, T., Gong, Y.: Locality-constrained linear coding for image classification. In: Proceeding IEEE Conference on Computer Vision and Pattern Recognition, pp. 3360–3367 (2010)
7. Cai, T.T., Wang, L.: Orthogonal matching pursuit for sparse signal recovery with noise. IEEE Trans. Inf. Theor. **57**, 4680–4688 (2011)
8. Figueiredo, M.A., Nowak, R.D., Wright, S.J.: Gradient projection for sparse reconstruction: application to compressed sensing and other inverse problems. IEEE J. Sel. Top. Sign. Proces. **1**, 586–597 (2007)
9. Wright, S.: Primal-Dual Interior-Point Methods. SIAM, Philadelphia (1997)
10. Xu, B., Liu, Q.: Iterative projection based sparse reconstruction for face recognition. Neurocomputing **284**, 99–106 (2018)
11. Xu, B., Liu, Q., Huang, T.: A discrete-time projection neural network for sparse signal reconstruction with application to face recognition. IEEE Trans. Neural Netw. Learn. Syst. (2018). https://doi.org/10.1109/TNNLS.2018.2836933
12. Tank, D., Hopfield, J.: Simple neural optimization networks: an A/D converter, signal decision circuit, and a linear programming circuit. IEEE Trans. Circ. Syst. **33**, 533–541 (1986)
13. Wang, J.: Analysis and design of a recurrent neural network for linear programming. IEEE Trans. Circ. Syst.-I **40**, 613–618 (1993)
14. Xia, Y., Wang, J.: A general projection neural network for solving monotone variational inequalities and related optimization problems. IEEE Trans. Neural Netw. **15**, 318–328 (2004)

15. Liu, Q., Wang, J.: A one-layer projection neural network for nonsmooth optimization subject to linear equalities and bound constraints. IEEE Trans. Neural Netw. Learn. Syst. **24**, 812–824 (2013)
16. Liu, Q., Yang, S., Wang, J.: A collective neurodynamic approach to distributed constrained optimization. IEEE Trans. Neural Netw. Learn. Syst. **28**, 1747–1758 (2017)
17. Liu, Q., Yang, S., Hong, Y.: Constrained consensus algorithms with fixed step size for distributed convex optimization over multiagent networks. IEEE Trans. Autom. Control **62**, 4259–4265 (2017)
18. Xia, Y., Kamel, M.S.: Novel cooperative neural fusion algorithms for image restoration and image fusion. IEEE Trans. Image Process. **16**, 367–381 (2007)
19. Xia, Y., Wang, J.: A one-layer recurrent neural network for support vector machine learning. IEEE Trans. Syst. Man Cybern.- Part B: Cybern. **34**, 1261–1269 (2004)
20. Zhang, Y., Wang, J., Xu, Y.: A dual neural network for bi-criteria kinematic control of redundant manipulators. IEEE Trans. Robot. Autom. **18**, 923–931 (2002)
21. Zhang, L., Yang, M., Feng, X.: Sparse representation or collaborative representation: which helps face recognition? In: Proceedings 2011 IEEE International Conference on Computer Vision, pp. 471–478 (2011)
22. Kinderlehrer, D., Stampacchia, G.: An Introduction to Variational Inequalities and Their Applications. Academic, New York (1982)
23. LaSalle, J.: The Stability of Dynamical Systems. SIAM, Philadelphia (1976)
24. Samaria, F.S., Harter, A.C.: Parameterisation of a stochastic model for human face identification. In: Proceedings of the Second IEEE Workshop on Applications of Computer Vision, pp. 138–142 (1994)
25. Duda, R.O., Hart, P.E., Stork, D.G.: Pattern Classification. Wiley, Hoboken (2012)
26. Georghiades, A.S., Belhumeur, P.N., Kriegman, D.J.: From few to many: illumination cone models for face recognition under variable lighting and pose. IEEE Trans. Pattern Anal. Mach. Intell. **23**, 643–660 (2001)
27. Sim, T., Baker, S., Bsat, M.: The CMU pose, illumination, and expression database. IEEE Trans. Pattern Anal. Mach. Intell. **25**, 1615–1618 (2003)

Cross-Database Facial Expression Recognition with Domain Alignment and Compact Feature Learning

Lan Wang[1], Jianbo Su[1(✉)], and Kejun Zhang[2]

[1] Shanghai Jiao Tong University, Shanghai, China
{wang.lan,jbsu}@sjtu.edu.cn
[2] Shanghai Lingzhi Hi-Tech Corporation, Shanghai, China

Abstract. Expression recognition has achieved impressive success in recent years. Most methods are based on the assumption that the training and testing databases follow the same feature distribution. However, distribution discrepancy among datasets is pretty common in practical scenarios. Thus, the performance of these methods may drop sharply on target datasets. To address this issue, we aim to learn a facial expression classification model from several labeled source databases and generalize it to target databases. This is achieved by integrating domain alignment and class-compact features learning across source domains. Domain alignment paves the way to involve more expression-related representations. Learning compact features can signicantly diminish the intra-class divergence, which is beneficial to both domain alignment and expression recognition. Experimental results demonstrate that the proposed model has a more promising performance compared with other cross-database expression recognition methods.

Keywords: Facial expression recognition · Domain alignment ·
Class-compact feature learning

1 Introduction

Facial expression recognition is an essential research field to comprehend human emotions. Autonomous recognition of expressions has drawn considerable attention of researchers. Extensive studies have been performed on facial expression recognition [1,10] and facial feature extraction [5–7]. Usually, the existing methods depend on supervised learning with the assumption that the distribution remains unchanged between training and testing face images. However, in many applications, biases in collected databases are common primarily due to different collecting conditions and personal characteristics. In this condition, the aforementioned distribution consistency between the training and testing instances is not satisfied. Consequently, the performance of the existing methods often degrades significantly. Hence, cross-database expression recognition is an important and challenging research topic.

© Springer Nature Switzerland AG 2019
H. Lu et al. (Eds.): ISNN 2019, LNCS 11555, pp. 341–350, 2019.
https://doi.org/10.1007/978-3-030-22808-8_34

To alleviate this issue, many studies have been conducted recently. Na Liu et al. [9] propose a regression network to solve the unsupervised cross-database expression recognition problem. Zheng et al. [25] introduce a transductive transfer subspace learning framework to predict the expression categories of the target samples. Yan et al. [19] present a domain adaptive dictionary learning (DADL) method to build a mutual relation between the source and target instances. The fundamental idea of these approaches is to adapt the learned model to unlabeled images of the corresponding target dataset. Nevertheless, gathering sufficient samples for each new target domain is often time-wasting and even impossible. Hence Marcus et al. [21] perform cross-database expression recognition based on the fine-tuned deep convolutional network without access to the target sets.

Though recent research on cross-database expression recognition has shown positive results, there remain two limitations: (1) Due to the unmatched feature distributions of source domains, irrelevant domain information is involved in the expression representations. Thus the generalization ability of the network in [21] may be poor. (2) Researches [4,10,14] reported that the performance of expression recognition systems on unseen subjects often decreases, mainly due to the intra-expression variations triggered by personal characteristics. Such variations are even more considerable among multi-source databases, which requires the model to extract compact features. However, in previous studies, the learned features are only separable but not compact enough.

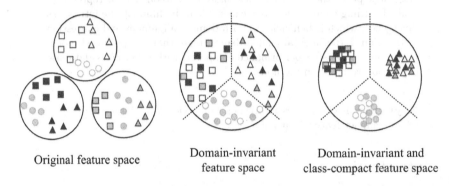

Original feature space   Domain-invariant   Domain-invariant and
                                feature space   class-compact feature space

Fig. 1. The necessity of integrating domain alignment and compact feature learning. Three source domains are shown, circle, square, triangle indicate three different expression categories, respectively. The features need to be generalized across domains and compact enough.

In order to address the aforementioned two problems, this paper proposes to joint domain alignment and class-compact representations learning in one optimization framework. As described in Fig. 1, minimizing the private domain characteristics is beneficial to construct a latent feature space shared by all source datasets, thus improving the generalization of the model. Additionally, enforcing the intra-expression samples with better compactness enhances the discriminability of representations. Specifically, to reduce the distribution discrepancy across source datasets, we approach cross-database expression recognition from the perspective of domain generalization (DG). A deep convolution network with domain-adversarial training is developed with the manner proposed in [2]. Simultaneously, a sample-based compact feature learning method is proposed. More importantly, the combination of these two strategies realizes a complementary advantage: (1) Along with the reduction of domain shift, the interference of domain-specific features is decreased, making the representations in each cluster more relevant to expressions. (2) Since the intra-class features are better clustered, more emphasis is put on the inconsistency across domains when performing domain alignment.

To validate the effectiveness of our method, we conducted experiments on the following widely used databases: the Extended Cohn-Kanade Database database (CK+) [11], the Japanese Female Facial Expression database (JAFFE) [13], the MMI database [15], the Karolinska Directed Emotional Faces database (KDEF) [12]. These databases comprise images from subjects of various genders, ages, and ethnicities in different environments. The results indicate the superior performance of our method.

2   Joint Domain Alignment and Compact Feature Learning

With the aim of obtaining a latent feature space that is also effective for the related target domain, we expect the following conditions to be satisfied in this space:

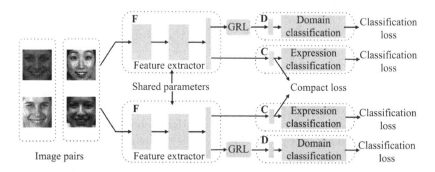

Fig. 2. The proposed network for cross-database expression recognition. We use two streams of CNN with shared weights. It consists of three components: feature extractor F which transforms the given image into corresponding feature maps, expression recognition network C which classifies the expressions with softmax loss, domain recognition network D which discriminates domains with softmax loss.

- Let x and y denote the expression features and the corresponding labels, the conditional distribution $p(y|x)$ should be aligned. Since if the distributions of diverse datasets differ greatly, which means there are more domain-related characteristics involved in the final classification, the generalization ability of the model would be poor.
- The learned feature representations need to form compact intra-class clusters to classify the target samples more precisely.

Based on the above ideas, we integrate domain alignment and class-compact representations learning into a unified framework. The architecture is illustrated in Fig. 2. Specifically, during the training process, each sample of the image pair is fed into one CNN stream. The expression classification loss is computed for both images to ensure the learned features are meaningful for expression recognition. The domain classification loss is calculated to help the feature extractor reduce the interference of domain-specific features. Simultaneously, the compact loss is computed using the expression related features to make the within-class samples closer in the latent space. Additionally, during testing, an input image is fed into one CNN stream, and the expression relevant features are used for prediction.

2.1   Domain Alignment with Adversarial Training

To match the conditional distribution across domains, we introduce an auxiliary task. It prevents the discriminator of distinguishing the domains, which benefits the main recognition task.

Assume we observe $i = 1...S$ source domains, the ith domain contains N_i labeled instances $\{x_j^i, y_j^i\}$, where x_j^i represents the input image and y_j^i denotes the expression label, $y = 1...K$. The domain classification loss can be formulated as:

$$L_d = \frac{1}{S} \sum_{i=1}^{S} \frac{1}{N_i} \sum_{j=1}^{N^i} I(z_j^i, \hat{z}_j^i), \tag{1}$$

where $I(\cdot)$ denotes the cross-entropy loss function between the probability distribution of predicted domain category $q(\hat{z})$ and the probability distribution of real category $p(z)$. Namely, $I(z, \hat{z}) = -p(z)log(q(\hat{z}))$.

We propose a gradient-reversal layer (GRL) to minimize the distribution divergence across domains by following [2]. This gradient-reversal layer feeds the input to the following layer during forward propagation and reverses the gradient during the back propagation to obtain an opposite optimization direction from the domain classifier. Such domain-adversarial training can guide the feature extractor to be less sensitive to the domain-specific information, which is beneficial to the recognition of unknown target samples.

2.2   Learning Compact Features

The expression recognition task will be executed by the classifier C. Thus the loss function for the classification of the labeled instances in source domains can be described as:

$$L_c = \frac{1}{S}\sum_{i=1}^{S}\frac{1}{N_i}\sum_{j=1}^{N_i}l(y_j^i, \hat{y}_j^i), \tag{2}$$

where $l(\cdot)$ denotes the cross-entropy loss function between the probability distribution of predicted expression category $q(\hat{y})$ and the probability distribution of real expression category $p(y)$. Namely, $I(y, \hat{y}) = -p(y)log(q(\hat{y}))$.

Furthermore, the discriminative power of feature space is of vital importance for the classification task. Since images of the same expression can be quite different due to various factors, there is great intra-class variation in multi-source expressions. To minimize the distance between samples in the same expression class, inspired by the contrastive loss [3], we propose the compact feature learning as:

$$d_{cp}^{a,b} = max(0, \|g(x_a) - g(x_b)\|^2), \tag{3}$$

$$L_{cp} = \frac{1}{2}\sum_{a,b} d_{cp}^{a,b}, \tag{4}$$

where $g(x_a)$ and $g(x_b)$ are d-dimensional vectors representing features of the intra-class samples x_a and x_b, squared Euclidean distance $\|\cdot\|$ is used to measure the feature distribution distance between the paired samples. Clearly, by minimizing the compact loss, we highly enhance the discriminative power of the feature space.

We integrate all the mentioned losses into a deep learning framework. The overall loss is defined as follows:

$$L = L_c - \lambda_1 L_d + \lambda_2 L_{cp}, \tag{5}$$

where parameters λ_1 and λ_1 are weights to balance the contribution of each loss. For λ_1 we adopt the rule introduced by [2], which increases the weight of domain discriminator from 0 to 1 with training epochs: $\lambda_1 = \frac{2}{1+exp(-10w)} - 1$, where $w = \frac{current\ epoch}{total\ epochs}$.

During minimization of the overall loss L, the feature extractor tends to learn both domain-invariant and class-compact representations for source domains, making the model better generalized to the target domains.

3   Experimental Results

To demonstrate the validity of our method, cross-database facial expression recognition experiments have been conducted on the following widely used expression databases: the CK+ database [11], the JAFFE database [13], the MMI database [15], and the KDEF database [12].

The Extended Cohn-Kanade (CK+) database [11] contains 327 video sequences from 118 subjects labeled with six basic expressions (disgust, fear, anger, happiness, sadness, surprise) and contempt. Each sequence starts with a neutral frame and gradually reaches the peak expression. The first and last three frames are extracted in our experiment.

The Japanese Female Facial Expression (JAFFE) database [13] consists of 213 images from 10 Japanese females. Each subject has six basic expressions (disgust, fear, anger, happiness, sadness, surprise) plus neutral.

The Karolinska Directed Emotional Faces (KDEF) database [12] comprises 980 images from 70 subjects with six basic facial expressions plus neutral.

The Oulu-CASIA database [24] contains 2880 sequences from 80 subjects. The sequences start from neutral frames and end with peak expressions. The last three peak frames from the 480 videos collected by the VIS system labeled with six basic expressions are selected, resulting in 1440 images.

The MMI database [15] contains 213 sequences labeled with six basic expressions. These sequences start from a neutral expression and reach a peak state in the middle, then back to the neutral expression. Three peak frames of each selected image sequence are collected, resulting in 323 images.

Pre-Processing: Pre-processing is required to reduce variations that are irrelevant to expressions, such as different scales, backgrounds, and poses [8]. Thus the multi-task convolutional neural network (MTCNN) [22] is used to detect landmarks of facial images. Further face alignment using the coordinates of landmarks is conducted, then the face regions are cropped and scaled to a size of 80×80. Finally, the images are transformed into grayscale to minimize the data bias among databases.

Data Augmentation: Deep neural networks demand large-scale training data to gain better generalization ability. However, most available databases for expression recognition do not have sufficient samples. Hence data augmentation is embedded to improve the generalizability of our model. Precisely, during the training phase, input images are randomly cropped and flipped horizontally. During the test phase, only the center patch of each instance is used for expression prediction.

Implementation Details: VGG-11 [18] is used to recognize facial expressions, it consists of 8 convolutional layers, followed by 3 FC layers. As mentioned above, directly training with insufficient expression datasets is easy to cause overfitting. Hence VGG-11 with softmax loss is pre-trained using the CASIA-WebFace dataset [20]. The architecture of our network is as follows: the convolutional layers of VGG-11 followed by an FC layer is used as the feature extractor F, attached a shallow network as expression classifier C and domain discriminator D. Units of the last FC layer are set to the number of categories. The GRL layer is connected to the feature extractor.

For all networks used in experiments, parameters of the convolution layers are initialized by weights of the pre-trained VGG-11 model, parameters of FC layers are randomly initialized. Therefore, the learning rate of FC layers is set to 10 times that of convolution layers. SGD with a momentum of 0.9 is used, and dropout is set at 0.5. Source domains are divided into a training set (90%) and a validation set (10%). To create positive and negative image pairs, we randomly selected two samples from all available source domains. All of the experiments are repeated five times to decrease the impact of random factors, and the average classification accuracies are reported.

We compare our proposed cross-database expression recognition method with the following methods in terms of accuracy on target domains.

- **Combine sources:** using all samples in source datasets to train the network with only expression classification loss.
- **Domain alignment:** using all samples in source datasets to train the network with expression classification loss and domain classification loss.
- **Domain alignment and compact feature learning (DACFL):** using all samples in source datasets to train the network with expression classification loss, domain classification loss, and compact loss.

Table 1. Performance comparison in terms of accuracy (%) on expression database

Sources	Target	Combine sources	Domain alignment	DACFL
KDEF, JAFFE, Oulu-CASIA	CK+	85.54	**85.77**	85.75
KDEF, JAFFE, Oulu-CASIA	MMI	63.13	63.60	**64.33**
CK+, JAFFE, Oulu-CASIA	KDEF	73.45	75.68	**76.45**
CK+, JAFFE, Oulu-CASIA	MMI	59.22	59.50	**61.32**
KDEF, CK+, Oulu-CASIA	JAFFE	42.67	46.60	**48.13**
KDEF, CK+, Oulu-CASIA	MMI	62.85	64.84	**65.06**
Avg		64.48	66.00	**66.84**

The experimental results of the aforementioned deep-learning-based methods are summarized in Table 1. DACFL achieves the best performance when testing on different target domains except for the CK+ database. The domain-aligned network gets an average accuracy of 66.00%, which is 1.52% higher than that for the original combine sources method. Additionally, learning compact features in the shared feature space enhances the average performance by 0.84%. Notably, our approach improves the accuracy of the JAFFE database significantly from 42.67% to 48.13%. For the KDEF database, our model achieves 76.45% of accuracy, which is 3.00% higher than the baseline (73.45%). These results convincingly emphasize the vital of constructing both domain alignment and class-compact representations learning.

Table 2. Results of different methods for cross-database expression recognition.

Target	Sources	Method	Accuracy
CK+	BOSPHORUS	Da Silva et al. [17]	57.60
	MMI	MKL framework [23]	61.20
	MMI	IACNN [14]	71.29
	FERA, MMI	Inception-ResNet with CRF [4]	73.91
	6 databases*	**Marcus et al.** [21]	**88.58**
	3 databases[†]	DACFL	85.75
KDEF	6 databases*	Marcus et al. [21]	72.55
	3 databases[‡]	**DACFL**	**76.45**
JAFFE	CK	SVM (RBF) [16]	41.30
	CK	Da Silva et al. [17]	42.30
	6 databases*	Marcus et al. [21]	44.32
	3 database°	**DACFL**	**48.13**
MMI	CK+	IACNN [14]	55.41
	3 databases°	**DACFL**	**65.06**

*Trained with CK+, BU3DFE, JAFFE, RaFD, MMI, KDEF, ARFace
except the target dataset.
[†]Trained with KDEF, JAFFE, Oulu-CASIA.
[‡]Trained with CK+, JAFFE, Oulu-CASIA.
°Trained with KDEF, CK+, Oulu-CASIA.

We further compare the performance of our model with other methods, the results are shown in Table 2. Notice that the accuracy of the JAFFE database is the lowest among these databases, mainly due to the high divergence in ethnic with other databases. As can be seen, DACFL achieves superior performance over competitive works. For the KDEF database, DACFL reports 76.45% of accuracy, which is 3.90% higher than that in [21]. For the JAFFE database, the result of DACFL is 3.81% higher than the best result of previous studies. For the MMI database, DACFL increases the performance by 9.65%. Hence, DACFL is an excellent feature extraction method for the expression classification task.

4   Conclusion

In this paper, we propose a novel framework for cross-database facial expression recognition. This framework outperforms previous methods because it combines domain alignment and class-compact representations learning. Domain-invariant adversarial training makes the discriminator cannot distinguish the source of images. Furthermore, a sample-based compact feature learning method is proposed to guarantee the discriminative ability of the generalized features. We evaluate the method on several widely used databases, and the results demonstrate the effectiveness of our model.

Acknowledgments. This work was partially financially supported by National Natural Science Foundation of China under grant 61533012 and 91748120.

References

1. Ding, H., Zhou, S.K., Chellappa, R.: FaceNet2ExpNet: regularizing a deep face recognition net for expression recognition. In: 12th IEEE International Conference on Automatic Face and Gesture Recognition, pp. 118–126 (2017)
2. Ganin, Y., et al.: Domain-adversarial training of neural networks. J. Mach. Learn. Res. **17**(59), 1–35 (2016)
3. Hadsell, R., Chopra, S., LeCun, Y.: Dimensionality reduction by learning an invariant mapping. In: IEEE Computer Society Conference on Computer Vision and Pattern Recognition, pp. 1735–1742 (2006)
4. Hasani, B., Mahoor, M.H.: Spatio-temporal facial expression recognition using convolutional neural networks and conditional random fields. In: 12th IEEE International Conference on Automatic Face and Gesture Recognition, pp. 790–795 (2017)
5. Jian, M., Lam, K.M.: Face-image retrieval based on singular values and potential-field representation. Signal process. **100**, 9–15 (2014)
6. Jian, M., Lam, K.M.: Simultaneous hallucination and recognition of low-resolution faces based on singular value decomposition. IEEE Trans. Circ. Syst. Video Technol. **25**(11), 1761–1772 (2015)
7. Jian, M., Lam, K.M., Dong, J.: A novel face-hallucination scheme based on singular value decomposition. Pattern Recognit. **46**(11), 3091–3102 (2013)
8. Jian, M., Lam, K.M., Dong, J.: Facial-feature detection and localization based on a hierarchical scheme. Inform. Sci. **262**, 1–14 (2014)
9. Liu, N., et al.: Super wide regression network for unsupervised cross-database facial expression recognition. In: IEEE International Conference on Acoustics, Speech and Signal Processing, pp. 1897–1901 (2018)
10. Liu, X., Kumar, B.V., You, J., Jia, P.: Adaptive deep metric learning for identity-aware facial expression recognition. In: CVPR Workshops, pp. 522–531 (2017)
11. Lucey, P., Cohn, J.F., Kanade, T., Saragih, J., Ambadar, Z., Matthews, I.: The extended cohn-kanade dataset (ck+): a complete dataset for action unit and emotion-specified expression. In: IEEE Computer Society Conference on Computer Vision and Pattern Recognition Workshops, pp. 94–101 (2010)
12. Lundqvist, D., Flykt, A., Öhman, A.: The Karolinska Directed Emotional Faces (KDEF). CD ROM from Department of Clinical Neuroscience, Psychology section, Karolinska Institutet 91, p. 630 (1998)
13. Lyons, M., Akamatsu, S., Kamachi, M., Gyoba, J.: Coding facial expressions with Gabor wavelets. In: Proceedings of the IEEE International Conference on Automatic Face and Gesture Recognition, pp. 200–205 (1998)
14. Meng, Z., Liu, P., Cai, J., Han, S., Tong, Y.: Identity-aware convolutional neural network for facial expression recognition. In: 12th IEEE International Conference on Automatic Face and Gesture Recognition, pp. 558–565 (2017)
15. Pantic, M., Valstar, M., Rademaker, R., Maat, L.: Web-based database for facial expression analysis. In: 2005 IEEE International Conference on Multimedia and Expo, pp. 317–321 (2005)
16. Shan, C., Gong, S., McOwan, P.W.: Facial expression recognition based on local binary patterns: a comprehensive study. Image Vis. Comput. **27**(6), 803–816 (2009)

17. da Silva, F.A.M., Pedrini, H.: Effects of cultural characteristics on building an emotion classifier through facial expression analysis. J. Electron. Imaging **24**(2), 1–9 (2015)
18. Simonyan, K., Zisserman, A.: Very deep convolutional networks for large-scale image recognition. arXiv preprint arXiv:1409.1556 (2014)
19. Yan, K., Zheng, W., Cui, Z., Zong, Y.: Cross-database facial expression recognition via unsupervised domain adaptive dictionary learning. In: International Conference on Neural Information Processing, pp. 427–434 (2016)
20. Yi, D., Lei, Z., Liao, S., Li, S.Z.: Learning face representation from scratch. arXiv preprint arXiv:1411.7923 (2014)
21. Zavarez, M.V., Berriel, R.F., Oliveira-Santos, T.: Cross-database facial expression recognition based on fine-tuned deep convolutional network. In: 30th SIBGRAPI Conference on Graphics, Patterns and Images, pp. 405–412 (2017)
22. Zhang, K., Zhang, Z., Li, Z., Qiao, Y.: Joint face detection and alignment using multitask cascaded convolutional networks. IEEE Signal Process. Lett. **23**(10), 1499–1503 (2016)
23. Zhang, X., Mahoor, M.H., Mavadati, S.M.: Facial expression recognition using l_p-norm MKL multiclass-SVM. Mach. Vis. Appl. **26**(4), 467–483 (2015)
24. Zhao, G., Huang, X., Taini, M., Li, S.Z., Pietikälnen, M.: Facial expression recognition from near-infrared videos. Image Vis. Comput. **29**(9), 607–619 (2011)
25. Zheng, W., Zong, Y., Zhou, X., Xin, M.: Cross-domain color facial expression recognition using transductive transfer subspace learning. IEEE Trans. Affect. Comput. **9**(1), 21–37 (2018)

An Adaptive Chirplet Filter for Range Profiles Automatic Target Recognition

Yifei Li and Zunhua Guo[✉]

School of Mechanical, Electrical and Information Engineering,
Shandong University, Weihai 264209, China
liyifei13@mail.sdu.edu.cn, gritty@sdu.edu.cn

Abstract. An adaptive Chirplet filter approach is proposed to deal with the aircraft recognition problem based on high-resolution range profiles. The Chirplet filter is a joint feature extraction and target identification method derived from the feed-forward neural networks, which consists of two layers: the Chirplet-atoms transform in the input layer is used for replacing the conventional sigmoid function, and the weights between the input and the output layer are taken as the linear classifier. The Chirplet-atoms parameters and the weights are adaptively adjusted by using the nonstationarity degree as the measurement of the features. The simulation results suggest that the adaptive Chirplet filter has advantages especially in noisy conditions.

Keywords: Automatic target recognition · Adaptive Chirplet filter ·
Neural network · Joint feature extraction and target identification ·
High-resolution range profiles

1 Introduction

Automatic target recognition (ATR) with one-dimensional high-resolution radar (HRR) range profiles is a research focus in the signal processing areas. The HRR range profiles can be obtained by transmitting a burst of broadband pulses (e.g. chirp signal and the stepped-frequency signal). Thus, an aircraft is separated into several range resolution units along the radar line-of-sight [1]. The range profiles reflect the distribution of the scattering strength of an aircraft's scatterers (e.g. the wings, engine intakes and the tail) at a certain aspect. The HRR range profiles can also avoid the scaling and rotation problems in the image recognition [2]. Therefore, it is efficient to use range profiles as the raw signatures for target recognition. Over the past few decades, many universities and institutes have made great achievements in range profiles target recognition [1–10].

The difficulty of HRR range profiles based ATR is to acquire effective information from the original data to characterize the target and to classify it accurately and timely. A target recognition system is commonly divided into two separated stages: feature extraction and classification. The classification performance of an ATR system is closely associated with the features extracted from the targets. Thus, it is necessary to design an ATR system considering the most feasible and suitable feature extraction approaches. The popular HRR range profiles features include Fourier transform, power

© Springer Nature Switzerland AG 2019
H. Lu et al. (Eds.): ISNN 2019, LNCS 11555, pp. 351–360, 2019.
https://doi.org/10.1007/978-3-030-22808-8_35

spectrum, Gabor transform, wavelet transform, etc. [3–5]. In the classification stage, the unknown targets are identified by pattern recognition methods, such as neural networks, support vector machine, and so on [5–10]. However, the parameters of the features and classifiers are not optimized collaboratively in such ATR system. If all parameters are adjusted concurrently, the classification performance can be improved [4].

The Chirplet transform is proved to be a more adequate representation of the original signals [11]. Furthermore, an adaptive algorithm was presented in reference [12], which has two distinct advantages: a compact description of the signal and a possible means of classification. It is thus a highly promising feature extraction method for the target recognition. The Chirplet-atoms network demonstrates these advantages in target recognition especially in noisy conditions [13]. Combined the Chirplet transform with the adaptive filter and the feed-forward neural networks, we proposed an adaptive Chirplet filter for joint feature extraction and classification. The Chirplet filter adopts the nonstationarity degree as the dynamic property measurement of the features to select the optimal Chirplet-atoms parameters as well as the tap-weights of the adaptive filter simultaneously.

2   Chirplet Filter

2.1   Chirplet-Atoms Transform

The Chirplet transform can be generalized from short-time Fourier transform (STFT) as well as wavelet transform [14]. In terms of the parameters space, the Chirplet transform is considered as a generalization of the three-dimensional space (time-frequency-scale) to four-dimensional space (time-frequency-scale-chirprate). The Chirplet atom is defined as [15]:

$$g_{u,\xi,s,c}(t) = \frac{1}{\sqrt{s}} g(\frac{t-u}{s}) e^{j\left[\xi(t-u) + \frac{c}{2}(t-u)^2\right]}, \tag{1}$$

where

$$g(t) = 2^{1/4} e^{-\pi t^2} \tag{2}$$

is the Gaussian window function, the parameters of shift u, frequency ξ, scale s and chirprate c are related to an aircraft's scatterers, geometric shape and size, material coatings, etc.

2.2   The Structure of Chirplet Filter

The Chirplet filter is a two-layer network structure, which reduces the computational complexity of the well-known multilayer feedforward neural network through single-layer linear combination. As shown in Fig. 1, the input layer uses Chirplet transform to extract range profiles features, the tap-weights and output layer constitute the adaptive filter for classification.

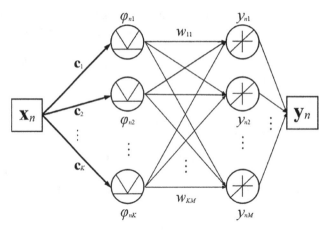

Fig. 1. The two-layer network structure of Chirplet filter

The input vector $\mathbf{x}_n = [x_{n1}, x_{n2}, \ldots, x_{nL}]^T$ is the range profile samples, where $n = 1, 2, \ldots, N$ denotes the number of the samples, and L is the number of range profile units. The range profiles are input to K Chirplet atoms $\mathbf{c}_1 \sim \mathbf{c}_K$ to extract the features $\varphi_{n1} \sim \varphi_{nK}$. The bold connected lines indicate to calculate the inner product. The weights are $w_{11} \sim w_{KM}$, and $\mathbf{y}_n = [y_{n1}, y_{n2}, \ldots, y_{nM}]^T$ represents the outputs, where M is the number of output nodes.

The HRR range profiles features are extracted by calculating the absolute value of the inner product of the input signals and Chirplet-atoms in the input layer:

$$\varphi = |\langle \mathbf{c}, \mathbf{x} \rangle| = \left| \int_t g^*\left(\frac{t - u}{s}\right) e^{-j\left[\xi(t-u) + \frac{\zeta}{2}(t-u)^2\right]} x(t) dt \right|. \tag{3}$$

The HRR range profiles are discrete-time real signals, so the integral in (3) can be replaced by the sum, namely, the kth feature of nth range profile φ_{nk} can be given by:

$$\varphi_{nk} = \left| \sum_{l=1}^{L} x_{nl} \frac{2^{1/4}}{\sqrt{s_k}} e^{-\pi\left(\frac{t_l - u_k}{s_k}\right)^2} \cos\left[\xi_k(t_l - u_k) + \frac{c_k}{2}(t_l - u_k)^2\right] \right|. \tag{4}$$

In the output layer, the feature φ_{nk} and the weights w_{km} are used to obtain the output y_{nm}:

$$y_{nm} = \sum_{k=1}^{K} \varphi_{nk} w_{km}, m = 1, \ldots, M \tag{5}$$

During the training phase, the Chirplet filter searches for the optimal Chirplet-atoms parameters and the weights concurrently. The parameters of the Chirplet atoms are tuned with the classical gradient descent algorithm, whereas the weights are updated by adaptive filter algorithms. We select the recursive least square (RLS) algorithm to adjust the weights for its faster convergence rate and smaller steady value of the ensemble-average squared error.

2.3 Nonstationarity Degree

Due to the uncertainty of the training process, the adjustment of the Chirplet-atoms parameters can be very complex. In order to effectively train the Chirplet filter, we use the nonstationarity degree as the measurement of the features [16]:

$$\alpha = \left[\frac{\sum_{n=1}^{N} |\omega^H(n)\varphi(n)|^2}{\sum_{n=1}^{N} |v(n)|^2} \right]^{1/2}, \tag{6}$$

where $\omega(n)$ is the process noise vector, $\varphi(n)$ is the input vector of the filter; $v(n)$ is the measurement noise assumed to be white with zero mean.

The nonstationarity degree α is used to regulate the update step size according to the following principles: If α decreases in a training period, which means the features are suitable for the classifier, the weights will be constantly updated until the filter converge; If α increases, the atoms parameters should be updated to start a new training period. In addition, it can be applied to regulate the learning rate for the atoms-parameters tuning.

2.4 Update the Chirplet-Atoms Parameters

Let $\beta_k = \{u_k, \xi_k, s_k, c_k\}$ indicate the parameters' set of the kth Chirplet-atoms node, the updated β_k by delta rule is defined as:

$$\beta_k(n) = \beta_k(n-1) - \alpha^2 \eta_0 \frac{\partial E}{\partial \beta_k}, \tag{7}$$

where α is the nonstationarity degree, η_0 is the initial learning rate, and E is the mean square error (MSE), which is calculated by the expected outputs d_{nm} and the actual outputs y_{nm} of the Chirplet filter:

$$E = \frac{1}{2} \sum_{n=1}^{N} \sum_{m=1}^{M} (d_{nm} - y_{nm})^2. \tag{8}$$

Accordingly, the partial derivative of the MSE with respect to the parameters u_k, ξ_k, s_k and c_k of the Chirplet-atoms is derived as follows:

$$\frac{\partial E}{\partial u_k} = -\sum_{n=1}^{N}\sum_{m=1}^{M}\sum_{l=1}^{L} x_{nl}(d_{nm}-y_{nm})w_{km}\frac{\theta_{nk}}{\varphi_{nk}}\frac{2^{1/4}}{\sqrt{s_k}}e^{-\pi\left(\frac{t_l-u_k}{s_k}\right)^2}$$

$$\times\left\{\frac{2\pi}{s_k}(t_l-u_k)\cos\left[\xi_k(t_l-u_k)+\frac{c_k}{2}(t_l-u_k)^2\right]\right.$$

$$\left. + [\xi_k+c_k(t_l-u_k)]\sin\left[\xi_k(t_l-u_k)+\frac{c_k}{2}(t_l-u_k)^2\right]\right\} \quad (9)$$

$$\frac{\partial E}{\partial \xi_k} = -\sum_{n=1}^{N}\sum_{m=1}^{M}\sum_{l=1}^{L} x_{nl}(d_{nm}-y_{nm})w_{km}\frac{\theta_{nk}}{\varphi_{nk}}\frac{2^{1/4}}{\sqrt{s_k}}e^{-\pi\left(\frac{t_l-u_k}{s_k}\right)^2}$$

$$\times (u_k-t_l)\sin\left[\xi_k(t_l-u_k)+\frac{c_k}{2}(t_l-u_k)^2\right] \quad (10)$$

$$\frac{\partial E}{\partial s_k} = -\sum_{n=1}^{N}\sum_{m=1}^{M}\sum_{l=1}^{L} x_{nl}(d_{nm}-y_{nm})w_{km}\frac{\theta_{nk}}{\varphi_{nk}}\frac{2^{1/4}}{\sqrt{s_k}}e^{-\pi\left(\frac{t_l-u_k}{s_k}\right)^2}$$

$$\times\left[-\frac{1}{2s_k}+\frac{2\pi}{s_k^3}(t_l-u_k)^2\right]\cos\left[\xi_k(t_l-u_k)+\frac{c_k}{2}(t_l-u_k)^2\right] \quad (11)$$

$$\frac{\partial E}{\partial c_k} = -\sum_{n=1}^{N}\sum_{m=1}^{M}\sum_{l=1}^{L} x_{nl}(d_{nm}-y_{nm})w_{km}\frac{\theta_{nk}}{\varphi_{nk}}\frac{2^{1/4}}{\sqrt{s_k}}e^{-\pi\left(\frac{t_l-u_k}{s_k}\right)^2}$$

$$\times\frac{-(t_l-u_k)^2}{2}\sin\left[\xi_k(t_l-u_k)+\frac{c_k}{2}(t_l-u_k)^2\right] \quad (12)$$

where

$$\theta_{nk} = \sum_{l=1}^{L} x_{nl}\frac{2^{1/4}}{\sqrt{s_k}}e^{-\pi\left(\frac{t_l-u_k}{s_k}\right)^2}\cos\left[\xi_k(t_l-u_k)+\frac{c_k}{2}(t_l-u_k)^2\right]. \quad (13)$$

2.5   Update the Tap-Weights

Let $\mathbf{w}_m = [w_{1m}, w_{2m}, \ldots, w_{Km}]^{\mathrm{T}}, (m = 1, 2, \ldots, M)$ denote the mth weight-vector, initialize the weight-vector $\mathbf{w}_m(0) = [1 \ldots 1]^{\mathrm{T}}$ and the inverse correlation matrix vector $\mathbf{B}_m(0) = (\delta\mathbf{I}_{K\times K})^{-1}$, where δ is the regularization factor. According to RLS algorithm, for each instant of time $n = 1, 2, \ldots, N$, the weight-vector can be updated by [17]:

$$\mathbf{k}_m(n) = \frac{\mathbf{B}_m(n-1)\boldsymbol{\varphi}(n)}{\lambda+\boldsymbol{\varphi}^{\mathrm{T}}(n)\mathbf{B}_m(n-1)\boldsymbol{\varphi}(n)}, \quad (14)$$

$$e_m(n) = d_m(n) - \hat{\mathbf{w}}_m^{\mathrm{T}}(n-1)\boldsymbol{\varphi}(n), \quad (15)$$

$$\hat{\mathbf{w}}_\mathrm{m}(n) = \hat{\mathbf{w}}_\mathrm{m}(n-1) + \mathbf{k}_m(n)e_m(n), \tag{16}$$

$$\mathbf{B}_m(n) = \lambda^{-1}\left[\mathbf{B}_m(n-1) - \mathbf{k}_m(n)\boldsymbol{\varphi}^\mathrm{T}(n)\mathbf{B}_m(n-1)\right], \tag{17}$$

where $\boldsymbol{\varphi}(n)$ is the feature vector, λ denotes the forgetting factor.

2.6   Training Algorithm and Testing Algorithm

The Chirplet filters of P classes can be trained in a parallel way. For the pth class of aircrafts, the training algorithm is shown in Algorithm 1. After the Chirplet-atoms parameters and the filter weights are obtained, the testing samples will be input into the Chirplet filter to identify the aircrafts (see Algorithm 2).

Algorithm 1. Training algorithm of Chirplet filter

Initialization: The nodes' number K and M, initial learning rate η_0, regularization factor δ, forgetting factor λ; Initialize Chirplet-atoms parameters set β_k and weights vector $\mathbf{w}_m$ randomly.

Repeat:

1: Calculate the features φ_{nk} by (4);

2: **Do**

3:      Calculate the outputs y_{nm} by (5);

4:      Calculate the mean square error E by (8);

5:      **If** $E <$ the chosen threshold **then**

6:          Return β_k and $\mathbf{w}_m$;

7:          **Exit;**

8:      **Else**

9:          **Do**

10:             Update $\mathbf{w}_m$ by (14) - (17);

11:             Calculate nonstationarity degree α by (6);

12:         **While** α is first time obtained in the period

13:     **End if;**

14: **While** α decreases

15: Update β_k by (7) - (13);

Until the number of training epochs exceeds the set value

Return β_k and $\mathbf{w}_m$.

Algorithm 2. Testing algorithm of Chirplet filter

Input: HRR range profiles testing samples $\mathbf{x}_n$, Chirplet-atoms parameters

sets $\beta_k^{(p)}$, weights vectors $\mathbf{w}_m^{(p)}$.

For p=1 to P **do**

    1: Calculate the features φ_{nk} by (4);

    2: Calculate the actual outputs y_{nm} by (5);

    3: Compare y_{nm} with the set value to judge whether the samples be-
       long to the pth class of the targets.

End for

Return classification results.

---

3 Simulations

The range profiles are collected from the scaled aircraft (YF22, F117, J6 and B2) models in a microwave anechoic chamber, which are obtained from the aspect range of $0° \sim 180°$ with an equal interval of $0.6°$. The training set includes the 1st, 7th, . . . , 295th HRR range profile of each aircraft, and the others are used for testing. Thus, the training set contains 200 samples and the testing set contains 1000 range profiles. The signal-noise ratio (SNR) of the simulation data is 30 dB. In order to reduce the influence of initialization in weights training, the initial value of the correlation matrix should be relatively small, we set the regularization factor $\delta = 0.01$. The other parameters of the Chirplet filter acquired through numerous simulations are shown in Table 1. For convenience, the number of Chirplet filter's output layer M is set to 1.

Table 1. The parameters of the Chirplet filter

Parameters	Targets			
	YF22	J6	F117	B2
The number of atom nodes	45	40	50	45
Forgetting factor	0.999	0.99	0.9999	0.9999
Initial learning rate	0.001	0.001	0.0001	0.0001

In order to evaluate the recognition performance of the Chirplet filter, we investigate the HRR range profiles, the magnitude of the FFT, the Chirplet transform coefficients, and the Gabor filter for comparison. The recognition results are shown in Table 2.

According to the recognition results, since the time-domain range profiles and the frequency-domain feature FFT describe the targets in a single domain, i.e., the

Table 2. The recognition rate (%) with SNR = 30 dB

Methods	Targets			
	YF22	J6	F117	B2
Range profiles	72.9	79.5	72.4	79.5
FFT	73.8	84.2	73.5	96.1
Chirplet transform	82.4	91.8	79.4	95.4
Gabor filter	83.0	91.3	82.7	95.6
Chirplet filter	84.3	92.2	84.7	95.3

representations are insufficient for classification, so their recognition rates are less than the joint time-frequency features Chirplet transform. Due to the optimal parameter sets, the Gabor filter and the Chirplet filter obtain better recognition performance. The Chirplet filter achieves the best recognition results based on the two advantages: Chirplet transform with more complete description of the aircrafts, and the collaborative optimization of all parameters.

To assess the noise immunity, we examine the average recognition rates of four aircrafts in different SNR conditions as shown in Fig. 2. In addition, Table 3 shows the detailed recognition rates of the Chirplet filter.

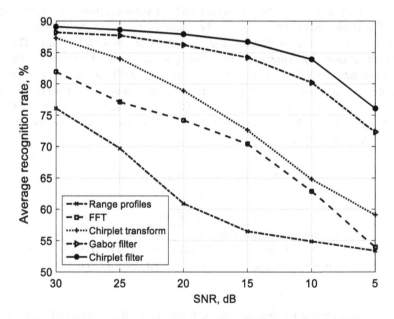

Fig. 2. The average recognition rate with the SNR from 30 dB to 5 dB

As Fig. 2 showed, with the noise increasing, the performance of the HRR range profiles, FFT and the Chirplet transform methods is rapidly getting worse. In the contrast, the performance of the adaptive filter methods still remains at a relevant high

Table 3. The recognition rate (%) of the Chirplet filter

SNR	Targets			
	YF22	J6	F117	B2
30 dB	84.3	92.2	84.7	95.3
25 dB	83.9	92.2	84.1	94.3
20 dB	83.1	91.6	83.7	93.0
15 dB	82.0	90.9	82.7	91.3
10 dB	80.4	88.1	79.3	87.7
5 dB	75.7	81.9	73.0	73.7

level due to the property of joint feature extraction and target identification. Even if the SNR is 5 dB, the average recognition rates of the Chirplet filter are over 75%. The Chirplet filter with good noisy immunity suggests that the four-dimensional Chirplet transform contains more useful information for target identification. Meanwhile the feature parameter sets and the weights of the classifier are optimized concurrently is effective to improve the recognition performance. Moreover, the initialization and the tuning of all parameters can be further investigated to obtain higher recognition rates.

4   Conclusion

This paper proposes an adaptive Chirplet filter approach for aircrafts recognition. Based on the feed-forward neural network structure, the Chirplet filter combines Chirplet-atoms transform with the adaptive filter to extract features and identify the aircrafts adaptively and cooperatively. The Chirplet atoms contain more effective characteristics information, and the training process is regulated by the nonstationarity degree to ensure the performance of the classifier. The simulation results show that the Chirplet filter has better recognition performance even in the lower SNR conditions. In the future works, it needs to collect more types of targets data and conduct the experiments with the HRRPs measured from actual targets in the outfield.

Acknowledgements. This work was supported by National Natural Science Foundation of China (61401252).

References

1. Guo, Z.H., Li, D., Zhang, B.Y.: Survey of radar target recognition using one-dimensional high range resolution profiles (in Chinese). Syst. Eng. Electron. **35**(1), 53–60 (2013)
2. Li, H.J., Yang, S.H.: Using range profiles as feature vectors to identify aerospace objects. IEEE Trans. Antennas Propag. **41**(3), 261–268 (1993)
3. Zhang, X.D.: Modern Signal Processing (in Chinese), pp. 349–443. Tsinghua University Press, Beijing (2002)
4. Zhu, F., Zhang, X.D., Hu, Y.: Gabor filter approach to joint feature extraction and target recognition. IEEE Trans. Aerosp. Electron. Syst. **45**(1), 17–30 (2009)

5. Guo, Z.H., Li, S.H.: One-dimensional frequency-domain features for aircraft recognition from radar range profiles. IEEE Trans. Aerosp. Electron. Syst. **46**(4), 1880–1892 (2010)
6. Shaw, A.K., Paul, A.S., Williams, R.: Eigen-template-based HRR-ATR with multi-look and time-recursion. IEEE Trans. Aerosp. Electron. Syst. **49**(4), 2369–2385 (2013)
7. Duin, R.P.W., Pekalska, E.: The science of pattern recognition. Achievements and perspectives. In: Duch, W., Mańdziuk, J. (eds.) Challenges for Computational Intelligence. SCI, vol. 63, pp. 221–259. Springer, Heidelberg (2007). https://doi.org/10.1007/978-3-540-71984-7_10
8. Liu, H., Feng, B., Chen, B., Du, L.: Radar high-resolution range profiles target recognition based on stable dictionary learning. IET Radar Sonar Navig. **10**(2), 228–237 (2016)
9. Liu, W.B., Wang, Z.D., Liu, X.H., et al.: A survey of deep neural network architectures and their applications. Neurocomputing **234**, 11–26 (2017)
10. Yin, H.Y., Guo, Z.H.: Radar HRRP target recognition with one-dimensional CNN (in Chinese). Telecommun. Eng. **58**(10), 1121–1126 (2018)
11. Mann, S., Haykin, S.: The Chirplet transform: a generalization of Gabor's logon transform. In: Vision Interface 1991, pp. 205–212. University of Calgary, Alberta (1991)
12. Mann, S., Haykin, S.: Adaptive "Chirplet" transform: an adaptive generalization of the wavelet transform. Opt. Eng. **31**(6), 1243–1256 (1992)
13. Li, Y.F., Guo, Z.H.: Chirplet-atoms network approach to high-resolution range profiles automatic target recognition. In: 11th International Congress on Image and Signal Processing, Bio Medical Engineering and Informatics (CISP-BMEI), pp. 1–5. IEEE Press, Beijing (2018)
14. Mann, S., Haykin, S.: The Chirplet transform: physical considerations. IEEE Trans. Sig. Process. **43**(11), 2745–2761 (1995)
15. Zhu, M., Jin, W.D., Pu, Y.W., Hu, L.Z.: Feature extraction of radar emitter signals based on Gaussian Chirplet atoms (in Chinese). J. Infrared Millimeter Waves **26**(4), 302–306 (2007)
16. Macchi, O.: Optimization of adaptive identification for time-varying filters. IEEE Trans. Autom. Control **31**(3), 283–287 (1986)
17. Haykin, S.: Adaptive Filter Theory, pp. 324–405. Prentice-Hall, Upper Saddle River (2013)

An Unsupervised Spiking Deep Neural Network for Object Recognition

Zeyang Song, Xi Wu, Mengwen Yuan, and Huajin Tang[✉]

Neuromorphic Computing Research Center, College of Computer Science,
Sichuan University, Chengdu 610065, China
huajin.tang@gmail.com

Abstract. In this paper, we propose an unsupervised HMAX-based Spiking Deep Neural Network (HMAX-SDNN) for object recognition. HMAX is a biologically plausible model based on the hierarchical activity of object recognition in visual cortex. In HMAX-SDNN, input layer with HMAX structure is followed by a stacked convolution-pooling structure, in which convolutional layers are hierarchically trained with STDP. After that, a linear SVM is used for classification. Then, we demonstrate that the firing threshold has positive correlation with receptive fields size in convolutional layers, and optimize HMAX-SDNN with this conclusion. With the optimized structure, we validate HMAX-SDNN on Caltech dataset, and HMAX-SDNN outperforms other SNNs by reaching 99.2% recognition accuracy. Furthermore, the experiments show that HMAX-SDNN is robust to different kinds of objects.

Keywords: HMAX · Spiking Deep Neural Network · STDP · Deep learning · Object recognition

1 Introduction

Recently, development and achievement of deep neural networks have brought much attention in artificial intelligence. Deep Convolutional Neural Networks (DCNNs) [1,2] perform well in wide range of applications such as image recognition. Though DCNNs perform high recognition accuracy, there are still some problems: (1) in the training process, large amount of training data is needed, limiting the applications with small-scale training data; (2) the bulk computation cost in training and testing process limits the applications in the hardware.

Spiking Neural Networks (SNNs) composed of biologically inspired neuron models can save computation cost and energy consumption through the event-driven computation. A straightforward idea is to convert CNNs to SNNs by replacing their computing nodes into spiking neurons [3–5]. In order to achieve the learning ability based on a biological mechanism, Spike Timing Dependent Plasticity (STDP) is provided to adjust connections between cortex neurons. STDP has been converted to learning rule in unsupervised forward-connected

© Springer Nature Switzerland AG 2019
H. Lu et al. (Eds.): ISNN 2019, LNCS 11555, pp. 361–370, 2019.
https://doi.org/10.1007/978-3-030-22808-8_36

SNNs [6,7]. Shallow spiking neural networks have scalability problem in processing high dimensional images [8]. To solve this problem, the hierarchical structure of DCNN has been applied in SNNs to construct deeper SNNs [9–13]. In [9], a STDP-based spiking deep neural network (SDNN) is proposed, which is comprised of a temporal-coding layer followed by a cascade of consecutive convolutional (feature extractor) and pooling layers.

In this paper, we proposed an HMAX-SDNN model, comprising HMAX layer, stacked convolutional and pooling layers for feature extraction and dimensionality reduction. HMAX model [14] is a biologically inspired model based on ventral stream of visual cortex. It enables HMAX-SDNN to mimic image reception process of human brain in image recognition tasks. Different from DoG filter in SDNN model [9], HMAX is more likely to detect features within more complex objects like face or motorbike in Caltech dataset, and keep more feature information such as position, scale and viewpoint. Furthermore, HMAX can extract precise direction features which contributes to robust recognition, whereas DoG filters cannot guarantee stable feature extraction.

2   Integrate-and-Fire Neuron

In a biologically inspired network, neurons can be excitatory or inhibitory and deliver spikes throughout the whole SNN. There are several types of spiking neuron model, such as integrate-and-fire (IF) model [15], the Hodgkin-Huxley (HH) model [16], and the Izhikevich (IM) model [17]. In HMAX-SDNN, we use the IF neuron model. As shown in Fig. 1(a) and (b), the membrane potential of the postsynaptic neuron increases as it receives weighted presynaptic current I_{syn}^i from the i_{th} neuron. After the neuron emits a spike, its membrane potential will gradually return to rest state. (Implementation of the neuron model is demonstrated in Sect. 3.2)

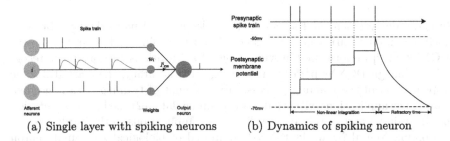

(a) Single layer with spiking neurons   (b) Dynamics of spiking neuron

Fig. 1. (a) Output neuron connects to afferent neurons with weighted synapses, receiving spike trains from them. (b) The membrane potential of postsynaptic neuron dynamically increases when received presynaptic spikes. The output neuron will fire a spike when the membrane potential reaches the firing threshold −50 mv, then the membrane potential will return back to the rest level.

3 HMAX-SDNN Model

As shown in Fig. 2, HMAX-SDNN model is composed of 1 HMAX layer, two convolutional layers each followed by a max pooling layer, and a global pooling layer. HMAX layer contains HMAX structure and temporal coding. In the convolutional layer, we use "same" mode of convolution, indicating the output does not differ in size as the input. For instance, $Pooling_1$ and $Conv_2$ are same in size (27×42), but differ in depth (8 and 20). Deeper layers with more filters detect more complex features by combining simpler features extracted from previous layers. Pooling layers compress redundant information from previous layers, and provide visual invariance as visual cortex.

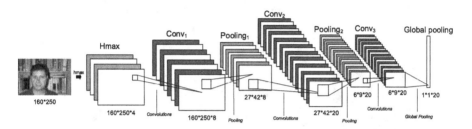

Fig. 2. Proposed HMAX-SDNN comprises three convolutional layers, each of which followed by a pooling layer. The ith convolutional layer and the ith pooling layer are separately denoted by $Conv_i$ and $Pool_i$. HMAX is applied on the first layer, the HMAX layer, to detect edges in four different orientations and encode input image information into spikes. The sizes of neuron maps are marked below each layer.

3.1 HMAX Layer

The role of HMAX layer is to extract edge features in images with HMAX structure and encode input signal into spike output. HMAX is a biologically plausible visual recognition system that closely follows organization of visual cortex and has been widely used in visual recognition, containing a series of position-tolerant and scale-tolerant feature detectors. We applied two S_1 and C_1 parts in HMAX layer. Within S_1, input images are filtered by Gabor filters in different scales and four orientations $-\frac{\pi}{4}$, 0, $\frac{\pi}{4}$ and $\frac{\pi}{2}$. We apply these filters in pyramid shape spanning in sizes from 7×7 to 37×37 pixels. Then we get 64 output results in S_1 (16 different sizes $\times$ 4 orientations in each band). The reason that we chose S_1 in HMAX layer is that in previous experiments we found that the weight matrices in the first convolutional layer converged to edge detectors in four orientations, $-\frac{\pi}{4}$, 0, $\frac{\pi}{4}$ and $\frac{\pi}{2}$. Hence, as the function of the first convolutional layer is the same as that of S_1, we use S_1 as the first part of HMAX layer. Figure 3 demonstrates the HMAX processing for two samples from Caltech and MNIST. Our experimental results show that S_1 has better capability in edge detection task and there is no trainable parameter.

As for C_1, it is used for translation invariance. In C_1, MAX pooling operations are performed over two adjacent S1 bands, thus 4 scale groups are formed. Figure 3(c) shows the first map out of 4 maps within a C_1 band. Thus, output from HMAX layer contains 4 maps indicating neurons' responses to 4 oriented edge features in different scales. The responses are temporally encoded and the firing time is inversely proportional to the activation extent.

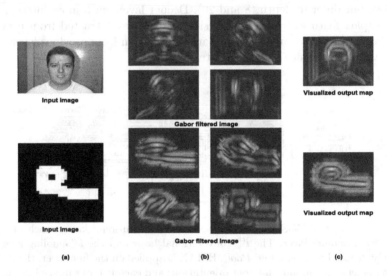

Input image   Gabor filtered image   Visualized output map

Input image   Gabor filtered image   Visualized output map

(a)   (b)   (c)

Fig. 3. (a) Two samples from Caltech and MNIST. (b) Filtered images in 4 different orientations within a C_1 scale band. (c) The first map of 4 maps varied in scale.

3.2   Convolutional Layer

Convolutional layers consist of several neuron maps. Each neuron map detects its preferred feature in different locations within images. At each time step, neurons in the convolutional layer perform convolution operations on the previous layer by weight kernels. When a neuron detects the appearance of its preferred feature, the neuron is more likely to fire.

Neuron maps consist of many IF neurons. As receptive field sliding over input activities (1 means firing a spike and 0 means no spike), the membrane potential of a neuron in the neuron map will update as follows:

$$\Delta P_i(t) = \sum_j W_{ij} S_j(t - 1), \tag{1}$$

where $\Delta P_i(t)$ is the change of the membrane potential of the ith neuron at time t, and W_{ij} is the synaptic weight between the ith neuron in convolutional layer and the jth neuron in previous layer. $S_j(t-1)$ refers to the activity of the jth neuron at time $t-1$, and $S_j(t-1) = 1$ when the jth neuron fires at time $t-1$, otherwise $S_j(t-1) = 0$. If $P_i(t)$ reaches the threshold Thr_i, then the ith neuron will fire a spike, and its membrane potential $P_i(t)$ will reset to 0 afterwards.

$$S_i(t) = \begin{cases} 0 & , if P_i(t-1) + \Delta P_i(t) < Thr_i, \\ 1 & , if P_i(t-1) + \Delta P_i(t) \geq Thr_i, \end{cases} \tag{2}$$

Within neuron maps, lateral inhibition is introduced in convolutional layers. Lateral inhibition is a mechanism to prevent action potential from spreading to neighbor neurons. When a neuron has fired, it is not allowed to fire again, and lateral inhibition will inhibit neurons nearby and neurons in same position but other maps to fire until the appearance of next image. Because neighbor neurons have highly overlapped receptive field, thus they may receive intensive stimuli and fire once the appearance of a same feature is detected. Once a feature is detected from one location, other features do not likely exist in this place. Thus, we use lateral inhibition to encode the output from convolutional layers in a sparse but highly informative way. Also, in each neuron map, all neurons detect the same visual feature because they share the same weight matrix. The sharing of weights can largely reduce trainable parameters in convolutional layers.

3.3 STDP Learning in the Convolutional Layer

As mentioned above, convolutional layers consisting of IF neurons are trained by STDP method that is derived from Hebbian learning. Hebbian theory claims that with repeated stimuli from presynaptic neurons to postsynaptic neuron, synaptic efficacy between them will be enhanced. Similar to Hebbian learning, simplified STDP learning rule used in HMAX-SDNN model can be described as:

$$\Delta w_{ij} - \begin{cases} a^+ w_{ij}(1-w_{ij}) & , if \ t_j \leq t_i, \\ a^- w_{ij}(1-w_{ij}) & , if \ t_j > t_i, \end{cases} \tag{3}$$

where a^+ and a^- are two parameters separately in excitation and inhibition, and t_i refers to the firing time of ith neuron. This equation, in which j^{th} and i^{th} neurons are respectively pre- and post-synaptic neuron, implies that the sign of weight change has no direct relation with exact firing times of jth and i^{th} neurons but with chronological order of their firing times. If the presynaptic neuron fires a spike just before the postsynaptic neuron, the presynaptic neuron contributes to the firing of the postsynaptic neuron. In this case, the connection between pre- and postsynaptic neuron will be strengthened. In addition, polynomial $w_{ij}(1-w_{ij})$ can ensure w_{ij} in range $[0,1]$ when updating weights.

a^+ and a^- are set to 0.003 and 0.004 initially and will be updated in STDP learning. If learning parameters are large at first, it will cause declination in learning memory, i.e. the network can memorize the last shown image but forget that of previous images. Whereas tiny value of the learning parameter will slower the learning speed.

As weights updated with iterations of STDP learning, they will gradually converge to be selective to one certain visual feature. When they reach the convergence point, it is needless to train the neuron maps with STDP continuously.

To this end, we use convergence index C_l to quantify the learning convergence at l^{th} layer:

$$C_l = \sum_f \sum_i \frac{w_{f,i}(1 - w_{f,i})}{n_w}, \tag{4}$$

where $w_{f,i}$ is the ith weight of the fth weights matrix, and n_w is the total number of weights. In our experiments, weights are randomly chosen from a normal distribution with $\mu = 0.8$ and $\sigma = 0.05$, and we set them in range $[0, 1]$. Weights will converge from the random numbers in $[0, 1]$ to zero or one, as C_l tends to be 0.

4   Experiment Results

The simulation experiments are in two phases, training phase and testing phase. In training phase, namely, HMAX-SDNN is trained with object training dataset. In the testing phase, the trained HMAX-SDNN is first used to extract features of training images and testing images without knowing labels. Then SVM classifier is trained with the training features and tested with the testing features. Testing images are not seen in the training phase.

We evaluated our proposed HMAX-SDNN on face and motorbike categories in Caltech dataset, containing 435 face images and 798 motorbike images. In our experiment, we randomly selected 200 images separately from face and motorbike into training set. And remaining images form testing set to evaluate performance of our model. This cross-validation method can measure the network's ability in generalization.

Here, the HMAX-SDNN structure we used is similar to the sample network mentioned in Sect. 3, with 1 HMAX layer, 3 convolutional layers, 2 pooling layers and a global pooling layer. Input images are of size 160×250 pixels. Processed by convolution and pooling layer by layer, 20 features are extracted from global pooling layer as a vector, which will be used in the classification afterward. Detailed parameters of the network are demonstrated in Table 1.

To start with, we tried a relatively shallower HMAX-SDNN (HMAX-2C-P-G-SDNN) composed of 1 HMAX layer, 2 convolutional layers, 1 pooling layer and a global pooling layer. In the experiment, recognition accuracy of HMAX-2C-P-G-SDNN fluctuated in multiple trials with cross-validation method, and its variance of accuracy was up to 0.02. The unstable performance of the HMAX-2C-P-G-SDNN can be attributed to its shallow structure. In maps of shallower convolutional layers, the preferred features are simple. More abstract features like object contour that is important in object recognition lack in HMAX-2C-P-SDNN. For the purpose of increasing accuracy and stability, we applied stacked convolution-pooling structure in HMAX-SDNN. We used an HMAX-3C-2P-SDNN, namely, comprising 1 HMAX layer, 3 convolutional layers, 2 pooling layer and a global pooling layer. HMAX-3C-2P-SDNN outperforms HMAX-2C-P-SDNN on recognition accuracy by up to 10%. The parameter in Table 1 is set by trial and error to get the lowest recognition error rate. Also, HMAX-3C-2P-SDNN performs well in stability by reducing the variance of accuracy to 0.003.

Table 1. Parameters of HMAX-3C-2P-G-SDNN.

	$Conv_1$	$Pool_1$	$Conv_2$	$Pool_2$	$Conv_3$
Number of filters	8	8	20	20	20
Size of filter	7	7	17	5	5
Convolution stride	1	/	1	/	1
Pooling stride	/	6	/	5	/
Threshold of firing	20	0	60	0	2

To optimize structure of HMAX-SDNN, we explored HMAX-SDNN with 3×3 and 7×7 filters in $Conv_1$. The result is shown in Fig. 4(a). With larger convolution filters, neurons will receive more spikes. Therefore, we correspondingly adjust the threshold of firing in each convolutional layer, ensuring receiver neurons can and only can fire when features appear in the receptive field. If we set a relatively low threshold and a larger filter, neurons are likely to fire even when no clear visual feature exists in receptive field. Thus, we explore different network structures that vary in size of convolution filters and thresholds. As shown in Fig. 4(a), HMAX-SDNN achieves high accuracy only when thresholds are moderate. Also, in this experiment, the 5×5 filter performs better than 3×3 and 7×7 filters by achieving 99.2% recognition accuracy (Table 2).

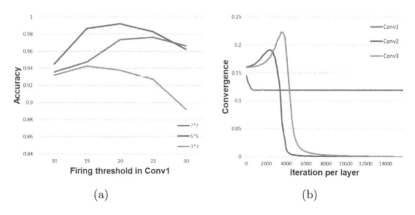

(a)   (b)

Fig. 4. (a) Recognition accuracy of HMAX-SDNN varies with filter size and firing threshold. (b) Each line shows the convergence variation of different convolutional layers in STDP learning process.

In Fig. 4(b), learning process of each layer is demonstrated by its extent of convergence. As it can be seen in Fig. 4(b), weight change in lower layer with fewer filters is much faster than that in higher layer. One direct explanation to this is that there is fewer and simpler features that need to be detected in lower layer. $Conv_2$ and $Conv_3$ have the same amount of filters but differ in convergence speed. This is mainly due to more intensive competition among neurons in larger filters. Because of lateral inhabitation, when a neuron won the competition in STDP learning, it inhibited excitation of other neurons. Neurons in higher layers, with larger filters, are likely to prevent more neurons from firing because they will compete with more neurons in these layers. In this case, only a few neurons can update their wights in an iteration of STDP learning. Thus more iterations are needed in training process.

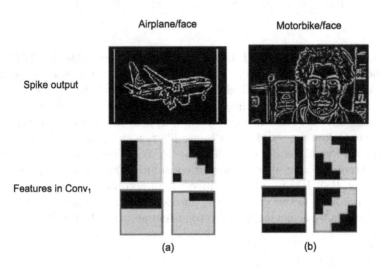

Fig. 5. The figure shows spike output and visualized features of SDNN separately extracted from airplane/face dataset and motorbike/face dataset. With those edges features in (a) extracted in $Conv_1$, complex features in deeper layers cannot be formed as that in [9].

We also tested SDNN and HMAX-SDNN with other categories in Caltech dataset. In airplane/face recognition task, accuracy of SDNN fluctuates between 92% and 96%, while HMAX-SDNN outperforms SDNN with 96% to 98% accuracy. As shown in Fig. 5, the filters of SDNN can extract edges in four orientations, $-\frac{\pi}{4}$, 0, $\frac{\pi}{4}$ and $\frac{\pi}{2}$, in motorbike/face recognition task (shown in Fig. 5(b)). While they did not perform stably in airplane/face recognition task. $Conv_1$ in SDNN preferred to learn horizontal features ($Conv_1$ features in Fig. 5(a)), which appear more frequently in airplane images. The same problem occurred in other experiments with motorbike/airplane, motorbike/watch, face/watch. As is shown in experiments above, the edge detectors in $Conv_1$ are not stable enough in all circumstances. The features in deeper convolutional layers combined from simple features in previous layers, are not shaped perfectly, hence reducing recognition accuracy. While in HMAX-SDNN, its hand-crafted Gabor filters enable HMAX-SDNN to extract features in four orientations stably, ensuring its robustness to different shape of objects.

Table 2. Recognition accuracy of SNNs on Caltech face/motor dataset

Architecture	Neural coding	Learning type	Learning rule	Accuracy
Two layer network [8]	Spike-based	Supervised	Tempotron rule	95
SDNN [9]	Spike-based	Unsupervised	STDP	99.1
SpiCNN [18]	Spike-based	Unsupervised	STDP	91.1
Proposed HMAX-SDNN	**Spike-based**	**Unsupervised**	**STDP**	**99.2**

5 Conclusion

In this paper, we proposed an HMAX-SDNN model that consists of HMAX layer, stacked convolutional and pooling layers. HMAX layer is used to extract features of input images. In convolutional layers, weight matrices are hierarchically trained with STDP learning rule. Comparing with DoG filter in the existing SDNN model, HMAX layer can keep more feature information. Furthermore, we also studied the relationship between the firing threshold and receptive field size to provide optimal values of threshold for each layer. We validated our proposed HMAX-SDNN on Caltech face/motorbike dataset. Experimental results show that the proposed HMAX-SDNN model can achieve 99.2% recognition accuracy. Comparing to SDNN, HMAX structure enhances robustness to various objects of IIMAX-SDNN by presetting edge detectors before the first convolutional layer.

Acknowledgements. This work was supported by the National Key R&D Program of China under Grant 2017YFB1300201 and the National Natural Science Foundation of China under Grant 61673283.

References

1. Krizhevsky, A., Sutskever, I., Hinton, G.E.: ImageNet classification with deep convolutional neural networks. In: Advances in Neural Information Processing Systems, pp. 1097–1105 (2012)
2. Zeiler, M.D., Fergus, R.: Visualizing and understanding convolutional networks. In: Fleet, D., Pajdla, T., Schiele, B., Tuytelaars, T. (eds.) ECCV 2014. LNCS, vol. 8689, pp. 818–833. Springer, Cham (2014). https://doi.org/10.1007/978-3-319-10590-1_53
3. Cao, Y., Chen, Y., Khosla, D.: Spiking deep convolutional neural networks for energy-efficient object recognition. Int. J. Comput. Vis. **113**(1), 54–66 (2015)
4. Diehl, P.U., Zarrella, G., Cassidy, A., Pedroni, B.U., Neftci, E.: Conversion of artfiicial recurrent neural networks to spiking neural networks for low-power neuromorphic hardware. In: IEEE International Conference on Rebooting Computing (2016)
5. Lee, J.H., Delbruck, T., Pfeiffer, M.: Training deep spiking neural networks using backpropagation. Front. Neurosci. **10**, 508 (2016)
6. Bliss, T.V., Collingridge, G.L.: A synaptic model of memory: long-term potentiation in the hippocampus. Nature **361**, 31–39 (1993)
7. Shiyong, H., et al.: Associative Hebbian synaptic plasticity in primate visual cortex. J. Neurosci. Off. J. Soc. Neurosc. **34**(22), 7575–7579 (2014)

8. Diehl, P.U., Cook, M.: Unsupervised learning of digit recognition using spike-timing-dependent plasticity. Front. Comput. Neurosci. **9**, 99 (2015)

9. Kheradpisheh, S.R., Ganjtabesh, M., Thorpe, S.J., Masquelier, T.: STDP-based spiking deep convolutional neural networks for object recognition. Neural Netw. **99**, 56–67 (2018)

10. Xu, X., Jin, X., Yan, R., Fang, Q., Lu, W.: Visual pattern recognition using enhanced visual features and PSD-based learning rule. IEEE Trans. Cogn. Dev. Syst. **10**(2), 205–212 (2018)

11. Yu, Q., Tang, H., Tan, K.C., Li, H.: Rapid feedforward computation by temporal encoding and learning with spiking neurons. IEEE Trans. Neural Netw. Learn. Syst. **24**(10), 1539–1552 (2013)

12. Xu, Q., Qi, Y., Yu, H., Shen, J., Tang, H., Pan, G.: CSNN: an augmented spiking based framework with perceptron-inception. In: IJCAI, pp. 1646–1652 (2018)

13. Xi, W., Yixuan, W., Huajin, T., Rui, Y.: A structure-time parallel implementation of spike-based deep learning. Neural Netw. **113**, 72–78 (2019)

14. Riesenhuber, M., Poggio, T.: Hierarchical models of object recognition in cortex. Nat. Neurosci. **2**, 1019–1025 (1999)

15. Gerstner, W.: Spiking neuron models: single neurons, populations, plasticity. Kybernetes **4**(7/8), 277–280 (2002)

16. Hodgkin, A.L., Huxley, A.F.: A quantitative description of membrane current and its application to conduction and excitation in nerve. Bull. Math. Biol. **52**(1–2), 25–71 (1989)

17. Izhikevich, E.M.: Simple model of spiking neurons. IEEE Trans. Neural Netw. **14**(6), 1569–72 (2003)

18. Lee, C., Srinivasan, G., Panda, P., Roy, K.: Deep spiking convolutional neural network trained with unsupervised spike timing dependent plasticity. IEEE Trans. Cogn. Dev. Syst. (2018). https://doi.org/10.1109/TCDS.2018.2833071

Procedural Synthesis of Remote Sensing Images for Robust Change Detection with Neural Networks

Maria Kolos[2], Anton Marin[2], Alexey Artemov[1], and Evgeny Burnaev[1(✉)]

[1] ADASE, Skolkovo Institute of Science and Technology, Moscow, Russia
{a.artemov,e.burnaev}@skoltech.ru
[2] Aeronet Groups, Skolkovo Institute of Science and Technology, Moscow, Russia
maria.kolos@skoltech.ru, anton.marin@skolkovotech.ru

Abstract. Data-driven methods such as convolutional neural networks (CNNs) are known to deliver state-of-the-art performance on image recognition tasks when the training data are abundant. However, in some instances, such as change detection in remote sensing images, annotated data cannot be obtained in sufficient quantities. In this work, we propose a simple and efficient method for creating realistic targeted synthetic datasets in the remote sensing domain, leveraging the opportunities offered by game development engines. We provide a description of the pipeline for procedural geometry generation and rendering as well as an evaluation of the efficiency of produced datasets in a change detection scenario. Our evaluations demonstrate that our pipeline helps to improve the performance and convergence of deep learning models when the amount of real-world data is severely limited.

Keywords: Remote sensing · Deep learning · Synthetic imagery

1 Introduction

Remote sensing data is utilized in a broad range of industrial applications including emergency mapping, deforestation, and wildfire monitoring, detection of illegal construction and urban growth tracking. Processing large volumes of the remote sensing imagery along with handling its high variability (*e.g.*, diverse weather/lighting conditions, imaging equipment) provides a strong motivation for developing automated and robust approaches to reduce labor costs.

Recently, data-driven approaches such as deep convolutional neural networks (CNNs) have seen impressive progress on a number of vision tasks, including semantic segmentation, object detection, and change detection [7,17,21,22,34,44,45]. Such methods offer promising tools for remote sensing applications as they can achieve high performance by leveraging the diversity

The work was supported by The Ministry of Education and Science of Russian Federation, grant No. 14.615.21.0004, grant code: RFMEFI61518X0004.

H. Lu et al. (Eds.): ISNN 2019, LNCS 11555, pp. 371–387, 2019.
https://doi.org/10.1007/978-3-030-22808-8_37

of the available imagery [11,19,41]. However, in order to successfully operate, most data-driven methods require large amounts of high-quality annotated
data [24,28,29,41]. Obtaining such data in the context of remote sensing poses
a significant challenge, as (1) aerial imagery data are expensive, (2) collection of
raw data with satisfactory coverage and diversity is laborious, costly and error-
prone, as is (3) manual image annotation; (4) moreover, in some instances such as
change detection, the cost of collecting a representative number of rarely occurring cases can be prohibitively high. Unsurprisingly, despite public real-world
annotated remote sensing datasets exist [8,19,27,38,48], these challenges have
kept them limited in size, compared to general-purpose vision datasets such as
the ImageNet [23].

The alternatives considered to avoid dataset collection issues suggest producing synthetic annotated images with the aid of game development software such
as *Unity* [4], *Unreal Engine 4* [5], and *CRYENGINE* [1]. This approach has been
demonstrated to improve the performance of computer vision algorithms in some
instances [20,24,28,46,49,50,54]. Its attractive benefits include (1) flexibility in
scene composition, addressing class imbalance issue, (2) pixel-level precise automated annotation, and (3) the possibility to apply transfer learning techniques
for subsequent "fine-tuning" on real data. However, little work has been done in
the direction of using game engines to produce synthetic datasets in the remote
sensing domain. Executing procedural changes on large-scale urban scenes is
computationally demanding and requires smart optimization of rendering or the
object-level reduction (number of polygons, textures quality). Levels of realism
rely heavily on the amount of labor on scene design and optimization. Thus,
research on the procedural construction of synthetic datasets would contribute
to the wider adoption of data-driven methods in the remote sensing domain.

In this work, we focus on the task of change detection, however, it is straightforward to adapt our method to other tasks such as semantic segmentation. We leverage game development tools to implement a semi-automated pipeline for procedural generation of realistic synthetic data. Our approach uses publicly available cartographic data and produces realistic 3D scenes of real territory (*e.g.*, relief, buildings). These scenes are rendered using *Unity* engine to produce high-resolution
synthetic RGB images. Taking advantage of real cartographic data and emula-
tion of image acquisition conditions, we create a large and diverse dataset with a
low *simulated-to-real* shift, which allows us to efficiently apply deep learning methods. We validate our data generation pipeline on the change detection task using
a state-of-the-art deep CNN. We observe consistent improvements in performance
and convergence of our models on this task with our synthetic data, compared to
when using (scarce) real-world data only.

In summary, our contributions in this work are:

- We describe a semi-automatic pipeline for procedural generation of realistic
  synthetic images for change detection in the domain of remote sensing.
- We demonstrate the benefits of large volumes of targeted synthetic images for
  generalization ability of CNN-based change detection models using extensive
  experiments and a real-world evaluation dataset.

The rest of this paper is organized as follows. In Sect. 2, we review prior work on change detection in the remote sensing domain, including existing image datasets, data generation tools, computational models, and transfer learning techniques. Section 3 presents our data generation pipeline. Section 4 poses three experiments investigating the possible benefits of our approach and presents their results. We conclude with a discussion of our results in Sect. 5.

2   Related Work

2.1   Computational Models For change Detection

Change detection in multi-temporal remote sensing images has attracted considerable interest in the research community, where approaches have been proposed involving anomaly detection on time series and spectral indices [18], Markov Random Fields and global optimization on graphs [32,58,65], object-based segmentation followed by changes classification [36,40,60], and Multivariate Alteration Detection [39,61] (cf. [57] for a broader review). These approaches generally work with low-resolution imagery (*e.g.*, 250–500 m/px) and require manual tuning of dozens of hyperparameters to handle variations in data such as sensor model, seasonal variations, image resolution, and calibration.

Recently, deep learning and CNNs have been extensively studied for classification, segmentation, and object detection in remote sensing images [13,66]. However, only a handful of CNN-based change detection approaches exist. Due to the lack of training data [26,53] use ImageNet pre-trained models to extract deep features and use super-pixel segmentation algorithms to perform change detection. We only study the influence of pre-training on ImageNet in one of the experiments; otherwise, we train our deep CNN from scratch using our synthetic dataset. [25,27] proposed a CNN-based method for binary classification of changes given a pair of two high-resolution satellite images. In contrast, we focus on predicting a dense mask of changes from the two registered images. The closest to our work are [11,62], which predict pixel-level mask of changes from the two given images; additionally, [11] uses a U-Net-like architecture as we do. Nevertheless, their models are different from ours, which is inspired by [17].

Due to the scarcity of the available data in some instances, transfer learning techniques have been extensively investigated in many image analysis tasks, including image classification [63,64], similarity ranking [67] and retrieval [10, 55]. Additionally, transfer from models pre-trained on RGB images to a more specialized domain, such as magnetic resonance images or multi-spectral satellite images, has been studied for automated medical image diagnostics [33,43,59] and remote sensing image segmentation [37]. In the context of the present work, of particular interest is the transfer learning from synthetic data to real-world data, that has proven effective for a wide range of tasks [20,24,28,46,49,50,54]. In the remote sensing domain, however, synthetic data has been only employed in the context of semantic segmentation [41]. However, their data generation method relies on pre-created scene geometry, while our system generates geometry based on the requested map data.

2.2 Image Datasets For change Detection

Real-World Datasets. Datasets for change detection are commonly structured in pairs of registered images of the same territory, made in distinct moments in time, accompanied by image masks per each of the annotated changes. With the primary application being emergency mapping, most datasets typically feature binary masks annotating damaged structures across the mapped areas [8,12,27,48]. L'Aquila 2009 earthquake dataset [8] contains data spanning $1.5 \times 1.5\,km^2$ annotated with masks of damaged buildings during the 2009 earthquake. California wildfires [48] contains $2.5 \times 2.5\,km^2$ and $5 \times 8\,km^2$ images, representing changes after a 2017 wildfire in California, annotated with masks of burnt buildings. ABCD dataset [27] is composed of patches for $66\,km^2$ of tsunami-affected areas, built to identify whether buildings have been washed away by the tsunami. Besides, OSCD dataset [19] addresses the issue of detecting changes between satellite images from different dates, containing 24 pairs of images from different locations, annotated with pixel-level masks of new buildings and roads. All these datasets are low in diversity and volume, while only providing annotations for a limited class of changes (*e.g.*, urban changes). In contrast, our pipeline can produce massive amounts of highly diverse synthetic images with flexible annotation, specified by the user. Other known datasets, such as the Landsat ETM/TM datasets [31], are of higher volume, but have an extremely low spatial resolution (on the order of $10\,m$), while featuring no annotation.

Synthetic Datasets and Visual Modelling Tools. To the best of our knowledge, AICD dataset [12] is the only published synthetic dataset on change detection in remote sensing domain. It consists of 1000 pairs of 800×600 images. It is a synthetic dataset in which the images are generated using a realistic rendering engine of a computer game, with ground truth generated automatically. The drawbacks of this dataset are low diversity in target and environmental changes and low graphics quality.

Despite tools for creating synthetic datasets are actively developed and studied in the domain of computer vision [4,5,49], research on the targeted generation of synthetic datasets for remote sensing applications is still in its infancy. Modern urban modeling packages require manual creation of assets and laborious tuning of rendering parameters (*e.g.*, lighting) [2,3,6]. This could be improved by leveraging extensive opportunities offered by game development engines, that combine off-the-shelf professional rendering presets, realistic shaders, and rich scripting engines for fine customization. In our pipeline, procedural generation of geometry and textures is followed by a rendering script, leveraging rich rendering opportunities. Other tools such as DIRSIG [30] allow simulating realistic multispectral renders of the scenes but rely on the existing geometry. Our pipeline, in contrast, enables us to create both geometry-based on real-world map data and realistic renders using a game engine.

3 Synthesis of Territory-Specific Remote Sensing Images

3.1 Data Requirements and the Design Choices of Our Pipeline

A good change detection dataset should contain application-specific target changes, such as, *e.g.*, deforestation or illegal construction, as well as high variability, which can be viewed as non-target changes. Such variability in data commonly involves appearance changes, *e.g.*, lighting and viewpoint variations, diverse directions of shadows, and random changes of scene objects. However, implementing an exhaustive list of non-target changes is too laborious. Thus, we restricted ourselves to the following *general requirements* to the synthetic data:

– *Visual scene similarity.* To reduce the omnipresent *simulated to real* shift, it is necessary that the modeled scenes have a high visual resemblance to the actual scenes. We approach the target territory modeling task by imitating the visual appearance of structures and environment.
– *Scale.* To match the largest known datasets in spatial scale, we have chosen to model scenes with large spatial size. In our dataset, scenes are generated with spatial extents of square kilometers, a spatial resolution of less than $1\,\mathrm{m}$, and image resolution of tens of Megapixels.
– *Target changes.* We have chosen to only model damaged buildings as target class as they are of general interest in applications such as emergency mapping (see, *e.g.*, [11,25,26,36,39,40,53,57,60,62]). They are also straightforward to implement in our pipeline with procedural geometry generation and rendering scripts.
– *Viewpoint variations.* Real-world multi-temporal remote sensing images for the same territory are commonly acquired using varying devices (*e.g.*, devices with different field of view) and viewpoints. The acquisition is commonly performed at angles not exceeding $25°$, and the data are then post-processed by registration and geometric correction. In our pipeline, we imitate the precession of a real satellite by randomly changing the image acquisition angle and the field of view.
– *Scene lighting changes.* Scene illumination is commonly considered to consist of a point-source illumination (produced by the Sun) and an ambient illumination due to the atmospheric scattering of the solar rays. In our work, we consider the changes in the Sun's declination angle and model both components of the illumination.
– *Shadows.* Real-world objects cast shadows that are irrelevant variations and should be ignored. We model realistic shadows again by varying the Sun's declination angle.

To meet these requirements, we have developed a two-stage pipeline consisting of geometry generation and rendering steps. The entire routine is semi-automatic and involves two widely used 3D engines. Specifically, we use *Esri CityEngine* [2] to procedurally build geometry from real-world map data and *Unity* [4] to implement the logic behind dataset requirements and leverage rendering capabilities. The reasons behind our choice of *CityEngine* as our geometry manipulation tool are its flexibility in the procedural geometric modeling

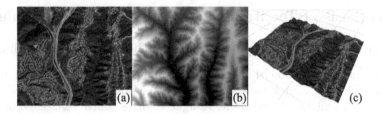

Fig. 1. Terrain reconstruction: (a) satellite imagery, (b) altitude data, (c) 3d model.

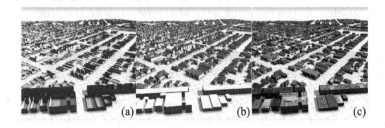

Fig. 2. Geometry generation: (a) polygon extrusion, (b) adding roofs, (c) texturing.

and built-in UV/texturing capabilities. Other tools, commonly implemented as plugins for *Unreal Engine 4* (*e.g.*, *StreetMap*[1]) and *Unity* (*e.g.*, *Mapbox Unity SDK*[2]), offered significantly less freedom. In these tools, either non-textured or textured but oversimplified shapes (*e.g.*, as simple as boxes) are the only objects available for urban geometry generation.

Unity game engine was selected to execute a generation procedure of change detection dataset. Compared to CAD software often used for the production of datasets, game engines offer advantages such as powerful lighting/shadows out of the box and scripting possibilities. Additionally, *Unity* allows implementing target changes, controlling lightning and viewpoint changes, and adjusting change rate in the dataset. Certain features in *Unity* are more suited to our needs, compared to *Unreal Engine 4*. For instance, we have found shadows in *Unity* to be more stable while rendering scenes from large distances, and the *Layers* feature[3] to add more flexibility by allowing to exclude objects from rendering or post-processing. It is natural, however, that we had to execute some initial settings of *Unity* before the generation, as the typical requirements of the remote sensing domain differ from those of 3D games. We describe these settings in Sect. 4.

3.2   Geometry Generation

To procedurally generate geometry in *CityEngine*, we obtain cartographic data (vector layers) from *OpenStreetMap* and elevation data (a geo-referenced Geo-TIFF image) from *Esri World Elevation*. Vector data contains information about

---

[1] https://github.com/ue4plugins/StreetMap.

[2] https://www.mapbox.com/unity.

[3] https://docs.unity3d.com/Manual/Layers.html.

the geometry of buildings and roads along with semantics of land cover (forests, parks, *etc.*), while elevation data is used to reconstruct terrain.

First, we reconstruct terrain using the built-in functionality in *CityEngine*, obtaining a textured 3D terrain mesh (see Fig. 1), to which we z-align flat vector objects; second, we run our geometry generation procedure implemented in the engine's rule-based scripting language; last, we apply textures to the generated meshes. Our implementation of the geometry generation involves extruding the polygon of a certain height and selecting a randomly textured roof (architectural patterns such as rooftop shapes are built into *CityEngine*), see Fig. 2. We have experimented with more complex geometry produced by operations such as polygon splitting and repeating; however, such operations (*e.g.*, splitting) increase the number of polygons significantly without adding much to the scene detail, render quality, or performance of change detection models. Focusing on our scalability and flexibility requirements, we avoid overloading our scenes with objects of redundant geometry.

We select two types of buildings to construct our scenes: small buildings with colored gable roofs and concrete industrial-looking structures with flat roofs, see Fig. 4. A set of roof textures has been selected manually, *CityEngine* built-in packs of textures for *OpenStreetMap* buildings were used for facades. We approach the emergency mapping use-case by texturing buildings footprints to imitate damaged appearance, see Fig. 3.

Fig. 3. Addressing the case of emergency mapping: scene model (a) before and (b) after the emergency situation (note that some structures are missing).

Fig. 4. The two models of buildings used in our pipeline: (a) photo and (c) our render of *residential houses*, (b) photo and (d) our render of *industrial structures*. (Color figure online)

3.3   Rendering

We construct the synthetic dataset by rendering the generated geometry using built-in functionality in *Unity*, obtaining high-definition RGB images.

To achieve a high degree of variations, we leverage rich scripting capabilities in *Unity*, that allow flexible scene manipulation via scripts written in C#. As shadow casting and lighting algorithms are built in, we only adjust their parameters, as indicated in Sect. 4.1. We implement target changes by randomly selecting object meshes and placing them onto the separate layer: the camera will not render these meshes, rendering corresponding damaged footprints instead.

Non-target variations are further added to scenes modified by target changes, by random alterations of lighting and camera parameters, resulting in m different image acquisition conditions per each target change.

Each element in the dataset was obtained after three rendering runs: first, we render the original scene; second, we apply both target and non-target changes, moving the changed objects in the separate layer; finally, only the layer with the changed objects is rendered to obtain annotations.

4   Experiments

We demonstrate the effectiveness of our pipeline in an emergency mapping scenario, where the goal is to perform a rapid localization and assessment of incurred damage with extremely limited amounts of annotated data [48]. To this end, we produce a synthetic training dataset in Sect. 4.1 and design a series of experiments to investigate the influence of training strategy on the change detection performance with different data volumes in Sects. 4.4 and 4.5.

4.1   Datasets

California Wildfires (CW). The dataset contains high-resolution satellite images depicting cases of wildfires in two areas of Ventura and Santa Rosa counties, California, USA. The annotation has been created manually [48].

Synthetic California Wildfires (SynCW) Dataset. Using our pipeline, we created a synthetic training dataset for the California wildfires case study. We collected *OpenStreetMap* and *ESRI world elevation* data from the area of interest in Ventura and Santa Rosa counties, California, USA. Two major meta-classes imported from *OpenStreetMap* data were *building* (including *apartments, garage, house, industrial, residential, retail, school,* and *warehouse*) and *highway* (road structures including *footway, residential, secondary, service,* and *tertiary*). We generate geometry using our pipeline configured according to Fig. 5. We add target changes by randomly selecting 30–50% of *buildings* geometry. To introduce non-target changes in the dataset, we select $m = 5$ different points of view, positioning camera at zenith and at four other locations determined by an inclination angle α from an axis pointing to zenith (we select α uniformly at random from $[5°, 10°]$), and orienting it to the center of the scene. To model daylight variations, we select Sun's declination angle uniformly at random from $[30°, 140°]$. In the resulting dataset, the generated scenes of 4 distinct locations have spatial extents of approximately $2 \times 2\,\mathrm{km}^2$ with spatial resolution of $0.6\,\mathrm{m}$ and image resolution of $3072 \times 3072\,\mathrm{px}$.

4.2   Our Change Detection Model and Training Procedure

When designing our change detection architecture, we take inspiration from recent progress in semantic segmentation [17,37,51]. Our architecture is a

Geometry generation parameters *(CityEngine)*	
Export format	Autodesk FBX
Terrain	Export all terrain layers
Mesh granularity	One model per start shape
Merge normals	Force separated normals
Center	Checked

Rendering parameters *(Unity)*	
Shadows	Hard and soft shadows
Shadow distance	1100
Shadow resolution	Very high
AntiAliasing	8× multi-sampling

Camera settings *(Unity)*	
Clipping planes	0, 1200
Field of view	30.0

Fig. 5. Configuration of geometry generation and rendering parameters in our evaluation.

Dataset	Large-scale	Real-world	Target domain	CD annotation
ImageNet [52]	✓	✓	–	–
CW [48]	–	✓	✓	✓
Our SynCW	✓	–	✓	✓

Fig. 6. Datasets considered in our evaluation.

Siamese U-Net [51] with residual units [35] in encoder blocks and upsampling units in decoder blocks, which can be viewed as a Siamese version of a segmentation model from [17] (see Fig. 7). While our model is composed of well-known building blocks, to the best of our knowledge, we are the first to study its Siamese version in a change detection setting. While selecting the top-performing architecture is beyond the scope of this paper, we have found our architecture to consistently outperform those examined previously [37,51] in all settings we have considered.

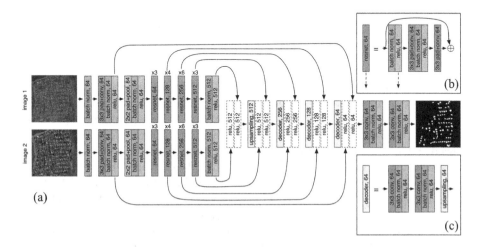

Fig. 7. Our siamese neural network for change detection: (a) overall architecture, (b) ResNet units (dashed arrows indicate skip-connections, if present), (c) decoder blocks.

To study the benefits of pre-training on a large image dataset, we have kept the encoder architecture a replica of ResNet-34 architecture [35]. In all settings, the models were trained using 352×352 patches for 20 epochs. Adam optimizer [42] was used with a batch size of 8 and initial learning rate of 10^{-4}.

4.3 Metrics

To evaluate our models, we chose two performance measures standard for segmentation tasks: *Intersection over Union* (IoU) and F1-measure, both obtained by applying the threshold 0.5 to the confidence output. Given a pair of binary

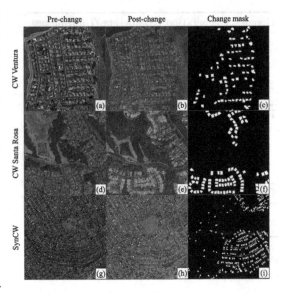

Fig. 8. Visual comparison of our training and evaluation data: (a)–(c) CW, Ventura area, (d)–(f) CW, Santa Rosa area, (g)–(i) synthetic California wildfires (SynCW).

masks, IoU can be interpreted as a pixel-wise metric that corresponds to localization accuracy between these two samples, $\text{IoU}(A, B) = \frac{|A \cap B|}{|A \cup B|} = \frac{|A \cap B|}{|A| + |B| - |A \cap B|}$. F1-measure is the harmonic mean of precision and recall values between the predicted and ground truth masks: $\text{F1} = 2 \cdot (\text{Precision}^{-1} + \text{Recall}^{-1})$.

4.4 The Evaluation Setup

When training a deep learning-based change detection model, an annotated real-world remote sensing dataset (*e.g.*, CW [48]), would be a natural choice; however, its volume does not allow training an architecture such as ours from scratch (*i.e.*, starting from randomly-initialized weights). A stronger initialization is commonly obtained with models pre-trained on ImageNet [52], a large-scale and real-world dataset. Unfortunately, ImageNet contains images from a completely different domain; thus, it is unclear whether features trained on ImageNet would generalize well for the change detection scenario. Furthermore, the decoder cannot be initialized and must still be trained. In our setting, SynCW, which is a target-domain large-scale dataset with change annotations, could be employed, but would training on synthetic images lead to good generalization? Thus, there exists no definitive choice of a training data source (*cf*. Fig. 6); as we demonstrate further, the choice of training strategy is crucial for achieving high performance.

We design seven training strategies for our task, summarized in Table 1. Strategy *A* would be a standard setting with excessive amounts of data. In strategies *B-1* and *B-2*, we attempt to model the *synthetic-to-real* transfer scenario. During pre-training, we either randomly initialize and freeze the encoder

Table 1. Statistical change detection results with models trained using strategies A–E. When using CW as a fine-tuning target, we only select a training subset; (*) indicates frozen encoder, (**) indicates augmentations (see Sect. 4.4).

	Training datasets			Ventura full		Ventura 1/16		SR full		SR 1/16	
	Init.	Pre	Fine	IoU	F1	IoU	F1	IoU	F1	IoU	F1
A	–	CW	–	0.695	0.820	0.310	0.460	0.487	0.504	0.337	0.639
B-1	–	SynCW*	CW	0.713	0.832	0.312	0.476	0.487	0.549	0.392	0.563
B-2	–	SynCW	CW	0.702	0.825	0.327	0.490	0.487	0.629	0.331	0.497
C	ImageNet	–	CW	0.716	0.835	0.499	0.704	0.626	0.770	0.435	0.607
D-1	ImageNet	SynCW*	CW	0.714	0.833	**0.572**	**0.735**	**0.680**	**0.800**	**0.649**	**0.787**
D-2	ImageNet	SynCW	CW	0.718	0.835	**0.580**	**0.744**	**0.684**	**0.812**	**0.631**	**0.774**
E	ImageNet	–	CW**	**0.724**	**0.840**	0.317	0.458	0.034	0.066	0.044	0.084

(*i.e.*, set its learning rate to zero, B-1) or train it (B-2). Strategies C, D-1, D-2, and E all initialize the encoder with ImageNet-pretrained weights, a widely used initialization: C realizes a common transfer learning setting, D-1 and D-2 proceed in two fine-tuning stages and use the synthetic, then the target training set, either training decoder only (D-1, similarly to B-1) or the entire model (D-2). E is a common transfer learning setting widely used in, *e.g.*, Kaggle[4] competitions, leveraging strong augmentations (*e.g.*, rotations, flips, brightness changes, blur, and noise).

Following [48], we use a pair of Ventura train images (4573×4418 px) for training or fine-tuning our models. As our goal is to study the effect of decreasing volumes of real-world data, we crop a random patch from these images, setting the ratio of patch area to the full image area to be 1, 1/2, 1/4, 1/8, and 1/16. A non-overlapping pair of Ventura test images (1044×1313 px) and a pair of visually distinct Santa Rosa images (2148×2160 px) are held out for testing. Note that when testing on Santa Rosa, we do not fine-tune on the same data to test generalization ability. In all experiments, we preserve the same architectural and training details as described in Sect. 4.2. We release the code used to implement and test our models[5].

4.5 Results

We present the statistical results of our evaluation in Table 1. As expected, when training data is present in large volumes (*e.g.*, using augmentations in strategy E), models pre-trained on ImageNet perform well. However, when the volume of real-world data supplied for fine-tuning decreases (up to 1//16), such strategies lead to unpredictable results (*e.g.*, for strategy E on Ventura test images IoU

---

[4] https://www.kaggle.com.
[5] https://github.com/mvkolos/siamese-change-detection.

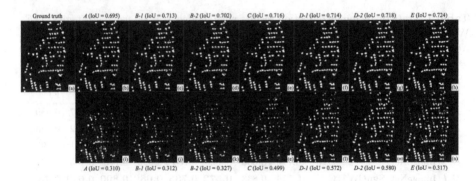

Ground truth   A (IoU = 0.695)   B-1 (IoU = 0.713)   B-2 (IoU = 0.702)   C (IoU = 0.716)   D-1 (IoU = 0.714)   D-2 (IoU = 0.718)   E (IoU = 0.724)

A (IoU = 0.310)   B-1 (IoU = 0.312)   B-2 (IoU = 0.327)   C (IoU = 0.499)   D-1 (IoU = 0.572)   D-2 (IoU = 0.580)   E (IoU = 0.317)

Fig. 9. Qualitative change detection results: (a) ground truth change mask and (b)–(n): change masks predicted by models A–E, trained on full real-world dataset (upper row (b)–(h)) and 1/16 of real-world data (lower row (i)–(n)).

measure drops by a factor of 2.3 from 0.724 to 0.317). In contrast, fine-tuning using our synthetic images helps to retain a significant part of the efficiency and leads to a more predictable change in the quality of the resulting model (*e.g.*, in strategy D-1 we observe a decrease in IoU by 20% only from 0.714 to 0.572). We note how for Santa Rosa images, the performance of models trained without synthetic data degrades severely, while fine-tuning using our synthetic images helps to suffer almost no drop in performance. We plot IoU/F1 vs. volume of used data in Fig. 10 to visualize this. We also display qualitative change-detection results in Fig. 9. Note how the output change masks tend to become noisy for strategies A, C, and E, and less so for D-1 and D-2. Our synthetic data also leads to faster convergence (see Fig. 11).

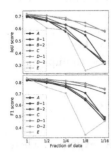

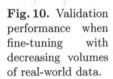

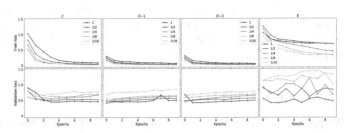

Fig. 10. Validation performance when fine-tuning with decreasing volumes of real-world data.

Fig. 11. Learning progress of our models trained with different fractions of real-world data. Note that models D-1 and D-2 pre-trained on our synthetic dataset converge faster and remain robust to train dataset reduction.

5 Conclusion

We have developed a pipeline for producing realistic synthetic data for the remote sensing domain. Using our pipeline, we have modeled the emergency mapping scenario and created 3D scenes and change detection image datasets of two real-world areas in California, USA. Results of the evaluation of deep learning models trained on our synthetic datasets indicate that synthetic data can be efficiently used to improve performance and robustness of data-driven models in real-world resource-poor remote sensing applications. We could further increase overall computational efficiency thanks to sparse CNNs [47], detection accuracy by using approaches to utilizing multi-modal data [14], imbalanced classification [15,56] and a loss, tailored for change detection in sequences of events [9,16].

References

1. Cryengine. https://www.cryengine.com. Accessed 30 Jan 2019
2. Esri cityengine. https://www.esri.com/en-us/arcgis/products/esri-cityengine/overview. Accessed 30 Jan 2019
3. Osm2xp. https://wiki.openstreetmap.org/wiki/Osm2xp. Accessed 30 Jan 2019
4. Unity. https://unity3d.com. Accessed 30 Jan 2019
5. Unreal engine 4. https://www.unrealengine.com/en-US/what-is-unreal-engine-4. Accessed 30 Jan 2019
6. World machine. http://www.world-machine.com/. Accessed 30 Jan 2019
7. Alcantarilla, P.F., Stent, S., Ros, G., Arroyo, R., Gherardi, R.: Street-view change detection with deconvolutional networks. Auton. Robots **42**(7), 1301–1322 (2018)
8. Anniballe, R., et al.: Earthquake damage mapping: an overall assessment of ground surveys and VHR image change detection after l'aquila 2009 earthquake. Remote Sens. Environ. **210**, 166–178 (2018)
9. Artemov, A., Burnaev, E.: Ensembles of detectors for online detection of transient changes. In: Proceedings of SPIE, vol. 9875, pp. 9875–9875-5 (2015). https://doi.org/10.1117/12.2228369
10. Babenko, A., Slesarev, A., Chigorin, A., Lempitsky, V.: Neural codes for image retrieval. In: Fleet, D., Pajdla, T., Schiele, B., Tuytelaars, T. (eds.) ECCV 2014. LNCS, vol. 8689, pp. 584–599. Springer, Cham (2014). https://doi.org/10.1007/978-3-319-10590-1_38
11. Bai, Y., Mas, E., Koshimura, S.: Towards operational satellite-based damage-mapping using U-net convolutional network: a case study of 2011 Tohoku Earthquake-Tsunami. Remote Sens. **10**(10), 1626 (2018)
12. Bourdis, N., Denis, M., Sahbi, H.: Constrained optical flow for aerial image change detection. In: 2011 IEEE International Geoscience and Remote Sensing Symposium (IGARSS), pp. 4176–4179 (2011)
13. Bruzzone, L., Demir, B.: A review of modern approaches to classification of remote sensing data. In: Manakos, I., Braun, M. (eds.) Land Use and Land Cover Mapping in Europe. RDIP, vol. 18, pp. 127–143. Springer, Dordrecht (2014). https://doi.org/10.1007/978-94-007-7969-3_9
14. Burnaev, E., Cichocki, A., Osin, V.: Fast multispectral deep fusion networks. Bull. Pol. Acad. Sci.: Techn. Sci. **66**(4), 875–880 (2018)

15. Burnaev, E., Erofeev, P., Papanov, A.: Influence of resampling on accuracy of imbalanced classification. In: Proceedings of SPIE, vol. 9875, pp. 9875–9875-5 (2015). https://doi.org/10.1117/12.2228523

16. Burnaev, E., Koptelov, I., Novikov, G., Khanipov, T.: Automatic construction of a recurrent neural network based classifier for vehicle passage detection, vol. 10341, pp. 10341–10341-6 (2017). https://doi.org/10.1117/12.2268706

17. Buslaev, A., Seferbekov, S., Iglovikov, V., Shvets, A.: Fully convolutional network for automatic road extraction from satellite imagery. CoRR, abs/1806.05182 (2018)

18. Cai, S., Liu, D.: Detecting change dates from dense satellite time series using a sub-annual change detection algorithm. Remote Sens. **7**(7), 8705–8727 (2015)

19. Caye Daudt, R., Le Saux, B., Boulch, A., Gousseau, Y.: Urban change detection for multispectral earth observation using convolutional neural networks. In: IEEE International Geoscience and Remote Sensing Symposium (IGARSS), July 2018

20. Chen, C., Seff, A., Kornhauser, A., Xiao, J.: DeepDriving: learning affordance for direct perception in autonomous driving. In: Proceedings of the IEEE International Conference on Computer Vision, pp. 2722–2730 (2015)

21. Chen, L.C., Papandreou, G., Kokkinos, I., Murphy, K., Yuille, A.L.: DeepLab: semantic image segmentation with deep convolutional nets, atrous convolution, and fully connected CRFs. IEEE Trans. Pattern Anal. Mach. Intell. **40**(4), 834–848 (2017)

22. Cheng, G., Han, J., Lu, X.: Remote sensing image scene classification: benchmark and state of the art. Proc. IEEE **105**(10), 1865–1883 (2017)

23. Deng, J., Dong, W., Socher, R., Li, L.J., Li, K., Fei-Fei, L.: ImageNet: a large-scale hierarchical image database. In: 2009 IEEE Conference on Computer Vision and Pattern Recognition, CVPR 2009, pp. 248–255. IEEE (2009)

24. Dosovitskiy, A., Ros, G., Codevilla, F., Lopez, A., Koltun, V.: CARLA: an open urban driving simulator. In: Conference on Robot Learning, pp. 1–16 (2017)

25. El Amin, A.M., Liu, Q., Wang, Y.: Convolutional neural network features based change detection in satellite images. In: First International Workshop on Pattern Recognition, vol. 10011, p. 100110W. International Society for Optics and Photonics (2016)

26. El Amin, A.M., Liu, Q., Wang, Y.: Zoom out CNNs features for optical remote sensing change detection. In: 2017 2nd International Conference on Image, Vision and Computing (ICIVC), pp. 812–817. IEEE (2017)

27. Fujita, A., Sakurada, K., Imaizumi, T., Ito, R., Hikosaka, S., Nakamura, R.: Damage detection from aerial images via convolutional neural networks. In: IAPR International Conference on Machine Vision Applications, Nagoya, Japan, pp. 08–12 (2017)

28. Gaidon, A., Lopez, A., Perronnin, F.: The reasonable effectiveness of synthetic visual data. Int. J. Comput. Vis. **126**(9), 899–901 (2018). https://doi.org/10.1007/s11263-018-1108-0

29. Gaidon, A., Wang, Q., Cabon, Y., Vig, E.: Virtual worlds as proxy for multi-object tracking analysis. In: Proceedings of the IEEE Conference on Computer Vision and Pattern Recognition, pp. 4340–4349 (2016)

30. Goodenough, A.A., Brown, S.D.: DIRSIG 5: core design and implementation. In: Algorithms and Technologies for Multispectral, Hyperspectral, and Ultraspectral Imagery XVIII, vol. 8390, p. 83900H. International Society for Optics and Photonics (2012)

31. Goward, S.N., Masek, J.G., Williams, D.L., Irons, J.R., Thompson, R.: The Landsat 7 mission: terrestrial research and applications for the 21st century. Remote Sens. Environ. **78**(1–2), 3–12 (2001)

32. Gu, W., Lv, Z., Hao, M.: Change detection method for remote sensing images based on an improved markov random field. Multimed. Tools Appl. **76**(17), 17719–17734 (2017)

33. Haarburger, C., et al.: Transfer learning for breast cancer malignancy classification based on dynamic contrast-enhanced MR images. Bildverarbeitung für die Medizin 2018. INFORMAT, pp. 216–221. Springer, Heidelberg (2018). https://doi.org/10.1007/978-3-662-56537-7_61

34. He, K., Gkioxari, G., Dollár, P., Girshick, R.: Mask R-CNN. In: 2017 IEEE International Conference on Computer Vision (ICCV), pp. 2980–2988. IEEE (2017)

35. He, K., Zhang, X., Ren, S., Sun, J.: Deep residual learning for image recognition. In: Proceedings of the IEEE Conference on Computer Vision and Pattern Recognition, pp. 770–778 (2016)

36. Huang, L., Fang, Y., Zuo, X., Yu, X.: Automatic change detection method of multitemporal remote sensing images based on 2D-OTSU algorithm improved by firefly algorithm. J. Sens. **2015** (2015)

37. Iglovikov, V., Shvets, A.: TernausNet: U-net with VGG11 encoder pre-trained on imagenet for image segmentation. arXiv preprint arXiv:1801.05746 (2018)

38. Ignatiev, V., Trekin, A., Lobachev, V., Potapov, G., Burnaev, E.: Targeted change detection in remote sensing images, vol. 110412H (2019). https://doi.org/10.1117/12.2523141

39. Jabari, S., Zhang, Y.: Building change detection using multi-sensor and multi-view-angle imagery. In: IOP Conference Series: Earth and Environmental Science, vol. 34, p. 012018. IOP Publishing (2016)

40. Jianya, G., Haigang, S., Guorui, M., Qiming, Z.: A review of multi-temporal remote sensing data change detection algorithms. Int. Arch. Photogramm. Remote Sens. Spat. Inf. Sci. **37**(B7), 757–762 (2008)

41. Kemker, R., Kanan, C.: Deep neural networks for semantic segmentation of multispectral remote sensing imagery. CoRR, vol. abs/1703.06452 (2017)

42. Kingma, D.P., Ba, J.: Adam: a method for stochastic optimization. In: International Conference on Learning Representations (ICLR) (2015)

43. Lee, H., et al.: An explainable deep-learning algorithm for the detection of acute intracranial haemorrhage from small datasets. Nat. Biomed. Eng. 1 (2018)

44. Lin, T.Y., Dollár, P., Girshick, R., He, K., Hariharan, B., Belongie, S.: Feature pyramid networks for object detection. In: CVPR. vol. 1, p. 4 (2017)

45. Lin, T.Y., Goyal, P., Girshick, R., He, K., Dollár, P.: Focal loss for dense object detection. IEEE Trans. Pattern Anal. Mach. Intell. (2018)

46. Müller, M., Casser, V., Lahoud, J., Smith, N., Ghanem, B.: Sim4CV: a photorealistic simulator for computer vision applications. Int. J. Comput. Vis. **128**, 902–919 (2018)

47. Notchenko, A., Kapushev, Y., Burnaev, E.: Large-scale shape retrieval with sparse 3D convolutional neural networks. In: van der Aalst, W.M.P., et al. (eds.) AIST 2017. LNCS, vol. 10716, pp. 245–254. Springer, Cham (2018). https://doi.org/10.1007/978-3-319-73013-4_23

48. Novikov, G., Trekin, A., Potapov, G., Ignatiev, V., Burnaev, E.: Satellite imagery analysis for operational damage assessment in emergency situations. In: Abramowicz, W., Paschke, A. (eds.) BIS 2018. LNBIP, vol. 320, pp. 347–358. Springer, Cham (2018). https://doi.org/10.1007/978-3-319-93931-5_25

49. Qiu, W., Yuille, A.: UnrealCV: connecting computer vision to unreal engine. In: Hua, G., Jégou, H. (eds.) ECCV 2016. LNCS, vol. 9915, pp. 909–916. Springer, Cham (2016). https://doi.org/10.1007/978-3-319-49409-8_75

50. Richter, S.R., Vineet, V., Roth, S., Koltun, V.: Playing for data: ground truth from computer games. In: Leibe, B., Matas, J., Sebe, N., Welling, M. (eds.) ECCV 2016. LNCS, vol. 9906, pp. 102–118. Springer, Cham (2016). https://doi.org/10.1007/978-3-319-46475-6_7

51. Ronneberger, O., Fischer, P., Brox, T.: U-Net: convolutional networks for biomedical image segmentation. In: Navab, N., Hornegger, J., Wells, W.M., Frangi, A.F. (eds.) MICCAI 2015. LNCS, vol. 9351, pp. 234–241. Springer, Cham (2015). https://doi.org/10.1007/978-3-319-24574-4_28

52. Russakovsky, O., et al.: ImageNet large scale visual recognition challenge. Int. J. Comput. Vis. **115**, 211–252 (2015)

53. Saha, S., Bovolo, F., Brurzone, L.: Unsupervised multiple-change detection in VHR optical images using deep features. In: IGARSS 2018, 2018 IEEE International Geoscience and Remote Sensing Symposium, pp. 1902–1905. IEEE (2018)

54. Shah, S., Dey, D., Lovett, C., Kapoor, A.: AirSim: high-fidelity visual and physical simulation for autonomous vehicles. In: Hutter, M., Siegwart, R. (eds.) Field and Service Robotics. SPAR, vol. 5, pp. 621–635. Springer, Cham (2018). https://doi.org/10.1007/978-3-319-67361-5_40

55. Sharif Razavian, A., Azizpour, H., Sullivan, J., Carlsson, S.: CNN features off-the-shelf: an astounding baseline for recognition. In: Proceedings of the IEEE Conference on Computer Vision and Pattern Recognition Workshops, pp. 806–813 (2014)

56. Smolyakov, D., Korotin, A., Erofeev, P., Papanov, A., Burnaev, E.: Meta-learning for resampling recommendation systems, vol. 11041 (2019). https://doi.org/10.1117/12.2523103

57. Tewkesbury, A.P., Comber, A.J., Tate, N.J., Lamb, A., Fisher, P.F.: A critical synthesis of remotely sensed optical image change detection techniques. Remote Sens. Environ. **160**, 1–14 (2015)

58. Vakalopoulou, M., Karantzalos, K., Komodakis, N., Paragios, N.: Simultaneous registration and change detection in multitemporal, very high resolution remote sensing data. In: Proceedings of the IEEE Conference on Computer Vision and Pattern Recognition Workshops, pp. 61–69 (2015)

59. Van Ginneken, B., Setio, A.A., Jacobs, C., Ciompi, F.: Off-the-shelf convolutional neural network features for pulmonary nodule detection in computed tomography scans. In: 2015 IEEE 12th International Symposium on Biomedical Imaging (ISBI), pp. 286–289. IEEE (2015)

60. Vittek, M., Brink, A., Donnay, F., Simonetti, D., Desclée, B.: Land cover change monitoring using landsat MSS/TM satellite image data over West Africa between 1975 and 1990. Remote Sens. **6**(1), 658–676 (2014)

61. Wang, B., Choi, J., Choi, S., Lee, S., Wu, P., Gao, Y.: Image fusion-based land cover change detection using multi-temporal high-resolution satellite images. Remote Sens. **9**(8), 804 (2017)

62. Wiratama, W., Lee, J., Park, S.E., Sim, D.: Dual-dense convolution network for change detection of high-resolution panchromatic imagery. Appl. Sci. **8**(10), 1785 (2018)

63. Yao, Y., Doretto, G.: Boosting for transfer learning with multiple sources. In: 2010 IEEE Conference on Computer vision and pattern recognition (CVPR), pp. 1855–1862. IEEE (2010)

64. Yosinski, J., Clune, J., Bengio, Y., Lipson, H.: How transferable are features in deep neural networks? In: Advances in Neural Information Processing Systems, pp. 3320–3328 (2014)

65. Yu, H., Yang, W., Hua, G., Ru, H., Huang, P.: Change detection using high resolution remote sensing images based on active learning and Markov random fields. Remote Sens. **9**(12), 1233 (2017)
66. Zhang, L., Zhang, L., Du, B.: Deep learning for remote sensing data: a technical tutorial on the state of the art. IEEE Geosci. Remote Sens. Mag. **4**(2), 22–40 (2016)
67. Zhang, R., Isola, P., Efros, A.A., Shechtman, E., Wang, O.: The unreasonable effectiveness of deep features as a perceptual metric. In: CVPR (2018)

Boundary Loss for Remote Sensing Imagery Semantic Segmentation

Alexey Bokhovkin[1] and Evgeny Burnaev[2(✉)]

[1] Aeronet, Skoltech, Moscow, Russia
a.bokhovkin@skoltech.ru
[2] ADASE, Skoltech, Moscow, Russia
e.burnaev@skoltech.ru

Abstract. In response to the growing importance of geospatial data, its analysis including semantic segmentation becomes an increasingly popular task in computer vision today. Convolutional neural networks are powerful visual models that yield hierarchies of features and practitioners widely use them to process remote sensing data. When performing remote sensing image segmentation, multiple instances of one class with precisely defined boundaries are often the case, and it is crucial to extract those boundaries accurately. The accuracy of segments boundaries delineation influences the quality of the whole segmented areas explicitly. However, widely-used segmentation loss functions such as BCE, IoU loss or Dice loss do not penalize misalignment of boundaries sufficiently. In this paper, we propose a novel loss function, namely a differentiable surrogate of a metric accounting accuracy of boundary detection. We can use the loss function with any neural network for binary segmentation. We performed validation of our loss function with various modifications of UNet on a synthetic dataset, as well as using real-world data (ISPRS Potsdam, INRIA AIL). Trained with the proposed loss function, models outperform baseline methods in terms of IoU score.

Keywords: Semantic segmentation · Deep learning · Aerial imagery · CNN · Loss function · Building detection · Computer vision

1 Introduction

Semantic segmentation of remote sensing images is a critical process in the workflow of object-based image analysis, which aim is to assign each pixel to a semantic label [13,20]. It has applications in environmental monitoring, urban planning, forestry, agriculture, and other geospatial analysis. Although a general image segmentation problem is relatively well investigated, we consider a particular type of this problem related to buildings segmentation.

The common aspect of urban aerial imagery is its high resolution (from 0.05 to 1.0 m). Higher GSD brings a lot of small details and structures, but also increases intra-class variance and decreases inter-class differences. This applies

The work was supported by The Ministry of Education and Science of Russian Federation, grant No. 14.615.21.0004, grant code: RFMEFI61518X0004.

© Springer Nature Switzerland AG 2019
H. Lu et al. (Eds.): ISNN 2019, LNCS 11555, pp. 388–401, 2019.
https://doi.org/10.1007/978-3-030-22808-8_38

particularly to the class of buildings, including an enormous number of shapes and patterns. After the stage of segmenting buildings, further analysis could be done, such as distinct building classification and height estimation, population prediction, economic forecasting, etc. For most of the tasks, it is crucial to separate all buildings and minimize the number of false positive/false negative instances. Even with high-resolution imagery, it is difficult to segment every object separately, especially in high-density urban areas, although we get more accurate information about the boundaries of buildings. Therefore it is necessary to construct a method that increases the attention of a neural network, used for semantic segmentation, to borders of closely located objects.

CNNs for semantic segmentation typically use such loss functions as cross-entropy, dice score and direct IoU optimization while training the network, but they are not sensitive enough to some misalignment of boundaries. Even being deviated from the ground truth by 5–10 pixels, predicted boundary does not significantly contribute to the value of the loss functions, mentioned above, and scores of common pixel-wise metrics. To better account for boundary pixels we select BF_1 metric (see original work [9] and its extension [12]) to construct a differentiable surrogate and use it in training. The surrogate is not used alone for training, but as a weighted sum with IoU loss (from direct IoU optimization). We found that the impact of the boundary component of the loss function should be gradually increased during training, and so we proposed a policy for the weight update. The assumption why this works is that for the first epochs network learns to label instances as shapeless blobs with very uncertain boundaries, and after finding an instance we only need to adjust its borders because they define the whole segments explicitly.

In experiments we also found that network trained with a combination of IoU and boundary loss converges faster: we increase the confidence of the network in its predictions on the border as fast as for intra-segment pixels; with other losses, the mask near the borders is very blurred. The next point is that network better handles edge effects while baseline models tend to miss or distort masks on the edge of an input image. Besides, if it is highly important to separate neighboring buildings, methods of instance segmentation are often applied, which can require extra data preparation from raw masks and they are commonly multistage. However, binary segmentation with boundary loss is end-to-end. In experiments on the challenging datasets ISPRS Potsdam and INRIA AIL, we obtain >93.8% and >74.3% pixel-wise IoU score respectively. There is also a comparison with another method for accurate delineation of curvilinear structures [17], which encounters difficulties in remote sensing tasks. Our results show a consistent increase in the performance for all models on both datasets in comparison to various loss functions.

The remainder of the paper is organized as follows. Section 2 briefly reviews related work. The proposed boundary loss, as well as the steps to construct its surrogate, are presented in Sect. 3. Experiments and results are discussed in Sect. 4, followed by concluding remarks in Sect. 5.

2   Related Work

The subject of our work intersects with two branches of research, which are Deep Neural Networks (DNNs) and direct optimization of performance measures or their differentiable versions also called surrogates. Besides, we should mention methods for accurate boundary delineation, as one of them (see [17]) was compared to our approach.

DNNs and Semantic Segmentation. Semantic segmentation based on DNNs is a pixel-wise mapping to semantic labels. Since 2012 a lot of neural networks [26] for solving this task were constructed. The success of AlexNet [15] in ImageNet challenge marked the beginning of neural networks application in computer vision tasks such as classification, object detection, and semantic segmentation. Initially, for the latter problem, they adapted CNNs, used for classification. Namely, they remove fully connected layers, and use initial layers as an encoder to downsample an image and derive features that have stronger semantic information but low spatial resolution. Another part of the network called decoder obtains a high-resolution mask of labels, which upsamples features into the resulting mask.

For the last few years, DNNs have shown to be very effective for semantic segmentation task. There are several models now that are regarded as state-of-the-art. First of all, it is SegNet architecture [2] which fits online video segmentation very well. DeepLab network [6] uses atrous convolutions and CRFs at postprocessing step to refine small details of the segmentation. The FRRN [22] is an example of a model with multi-scale processing technique; it is based on two separate streams of data to better handle the high-level semantic information and the low-level pixel information simultaneously. Finally, UNet architecture [23] is very popular nowadays for its enormous flexibility. Among aforementioned models UNet remains preferable because of the high efficiency of feature extraction and ability to apply various neural network architectures as backbones. UNet is an asymmetric FCNN (Fully Convolutional Neural Network) with skip connections between the downsampling and upsampling paths. UNet has shown its high performance in competitions and research projects, and we use it for all experiments in the paper.

Performance Measure Optimization. There already exist papers concerning direct optimization of metrics or their differentiable versions. In various applications such measures as IoU, F_1-score, ROC-area, mAP are widely used. Constructing surrogates of these measures is the most tricky part because it can be very hard or even impossible to replace non-differentiable operations with differentiable ones and keep computational efficiency at the same time. For the task of segmentation, IoU is usually used to measure the performance of any segmentation approach. As a result, there exists a lot of its surrogates, and the goal is to minimize the gap between the actual IoU value and its differentiable approximation. In paper [18] they proposed $NeuroIoU$ loss, which approximates naive IoU-loss with a neural network. Another approach [3] called Lovasz-Softmax loss, is based on the convex Lovasz extension of submodular losses. A critical property for this surrogate is that it effectively captures the absolute minimum of the original loss.

In this paper, we construct a differentiable version of the metric, presented in [9]. The motivation is that boundaries of segments explicitly define them and their extraction is necessary for accurate building segmentation. Authors of [9] proposed a novel boundary metric BF_1 that accounts for accuracy of contour extraction and overcomes the weaknesses of mainstream performance measures. For BF_1 we will construct a differentiable surrogate and show how it increases the accuracy of segmenting borders.

Below we provide a list of metrics and loss functions to be used for comparative analysis:

- Direct IoU loss: $IoU = \frac{TP}{FP+TP+FN}$, $L_{IoU} = 1 - IoU$,
- $Dice$ loss: $F_1 = \frac{2TP}{2TP+FN+FP}$, $L_{F_1} = 1 - F_1$,
- SS loss: $SS = \lambda\frac{TN}{TN+FP} + (1-\lambda)\frac{TP}{TP+FN}$, $L_{SS} = 1 - SS$, where λ is a weight to balance two components,
- VGG loss: implemented in [17].

The terms TP, FP, TN, FN denote pixel sets of true positive, false positive, true negative, false negative classes on a predicted binary map. For all loss functions, notations of corresponding metrics or approaches, inducing them, are provided in a subscript. By $L_{BF_1,IoU}$ we denote a weighted sum

$$wL_{BF_1} + (1 - w)L_{IoU}$$

for an arbitrary weight $w \in [0, 1]$.

Accurate Boundary Delineation. One possible solution to extract borders accurately is to use CRF (Conditional Random Fields). It works as an extra layer above the output of the original neural network. CRF is applied to capture additional contextual information and to produce much more refined prediction [1]. Authors of [21] used CRF together with Bayesian decision theory and proposed a heuristic to maximize the value of $EIoEU$. Another approach, a combination of CRF and superpixels [29], takes advantage of superpixel edge information and the constraint relationship among different pixels. Proposed algorithm of boundary optimization made it possible to improve IoU score by 3% on PASCAL VOC 2012 [11] and Cityscapes [8] datasets compared to the performance of plain FCN.

We compare an approach of [17] to the one, proposed in our paper. Authors claim that pixel-wise losses alone such as binary cross-entropy are unsuitable for predicting curvilinear structures. For this problem, they developed a new loss term based on features extracted with VGG19 network [27]. Conceptually their approach is similar to ours because we add our surrogate to the weighted conventional loss function. In contrast to VGG19 features, using our surrogate, we manually extract boundaries which are features too and encourage a neural network to draw attention to them much stronger.

3 Our Method

In this work, we use deep neural network UNet as a base model, because it is one of the most reliable and universal models. As we already pointed out in the introduction, there are a lot of different types of encoders and decoders that we can use as backbones. Of course, this network can be extended to two-headed version (UNet with two decoders), where the first head predicts whole segments and the other predicts only their boundaries. However, this approach requires additional computational resources and data preparation compared to the original UNet. Our interest is to propose a method that would not increase implementation and computational complexities. Further, we first introduce BF_1 metric from [9] and then describe how to construct its differentiable surrogate.

3.1 Boundary Metric (BF_1)

Let S_{gt}^c, S_{pd}^c be the binary maps of class c in the ground truth and predicted segmentation respectively, B_{gt}^c, B_{pd}^c—the boundaries for these binary maps. Then the precision and the recall for class c are defined as follows:

$$P^c = \frac{1}{|B_{pd}^c|} \sum_{x \in B_{pd}^c} [[d(x, B_{gt}^c) < \theta]], \ R^c = \frac{1}{|B_{gt}^c|} \sum_{x \in B_{gt}^c} [[d(x, B_{pd}^c) < \theta]],$$

where brackets $[[\cdot]]$ denote indicator function of a logical expression, $d(\cdot)$ is Euclidean distance measured in pixels, θ - some predefined threshold on a distance; in all experiments we set θ to 3 or 5. Here distance is calculated from a point to one of two sets (ground truth or predicted boundary) as the shortest distance from the point to the point of the boundary. The BF_1^c measure for class c is defined as:

$$BF_1^c = \frac{2P^c R^c}{P^c + R^c}.$$

3.2 Surrogate Construction (L_{BF_1})

To construct the differentiable version of the metric BF_1, we need to extract the boundary of any segment somehow. There are several ways to do this. Let us denote by y_{pd} a binary map for an arbitrary class c for some image I, predicted by a neural network, y_{gt} a ground truth map for the same class and the same image. We expect values of the predicted map to be distributed in $[0, 1]$, and values of the ground truth map to be in $\{0, 1\}$. The boundaries can be defined as

$$y_{gt}^b = pool(1 - y_{gt}, \theta_0) - (1 - y_{gt}), \ y_{pd}^b = pool(1 - y_{pd}, \theta_0) - (1 - y_{pd}). \tag{1}$$

Here $pool(\cdot, \cdot)$ applies a pixel-wise max-pooling operation to the inverted predicted or ground truth binary map with a sliding window of size θ_0 (hyperparameter θ_0 must be as small as possible to extract vicious boundary; usually we set $\theta_0 = 3$). Value $(1 - y_{gt,pd})$ corresponds to inversion of any pixel of the map. To compute Euclidean distances from pixels to boundaries a supporting map should be obtained, which is the map of the extended boundary:

$$y_{gt}^{b,ext} = pool(y_{gt}^b, \theta), \quad y_{pd}^{b,ext} = pool(y_{pd}^b, \theta). \tag{2}$$

The value of hyperparameter θ can be determined as not greater than the minimum distance between neighboring segments of the binary ground truth map. After that precision and recall can be computed as follows:

$$P^c = \frac{sum(y_{pd}^b \circ y_{gt}^{b,ext})}{sum(y_{pd}^b)}, \quad R^c = \frac{sum(y_{gt}^b \circ y_{pd}^{b,ext})}{sum(y_{gt}^b)}, \tag{3}$$

where operation $\circ$ denotes pixel-wise multiplication of two binary maps and operation $sum(\cdot)$—pixel-wise summation of a binary map. Finally, the reconstructed metric and a corresponding loss function are defined as:

$$BF_1^c = \frac{2P^c R^c}{P^c + R^c}, \quad L_{BF_1^c} = 1 - BF_1^c. \tag{4}$$

All operations above are differentiable in terms of either derivative or sub-derivative. In Fig. 1 there is an example of how the proposed surrogate works. It is worth to mention that of course we can use convolutions with edge detection filters, such as Sobel, but in experiments, we found that this approach is not as effective as max-pooling. Nevertheless, both methods have their drawbacks. The disadvantages of both methods are that on the first step we extract boundary not narrower than 2–3 pixels in width, but with Sobel filters, boundaries are even wider and more blurred. The main drawback of the used max-pooling operation is that gradients are passed only in tensor cells with a maximum value within a current sliding window. Sobel operators do not have this problem and take into account all pixels of the sliding window.

4 Experiments

4.1 Example on Synthetic Dataset (AICD)

To demonstrate capabilities of the novel boundary loss and test the hypothesis that L_{BF_1} can assist L_{IoU} a synthetic dataset was used. It consists of 800 images with one primitive segment in each image. We trained on this dataset without augmentations and postprocessing a not deep fully convolutional neural network with four conv layers in the encoder and four conv layers in the decoder to show that $L_{BF_1,IoU}$ outperforms other losses under simple conditions.

First of all, we present a working principle of the new loss in Fig. 1. In the upper row (see (b), (d) and (f)) we depict ground truth, where (b) is a binary mask for the ground truth segment, (d) is a binary mask of the gt boundary after applying (1). Then we compute expanded boundary (f) using pooling operations

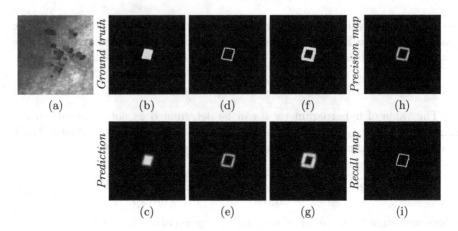

Fig. 1. (a) original orthophoto; (b) ground truth segment (`gt`); (c) predicted segment (`pred`); (d) boundary of `gt`; (e) boundary of `pred`; (f) expanded boundary of `gt`; (g) expanded boundary of `pred`; (h) pixel-wise multiplication of masks (d) and (g); (i) pixel-wise multiplication of masks (e) and (f)

(2). The same pipeline is applied to the predicted mask (color intensity represents the probability that pixel belongs to a foreground). After that we obtain `Precision map`, which represents pixel-wise multiplication of maps corresponding to (d) and (g), and `Recall map`, which is a multiplication of maps (e) and (f). `Precision` and `Recall` are normalized pixel-wise sums of these maps.

Some examples of how $L_{BF_1,IoU}$ loss outperforms L_{IoU} loss are presented in Fig. 2. It is clear that in comparison with L_{IoU} training with $L_{BF_1,IoU}$ helps better delineate segment borders: they are very similar to ground truth boundaries and even meet at the corners in places where they should be. The second advantage is that the loss better handles edge effects. The third column clearly explains this kind of a problem when a building is located near the edge of the image. L_{IoU} loss is unable to segment the building near the edge, however $L_{BF_1,IoU}$ does it accurately. Finally, the boundary loss converges faster on this synthetic dataset, and such behavior is expected not only for this simple data distribution. The intuition behind is that network keeps attention mainly on a boundary, which consists of fewer pixels in comparison with the entire segment.

4.2 INRIA AIL Dataset Segmentation

The dataset [16] consists of 180 tiles (RGB) of $405\,km^2$ area in total with GSD of $0.3\,m$. There are only two classes: *building* and *not building*.

Training on this dataset is more complex than on AICD. As for preprocessing, very strong augmentations were applied including vertical, horizontal flips, hue/saturation adjustments, many kinds of noises and blurs. Every orthophoto was split into patches of size 512×512 and UNet with Inception-ResNet-v2 [30] as backbone was trained on these patches. After that, all predicted masks for

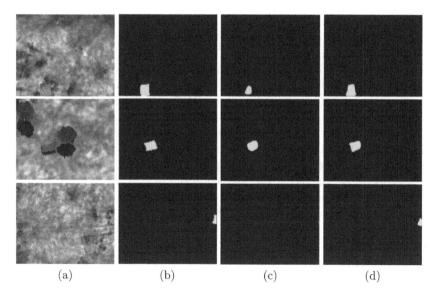

Fig. 2. (a) original orthophotos; (b) binary maps of ground truth segments; (c) predicted binary maps of segments (L_{IoU} loss); (d) predicted binary maps of segments ($L_{BF_1,IoU}$ loss)

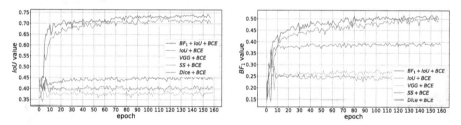

Fig. 3. Evolution of IoU value during training for various loss functions (INRIA AIL)

Fig. 4. Evolution of BF_1 value during training for various loss functions (INRIA AIL)

every patch were merged into one mask of the whole orthophoto. Also, test-time augmentation (flips, reflections) was applied during the inference step. As for the training process, first, the network was trained from ImageNet [10] weights with a frozen encoder for three epochs, and the learning rate was set to 3e−3. After that all batch-normalization layers were unfrozen, and the learning rate was decreased to 1e−3. Then all layers were unfrozen and trained for 100 epochs with the learning rate of 1e−4. Finally, the network was trained for 60 epochs with the learning rate of 1e−5. We used Adam optimizer with default Keras [7] settings. There were several experiments with different loss functions. As for L_{IoU}, L_{Dice}, L_{SS} and L_{VGG} losses, they did not require any adjustments while training: these losses were trained as a weighted sum with BCE one by one.

As for $L_{BF_1,IoU}$ loss, it requires an additional procedure for mini grid-search: after the 8th epoch for every 30 epochs and for every weight $w \in \{0.1, 0.3, 0.5, 0.7, 0.9\}$ in equation $(BCE + wL_{BF_1} + (1 - w)L_{IoU})$ a network was trained. Then the best weight is chosen, and the process repeats. In Fig. 3 we see the evolution of IoU metric while training with combinations of BCE and L_{IoU}, L_{Dice}, L_{SS}, L_{VGG}, or $L_{BF_1,IoU}$. BF_1 is also a very important metric for remote sensing image segmentation (see Fig. 4).

Finally, from figures, we see that the final segmentation is better than the baseline model trained with L_{IoU} in terms of IoU and BF_1. Now we want to understand what is better when training with the boundary loss. Below we provide an example comparing two segmentations.

In Fig. 5 on the right we see more accurate shapes of edges and corners. The important feature, which follows from Fig. 6, is that several buildings are standing in one row with a distance between adjacent constructions about 5 pixels. In comparison with L_{IoU}, $L_{BF_1,IoU}$ our loss managed to separate instances of buildings much better.

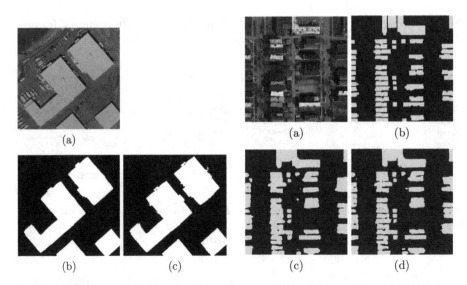

Fig. 5. (a) original orthophoto; (b) predicted segmentation trained with L_{IoU}; (c) predicted segmentation trained with $L_{BF_1,IoU}$

Fig. 6. (a) original orthophoto; (b) ground truth segmentation (**gt**); (c) predicted segmentation trained with L_{IoU}; (d) predicted segmentation trained with $L_{BF_1,IoU}$

4.3 ISPRS Potsdam Dataset Segmentation

The dataset [24] contains 38 patches, each consisting of a true orthophoto extracted from the larger mosaic. There are five channels (RGB+NIR+DSM) with GSD 0.2 m for each channel. As a model, UNet was trained with very strong augmentations similar to that used for INRIA AIL. We used test-time augmentation (flips, reflections), set the learning rate to 1e−3 and divided it by a factor of 10 if for 10 epochs there was no improvement in validation loss. We trained a neural network on patches of orthophotos with Adam optimizer.

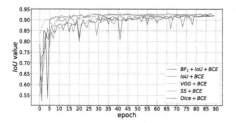 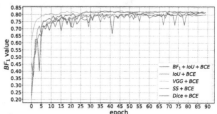

Fig. 7. Evolution of IoU value during training for various loss functions (ISPRS Potsdam)

Fig. 8. Evolution of BF_1 value during training for various loss functions (ISPRS Potsdam)

In Figs. 7 and 8 we see the evolution of IoU and BF_1 metric while training with combinations of BCE and L_{IoU}, L_{Dice}, L_{SS}, L_{VGG}, $L_{BF_1,IoU}$. Here we see slightly faster convergence of IoU for $L_{BF_1,IoU}$, the final segmentation is 1–2% better than baseline ($BCE + L_{IoU}$). Also there is an advantage that training with $L_{BF_1,IoU}$ loss is much less noisy.

Even though IoU value is the best with $L_{BF_1,IoU}$ training, boundary metric converges faster with VGG loss; authors of the original paper mentioned that this loss function is also efficient for boundaries delineation. In Figs. 9 and 10 there are several examples segmented with L_{IoU} and the boundary loss.

Direct IoU loss (Fig. 9) has poorer performance on edges of segments and complicated shapes, whereas boundary loss (Fig. 10) on the right retrieved almost all tricky shapes.

In Tables 1 and 2 we provide final results of different losses in the task of binary segmentation of buildings for different models and datasets. Backbones for comparison in these tables are chosen is such way that they showed the best performance over all we have tried (VGG16, VGG19, ResNet34, DenseNet121, DenseNet169, Inception-ResNet-v2).

Finally, here are some constraints and recommendations:

1. It is better to use boundary loss only for segments with precise edges and corners such as buildings, roads etc.
2. Let us discuss the choice of parameters θ_0 and θ (1) and (2) respectively. The best choice for the parameter θ_0 is as less as possible, but at the same time it

Fig. 9. Example of segmentation (loss—L_{IoU}). **Left**: original orthophoto, **top right**: ground truth segmentation, **bottom right**: predicted segmentation

Fig. 10. Example of segmentation (loss—$L_{BF_1,IoU}$). **Left**: original orthophoto, **top right**: ground truth segmentation, **bottom right**: predicted segmentation

Table 1. IoU values for different datasets and loss functions. ConvNN is a not very deep neural network with 5 convolutional blocks in both the encoder and the decoder. For ISPRS dataset UNet model with ResNet34 backbone was used, for INRIA dataset – UNet with Inception-ResNet-v2 backbone

	AICD	ISPRS Potsdam	INRIA AIL
	ConvNN	ResNet34	Inc.-ResNet-v2
$L_{BF_1,IoU}$	**86.91**	**93.85**	**74.30**
L_{IoU}	85.66	92.56	71.91
$L_{SS,IoU}$	61.14	92.47	43.66
$L_{VGG,IoU}$	85.76	92.53	45.89
$L_{Dice,IoU}$	85.13	92.61	44.90

Table 2. BF_1 values for different datasets and loss functions. For ISPRS dataset UNet model with *ResNet34* backbone was used, for INRIA dataset – UNet with *Inception-ResNet-v2* backbone

	AICD	ISPRS Potsdam	INRIA AIL
	ConvNN	ResNet34	Inc.-ResNet-v2
$L_{BF_1,IoU}$	**71.51**	82.48	**51.75**
L_{IoU}	68.01	81.02	50.75
$L_{SS,IoU}$	20.74	81.55	29.53
$L_{VGG,IoU}$	69.41	**82.77**	40.18
$L_{Dice,IoU}$	70.49	81.86	29.45

must be possible to extract solid boundaries of segments. The next constraint is that θ_0 and θ should be smaller than minimum distance between segments, otherwise the performance of segmenting current element is affected by other segments because of overlapped expanded boundaries. Via the trial and error process we set θ_0 to 3 and θ to 5–7 as a proper choice, because theses values deliver the most accurate boundaries in all experiments.

5 Conclusions

In this work, we introduced a novel loss function to encourage a neural network better taking into account segment boundaries. As motivation, we applied this approach to a synthetic dataset of binary segmentation and after that proved its validity on real-world data. Our experimental results demonstrate that optimizing IoU assisted by BF_1 metric surrogate leads to better performance and more accurate boundaries delineation compared to using other loss functions. As for future work, we would like to extend our approach to handle multi-class semantic segmentation. We are going to elaborate on this approach by increasing its computational capabilities with large scale sparse convolutional neural networks [19], using a new loss function, specially tailored for imbalanced classification [5,28], utilizing an approach for combining multi-modal data through CNNs features aggregation [4] and imposing a confidence measure on top of the segmentation model based on the non-parametric conformal approach [14,25,31].

References

1. Alam, F.I., Zhou, J., Liew, A.W., Jia, X., Chanussot, J., Gao, Y.: Conditional random field and deep feature learning for hyperspectral image classification. IEEE Trans. Geosci. Remote Sens. **57**(3), 1612–1628 (2019). https://doi.org/10.1109/TGRS.2018.2867679
2. Badrinarayanan, V., Kendall, A., Cipolla, R.: SegNet: a deep convolutional encoder-decoder architecture for image segmentation. IEEE Trans. Pattern Anal. Mach. Intell. **39**, 2481–2495 (2016)

3. Berman, M., Triki, A.R., Blaschko, M.B.: The Lovasz-Softmax loss: a tractable surrogate for the optimization of the intersection-over-union measure in neural networks. In: 2018 IEEE/CVF Conference on Computer Vision and Pattern Recognition, pp. 4413–4421 (2018)
4. Burnaev, E., Cichocki, A., Osin, V.: Fast multispectral deep fusion networks. Bull. Pol. Acad. Sci.: Techn. Sci. **66**(4), 875–880 (2018)
5. Burnaev, E., Erofeev, P., Papanov, A.: Influence of resampling on accuracy of imbalanced classification. In: Eighth International Conference on Machine Vision. Proceedings of SPIE, 8 December 2015, vol. 9875, p. 987525 (2015). https://doi.org/10.1117/12.2228523
6. Chen, L., Papandreou, G., Kokkinos, I., Murphy, K., Yuille, A.L.: DeepLab: semantic image segmentation with deep convolutional nets, atrous convolution, and fully connected CRFs. IEEE Trans. Pattern Anal. Mach. Intell. **40**(4), 834–848 (2018). https://doi.org/10.1109/TPAMI.2017.2699184
7. Chollet, F., et al.: Keras (2015). https://github.com/fchollet/keras
8. Cordts, M., et al.: The cityscapes dataset for semantic urban scene understanding. In: Proceedings of the IEEE Conference on Computer Vision and Pattern Recognition (CVPR) (2016)
9. Csurka, G., Larlus, D., Perronnin, F.: What is a good evaluation measure for semantic segmentation? IEEE PAMI **26**, 1–11 (2004)
10. Deng, J., Dong, W., Socher, R., Li, L.: ImageNet: a large-scale hierarchical image database. In: 2009 IEEE Conference on Computer Vision and Pattern Recognition, pp. 248–255 (2009)
11. Everingham, M., Van Gool, L., Williams, C.K.I., Winn, J., Zisserman, A.: The PASCAL visual object classes challenge 2012 (VOC 2012) results. www.pascal-network.org/challenges/VOC/voc2012/workshop/index.html
12. Fernandez-Moral, E., Martins, R., Wolf, D., Rives, P.: A new metric for evaluating semantic segmentation: leveraging global and contour accuracy. In: Workshop on Planning, Perception and Navigation for Intelligent Vehicles, PPNIV 2017, Vancouver, Canada, September 2017. https://hal.inria.fr/hal-01581525
13. Ignatiev, V., Trekin, A., Lobachev, V., Potapov, G., Burnaev, E.: Targeted change detection in remote sensing images. In: Eleventh International Conference on Machine Vision (ICMV 2018). Proceedings of SPIE, vol. 11041, p. 110412H (2019). https://doi.org/10.1117/12.2523141
14. Ishimtsev, V., Bernstein, A., Burnaev, E., Nazarov, I.: Conformal k-NN anomaly detector for univariate data streams. In: Proceedings of 6th Workshop COPA. PRML, vol. 60, pp. 213–227. PMLR (2017)
15. Krizhevsky, A., Sutskever, I., E. Hinton, G.: ImageNet classification with deep convolutional neural networks. In: Advances in Neural Information Processing Systems, pp. 1097–1105 (2012)
16. Maggiori, E., Tarabalka, Y., Charpiat, G., Alliez, P.: Can semantic labeling methods generalize to any city? The Inria aerial image labeling benchmark. In: 2017 IEEE International Geoscience and Remote Sensing Symposium (IGARSS), pp. 3226–3229 (2017). https://doi.org/10.1109/IGARSS.2017.8127684
17. Mosinska, A., Marquez-Neila, P., Kozinski, M., Fua, P.: Beyond the pixel-wise loss for topology-aware delineation, pp. 3136–3145, June 2018. https://doi.org/10.1109/CVPR.2018.00331
18. Nagendar, G., Singh, D., Balasubramanian, V.N., Jawahar, C.V.: Neuro-IoU: learning a surrogate loss for semantic segmentation. In: British Machine Vision Conference 2018, BMVC 2018, Northumbria University, Newcastle, UK, 3–6 September 2018, pp. 278–289 (2018). http://bmvc2018.org/contents/papers/1055.pdf

19. Notchenko, A., Kapushev, Y., Burnaev, E.: Large-scale shape retrieval with sparse 3D convolutional neural networks. In: van der Aalst, W.M.P., et al. (eds.) AIST 2017. LNCS, vol. 10716, pp. 245–254. Springer, Cham (2018). https://doi.org/10.1007/978-3-319-73013-4_23

20. Novikov, G., Trekin, A., Potapov, G., Ignatiev, V., Burnaev, E.: Satellite imagery analysis for operational damage assessment in emergency situations. In: Abramowicz, W., Paschke, A. (eds.) BIS 2018. LNBIP, vol. 320, pp. 347–358. Springer, Cham (2018). https://doi.org/10.1007/978-3-319-93931-5_25

21. Nowozin, S.: Optimal decisions from probabilistic models: the intersection-over-union case. In: 2014 IEEE Conference on Computer Vision and Pattern Recognition, pp. 548–555 (2014)

22. Pohlen, T., Hermans, A., Mathias, M., Leibe, B.: Full-resolution residual networks for semantic segmentation in street scenes. In: 2017 IEEE Conference on Computer Vision and Pattern Recognition (CVPR), pp. 3309–3318 (2017)

23. Ronneberger, O., Fischer, P., Brox, T.: U-Net: convolutional networks for biomedical image segmentation. In: Navab, N., Hornegger, J., Wells, W.M., Frangi, A.F. (eds.) MICCAI 2015. LNCS, vol. 9351, pp. 234–241. Springer, Cham (2015). https://doi.org/10.1007/978-3-319-24574-4_28

24. Rottensteiner, F., et al.: The ISPRS benchmark on urban object classification and 3D building reconstruction. In: Shortis, M., Paparoditis, N., Mallett, C. (eds.) ISPRS 2012 Proceedings of the XXII ISPRS Congress: Imaging a Sustainable Future, 25 August–01 September 2012, Melbourne, Australia, vol. I-7, pp. 293–298, August 2012. Peer Reviewed Annals, International Society for Photogrammetry and Remote Sensing (ISPRS)

25. Safin, A., Burnaev, E.: Conformal kernel expected similarity for anomaly detection in time-series data. Adv. Syst. Sci. Appl. **17**(3), 22–33 (2017)

26. Shelhamer, E., Long, J., Darrell, T.: Fully convolutional networks for semantic segmentation. IEEE Trans. Pattern Anal. Mach. Intell. **39**(4), 640–651 (2017). https://doi.org/10.1109/TPAMI.2016.2572683

27. Simonyan, K., Zisserman, A.: Very deep convolutional networks for large-scale image recognition. arXiv:1409.1556, September 2014

28. Smoliakov, D., Korotin, A., Erifeev, P., Papanov, A., Burnaev, E.: Meta-learning for resampling recommendation systems. In: Eleventh International Conference on Machine Vision (ICMV 2018); 110411S (2019). Proceedings of SPIE, vol. 11041 (2019). https://doi.org/10.1117/12.2523103

29. Sulimowicz, L., Ahmad, I., Aved, A.J.: Superpixel-enhanced pairwise conditional random field for semantic segmentation. 2018 25th IEEE International Conference on Image Processing (ICIP) pp. 271–275 (2018)

30. Szegedy, C., Ioffe, S., Vanhoucke, V.: Inception-v4, inception-resnet and the impact of residual connections on learning. In: AAAI. pp. 4278–4284 (2016)

31. Volkhonskiy, D., Burnaev, E., Nouretdinov, I., Gammerman, A., Vovk, V.: Inductive conformal martingales for change-point detection. In: Proceedings of 6th Workshop COPA. PRML, vol. 60, pp. 132–153. PMLR (2017)

Ship Segmentation and Orientation Estimation Using Keypoints Detection and Voting Mechanism in Remote Sensing Images

Mingxian Nie[1], Jinjie Zhang[1], and Xuetao Zhang[2]($\boxtimes$)

[1] School of Software Engineering, Xi'an Jiaotong University,
Xi'an 710049, Shaanxi, China
nmx136@stu.xjtu.edu.cn, xbzzhc@qq.com
[2] School of Electronic and Information Engineering, Xi'an Jiaotong University,
Xi'an 710049, Shaanxi, China
xuetaozh@xjtu.edu.cn

Abstract. Ship detection in remote sensing images is an important and challenging task in civil fields. However, the various types of ships with different scale and ratio and the complex scenarios are the main bottlenecks for ship detection and orientation estimation of the ship. In this paper, we propose a new method based on Mask R-CNN, which can perform ship segmentation and direction estimation on ships at the same time by simultaneously output the binary mask and the bow and sterns keypoints locations. We can achieve keypoints detection of the ship without significantly losing the accuracy of the mask. Finally, we regress the coordinates of the ship's bow and sterns to four quadrants and use the voting mechanism to determine which quadrant the bow keypoint locates. Then we combine the quadrant of bow keypoint with the minimum bounding box of the mask to determine the final orientation of the ship. Experiments on the datasets have achieved effective performance.

Keywords: Remote sensing images · Ship detection ·
Orientation estimation · Mask R-CNN

1 Introduction

Ship detection in remote sensing images has attracted more and more attention of researchers in recent years because of its vital role in port management, traffic surveillance, and ship search and rescue [1]. With the development of remote sensing technology, we can easily obtain high resolution images, which prompts us to make a finer segmentation of the ship. However, accurately segmenting ships in remote sensing images is not trivial. Firstly, the size of the ship varies greatly and the aspect ratio is quite different [2], making the ship detection become more difficult than other object, such as vehicles [3] or buildings [4].

© Springer Nature Switzerland AG 2019
H. Lu et al. (Eds.): ISNN 2019, LNCS 11555, pp. 402–413, 2019.
https://doi.org/10.1007/978-3-030-22808-8_39

Secondly, the color and texture of the inshore ship is similar to that of land area such as an anti-wave or residential building which can lead to many false alarms [5], and the detection of sailing ships is susceptible to fog and waves on the sea. Thirdly, the ship in remote sensing images is multi-directional, making the direction estimation of the ship much more difficult than direction estimation of other objects in the natural scene.

To handle the problems above, researchers have developed a lot of methods and have achieve amazing results in ship detection in the past few years. Before CNN-based methods [6], some traditional methods based on sea-land segment are widely used in ship detection. For example, [7] firstly realized the sea-land separation through the difference in characteristics of the texture and color between ports and sea so that only the sea area named ROI (region of interests) needed to be processed, and then they used the contrast box algorithm to obtain the candidates of targets. Finally, through post-processing, the false alarms were eliminated and the final boxes containing the ship were obtained. It required pre-process such as sea-land separation, and the detection accuracy was heavily dependent on the results of sea-land separation. Such kinds of methods perform well in a special application, but will get poor practicability in complex scenarios. Recently, the deep learning based methods have achieved good results in the field of object detection. The Faster-RCNN [8], which is one of the representative models, consists of RPN and Fast-RCNN [9]. It generates region proposals through the RPN, then executes pooling operation on the feature maps of the proposals to generate RoIs. The RPN feeds RoIs into the Fast-RCNN, which classifies the RoIs and outputs the class probability and the regression coordinates of the RoIs. However, Faster-RCNN is a horizontal region detection which is suitable for natural scene detection but not suitable for satellite remote sensing ship detection [10]. [11] utilized instance segmentation into ship detection framework via Mask R-CNN to achieve a more thorough segmentation between the ship and background, they used soft-NMS rather than standard NMS to improve the robustness of detecting inshore ships. This operation achieved a certain effect in detecting dense ships which are close to each other. What's more, by classifying each pixel of the ship, the accuracy of the object detection can be improved. And by combining the resolution of the satellite itself with the number of pixels of the ship in the image, we can learn the actual area, perimeter and size of the ship, which are important for ship type recognition. [10] adopted the rotational region detection framework to detect the dense objects by outputting rotated boxes rather than horizontal boxes and output the ship orientation simultaneously. It determined the ship orientation via selecting from the four sides of the rotated box. The orientation estimation of the ship is a meaningful job, especially for port management and maritime traffic. But the method in [10] is simple, and the applicability of ship orientation estimation is relatively poor in complex situations.

In this paper, we put forward a new proposal for estimating the direction of the bow. Inspired by the human pose estimation [12], we decide to apply the human body keypoints detection technology used in human pose estimation to that of the bow and stern of the ship. [13] achieves very good results by using multi-scale features to capture spatial position information of joint points in the human body. But it can only output the precise pixel-level positions of the keypoints of a single human, not suitable for keypoints detection of multi-person or multi-ships. [14] first uses the pedestrian detection algorithm to detect the pedestrian candidate boxes, then uses the CPN (Cascaded Pyramid Network) to perform coordinate regression of the human body keypoints for each detected pedestrian candidate box and output the final results. It performs well in dealing with multi-person keypoints in case of occlusion, invisibility or complex background. But it is a two-stage framework which is time-consuming and computationally intensive and relies heavily on the accuracy of the pedestrian detection module. Mask R-CNN [15] can also be used to estimate the pose of the human body. Since multiple ships do not have mutual occlusion, we incorporate ship keypoints detection into the Mask R-CNN basic framework to form a unified architecture that can simultaneously obtain the ship's mask and the three keypoints of ship. The angle of the ship is initially determined by the minimum bounding box of the mask. We propose a voting mechanism by combing the relative positions information between one bow and two sterns to determine the final orientation of the bow. When the ship is partially detected, we can still estimate the direction of the ship by the detected individual kepoints, which improve the robustness of the algorithm.

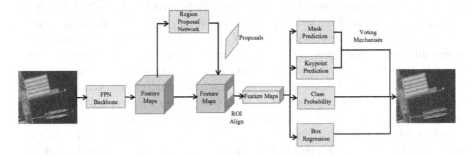

Fig. 1. Our framework with proposed method.

2   Methodology

The method we use is based on the Mask R-CNN architecture, and the overall architecture is shown in Fig. 1. Firstly, we use FPN [16] to extract the features of different scales from the input image. Then RPN generates the proposals on the feature maps. After performing the ROI Align operation on these proposals, RPN produces RoIs and feeds them into Fast-RCNN. Fast-RCNN implements the classification, coordinate regression, binary mask and keypoints output for each RoI. Our approach consists in adding keypoints prediction to Mask R-CNN original architecture and defining three keypoints for a bow and two sterns. By combining information such as the mask branch and the keypoints branch, we use a voting mechanism to obtain the orientation of the bow.

2.1 Three-Keypoints Ship Model

Drawing on the human skeleton keypoints model, we propose a ship keypoints model (named three-keypoints ship model) which consists of three keypoints, the bow ($b1$), the first stern ($s1$) and the second stern ($s2$). Each key point has three values: x, y, v, which represent the abscissa, ordinate and visibility of the keypoint on the image coordinate system respectively. We abstract the ship into a three-keypoints ship model because the bow and sterns are at the edge of the ship and are the corner points, which have obvious features. These three keypoints represent the commonality of most ships like civilian ships which all have a bow and two sterns so we use them for keypoint detection of the ships. Our three-keypoints ship model is not limited to three keypoints and can be extended to more than three keypoints according to the characteristics of the detected object. The three-keypoints ship model and some samples in application of our model are shown in Fig. 2.

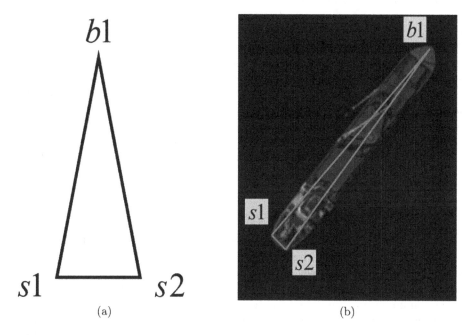

Fig. 2. Three-keypoints ship model and some samples. (a) three-keypoints ship model. (b) civilian ship.

2.2   A Unified Framework

Our method is based on the Mask R-CNN architecture using FPN as backbone. The idea of FPN is to fuse the adjacent two layers of the feature pyramid from top to bottom. This operation combines the low-level position information and high-level semantic information of the neural network and is useful to detect ships of different scales, especially for particularly small targets. FPN generates feature maps and sends it to RPN. The RPN generates region proposals and finally generates regions of interest (RoIs) based on the location of the region proposals on the feature maps. The RoIs are processed by the ROI Align pooling operation and sent to Fast-RCNN. The classifier produces the category probability of each RoI, the regression branch produces the refined bounding box coordinates and the mask branch encodes the probability each RoI belonging to each class.

Mask R-CNN has achieved good results in the detection of keypoints in the human body. In theory, the keypoints detection of the ship is feasible since the ship has some obvious characteristic points, for example, the bow and the stern are corner points. We define one bow and two sterns as the keypoints of the ship, and add the keypoint branch to the original Mask R-CNN framework including ship classification, regression and mask. For each RoI, the output of keypoints branch is a heatmap encoded by Km^2-dimensional array where K means the number of ship keypoints and m^2 is the resolution. By finding the maximum value on the m^2 array, we can determine the position coordinates of the bow and stern, which is important for estimating the orientation of the ship.

Our framework is built on the Mask R-CNN base framework, so the FPN and RPN module structures are the same. When RPN generates RoIs, we also perform ROI Align on the RoIs in our keypoints branch, and the generated feature maps are up-sampled. We add a deconvolution layer and perform a ReLU activation on the feature maps to obtain the final heatmaps whose size are four times the size of the original feature maps. Although we separately perform ROI Align on the RoIs in the keypoints branch, we share the parameters of the FPN and RPN modules with the classification, regression and mask branches, which may reduce the accuracy of the mask branch slightly.

2.3   Orientation Voting Mechanism

The mask branch outputs the probability that each pixel belonging to each class. The knowledge of the digital image processing is used to binarize the mask image, find the contour of the mask, and finally obtain the minimum bounding box of the mask. Firstly, we rotate a horizontal line through the center point of the minimum bounding box counterclockwise. When the horizontal line first coincides with the edges of the minimum bounding box, the side is the width side of the minimum bounding box, and the height side can be obtained naturally. We define the rotated angle ranging from $0°$ to $180°$ is the angle of the minimum bounding box, as shown in Fig. 3(a). Secondly, we divide the image coordinate system into four quadrants defined as $q1$, $q2$, $q3$ and $q4$ counterclockwise, and map $b1$, $s1$, $s2$ to the four quadrants according to their coordinates and visibility, just as shown in Fig. 3(b). Thirdly, using the prior knowledge that the ship is narrow and long, we can know that $b1$ is in the two short sides of the minimum bounding box. Corresponding to Fig. 4 (the angle of the minimum bounding box ranges from 0 to 45°), $b1$ is in the height sides, not in the width sides, so we integrate $q1$ and $q2$ into the right quadrant, recorded as qr, and integrate $q3$ and $q4$ into the left quadrant, recorded as ql. Thus, we only need to judge whether the keypoints are in ql or in qr. We vote on the quadrant of each keypoint. The quadrant of $b1$ is $q2$, which belongs to qr, so the votes of qr is increased by 1, and the quadrant of $s1$ is $q3$, according to the assumption that the bow and stern symmetry about the center of the ship, we can determine reversely that the $b1$ corresponding to $s1$ is in $q1$. Similarly, the $b1$ corresponding to $s2$ is in $q2$. Since $q1$ and $q2$ belong to qr, qr gets a final votes of three, and ql gets the votes of zero. Although this reverse judgment is not very accurate, we only consider weather the keypoints are in ql or in qr, it does not affect the final result that $b1$ is exactly in the right height side of the minimum bounding box. When the angle is between 135 and 180°, the judgment method is the same.

When the angle of the minimum bounding box is between 45 and 135°, the algorithm is similar. We need to integrate $q2$ and $q3$ into the top quadrant, recorded as qt, and integrate $q1$ and $q4$ into the bottom quadrant, recorded as qb, then map $b1$, $s1$ and $s2$ to qt and qb respectively. The detailed method is shown in Fig. 5.

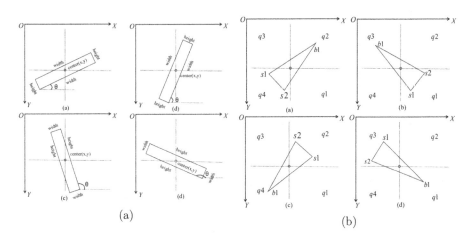

Fig. 3. (a) The width side of the minimum bounding box is different according to different angles. (b) Map $b1$, $s1$ and $s2$ to four quadrants.

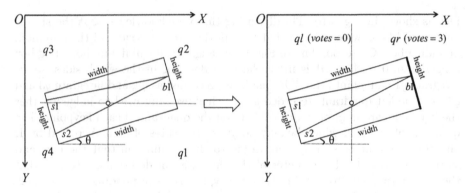

Fig. 4. Map $b1$, $s1$ and $s2$ to ql and qr when the angle is between 0 and 45° or between 135 and 180°.

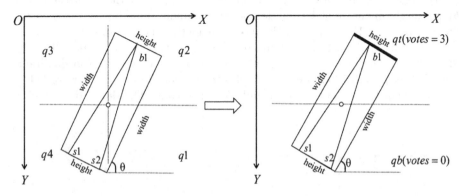

Fig. 5. Map $b1$, $s1$ and $s2$ to qt and qb when the angle is between 45 and 135°.

3   Experiments

3.1   Dataset

We have built a ship's keypoints detection dataset on remote sensing images. The original images come from the HRSC2016 dataset [17]. The dataset contains images from seven famous harbors, six of them come from Google Earth in different years and one comes from Mumansk harbor. The image resolutions are between 2-m and 0.4-m. The high-resolution images are clipped into smaller images ranging from 300 × 300 to 1500 × 900 including 226 sea images and 1445 sea-land images.

In order to label the keypoints of the ship, we removed some images that did not contain the ship object. For the remaining images, we label $b1$, $s1$ and $s2$ counterclockwise for each of the ships. After adding annotations to these samples, we finally get 1464 images with 5258 samples in total. We split the dataset into training set including 1172 images with 4187 samples and test set including 292 images with 1071 samples respectively.

We also created a separate dataset to test the applicability of our model. The images are part of the training set in the kaggle Airbus Ship Detection Challenge [18]. The training set has 104168 images whose size are 768 × 768. The target of each image is a merchant or civilian ship with a background in the sea, which is different from the HRSC2016 dataset where most of the backgrounds are the port. Since most of the images are background without a target, and labeling the mask dataset requires time and labor, we only select 122 images containing the ship from the 104168 images to create a dataset named KASDC2018.

3.2 Evaluation Indicators

Our experiments are mainly to verify the effectiveness of estimating the ship orientation by combining ship keypoints detection and ship segmentation techniques. To this end, we did a comparative experiment with [11]. Our architecture is very similar to the method mentioned in [11]. Both the classification, detection and segmentation of ships are based on the Mask R-CNN architecture. The difference is that we add the keypoints detection module to the Mask R-CNN architecture, and the final orientation of the ship is determined by integrating the mask information and the keypoints information. During the experiment, we have the same parameter settings for all the same parts of [11], the initial learning rate is 0.001 and the total number of iterations is 90,000. We adopt AP (average precision), AP_{50} and AP_{75} (average precision rates assessed at 0.5 and 0.75 IoU threshold), as well as AP on small (AP_S), medium (AP_M) and large (AP_L) targets as the evaluation indicators of the ship detection and segmentation. The experimental results in Tables 1 and 2 show that our method can guarantee the accuracy almost not reduced and output the orientation information of the ship in addition.

Table 1. Box detection task of our method.

Method	AP	AP_{50}	AP_{75}	AP_S	AP_M	AP_L
Method in [11]	0.766	0.927	0.888	0.635	0.723	0.836
Our method	0.756	0.941	0.883	0.720	0.720	0.825

We also validated the validity of the voting mechanism to estimate the ship's direction. Referring to [10], we use precision rate and recall rate with four components which are true positive (TP), false positive (FP), true negative (TN) and false negative (FN) as the quantitative indicators of direction estimation. We define the precision rate and recall rate as:

$$Precision = \frac{TP}{TP + FP} \qquad (1)$$

$$Recall = \frac{TP}{TP + FN} \qquad (2)$$

Table 2. Segmentation task of our method.

Method	AP	AP_{50}	AP_{75}	AP_S	AP_M	AP_L
Method in [11]	0.638	0.880	0.774	0.438	0.540	0.740
Our method	0.644	0.888	0.776	0.421	0.534	0.755

Table 3. The precision and recall rate of ship direction estimation.

Method	Precision	Recall
One-bow method	0.917	0.866
Two-sterns method	0.918	0.867
Three-keypoints method	0.957	0.904

The *Precision* refers to the ratio of the number of the correct directions in all detected directions and the *Recall* means the ratio of the number of correct directions in all target directions. We compare the segmentation result with ground truth, and only do the direction statistics for the segmentation results whose iou with ground truth is greater than 0.7 and whose category confidence is greater than 0.7. It is meaningless to make a direction estimation based on the segmentation result when the iou or the category confidence is too low. We have designed three different methods to estimate the direction of the ship according to the number of keypoints used, namely one-bow method (only using $b1$), two-sterns method (only using $s1$ and $s2$) and three-keypoints method (using both $b1$, $s1$ and $s2$). The experiment results in Table 3 show that the precision and recall rate of one-bow method and two-sterns method are almost the same, but much lower than that of three-keypoints method. It may be because one-bow and two-sterns method only work well in some conditions. Take the one-bow method as an example, once $b1$ is not detected (maybe the detection score of $b1$ is too low, or $b1$ is invisible), or falsely detected, it will directly lead to errors in the direction judgment. Even if the $s1$ or $s2$ is detected correctly, one-bow method chose to ignore it. The same problem also exists in the two-sterns method. If we use $b1$, $s1$ and $s2$, we can make up for each other's defects to circumvent these problems. The precision and recall rate of three-keypoints method are significantly improved, this is why we use the three-keypoints ship model as it improves the robustness of our algorithm greatly. The detection result is shown in Fig. 6.

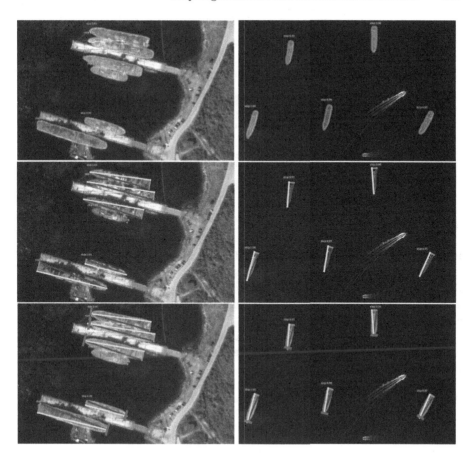

Fig. 6. Combining the minimum bounding box of the mask and the keypoints of bow and sterns to infer the orientation of the ship. Each row represents the mask results, keypoints results, and final results graph respectively. The red line in the third row indicates the direction of bow. (Color figure online)

Finally, we performed experiments on the KASDC2018 dataset. The ship detection, segmentation and direction estimation results are shown in Tables 4, 5, and 6 respectively. The experimental results show that our algorithm has generalization ability.

Table 4. Box detection task in KASDC2018 dataset.

AP	AP_{50}	AP_{75}	AP_S	AP_M	AP_L
0.756	0.942	0.883	0.720	0.713	0.825

Table 5. Segmentation task in KASDC2018 dataset.

AP	AP_{50}	AP_{75}	AP_S	AP_M	AP_L
0.644	0.888	0.776	0.421	0.534	0.755

Table 6. The results of ship direction estimation in KASDC2018 dataset.

Precision	Recall
0.804	0.765

4   Conclusion

In this paper, we present a new method for estimating ship orientation based on Mask R-CNN. We integrated the keypoints detection technology into the Mask R-CNN basic framework and achieved good results on our dataset. We adopted a voting mechanism to ensure that the ship's orientation is accurately estimated when some of the individual keypoints of the ship are not accurately detected, which improves the robustness of our algorithm. Through the orientation judgment, we realized a more detailed semantic understanding of the ship in the remote sensing image.

Acknowledgments. This work was supported by the National Science and Technology Major Project of China grant number 2018ZX01008103 and National Key Research and Development Program of China under Grant 2017YFC0803905.

References

1. Yang, X.: Automatic ship detection in remote sensing images from Google Earth of complex scenes based on multiscale rotation dense feature pyramid networks. Remote Sens. **10**(1), 132 (2018)
2. Zhang, R.: S-CNN ship detection from high-resolution remote sensing images. ISPRS - Int. Arch. Photogramm. Remote Sens. Spat. Inf. Sci. **XLI-(B7)**, 423–430 (2016)
3. Liu, W.: Automated vehicle extraction and speed determination from QuickBird satellite images. IEEE J. Sel. Top. Appl. Earth Obs. Remote Sens. **4**(1), 75–82 (2011)
4. Tong, X.: Building-damage detection using pre- and post-seismic high-resolution satellite stereo imagery: a case study of the May 2008 Wenchuan earthquake. ISPRS J. Photogramm. Remote Sens. **68**, 13–27 (2012)
5. Zhu, C.: A novel hierarchical method of ship detection from spaceborne optical image based on shape and texture features. IEEE Trans. Geosci. Remote Sens. **48**(9), 3446–3456 (2010)
6. He, K., Sun, J.: Convolutional neural networks at constrained time cost. In: 2015 IEEE Conference on Computer Vision and Pattern Recognition (CVPR), pp. 5353–5360. IEEE, Boston (2015)

7. Yu, Y.: Automated ship detection from optical remote sensing images. Key Eng. Mater. **500**, 785–791 (2012)
8. Ren, S., He, K., Girshick, R., Sun, J.: Faster R-CNN: towards realtime object detection with region proposal networks. In: Advances in Neural Information Processing Systems, vol. 28, pp. 91–99. Curran Associates Inc., Montreal (2015)
9. Girshick, R.: Fast R-CNN. In: 2015 IEEE International Conference on Computer Vision (ICCV), pp. 1440–1448. IEEE, Santiago (2015)
10. Yang, X.: Position detection and direction prediction for arbitrary-oriented ships via multitask rotation region convolutional neural network. IEEE Access **6**, 50839–50849 (2018)
11. Nie, S., Jiang, Z., Zhang, H., Cai, B., Yao, Y.: Inshore ship detection based on mask R-CNN. In: IGARSS 2018, 2018 IEEE International Geoscience and Remote Sensing Symposium, pp. 693–696. IEEE, Valencia (2018)
12. Andriluka, M., Roth, S., Schiele, B.: Pictorial structures revisited: people detection and articulated pose estimation. In: 2009 IEEE Conference on Computer Vision and Pattern Recognition, pp. 1014–1021. IEEE, Miami (2009)
13. Newell, A., Yang, K., Deng, J.: Stacked hourglass networks for human pose estimation. In: Leibe, B., Matas, J., Sebe, N., Welling, M. (eds.) ECCV 2016. LNCS, vol. 9912, pp. 483–499. Springer, Cham (2016). https://doi.org/10.1007/978-3-319-46484-8_29
14. Chen, Y., Wang, Z., Peng, Y., Zhang, Z., Yu, G., Sun, J.: Cascaded pyramid network for multi-person pose estimation. In: 2018 IEEE/CVF Conference on Computer Vision and Pattern Recognition, pp. 7103–7112. IEEE, Salt Lake City (2018)
15. He, K., Gkioxari, G., Dollar, P., Girshick, R.: Mask R-CNN. In: 2017 IEEE International Conference on Computer Vision(ICCV), pp. 2980–2988. IEEE, Venice (2017)
16. Lin, T., Dollar, P., Girshick, R., He, K., Hariharan, B., Belongie, S.: Feature pyramid networks for object detection. In: 2017 IEEE Conference on Computer Vision and Pattern Recognition (CVPR), pp. 936–944. IEEE, Honolulu (2017)
17. Liu, Z., Yuan, L., Weng, L., Yang, Y.: A high resolution optical satellite image dataset for ship recognition and some new baselines. In: 6th International Conference on Pattern Recognition Applications and Methods, pp. 324–331. SciTePress, Porto (2017)
18. Airbus Ship Detection Challenge. https://www.kaggle.com/c/airbus-ship-detection. Accessed 1 Jan 2019

A Deep Learning Approach to Detecting Changes in Buildings from Aerial Images

Bin Sun[1], Guo-Zhong Li[2], Min Han[1], and Qiu-Hua Lin[1(✉)]

[1] Faculty of Electronic Information and Electrical Engineering,
Dalian University of Technology, Dalian 116024, China
qhlin@dlut.edu.cn
[2] Geotechnical Engineering and Mapping Institute CO., LTD,
Dalian 116021, China

Abstract. Detecting building changes via aerial images acquired at different times is important in the urban planning and geographic information updating. Deep learning solutions have high potential in improving detection performance as compared with traditional methods. However, existing methods usually carry out detection for whole images. Non-building interferences involved may result in an increase of false alarm rate, a decrease in accuracy rate, and a heavy computational load. In addition, they mostly utilize supervised deep learning networks dependent highly on massive labeled samples. In this study, we present an unsupervised deep learning solution with detection only on segmented building areas. We first employ a masking technique based on building segmentation to remove non-building interferences. We then use a classification model combing an unsupervised deep learning network PCANet and linear SVM to realize building change detection. Experimental results show that our method achieves 34.96% higher accuracy rate, 45.18% lower missed detection rate, 37.92% lower false alarm rate, and 50.12% lesser computational time than the whole-image detection method without building segmentation.

Keywords: Building change detection · Masking ·
Unsupervised deep learning network · Non-building interferences

1 Introduction

Detection of building changes plays an important role in the urban planning, geographic information updating, and land resources management. With the development of remote sensing and aerial photography, building change is readily detected based on satellite or aerial images acquired at different times.

The existing change detection methods can be divided into two categories, traditional methods and newly developed deep learning methods. Compared to the traditional methods based on such as clustering [1], object-level image analysis [2, 3], and geographic information systems (GIS) [4], deep learning methods make features extraction more robust by applying neural networks. Deep learning methods can be further grouped into three categories, according to the need of labeled samples. They are supervised, semi-supervised, and unsupervised methods, respectively. Although the

© Springer Nature Switzerland AG 2019
H. Lu et al. (Eds.): ISNN 2019, LNCS 11555, pp. 414–421, 2019.
https://doi.org/10.1007/978-3-030-22808-8_40

supervised and semi-supervised methods [5–8] usually have higher accuracy of features extraction, these methods have a strong dependency on the amount of labeled samples. In practice, it is not easy to collect a large amount of labeled samples for satellite or aerial images. Therefore, unsupervised methods are more promising in detection of changes. Zhang et al. proposed a method applying the Gaussian-Bernoull Deep Boltzmann Machine (GDBM), which can extract deep features to achieve change detection [9]. Gao et al. proposed an automatic change detection algorithm; the change detection for images was realized by utilizing the excellent features extraction ability of PCANet [10].

However, existing unsupervised deep learning methods such as GDBM usually carry out detection for the whole image. In the aerial or satellite images, there are not only buildings, but also non-building environments, such as roads, vehicles and pedestrians. These inevitably produce different degrees of interference to the change detection, including the increase of false alarm rate and the computational load. In the meantime, the accuracy rate is also reduced. As such, we propose to detect building changes using unsupervised deep learning method together with removal of non-building interferences. Specifically, we employ masking technique based on building segmentation to remove non-building interferences, and use a classification model combing an unsupervised deep learning network PCANet [11] and linear support vector machine (SVM) to realize change detection. We carry out evaluations of the proposed method using the aligned aerial images provided by Dalian Geotechnical Engineering and Mapping Institute CO., LTD.

The remaining of this paper is organized as follows. Section 2 describes details of the proposed method. Section 3 presents experiments and results. The conclusion is given in Sect. 4.

2 Proposed Method

Given two aerial images $I^{(x)}(i,j) \in \mathbb{R}^{m \times n}$, $x = 1, 2$; $i = 1, \ldots, m$; $j = 1, \ldots, n$, $m \times n$ is the size of the images, which are acquired at different times. Figure 1 shows the framework of the proposed method. There are generally two stages, firstly, segmentation and masking stage, secondly, sample images generation and pixels classification stage.

In the segmentation and masking stage, the two input images $I^{(1)}$ and $I^{(2)}$ are first fed into an image segmentation algorithm to extract buildings, the two segmentation images are used to generate two binary masks and then a merged binary mask. Next, non-building interferences are removed by masking the original images with the merged mask, and the denoising images $I_p^{(1)}$ and $I_p^{(2)}$ are to be detected for changes.

In the second stage, sample images are generated based on pre-classification results for pixels in $I_p^{(1)}$ and $I_p^{(2)}$, and utilized for training PCANet and linear SVM. Note pre-classification results are obtained by the pixel-level operation, which will be described in detail in Subsect. 2.2. The trained linear SVM classify the intermediate pixels in pre-classification into changed and unchanged classes. The changed pixels detected by SVM classification and the pre-classification form the final change map.

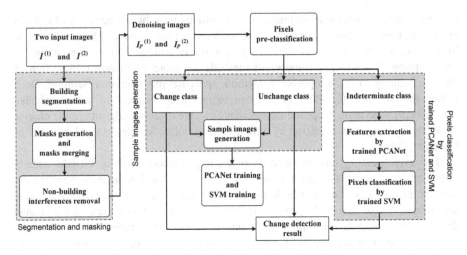

Fig. 1. Framework of the proposed method

2.1   Masking Technique Based on Building Segmentation

Figure 2 shows the main steps in the segmentation and masking on two input images. We first use a deep neural network U-Net [12] to realize the segmentation of buildings in the two input images $I^{(1)}$ and $I^{(2)}$. Two binary masks are then generated based on the segmentation images $I_s^{(1)}$ and $I_s^{(2)}$ as follows:

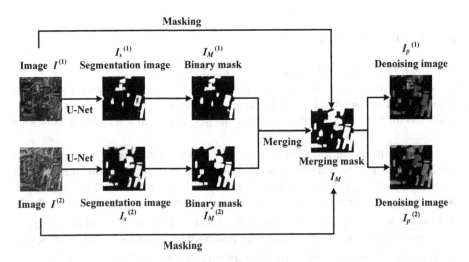

Fig. 2. Image segmentation and masking

$$I_M^{(x)}(i,j) = \begin{cases} 2^B - 1, & \text{buildings in } I_s^{(x)}(i,j) \\ 0, & \text{else} \end{cases} \tag{1}$$

where $I_M^{(x)} \in \mathbb{R}^{m \times n}$, $x = 1, 2$, denote the binary masks, and B is the color depth of single channel.

Next, we combine two binary masks $I_M^{(x)}$ to obtain a merging binary mask to cover whole buildings included in the two images. The merging mask $I_M \in \mathbb{R}^{m \times n}$ is generated as follows:

$$I_M(i,j) = \begin{cases} 2^B - 1, & I_M^{(1)}(i,j) = 2^B - 1, \ or \ I_M^{(2)}(i,j) = 2^B - 1 \\ 0, & I_M^{(1)}(i,j) = 0 \ and \ I_M^{(2)}(i,j) = 0 \end{cases} \tag{2}$$

Finally, we use the merging mask I_M to remove non-building interferences from the original images:

$$I_p^{(x)}(i,j) = \begin{cases} I^{(x)}(i,j), & I_M(i,j) = 2^B - 1 \\ 0, & I_M(i,j) = 0 \end{cases} \tag{3}$$

where $I_p^{(x)} \in \mathbb{R}^{m \times n}$, $x = 1, 2$, are the denoising images used for the sequential change detection.

2.2   Sample Images Generation and Pixels Classification

We use the method proposed in [10] to generate sample images for training PCANet. A log-ratio image $I_D(i,j) = |\log(I_p^{(1)}(i,j)/I_p^{(2)}(i,j))|$ is generated. Pixels in I_D are pre-classified (changed, unchanged, and intermediate classes) and labelled using Gabor wavelet representation and fuzzy c-means (FCM). Sample images are generated with centers of labelled pixels which having high probability of being changed or unchanged, the number of which is N_{sum}.

Specifically, given a labelled pixel (i, j), we generate image patches $p_{ij}^1 \in \mathbb{R}^{k \times k}$ and $p_{ij}^2 \in \mathbb{R}^{k \times k}$ of size $k \times k$ centered at pixel (i, j) from the two denoising images $I_p^{(1)}$ and $I_p^{(2)}$, and then vertically combine the two image patches to form a sample image $P_y \in \mathbb{R}^{2k \times k}$. At last, we obtain a set of sample images $\{P_y\}_{y=1}^{N_{sum}}$.

We train PCANet using sample images $\{P_y\}_{y=1}^{N_{sum}}$. The extracted features from PCANet are fed into linear SVM for training. The classification model is then constructed by combining the trained PCANet and SVM. This model is then employed to classify the pixels belonging to the intermediate class (obtained in pre-classification) into changed and unchanged classes. Finally, we combine the changed pixels detected by SVM classification and the pre-classification to output the final change map.

2.3   Performance Indexes

We use the following four performance indexes to evaluate the detection results, including accuracy rate ACC, missed detection rate P_{MD}, false alarm rate P_{FA}, and computational time.

The accuracy rate is defined as:

$$ACC = (N_{TUC} + N_{TC})/(N_{FUC} + N_{FC} + N_{TUC} + N_{TC}) \times 100\% \tag{4}$$

where N_{FC} denotes the number of unchanged pixels but are detected as changed, while N_{TC} represents the number of pixels being correctly detected as changed. In a similar way, N_{FUC} represents the number of changed pixels but are detected as unchanged, and N_{TUC} is the number of correct detection of unchanged pixels.

The missed detection rate is defined as:

$$P_{MD} = N_{FUC}/(N_{FUC} + N_{TUC}) \times 100\% \tag{5}$$

The false alarm rate is computed by:

$$P_{FA} = N_{FC}/(N_{FC} + N_{TC}) \times 100\% \tag{6}$$

Different from the whole-image detection method, the computational time of the proposed method include the time consumed by the image segmentation and masking, in addition to the time for pixels classification.

3   Experiments and Results

3.1   Datasets and Parameters

We tested the proposed method with 105 sets of aerial images collected in 2008 and 2013, respectively, by Dalian Geotechnical Engineering and Mapping Institute CO., LTD. The size of the test images is 512×512 with single channel depth of 8. We trained U-Net using the dataset collated and released by the CSU-DP laboratory [13]. There is a total of 137 aerial images of size 1500×1500, and their spatial resolution is 0.22 m. In order to adapt to the network training, we divided each image into nine small images of size 512×512, thus generate 1,233 small images. The super-parameters for training U-Net include a learning rate ($\alpha = 10^{-5}$) and the number of epochs ($epochs = 19$). The test platform is run on an Intel Core i7-4790 processor, with a 3.6 GHz frequency, 16 GB of memory, and the operating system is Windows 7. The programming software is Spyder (python 3.5) in the first stage, and MATLAB R2014a in the second stage.

When using PCANet and SVM to perform classification, the main parameters include stages of PCA filter convolutions in the PCANet, the number of filters for each stage, the penalty coefficient C for SVM, and the patch size $k \times k$ in the process of sample images generation. We used 2 stages of PCA filter convolutions for PCANet, with 8 PCA filters at each stage, and $C = 1$ for SVM. As for the patch size $k \times k$, we test the results of the odd numbers between 3 and 11 for k, and select $k = 9$ due to its better performance.

3.2 Experimental Results

Table 1 shows comparison of the proposed method with the whole-image detection method, in terms of four performance indexes averaged across 105 sets of test images. It can be observed that the proposed method can improve the accuracy rate, and reduce the missed detection rate, the false alarm rate, and the computational time. Specifically, the accuracy rate was improved by 34.96%, the missed detection rate was reduced by 45.18%, the false alarm rate was decreased by 37.92%, and the computational time was decreased by 50.12%. Note the improvement of the missed detection rate and the false alarm rate were limited by the performance of image segmentation.

Table 1. Comparison of the proposed method with the whole-image detection method for detecting changes in 105 groups of aerial images, in terms of ACC, P_{MD}, P_{FA}, and computational time. The average values are included, and the better results are shown in bold.

	ACC (%)	P_{MD} (%)	P_{FA} (%)	Time (s)
Proposed	**83.35**	**10.86**	**45.25**	**3748**
Whole-image	61.76	19.81	72.89	7514

The experimental results of the proposed method for five exampling groups of aerial images are shown in Fig. 3 and Table 2. Figure 3(a) and (b) display whole images of the five groups of aerial images, respectively. Figure 3(c) and (d) represent the corresponding segmented images for buildings. Figures 3(e) and (f) are binary masks generated using Eq. (1), and Fig. 3(g) illustrates the merging masks obtained using Eq. (2). With the merging masks, the resulting denoising images are shown in Fig. 3(h) and (i). The changes detected by the proposed method are shown in Fig. 3(j). For comparison, we also show ground-truth change maps marked manually in Fig. 3 (k), and results obtained by proposed method without masking of non-building interferences in Fig. 3(l). We can see that the proposed method achieved much closer results with the ground-truth, as compared with the whole-image detection method.

Table 2 shows the four performance indexes obtained by the proposed method and the whole-image detection method. The proposed method improved the accuracy rate by 26.43%, reduced the false alarm rate by 70.72%, and decreased the computational time by 47.99%. The missed detection rate was also decreased by 12.29% on average, but this index was increased for group 4 and group 5 because of the unsatisfying results of image segmentation.

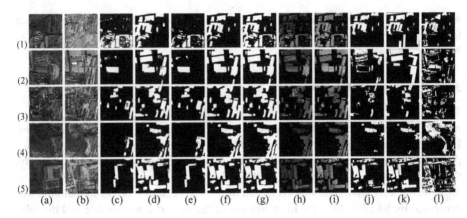

Fig. 3. The building change detection results. (a) Aerial images of 2008; (b) Aerial images of 2013; (c) Segmentation images of 2008; (d) Segmentation images of 2013; (e) Binary masks of 2008; (f) Binary masks of 2013; (g) Merging masks; (h) Denoising images of 2008; (i) Denoising images of 2013 (j) Results by the proposed method; (k) Ground-truth change maps; (l) Results by whole-image detection method.

Table 2. Comparison of the proposed method with the whole-image detection method for detecting changes in 5 exampling groups of aerial images, in terms of ACC, P_{MD}, P_{FA}, and computational time. The better results are shown in bold.

Test		ACC (%)	P_{MD} (%)	P_{FA} (%)	Time (s)
1	Proposed	**89.50**	**10.57**	**10.20**	**3244**
	Whole-image	81.91	15.78	29.68	5031
2	Proposed	**84.09**	**13.85**	**22.36**	**3459**
	Whole-image	66.60	18.26	55.06	7311
3	Proposed	**91.43**	**6.54**	**29.53**	**3532**
	Whole-image	75.05	11.78	86.11	7228
4	Proposed	**96.25**	2.66	**15.23**	**3035**
	Whole-image	72.24	**2.15**	75.94	6291
5	Proposed	**86.69**	16.35	**12.37**	**4185**
	Whole-image	58.49	**8.99**	59.63	7766
Average	Proposed	**89.59**	**9.99**	**17.94**	**3498**
	Whole-image	70.86	11.39	61.28	6725

4   Conclusion

We proposed a framework for building change detection based on unsupervised deep neural network combined with removal of non-building interferences. In contrast to the whole-image method, our method shows promising results in reducing the false alarm rate, decreasing the computational time, and improving the accuracy rate. Moreover, the unsupervised deep neural network does not rely on massive labeled samples. Therefore, this method is less demanding and more flexible. In the future, we will optimize the segmentation approach in an effort to retain full building information and to reduce both the false alarm rate and missed detection rate. In addition, new methods can be developed for generating sample images, thus the feature extraction accuracy can be further improved for the unsupervised deep neural network.

References

1. Sjahputera, O., et al.: Clustering of detected changes in satellite imagery using fuzzy C-means algorithm. In: International Geoscience and Remote Sensing Symposium, Hawaii, United States, pp. 468–471 (2010)
2. Argialas, D., Michailidou, S., Tzotsos, A.: Change detection of buildings in suburban areas from high resolution satellite data developed through object based image analysis. Surv. Rev. 45(333), 441–450 (2013)
3. Tabib Mahmoudi, F., Samadzadegan, F., Reinartz, P.: Context aware modification on the object based image analysis. J. Indian Soc. Remote Sens. 43(4), 709–717 (2015)
4. Sofina, N., Ehlers, M.: Building change detection using high resolution remotely sensed data and GIS. IEEE J. Sel. Top. Appl. Earth Obs. Remote Sens. 9(8), 3430–3438 (2016)
5. Bruzzone, L., Serpico, S.B.: An iterative technique for the detection of land-cover transitions in multitemporal remote-sensing images. IEEE Trans. Geosci. Remote Sens. 35(4), 858–867 (1997)
6. Fernández-Prieto, D., Marconcini, M.: A novel partially supervised approach to targeted change detection. IEEE Trans. Geosci. Remote Sens. 49(12), 5016–5038 (2011)
7. Roy, M., Ghosh, S., Ghosh, A.: A neural approach under active learning mode for change detection in remotely sensed images. IEEE J. Sel. Top. Appl. Earth Obs. Remote Sens. 7(4), 1200–1206 (2014)
8. Zagoruyko, S., Komodakis, N.: Learning to compare image patches via convolutional neural networks. In: Proceedings of the IEEE Computer Society Conference on Computer Vision and Pattern Recognition, Boston, United States, pp. 4353–4361 (2015)
9. Zhang, X., Chen, X., Li, F., Yang, T.: Change detection method for high resolution remote sensing images using deep learning. Acta Geodaetica Cartogr. Sin. 46(8), 999–1008 (2017)
10. Gao, F., Dong, J., Li, B., Xu, Q.: Automatic change detection in synthetic aperture radar images based on PCANet. IEEE Geosci. Remote Sens. Lett. 13(12), 1792–1796 (2017)
11. Chan, T.-H., Jia, K., Gao, S.: PCANet: a simple deep learning baseline for image classification? IEEE Trans. Image Process. 24(12), 5017–5032 (2015)
12. Ronneberger, O., Fischer, P., Brox, T.: U-Net: convolutional networks for biomedical image segmentation. In: Navab, N., Hornegger, J., Wells, W.M., Frangi, A.F. (eds.) MICCAI 2015. LNCS, vol. 9351, pp. 234–241. Springer, Cham (2015). https://doi.org/10.1007/978-3-319-24574-4_28
13. CSU-DP laboratory. https://github.com/gengyanlei/build_segmentation_dataset

Scene Recognition in User Preference Prediction Based on Classification of Deep Embeddings and Object Detection

Andrey V. Savchenko$^{(\boxtimes)}$ and Alexandr G. Rassadin

Laboratory of Algorithms and Technologies for Network Analysis,
National Research University Higher School of Economics, Nizhny Novgorod, Russia
avsavchenko@hse.ru, arassadin@hse.ru

Abstract. In this paper we consider general scene recognition problem for analysis of user preferences based on his or her photos on mobile phone. Special attention is paid to out-of-class detections and efficient processing using MobileNet-based architectures. We propose the three stage procedure. At first, pre-trained convolutional neural network (CNN) is used extraction of input image embeddings at one of the last layers, which are used for training a classifier, e.g., support vector machine or random forest. Secondly, we fine-tune the pre-trained network on the given training set and compute the predictions (scores) at the output of the resulted CNN. Finally, we perform object detection in the input image, and the resulted sparse vector of detected objects is classified. The decision is made based on a computation of a weighted sum of the class posterior probabilities estimated by all three classifiers. Experimental results with a subset of ImageNet dataset demonstrate that the proposed approach is up to 5% more accurate when compared to conventional fine-tuned models.

Keywords: Image recognition · Scene recognition ·
Convolutional neural network (CNN) · Object detection ·
Ensemble of classifiers · Classifier fusion

1 Introduction

Deep understanding of user characteristics by analyzing the photos on a mobile device is a challenging task due to the extreme variability of visual data gathered in uncontrolled environment [1]. Such user preference prediction engine is applicable for various personalized services. Unfortunately, the analytics engines should be preferably run on the embedded system, because most photos contain private information, for which the user does not permit the remote client-server processing. As a result, the state-of-the-art very deep convolutional neural networks (CNNs) [2] cannot be directly applied due to their enormous inference time and energy consumption. Hence, it is extremely important to seek the ways to improve the accuracy of MobileNet-based architectures [3,4].

© Springer Nature Switzerland AG 2019
H. Lu et al. (Eds.): ISNN 2019, LNCS 11555, pp. 422–430, 2019.
https://doi.org/10.1007/978-3-030-22808-8_41

In this paper we focus on one of the main parts of visual preference prediction engine, namely, scene recognition, in order to extract such interests as art and theater, nightlife, sport, etc. The scene is composed of parts and some of those parts can be named and correspond to objects [5]. It is known that the object detection models [6] are used to estimate the scene inconsistencies [7]. The region proposals with potentially most important objects can be detected with modern saliency detection methods [8,9]. In this paper we developed the novel approach to improve the accuracy of scene recognition by using scores and embeddings from the MobileNets combined in an ensemble with the object detector.

The rest of the paper is organized as follows. In Sect. 2 we introduce the proposed algorithm. Experimental results for 40 scenes from ImageNet datasets are presented in Sect. 3. Concluding comments are discussed in Sect. 4.

2   Materials and Methods

In this paper we consider scene recognition as a general image classification task, in which it is required to assign an input photo X to one of $C > 1$ categories (classes). The classes are defined with the aid of $N \geq C$ reference images $\{X_n\}, n \in \{1, ..., N\}$, and the class label $c_n \in \{1, ..., C\}$ of each n-th reference image is supposed to be given. We assume that the input image and all reference images are resized in order to have the same width W and height H.

If the training sample is rather small to train a deep CNN from scratch, the transfer learning or domain adaptation can be applied [2]. In these methods a large external dataset, e.g. ImageNet-1000 [10], is used to pre-train a deep CNN. As we pay special attention to offline recognition on mobile devices, it is reasonable to use such CNNs as MobileNet v1/v2 [3,4]. The final step in transfer learning is fine-tuning of this neural network using the training set from the limited sample of instances $\{X_n\}$. Hence, this step includes replacement of the last (logistic regression) layer of the pre-trained CNN to the new layer with Softmax activations and C outputs. During the classification process, the input image is fed to the fine-tuned CNN, and the scores (predictions at the last layer) $\mathbf{p} = [p_1, ..., p_C], \sum_{c=1}^{C} p_c = 1$ are used to make a final decision in favour of the most probable class $c^* = \underset{c \in \{1,...,C\}}{\operatorname{argmax}} p_c$ [11].

This procedure can be modified by replacing the logistic regression in the last layer to more complex classifier. In this case the off-the-shelf features [12] are extracted using the outputs of one of the last layers of pre-trained or fine-tuned CNN [13,14]. It is especially suitable for small training samples ($N \approx C$) when the results of the fine-tuning is not too accurate. Namely, the images X and X_n are fed to the CNN, and the outputs of the one-but-last layer are used as the D-dimensional feature vectors $\mathbf{x} = [x_1, ..., x_D]$ and $\mathbf{x}_n = [x_{n;1}, ..., x_{n;D}]$, respectively. Such deep learning-based feature extractors allow training of a general classifier, e.g., random forest (RF), support vector machine (SVM) or gradient boosting that performs nearly as well as if a large training dataset of images from these C classes is available [2].

In this paper we propose to extend such traditional approach with additional features. Namely, we use either classification of off-the-shelf CNN features or similar approach with classification of the *scores (predictions)* $\mathbf{p_n} = [p_{n;1}, ..., p_{n;C}]$ from the last layer of the pre-trained CNN. Moreover, we noticed that many scenes from the one category contains identical objects (e.g., ball in the football), which can be detected by contemporary CNN-based methods, i.e., SSDLite [4] or Faster R-CNN [6]. These methods detect the positions of several objects in the input image and predict the scores of each class from the predefined set of $K > 1$ types. We decided to completely ignore bounding boxes and only extract the sparse vector $\mathbf{o} = [o_1, ..., o_K]$ of scores for each type. If there are several objects of the same type, the maximal score is stored in this feature vector.

The complete pipeline is shown in Fig. 1 and the general idea of our approach is presented in Fig. 2.

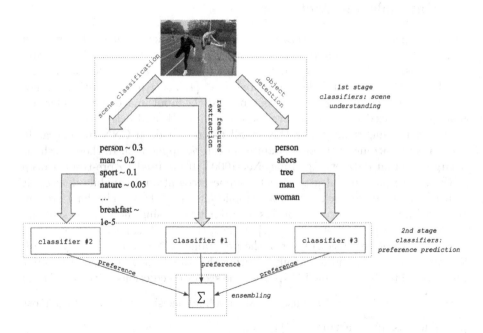

Fig. 1. Proposed pipeline of scene recognition

Here we implemented the most powerful classifier fusion technique, which consists of the features from the pre-trained model, scores from the fine-tuned CNN and predictions of the Faster R-CNN object detection model trained on large dataset, e.g. Open Image Dataset. The outputs of individual classifiers are combined with soft aggregation, i.e., the decision is taken in favor of the class with the highest weighted sum of outputs of individual classifier [15]. The weights can be chosen proportionally to the accuracy of individual classifier estimated on the special validation set [16]. However, if the RF classifier is applied, the out-of-bag (OOB) error can be used as its weight. As many objects will appear together

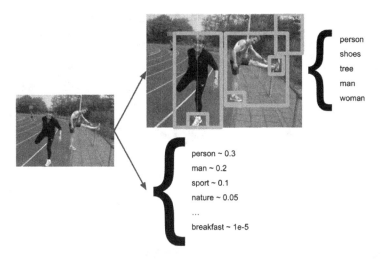

Fig. 2. Example of scene recognition using proposed pipeline

in a variety of scenarios and the object detection may be wrongly prejudged, we set the weight corresponding to the detector to be much smaller when compared to the weights of other fused classifiers.

It is important to emphasize that scene recognition usually requires rejection of all categories by making out-of-class decision. In order to deal with the out-of-class recognition, it is possible to gather additional training images from other categories and put them into the training set. This set can be used in various ways. For example, the new binary classifier can be created for the prediction vector p to verify that the decision c^* is reliable. However, we experimentally noticed that the most efficient approach is to train a separate (negative) class within basic image classification models rather than work over the output of models without negative class. Indeed, training such models can be extremely sophisticated because the positive classes are just a little fraction of the whole images variety. Hence, in this paper we decided to simply use additional $(C+1)$-th "background" category during the training of classifiers and fine-tuning of the CNNs. It was experimentally found that such approach still suffers from false negatives, though precision of out-of-class detection is approximately equal to 99%. The qualitative results for MobileNet v2 are shown in Fig. 3.

The proposed approach (Fig. 1) was implemented as a part of special mobile application for user preference prediction on Android. The application may operate in offline mode and does not require Internet connections. Several results of scene recognition together with the bounding boxes of detected objects are shown in Fig. 4. One can notice here the qualitative reasons for object detection to improve the scene recognition results. In fact, the detected objects (car/bus, food/drink/coffee cup and sport equipment) have high correlation with the scene labels "street" (Fig. 4a), "restaurant" (Fig. 4b) and "football" (Fig. 4c), respectively.

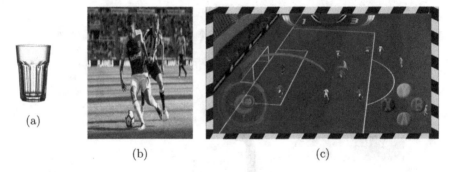

Fig. 3. The results of out-of-class detection: (a) True negative; (b) True positive; (c) False positive (soccer scene is detected as out-of-class sample).

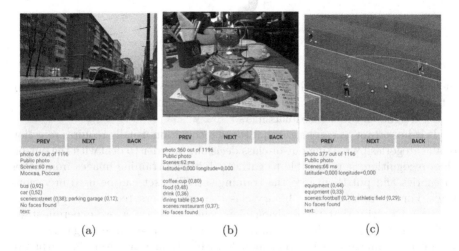

Fig. 4. Results of scene recognition in the mobile demo

3 Experimental Results

We downloaded an imbalanced set of 45 K image from preliminarily chosen $C = 40$ preference categories of scenes, which are presented as synsets in the ImageNet dataset (sport scenes, drugstore, gas station, beauty shop/salon, spa, department store, bookstore, grocery store, amusement park, gallery, art gallery, picture gallery, music hall, opera, cinema, alehouse, cabaret, nightclub, night club, club, nightspot, news magazine, comic book) [10]. Sample images are shown in Fig. 5. In order to implement the out-of-class detection, new "distractor" category with out-of-class images from Caltech101/Caltech-256 datasets was added.

This dataset was split into training/test sets with 85% of data (47983 images) in the training set and the rest 8490 images in the test set. The training set with appropriate data augmentation was used to fine-tune several CNNs, namely,

Fig. 5. Sample images of interest scenes from ImageNet

highly accurate state-of-the-art Inception v3 architecture and fast MobileNet v1 and MobileNet v2 with width multipliers $\alpha = 1$ and $\alpha = 1.4$. At first, all the layers of the pre-trained model were frozen, and the weights of the new fully connected layer were trained with the Adam optimizer (learning rate 0.001) during 10 epochs. Next, all weights were optimized during 40 epochs by the stochastic gradient descent method (learning rate 0.0001) with early stopping.

We examined several classification techniques for inclusion into our final pipeline. In particular, we focused on classification of: (1) features at the output of the last by one layer of the CNN, which was pre-trained on the ImageNet-1000 dataset, which does not intersect with our data; and (2) predictions (scores) extracted by the fine-tuned CNN. Moreover, we added the classification of object detection scores (Faster R-CNN trained on the Open Image Dataset v2) into the final ensemble. The linear SVM (LinearSVC from Scikit-learn library) was used as a classifier because it obtained the highest accuracy when compared to other classifiers (RF, SVM with radial basis functions) in our preliminarily experiments. In addition, we implemented the factorization machines (FM) [17], which learns the binary classification, i.e., prediction whether the image contains particular category or not. The negative examples in the FMs were generated by selecting the instances with the closest features. The FM is known to be able to estimate interactions even in problems with huge sparsity where RF or SVM fail [17]. The accuracy of all classifiers are presented in Table 1.

Table 1. Accuracy (%) of scene recognition models

	MobileNet v1	MobileNet v2, $\alpha = 1$	MobileNet v2, $\alpha = 1.4$	Inception v3
Fine-tuned CNN	76.84	78.66	80.02	82.22
Pre-trained features, FM	18.5	21.34	22.76	24.27
Pre-trained features, SVM	78.25	83.6	85.12	86.38
Fine-tuned scores, FM	24.11	27.8	28.92	29.0
Fine-tuned scores, SVM	72.76	76.25	77.9	77.91
Proposed ensemble, FM	48.35	50.8	52.56	53.0
Proposed ensemble, SVM	80.15	85.14	86.39	87.52

Here the SVM classifier for the features of the pre-trained CNN is slightly more accurate when compared to the complete CNN itself (i.e., with logistic regression of the same features). The features from the CNN are much more powerful over classification scores. It is important to emphasize that, in contrast to such conventional classifiers as SVM or Gradient Boosting, the replacement of features from the CNN scores to the embeddings from the last average pooling layer in the FM models *decreases* the accuracy.

Though single classification of the object detection scores has 56% error rate, its inclusion into the final ensemble slightly increases the validation accuracy. The gain is especially noticeable for the FM, which was claimed to be more appropriate for sparse data. Hence, their usage with sparse scores from object detection model caused an increase of accuracy up to 24%. Anyway, the classification scores are much more representative in this task: even SVM for simple scores extracted from the *pre-trained* MobileNet v1 achieved 10% higher accuracy. The proposed approach up to 5% more accurate when compared to conventional fine-tuned models. However, detection scores for the SVM classifier give no more than 0.5% to the overall accuracy. As the inference time in the Faster R-CNN with InceptionResNet backbone is approximately equal to 500 ms even on GTX1080 Ti GPU, the object detector can be excluded from the ensemble if the running time is critical. In such case the most appropriate for offline mobile applications will be the classification of the features extracted by the MobileNet v2 network [4].

4   Conclusion and Future Works

In this section we examined several ways to recognize scenes for visual user preference prediction engine. We experimentally evaluated known techniques to classify the whole scene. It was found that simple MobileNet v2 with $\alpha = 1.4$ is the best choice for offline mobile applications with strict constraints to the running time. However, the remote processing of public scenes should be better implemented with the proposed pipeline (Fig. 1) using fusion of classifiers based on the deep CNN (Inception v3 in our experiments). Our approach was implemented in a special mobile application (Fig. 4). It is important to highlight that the factorization machines cannot improve the accuracy when compared to simple CNN-based scene recognition.

In future research, we are planning to test the proposed approach on other datasets with large number of scenes, e.g., the Places2 dataset [5]. In addition, it is important to extend our solution for predicting the user preferences from a *set* of photos rather than process each photo independently [18]. If the number of classes is large, approximate nearest neighbor methods [13] and fast sequential analysis [11] can be applied. Finally, the object detector speed should be significantly increase by using either one-shot detectors or appropriate methods for detecting of salient objects [9].

Acknowledgements. The article was prepared within the framework of the Academic Fund Program at the National Research University Higher School of Economics (HSE University) in 2019 (grant No. 19-04-004) and by the Russian Academic Excellence Project "5–100".

References

1. Prince, S.J.: Computer Vision: Models Learning and Inference. Cambridge University Press, Cambridge (2012)
2. Goodfellow, I., Bengio, Y., Courville, A.: Deep Learning. MIT Press, Cambridge (2016)
3. Howard, A.G., et al.: MobileNets: efficient convolutional neural networks for mobile vision applications. arXiv preprint arXiv:1704.04861 (2017)
4. Sandler, M., Howard, A., Zhu, M., Zhmoginov, A., Chen, L.C.: Inverted residuals and linear bottlenecks: mobile networks for classification, detection and segmentation. arXiv preprint arXiv:1801.04381 (2018)
5. Zhou, B., Lapedriza, A., Khosla, A., Oliva, A., Torralba, A.: Places: a 10 million image database for scene recognition. IEEE Trans. Pattern Anal. Mach. Intell. **40**(6), 1452–1464 (2018)
6. Ren, S., He, K., Girshick, R., Sun, J.: Faster R-CNN: towards real-time object detection with region proposal networks. In: Advances in Neural Information Processing Systems (NIPS), pp. 91–99 (2015)
7. Bayat, A., Do Koh, H., Kumar Nand, A., Pereira, M., Pomplun, M.: Scene grammar in human and machine recognition of objects and scenes. In: Proceedings of the IEEE Conference on Computer Vision and Pattern Recognition Workshops, pp. 1992–1999 (2018)

8. Jian, M., et al.: Saliency detection based on background seeds by object proposals and extended random walk. J. Vis. Commun. Image Represent. **57**, 202–211 (2018)
9. Jian, M., et al.: Assessment of feature fusion strategies in visual attention mechanism for saliency detection. Pattern Recogn. Lett. (2018, in press). https://doi.org/10.1016/j.patrec.2018.08.022
10. Deng, J., Dong, W., Socher, R., Li, L.J., Li, K., Fei-Fei, L.: ImageNet: a large-scale hierarchical image database. In: Proceedings of the International Conference on Computer Vision and Pattern Recognition (CVPR), pp. 248–255. IEEE (2009)
11. Savchenko, A.V.: Sequential three-way decisions in multi-category image recognition with deep features based on distance factor. Inf. Sci. **489**, 18–36 (2019)
12. Sharif Razavian, A., Azizpour, H., Sullivan, J., Carlsson, S.: CNN features off-the-shelf: an astounding baseline for recognition. In: Proceedings of the International Conference on Computer Vision and Pattern Recognition Workshops (CVPRW), pp. 806–813. IEEE (2014)
13. Savchenko, A.V.: Maximum-likelihood approximate nearest neighbor method in real-time image recognition. Pattern Recogn. **61**, 459–469 (2017)
14. Savchenko, A.V., Belova, N.S.: Unconstrained face identification using maximum likelihood of distances between deep off-the-shelf features. Expert Syst. Appl. **108**, 170–182 (2018)
15. Rassadin, A., Gruzdev, A., Savchenko, A.: Group-level emotion recognition using transfer learning from face identification. In: Proceedings of the 19th International Conference on Multimodal Interaction (ICMI), pp. 544–548. ACM (2017)
16. Tarasov, A.V., Savchenko, A.V.: Emotion recognition of a group of people in video analytics using deep off-the-shelf image embeddings. In: van der Aalst, W.M.P., et al. (eds.) AIST 2018. LNCS, vol. 11179, pp. 191–198. Springer, Cham (2018). https://doi.org/10.1007/978-3-030-11027-7_19
17. Rendle, S.: Factorization machines. In: 10th International Conference on Data Mining (ICDM), pp. 995–1000. IEEE (2010)
18. Savchenko, A.V.: Probabilistic neural network with homogeneity testing in recognition of discrete patterns set. Neural Netw. **46**, 227–241 (2013)

Chinese Address Similarity Calculation Based on Auto Geological Level Tagging

Jing Liu[1], Jianbin Wang[2(✉)], Changqing Zhang[2], Xiubo Yang[2], Jianbo Deng[3], Ruihe Zhu[3], Xiaojie Nan[2], and Qinghua Chen[1(✉)]

[1] School of Systems Science, Beijing Normal University, Beijing 100875, China
qinghuachen@bnu.edu.cn
[2] Credit Harmony Research, Building 3 District 3 Hanwei International, Beijing 100071, China
jianbinwang@creditharmony.cn
[3] Swarma Club, 4059 Building 1, 1 Shuangqing Road, Beijing 100085, China

Abstract. How to quickly measure the similarity of addresses has become an urgent need in various fields including financial anti-fraud. Traditional string-based similarity calculation methods have not completed this task perfectly. Taking into account the hierarchical nature of addresses, we constructed a framework for calculating the similarity of Chinese addresses. First, the whole address strings are split and annotated with proper level by a LM-LSTM-CRF model, and then sub-string level similarities are calculated. Last, similarity scores are combining by BP neural networks. This framework has achieved good results in practice for financial anti-fraud tasks.

Keywords: Address similarity · Auto geological level tagging · LM-LSTM-CRF model

1 Introduction

Internet finance, the new business model, has brought new growth points but also new challenges to risk management. Among them, group fraud is particularly prominent [8]. How to quickly measure the similarity of addresses has become an urgent need in the field of financial anti-fraud, because some internet financial companies in China have found there existing similarities between addresses of these swindling gang [16]. Besides, such similarity comparison task also has applications in many fields such as logistics transportation, government management, etc.

Formally, an address is presented as a string and the similarity of two addresses should be performed by similarity computation of strings. A quantitative way to determine whether two strings are similar is based on some certain

This work is supported by joint project of Beijing Normal University and Credit Harmony Research, and in part by the National Natural Science Foundation of China under grant 71701018. Jing Liu, Jianbin Wang and Changqing Zhang contributed equally to this work.

© Springer Nature Switzerland AG 2019
H. Lu et al. (Eds.): ISNN 2019, LNCS 11555, pp. 431–438, 2019.
https://doi.org/10.1007/978-3-030-22808-8_42

similarity functions [7]. In general, it can be classified as string similarity metric and semantic similarity.

String similarity metric measures the similarity between two strings by matching or comparison. Levenshtein distance is very commonly used to calculate the similarity between the given two strings by counting the minimum number of operation needed to transform one string into another one [17]. Besides, the longest common subsequence (LCS) similarity measures the longest total length of all the matched substring between two string where this sub-string appear in the same order as they appear in the other string [9]. Kondrak et al. showed that edit distance and the length of the longest common subsequence are special cases of the so-called n-gram distance and similarity, respectively [11].

Semantic similarity, on the other hand, measures the similarity between texts or documents on the basis of their meaning rather than symbolic matching [1]. It uses the methods of corpus-based and knowledge-based measures [7]. Corpus-based similarity concerns the similarity between words on the basis of information obtained from a large corpus. Knowledge-based similarity sets the degree of similarity between words by using information derived from semantic networks. "WordNet" [4] and Natural language Toolkit (NLTK) [15] are popular semantic networks to measure the knowledge-based similarity between words.

Although these similarity methods are becoming more and more mature and have achieved good results in many aspects [14], they are not effective in comparing the similarity of addresses. Address data is different from general strings and it is hierarchical. For example, a typical Chinese address can be divided into provinces, county, city, town, district, village, road etc. and symbols at different levels cannot be compared directly [2]. Moreover, the contribution of different levels to the dissimilarity is different. How to integrate different levels of information is also an urgent problem to be solved.

In this paper, we propose a novel approach for address similarity calculation. First, the address strings are split into sub-strings at different levels, and then substrings are compared by string similarity metric and also in a semantic way. Finally, we integrate these similarity scores to get a similarity index. In detail, we apply a so-called LM-LSTM-CRF architecture to annotate addresses and divide them into different levels. We calculate the similarity belonging to the same level by word2vec and other methods. Then we synthesize them into a neural network at last. The remainder of the paper is organized as follows. Section 2 describes relative works including sequence tagging models used in this paper. Section 3 shows the whole framework and training procedure. Section 4 reports the results of the experiments. Section 5 discusses and draws conclusions.

2   Related Work

2.1   String Similarity Metric

Levenshtein Distance. Levenshtein distance is a way of quantifying the similarity of two strings by counting the minimum number of operations (including insertion, deletion and substitution) required to transform one string into the other [10]. For example, the Levenshtein distance between "北京市" and "天津市" is 2. We define Levenshtein distance of two strings A and B:

$$S_{\text{edit}}(A, B) = 1 - \frac{L(A, B)}{\max(|A|, |B|)}, \tag{1}$$

where the value of S_{edit} is constrained to the range of $[0, 1]$. Here, $|A|$ and $|B|$ are the length of A and B respectively, and $L(A, B)$ is the Levenshtein distance between them. $\max(\cdot)$ is a function which returns the maximum one.

Pinyin Edit Distance. At present, the input of Chinese characters is mainly based on pinyin input method, which is often mistaken for homonyms [3]. Although pinyin input methods adopt some automatic error correction technology, the final results also have some errors especially when one needs to type quickly at the counter. For example, "陕西省" and "山西省" are different provinces in China, but they have the same pinyin "shanxisheng" and there will be confusion at some points. We can also define the pinyin edit distance as follows

$$S_{\text{pyedit}}(A, B) = S_{\text{edit}}(P_A, P_B) = 1 - \frac{L(P_A, P_B)}{\max(|P_A|, |P_B|)}. \tag{2}$$

Here P_A, P_B are pinyin of A and B.

2.2 Semantic Similarity Methods

Word2vec Similarity. Word2vec is a tool based on shallow neural networks and released by Google in 2013. Through continuous bag-of-words (CBOW) or continuous skip-gram procedure, words are projected into a high dimensions space. Words with similar meanings are closer in space. Word2vec provides an easy comparison in the semantic similarity between words because each word is represented as a vector [18]. We define word2vec similarity as

$$S_{\text{w2v}}(A, B) = \frac{\mathbf{X_A} \cdot \mathbf{X_B}}{\|\mathbf{X_A}\| \|\mathbf{X_B}\|} = \cos(\mathbf{X_A}, \mathbf{X_B}). \tag{3}$$

Here $\mathbf{X_A}$ and $\mathbf{X_B}$ are the word vectors in semantic space and $\|\cdot\|$ donates the norm of vectors.

3 Methodology

3.1 LM-LSTM-CRF Model

Linguistic sequence labeling is a fundamental framework. It has been applied to a variety of tasks including part-of-speech (POS) tagging, noun phrase chunking and named entity recognition (NER) [12].

In the past few years, distributed representations, also known as word embeddings, have been broadly applied to natural language processing (NLP) problems with great success. Recently, recurrent neural networks (RNN) [6], together with its variants such as long-short term memory (LSTM) [5] have shown great success in modeling sequential data. But their performance drops rapidly when models solely depend on word embeddings [13].

In 2016, Ma etc. first use convolutional neural networks (CNNs) to encode character-level information of a word into its character-level representation. They combine character-level and word-level representations and feed them into bi-directional LSTM (BiLSTM) to model context information of each word and get significant performance [13]. After that, Liu etc. propose an effective sequence labeling framework called LM-LSTM-CRF, which leverages both word-level and character-level knowledge in a more efficient way [12]. In this paper, we will adapt the LM-LSTM-CRF architecture to make auto tagging.

3.2 Framework

The whole framework used in this paper is present as Fig. 1.

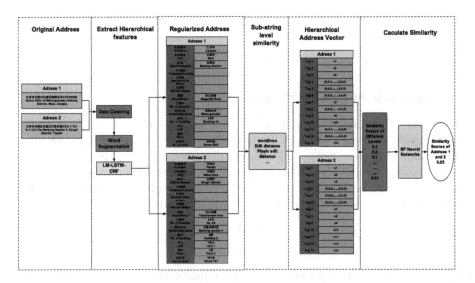

Fig. 1. The framework of our model. Two addresses are labeled by the LM-LSTM-CRF model.

As shown in this framework plot, after cleaning the address data, word segmentation is carried out, and the sequentially connected sub-strings are obtained. Then each sub-string is labeled by the LM-LSTM-CRF model at the proper address levels. The same level of substrings are compared by some mentioned methods including word2vec similarity, edit distance and search similarity. Finally, we synthesize these comparison results into a similarity index through BP neural networks.

Before running, the LM-LSTM-CRF model and BP neural networks need to be trained respectively.

3.3   Training Procedure

We established the LM-LSTM-CRF model with adaptation of source code from https://github.com/LiyuanLucasLiu/LM-LSTM-CRF. The procure of the model is shown in Fig. 2.

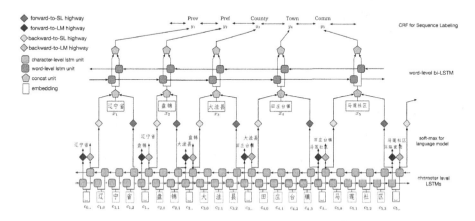

Fig. 2. The LM-LSTM-CRF model in our framework. The figure is adapted from [12].

For a sentence with annotations $\mathbf{y} = (y_1, ..., y_n)$, its word-level input is marked as $\mathbf{x} = (x_1, x_2, ..., x_n)$, where x_i is the ith word and its character-level input is recorded as $\mathbf{c} = (c_{0,\_}, c_{1,1}, c_{1,2}, ..., c_{1,\_}, c_{2,1}, ..., c_n, \_)$, where $c_{i,j}$ is the j-th character for word x_i and $c_{i,\_}$ is the space character after x_i.

Some necessary adjustments for Chinese address data are made, and then the following training operations are carried out as follows:

Data Collection and Cleaning. We first collected enough Chinese address string data, remove extraneous characters and then complete the address string data through standard address library completion, correction and other normalization processing.

Word Segmentation and Reintegration. Through word segmentation software and standard operation based on address-preserving words such as "province", "city", "district", "town" and "road", we decomposed address strings into sequentially substrings. Merging operation is also needed after exceeding constrain.

Manual Address Level Annotation. The sub-strings of the address string are labeled according to certain address administrative divisions standards. This step is to obtain absolutely accurate and reliable data.

Random Grouping and Training. Based on hierarchically labeled address substring data, immense amounts of the artificial complete sequence of address strings could be generated by random grouping. Finally we trained our LM-LSTM-CRF model.

After that, we constructed BP networks with one hidden layer. The number of input nodes corresponds to the results of similarity computation results. The last layer of neural networks output the final similarity score.

3.4   Address Levels

Comparing to American address, Chinese address is more complicated. In https://en.wikipedia.org/wiki/Administrative_divisions_of_China, there are 5-layered main administrative divisions of China. Some smaller address levels are discussed in related sources including "Industry standards of the People's Republic of China (CJJ/T 106-2010. No. J455-2010)" and literature [19]. Based on these research, we set the address levels of China shown as in Table 1. General addresses do not contain all levels, for example, there will be no 5 and 6 levels in the city address. The well trained LM-LSTM-CRF model will calibrate the corresponding levels for the input address substrings.

Table 1. Address levels of China.

Level	Name	Types	Sample
1	Province	Provinces (省), Municipalities (直辖市) Autonomous regions (自治区)	Jiangsu
2	Prefecture	Prefecture-level cities (地级市), Prefectures (地区) Leagues (盟), Autonomous prefectures (自治州)	Wuxi
3	County	Districts (市辖区), County-level cities (县级市) Counties (县), Leagues (盟), etc.	Beitang District
4	Township	Townships (乡), Towns (镇) Ethnic townships (民族乡), Subdistricts (街道)	
5	Administrative village	Administrative villages (行政村)	
6	Natural village	Natural villages (自然村)	
7	Road	Roads (街), Streets (路)	Qingshixi Road
8	No. of road	No. of road (门牌号)	
9	Community	Community (小区)	Wutongshuian
10	Building	Building (楼)	Building 43
11	Unit	Unit (单元)	
12	Floor	Floor (楼层)	
13	Room	Room (房间号)	2303

4 Results

We started with 1337 Chinese address with manual address level annotation. By random grouping process described in Sect. 3.3, we obtained about 1,400,000 tagged data for training the LM-LSTM-CRF model. Then with the well-trained LM-LSTM-CRF model, we trained the BP neural networks.

For the task of address similarity calculation, we deployed five servers with Centos 6.7, 32G RAM and 16 cores 32 threads processor. One of them is responsible for scheduling and four are responsible for calculations. It takes 2.5 h to complete the similarity calculation of 550 million pairs of address data. We then obtained the accuracy, recall rate and F_1 score on the test set, as shown in Table 2. Our model has performed well compared to traditional methods such as word2vec and edit distance. Besides, our framework deployed on the server can process about 60,000 pairs of address data in one second, which also meets the requirements in terms of efficiency.

Table 2. Our results for address similarity calculation.

Method	Accuracy	Recall	F_1 score
LM-LSTM-CRF + BP	87.42%	98.94%	87.44%
word2vec	72.23%	74.68%	72.91%
edit distance	65.81%	69.29%	66.98%

5 Conclusions and Discussion

Considering the hierarchical nature of Chinese address strings, in this paper we combined word segmentation, LM-LSTM-CRF model, sub-string level similarity calculation and BP neural networks to calculate the similarity of two Chinese address strings. With the well trained model, a large scale calculation can be carried out quickly and the accuracy rate can be guaranteed. Besides, the framework can be easily extended and we also applied our framework to other tasks such as recognition of company names. The applications of the model are helpful for companies and governments in various respects.

References

1. Budanitsky, A., Hirst, G.: Semantic distance in wordnet: an experimental, application-oriented evaluation of five measures. In: Workshop on WordNet and Other Lexical Resources **2**, 2–2 (2001)
2. Chang, C.H., Huang, C.Y., Su, Y.S.: On chinese postal address and associated information extraction. In: The 26th Annual Conference of the Japanese Society for Artificial Intelligence, pp. 1–7 (2012)

3. Chen, Z., Lee, K.F.: A new statistical approach to chinese pinyin input. In: Proceedings of the 38th Annual Meeting on Association for Computational Linguistics, pp. 241–247. Association for Computational Linguistics (2000)
4. Fellbaum, C.: WordNet. In: Poli, R., Healy, M., Kameas, A. (eds.) Theory and Applications of Ontology: Computer Applications, pp. 231–243. Springer, Dordrecht (2010). https://doi.org/10.1007/978-90-481-8847-5_10
5. Gers, F.A., Schmidhuber, J., Cummins, F.: Learning to forget: continual prediction with lstm. Neural Comput. 12(10), 2451–2471 (2000)
6. Goller, C., Kuchler, A.: Learning task-dependent distributed representations by backpropagation through structure. Neural Net. 1, 347–352 (1996)
7. Gomaa, W.H., Fahmy, A.A.: A survey of text similarity approaches. Int. J. Comput. Appl. 68(13), 13–18 (2013)
8. Hou, X., Gao, Z., Wang, Q.: Internet finance development and banking market discipline: evidence from china. J. Financ. Stab. 22, 88–100 (2016)
9. Julstrom, B.A., Hinkemeyer, B.: Starting from scratch: growing longest common subsequences with evolution. In: Runarsson, T.P., Beyer, H.-G., Burke, E., Merelo-Guervós, J.J., Whitley, L.D., Yao, X. (eds.) PPSN 2006. LNCS, vol. 4193, pp. 930–938. Springer, Heidelberg (2006). https://doi.org/10.1007/11844297_94
10. Jurafsky, D., Martin, J.H.: Speech and Language Processing. Pearson, London (2014)
11. Kondrak, G.: N-gram similarity and distance. In: Consens, M., Navarro, G. (eds.) SPIRE 2005. LNCS, vol. 3772, pp. 115–126. Springer, Heidelberg (2005). https://doi.org/10.1007/11575832_13
12. Liu, L., et al.: Empower sequence labeling with task-aware neural language model. arXiv preprint arXiv:1709.04109 (2017)
13. Ma, X., Hovy, E.: End-to-end sequence labeling via bi-directional lstm-cnns-crf. In: Proceedings of the 54th Annual Meeting of the Association for Computational Linguistics. pp. 1064–1074 (2016)
14. Navarro, G.: A guided tour to approximate string matching. ACM Comput. Surv. 33(1), 31–88 (2001)
15. Perkins, J.: Python Text Processing With NLTK 2.0 Cookbook. Packt Publishing Ltd, Birmingham (2010)
16. Ta, L.: The risk and prevention of internet finance. In: 2017 4th International Conference on Industrial Economics System and Industrial Security Engineering, pp. 1–5 (2017)
17. Yujian, L., Bo, L.: A normalized levenshtein distance metric. IEEE Trans. Pattern Anal. Mach. Intell. 29(6), 1091–1095 (2007)
18. Zhang, D., Xu, H., Su, Z., Xu, Y.: Chinese comments sentiment classification based on word2vec and svmperf. Expert Syst. Appl. 42(4), 1857–1863 (2015)
19. Zhao, Y., Wang, L., Chou, A.: A fusion method of marine sub-bottom acoustic spatial data based on features and applications. Sci. Surv. Map. 38(5), 74–76 (2013)

Enhancing Feature Representation for Saliency Detection

Tao Zheng, Bo Li$^{(\boxtimes)}$, and Haobo Rao

School of Electronics and Information Engineering,
South China University of Technology, Guangzhou, China
{eezhengtao,eerhb}@mail.scut.edu.cn, leebo@scut.edu.cn

Abstract. Current detectors for saliency detection adopt deep convolutional neural networks to continuously improve accuracy, but the results are still not satisfactory. We propose Multiple Receptive Field Aggregating Module (MRFAM) that can capture abundant context information to enhance feature representation. We assemble it into a novel network to predict saliency maps. Extensive experiments on six benchmark datasets demonstrate that the module is efficient and our proposed network can accurately capture salient objects with sharp boundaries in complex scene, performing favorably against the state-of-the-art methods in term of different evaluation metrics.

Keywords: Saliency detection · Deep learning · Feature representation

1 Introduction

Saliency detection aims to accurately localize the most prominent regions then separate them from complex scenes, which can be considered as pre-processing instrument in plenty of computer vision tasks, such as visual tracking [1], scene classification [2], image retrieval [3] and person re-identification [4].

In this visual task, abundant context information plays a crucial role. Traditional methods [5,6] design various hand-crafted features which are not robust enough to detect salient objects when they encounter complex and cluttered scenes. Because they are derived from the prior knowledge of previous datasets which can't be extended to be successfully useful in diverse situations.

Recently, significant accuracy gains in saliency detection have been obtained through the combination with Fully Convolutional Network (FCN) [7]. Although these models consist of various structures to extract features, the results are still far away from being satisfactory. Because most previous FCN-based saliency detection methods simply stack single-scale convolutions with limited receptive fields, this operation will result in the learned convolutional features not containing rich semantic information and texture information to precisely detect salient targets especially those having small scales, irregular shapes and sporadic locations. In addition, the receptive fields of these models are several sets of square

© Springer Nature Switzerland AG 2019
H. Lu et al. (Eds.): ISNN 2019, LNCS 11555, pp. 439–447, 2019.
https://doi.org/10.1007/978-3-030-22808-8_43

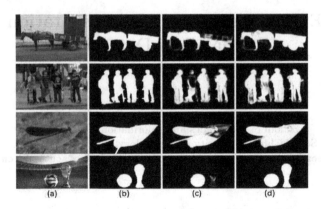

Fig. 1. Visual comparison of our proposed network and GBMPM [10]. From left to right: (a) Input images, (b) Ground truth, (c) GBMPM, (d) Our network.

due to the use of regular kernel shape, so the resulting features tend to be less distinctive when they aim to detect salient targets with complex boundaries such as flowers and animals (Fig. 1).

In this paper, we propose Multiple Receptive Field Aggregating Module (MRFAM) that can capture abundant information to enhance feature representation. We assemble it into a novel network to predict saliency maps. The contributions of our research can be summarized as follows:

(1) We propose a multiple receptive field aggregating module to extract abundant context information.
(2) We assemble MRFAM into a novel network for saliency detection. Quantitative and qualitative experiments on six benchmark datasets demonstrate the effectiveness of the module. In addition, the proposed network performs favorably against the state-of-the-art algorithms under different evaluation metrics.

2   Related Work

Szegedy [9] launches inception that contains multiple branch CNNs with different convolutional kernels to extract features from different receptive fields. Yu [13] proposes dilated convolution which support exponential expansion of the receptive field without loss of resolution or coverage. It applies four parallel dilated convolutions with different dilated rates to systematically aggregate multi-scale contextual information. Zhang [10] designs a MCFEM to extract multi-scale contextual features by stacking dilated convolutions with different receptive fields. Xie [8] illustrates that increasing branch is able to improve accuracy and is more effective than going deeper or wider when we increase the capacity, so it designs a homogeneous multi-branch architecture to extract feature. Although these FCN-based models have used different structures to promote accuracy, the effect is still not optimal. Because the receptive fields of these models are several sets of square due to the use of regular kernel shape, which is not robust enough to detect salient objects with small scales, irregular shapes and sporadic locations.

3   The Proposed Method

Multiple Receptive Field Aggregating Module. We attempt to decompose a complex object into multiple small parts and then separately detect them which is shown as Fig. 2. Firstly, we adopt dilated convolutions with rectangle kernel shape such as 1×5, 1×7 to captures structural information about object in a certain direction. Then we adopt dilated convolutions with different dilated rates to adjust receptive field, followed by a 1×1 convolution. The resulting feature maps are combined by cross-channel concatenation. In this way, we decompose a region into horizontal and vertical directions. The differences in shape and size of receptive fields are beneficial to detect complex edge regions. We integrate the above structure into MCFEM [10], proposing Multi Receptive Field Aggregating Module (MRFAM) to better extract features. In order to compare MCFEM and MRFAM fairly, we need to ensure that the feature dimension of output in MRFAM is the same as MCFEM, so we reduce the feature dimension of MCFEM in MRFAM to half of the original.

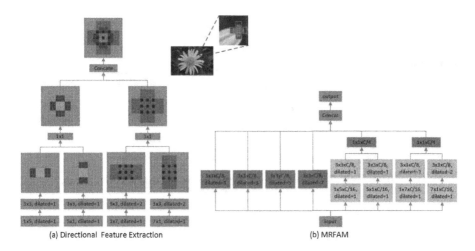

(a) Directional Feature Extraction   (b) MRFAM

Fig. 2. Left: directional feature extraction. Right: the structure of MRFAM. We integrate directional feature extraction with MCFEM, proposing our MRFAM. The block with (A × B × C, dilated = D) denotes dilated convolutions with kernel size A × B; C is the output channel, and D is the dilated rate. Concat denotes cross-channel concatenation.

Network Architecture. We assemble the MRFAM into a novel network to predict saliency maps. Our network applies pre-trained VGG16 [12] as the feature extraction network to produce a set of feature maps with different scales. Since our network is for pixel-wise dense prediction, we discard all the fully connected layers as well as the last pooling layer simultaneously. We resize the input image to 256×256. The revised VGG16 provided feature maps from different side output layers that can be represented as $F = \{f_i, i = 1, 2, 3, 4, 5\}$. We add the Multiple Receptive Field Aggregating Module (MRFAM) after each side output to capture information, obtaining f_i^c. Low-level spatial details in shallow layers can retain fine-grained object boundaries, while high-level semantic features in deep layers can locate where the salient objects are [14]. In order to take full advantages of multi-level features, we adopt a series of operations to combine different level features. This process can be performed by:

$$h_i^0 = Concat(f_i^c, Up(f_{i+1}^c)), i = 1, 2, 3, 4 \tag{1}$$

$$h_i^1 = Relu(Conv(h_i^0; \theta_i^1)), i = 1, 2, 3, 4 \tag{2}$$

$$h_4^2 = Relu(Conv(h_4^1; \theta_4^2)) \tag{3}$$

$$h_i^2 = Relu(Conv(Concat(h_i^1, Up(h_{i+1\sim4}^2)); \theta_i^2)), i = 3, 2, 1. \tag{4}$$

$$h^3 = Concat(h_1^2, Up(h_2^2), Up(h_3^2), Up(h_4^2)) \tag{5}$$

where $Up()$ is an up-sample operation to adjust the size of feature maps in deep layers; $Concat()$ denotes cross-channel concatenation and $Conv(*; \theta)$ represents a 3×3 convolution with parameter $\theta = \{W, b\}$; $Relu()$ denotes ReLU activation function. Finally, as previous methods, we adopt a convolution with kernel size 3×3 at the end of network to infer saliency maps. The structure of our network is shown in Fig. 3.

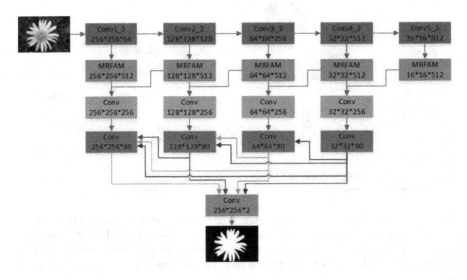

Fig. 3. The structure of our network.

To investigate the effectiveness of MRFAM, we will replace MRFAM with MCFEM, and then compare their results. The details of the experiment will be introduced in Sect. 4.2.

4   Experiments

4.1   Experimental Setup

Experimental setup consist of three parts: datasets, evaluation metrics and implementation details.

Datasets. We compare the proposed architecture and other previous algorithms on six benchmark datasets: ECSSD, PASCAL-S, HKU-IS, DUTS, DUT-OMRON and SOD. DUTS consists of two parts, DUTS-TR and DUTS-TE.

Evaluation metrics. Three metrics are utilized to better evaluate the performance of our architecture and other 11 state-of-the-art algorithms for salient object detection, including Precision-Recall (PR) curves, Maximum F-measure and Mean Absolute Error (MAE). The precision and recall are calculated by thresholding the predicted map and comparing the binary segmentation with the ground truth. F-measure represents the overall performance:

$$F_\beta = \frac{(1 + \beta^2) \times Precision \times Recall}{\beta^2 \times Precision + Recall}, \tag{6}$$

where we set β^2 to 0.3 as suggested in previous work. Maximum F-measure can well reflect the performance of the proposed model. The formula of MAE is as follows:

$$MAE = \frac{1}{W \times H} \sum_{x=1}^{W} \sum_{y=1}^{H} |S(x,y) - G(x,y)|. \tag{7}$$

Implementation details. In our proposed model, the parameters of the first 13 convolutional layers are initialized by pre-trained VGG16 net, while the weights of other convolutional layers are initialized by using truncated normal method. We utilize the DUTS-TR dataset to train the proposed model. For data augmentation, we extend the training set by horizontal flipping and vertical flipping. The architecture is trained until its training loss converges, without using validation. Adam is adopted to train the proposed network. The batch size N and the initial learning rate are set to 1 and 1e-6 respectively. The proposed architecture converges after 10 epochs by a NVIDIA TITAN X (Pascal) GPU that is adopted to accelerate the convergence speed.

4.2   Performance Comparison with State-of-the-art

We will demonstrates the effectiveness of MRFAM and our proposed network respectively.

(1) To better illustrate the effectiveness of the MRFAM, we replace MRFAM with MCFEM, doing additional experiment in six benchmark datasets. Comparison results are shown in Table 1, which demonstrates the performances of their results in term of the metrics of Maximum F-measure and MAE. MRFAM outperform MCFEM across all the datasets under different measurement.

(2) The proposed architecture is compared with existing 11 state-of-the-art algorithm (PiCANet [15], Amulet [16], GBMPM [10], DCL [17], DHS [18], DSS [14], ELD [19], NLDF [20], RFCN [21], SRM [22], UCF [23]). In order to fairly compare the performance between these algorithms, saliency prediction of different algorithms are provided by official source or generated by running available codes. Table 2 illustrates that our proposed architecture perform better than other algorithms across nearly all benchmark datasets in term of Maximum F-measure and MAE. From Fig. 4, it can be observed that the PR curves of our architecture outperform other algorithms on six datasets. Some saliency predictions randomly selected from testing set are listed in Fig. 5, showing that the proposed network can accurately capture salient objects with sharp boundaries in complex scene.

Table 1. We replace MRFAM with MCFEM in our network, and compare their effectiveness under the metrics of maximum F-measure and MAE.

Modules	SOD		HKU		OMRON		PASCAL		ECSSD		DUTS-TE	
	F_{max}	MAE	F_{max}	MAE	F_{max}	MAE	F_{max}	MAE	F_{max}	MAE	F_{max}	MAE
MRFAM	0.857	0.111	0.927	0.035	0.807	0.059	0.867	0.078	0.935	0.041	0.854	0.049
MCFEM	0.852	0.105	0.921	0.037	0.798	0.062	0.857	0.080	0.931	0.042	0.843	0.053

Table 2. The performances of different algorithms in term of the metrics of MAE and Maximum F-measure.

Methods	SOD		HKU		OMRON		PASCAL		ECSSD		DUTS-TE	
	F_{max}	MAE	F_{max}	MAE	F_{max}	MAE	F_{max}	MAE	F_{max}	MAE	F_{max}	MAE
Ours	0.857	0.111	0.927	0.035	0.807	0.059	0.867	0.078	0.935	0.041	0.854	0.048
PiCANet	0.855	0.108	0.921	0.042	0.794	0.068	0.880	0.088	0.931	0.047	0.851	0.054
GBMPM	0.851	0.106	0.920	0.038	0.774	0.063	0.862	0.074	0.928	0.044	0.850	0.049
Amulet	0.808	0.145	0.896	0.052	0.743	0.098	0.858	0.103	0.915	0.059	0.778	0.085
DCL	0.825	0.198	0.885	0.137	0.739	0.157	0.823	0.189	0.901	0.075	0.782	0.150
DHS	0.827	0.133	0.902	0.054	0.758	0.072	0.841	0.111	0.907	0.060	0.829	0.065
DSS	0.846	0.126	0.911	0.040	0.771	0.066	0.846	0.112	0.916	0.053	0.825	0.057
ELD	0.760	0.154	0.839	0.074	0.720	0.088	0.773	0.123	0.867	0.079	0.738	0.093
NLDF	0.837	0.123	0.902	0.048	0.753	0.080	0.845	0.112	0.905	0.063	0.812	0.066
RFCN	0.807	0.166	0.898	0.080	0.738	0.095	0.850	0.132	0.898	0.095	0.783	0.090
SRM	0.845	0.132	0.906	0.046	0.769	0.069	0.847	0.085	0.917	0.054	0.827	0.059
UCF	0.803	0.169	0.886	0.074	0.735	0.132	0.846	0.128	0.911	0.078	0.771	0.117

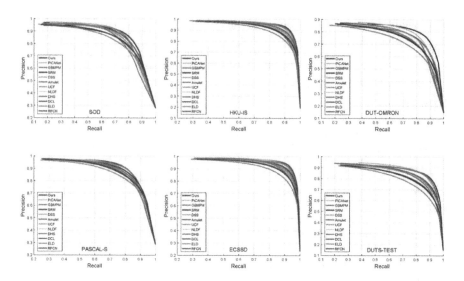

Fig. 4. Precision-Recall curves of the proposed network and 11 state-of-the-art algorithms on six benchmark datasets.

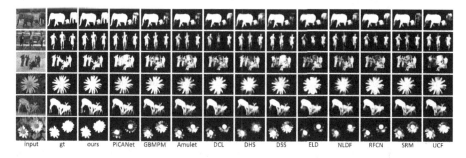

Fig. 5. Visual comparisons of the proposed network and previous state-of-the-art algorithms on different scenes.

5   Conclusion

In this paper, we propose Multiple Receptive Field Aggregating Module (MRFAM) that can capture abundant information to enhance feature representation. We assemble it into a novel network to predict saliency maps. Extensive experiments on six benchmark datasets demonstrate that the module is efficient and our proposed network can accurately capture salient objects with sharp boundaries in complex scene, performing favorably against other state-of-the-art algorithms in term of different evaluation metrics.

Acknowledgments. This research was supported by the National Natural Science Foundation of China (Grant Nos. 11627802, 51678249), by State Key Lab of Subtropical Building Science, South China University Of Technology (2018ZB33), and by the State Scholarship Fund of China Scholarship Council (201806155022)

References

1. Borji, A., Frintrop, S., Sihite, D.N., Itti, L.: Adaptive object tracking by learning background context. In: IEEE Computer Vision and Pattern Recognition Workshops, pp. 23–30 (2012)
2. Siagian, C., Itti, L.: Rapid biologically-inspired scene classification using features shared with visual attention. IEEE Trans. Pattern Anal. Mach. Intell. **29**(2), 300–312 (2007)
3. He, J., et al.: Mobile product search with bag of hash bits and boundary reranking. In: IEEE Computer Vision and Pattern Recognition, pp. 3005–3012 (2012)
4. Bi, S., Li, G., Yu, Y.: Person re-identification using multiple experts with random subspaces. J. Image Graph. **2**(2), 151–157 (2014)
5. Zhu, W., Liang, S., Wei, Y., Sun, J.: Saliency optimization from robust background detection. In: IEEE Computer Vision and Pattern Recognition, pp. 2814–2821 (2014)
6. Yang, C., Zhang, L., Lu, H., Ruan, X., Yang, M.H.: Saliency detection via graph-based manifold ranking. In: IEEE Computer Vision and Pattern Recognition, pp. 3166–3173 (2013)
7. Long, J., Shelhamer, E., Darrell, T.: Fully convolutional networks for semantic segmentation. In: IEEE Computer Vision and Pattern Recognition, pp. 3431–3440 (2015)
8. Xie, S., Girshick, R., Dollár, P., Tu, Z., He, K.: Aggregated residual transformations for deep neural networks. In: IEEE Computer Vision and Pattern Recognition, pp. 5987–5995 (2017)
9. Szegedy, C., et al.: Going deeper with convolutions. In: IEEE Computer Vision and Pattern Recognition, pp. 1–9 (2015)
10. Zhang, L., Dai, J., Lu, H., He, Y., Wang, G.: A bi-directional message passing model for salient object detection. In: IEEE Computer Vision and Pattern Recognition, pp. 1741–1750 (2018)
11. Krizhevsky, A., Sutskever, I., Hinton, G.E.: Imagenet classification with deep convolutional neural networks. In: Advances in Neural Information Processing Systems, pp. 1097–1105 (2012)
12. Simonyan, K., Zisserman, A.: Very deep convolutional networks for large-scale image recognition. arXiv preprint arXiv:1409.1556 (2014)
13. Yu, F., Koltun, V.: Multi-scale context aggregation by dilated convolutions. arXiv preprint arXiv:1511.07122 (2015)
14. Hou, Q., Cheng, M.M., Hu, X., Borji, A., Tu, Z., Torr, P.: Deeply supervised salient object detection with short connections. In: IEEE Computer Vision and Pattern Recognition, pp. 5300–5309 (2017)
15. Liu, N., Han, J., Yang, M.H.: PiCANet: learning pixel-wise contextual attention for saliency detection. In: IEEE Computer Vision and Pattern Recognition, pp. 3089–3098 (2018)
16. Zhang, P., Wang, D., Lu, H., Wang, H., Ruan, X.: Amulet: aggregating multi-level convolutional features for salient object detection. In: IEEE International Conference on Computer Vision, pp. 202–211 (2017)

17. Li, G., Yu, Y.: Deep contrast learning for salient object detection. In: IEEE Computer Vision and Pattern Recognition, pp. 478–487 (2016)
18. Liu, N., Han, J.: Dhsnet: deep hierarchical saliency network for salient object detection. In: IEEE Computer Vision and Pattern Recognition, pp. 678–686 (2016)
19. Lee, G., Tai, Y.W., Kim, J.: Deep saliency with encoded low level distance map and high level features. In: IEEE Computer Vision and Pattern Recognition, pp. 660–668 (2016)
20. Luo, Z., Mishra, A.K., Achkar, A., Eichel, J.A., Li, S., Jodoin, P.M.: Non-local deep features for salient object detection. In: IEEE Computer Vision and Pattern Recognition, pp. 6609–6617 (2017)
21. Wang, L., Wang, L., Lu, H., Zhang, P., Ruan, X.: Saliency detection with recurrent fully convolutional networks. In: Leibe, B., Matas, J., Sebe, N., Welling, M. (eds.) ECCV 2016. LNCS, vol. 9908, pp. 825–841. Springer, Cham (2016). https://doi.org/10.1007/978-3-319-46493-0_50
22. Wang, T., Borji, A., Zhang, L., Zhang, P., Lu, H.: A stagewise refinement model for detecting salient objects in images. In: IEEE International Conference on Computer Vision, pp. 4019–4028 (2017)
23. Zhang, P., Wang, D., Lu, H., Wang, H., Yin, B.: Learning uncertain convolutional features for accurate saliency detection. In: IEEE International Conference on Computer Vision, pp. 212–221 (2017)

Improving the Perceptual Quality of Document Images Using Deep Neural Network

Ram Krishna Pandey$^{(\boxtimes)}$ and A. G. Ramakrishnan

Department of Electrical Engineering, Indian Institute of Science, Bangalore, India
ramp@iisc.ac.in, agr@iisc.ac.in

Abstract. Given a low-resolution *binary document image*, we aim to improve its perceptual quality for *enhanced readability*. We have proposed a simple, deep learning based model, that uses convolution with transposed convolution and sub-pixel layers in the best possible way to construct the high-resolution image. The proposed architecture scales across the three different scripts tested, namely Tamil, Kannada and Roman. To show that the reconstructed output has enhanced readability, we have used the objective criterion of optical character recognizer (OCR) character level accuracy. The reported results by our CTCS architecture shows significant improvement in terms of the subjective criterion of human readability and objective criterion of OCR character level accuracy.

Keywords: Readability · Binary document image · Super-resolution · Deep learning · OCR

1 Introduction

The perceptual quality of binary document images can be ascertained in terms of the subjective criterion of human readability and the objective criterion of the character level accuracy (CLA) of an optical character recognizer (OCR). In this work, our primary goal is to improve the quality of low-resolution, binary document images for *better human readability*. OCR character level accuracy is used as a metric to objectively show that the quality of the document images has significantly improved. Thus, a secondary objective for the proposed technique can be as a preprocessing step before feeding a low resolution, document image as input to an OCR.

The performance of an OCR in terms of character and word level accuracies decreases when the documents are of very poor quality and resolution (see Table 1). Language models can be applied, as a post-processing step, on the text output by the OCR to correct the recognition errors. Alternately, a new classifier can be designed to operate on the low-resolution character images and still achieve good accuracy.

In this work, while dealing with our primary goal, we address three different tasks: (i) improving the human readability of the documents, (ii) improving the quality of the input image such that the existing OCRs perform well

© Springer Nature Switzerland AG 2019
H. Lu et al. (Eds.): ISNN 2019, LNCS 11555, pp. 448–459, 2019.
https://doi.org/10.1007/978-3-030-22808-8_44

(the character level accuracy of the OCR should be better than that on the input image), (iii) ensuring that the method works on multiple languages and resolutions. The advantages of this approach are: (i) low-resolution images stored on digital libraries are rendered better for direct human readability. (ii) We can avoid designing a new classifier to operate on such low-resolution document images. This reduces the burden of changing the design of the existing OCRs by training the classifier for each language independently, which in turn needs huge training data from each of these languages.

We have approached to solve the above-mentioned problem using techniques based on deep learning. Given an LR document image, the challenge is to generate a HR version of it, which should be perceived by a human as better readable, than the corresponding input image. Also the OCR should achieve higher CLA on the reconstructed output. This problem was earlier attempted in [4–7]. In [4], the authors have shown that the quality of the down-sampled version of the document image can be enhanced and brought to that of the input image. In [5], the authors have proposed an efficient convolutional neural network (CNN) architecture, coupled with bicubic interpolation, to achieve better performance in terms of OCR accuracy, starting from a down-sampled version of the same image. In [6], the authors have used the traditional interpolations (bicubic, bilinear and nearest neighbor), coupled with CNN, to achieve better performance in terms of WLA, when the input image is directly fed to the model. In [7], the authors have shown language dependent quality enhancement of document images. Motivated by these works, that aim to enhance the quality of low-resolution input images for better OCR recognition, here we propose an architecture which can upscale any low-resolution document image and improve the perceptual quality so that humans find it easy to read and also the generated output should have more recognition accuracy than the input. The architecture developed here is for an upscaling factor of 2. Our contributions can be summarized as follows:

- We have performed comprehensive experiments on a huge collection of document images. We found that the OCR character level accuracies can be increased by improving the resolution of poor resolution, binary document images as shown in Table 1.
- We *define, formulate and address* (to a good extent) this problem of increasing the perceptual quality of binary document images for better human readability.
- We have used state-of-art deep learning techniques to find the optimal and judicious combination of transposed convolution [10] and sub-pixel [11] layers to obtain super-resolution of binary document images.
- We have created a *unique dataset* that captures maximum possible variations in the input image space, particularly to address this challenging problem (details in Sec. 4).
- To our knowledge, this is the first report on the enhancement of the perceptual quality of document images for better human readability. (see Table 2).
- Our proposed algorithm enhances the quality of low-resolution binary document images and improves the OCR (CLA) by a good margin.

– Our algorithm works *independent of the three languages and the resolutions* it has been tested so far.

2   Related Work

Image super-resolution is a well-known problem for natural images. There are mainly two classes of image super-resolution, namely, single image super-resolution (SISR) and multi-image super-resolution (MISR). SISR has been the focus of many important publications [14–19]. The main assumption in all these papers is that the image in the LR space has the same local geometry as that of the image in the HR space. The researchers have endeavored to find out the representation common for both the spaces, using which the HR image can be formed. In [15], the authors have found out the common representation in the concatenated feature space of LR and HR dictionaries.

2.1   Traditional Approaches Not Based on Deep Learning

Some traditional approaches that deal with enhancing the quality of document images for better OCR recognition are mentioned below. Shi et al. [1] deal with the removal of noise (large blob or clutter noise, salt, and pepper noise) and non-text objects such as form line or rule lines from handwritten document images. They describe a region growing algorithm to fix salt and pepper noise. They also provide an approach to eliminate noisy artifacts that include multiple categories of degradation.

A non-parametric, unsupervised method is presented in [2] to deal with color or gray images e.g. camera captured or mobile document images. It Uses contrast limited, adaptive histogram equalization separately on HSV color space, an optimal conversion algorithm to transform the document image to the gray level and un-sharp masking to sharpen the useful information. The sharpened image is binarized by Otsu algorithm and fed to the OCR to obtain the final text.

Kumar et al. [3] have extended the application of sparse coding and dictionary learning techniques. Here, the basis/atoms of the dictionary for the binary document image are learnt by treating binary document images as distinct from natural images. They claim that their method restores degradations such as cuts, merge, blobs and erosion in the document. Besides these, there are reports in the literature that deal with noise removal from the document images with the sole aim of improving the OCR accuracy.

All the approaches mentioned above are used to improve the OCR accuracy for certain kinds of degradation (noise) associated with the input image. However, none of them deal with the improvement in the resolution of the input image or aim to enhance the readability of the document images, where, originally, a lot of missing pixels might have resulted in changes in the shape, structure and interpretation of the characters.

2.2   Approaches Based on Deep Learning

Since the advent of deep learning based models like [17], investigators have learned a multiple layer representation of the input images (called features) to capture maximum variations in the input image space. Once such a model is learned, they have used it to construct a HR from a single LR image. However, to our knowledge, the problem of super-resolution of binary document images has not been attempted prior to its conception in the recent papers [4–6]. In [4,5], the authors have shown that the downsampled version of a document image can be worked upon to reconstruct an image of the original resolution. However, these techniques fail to generalize, when one inputs a LR binary document image directly, and not a downsampled version of an existing HR image. The technique of nonlinear fusion of multiple interpolations (NFMI), proposed in [6], performs direct upscaling of a Tamil document image scanned at any resolution, to result in a better word level accuracy. The NFMI method uses multiple interpolations, together with a CNN, to learn a mapping function, which takes in a LR document image and produces the corresponding HR image.

In this work, instead of using interpolations to perform convolution in the high-resolution space, we have used two recent techniques, famous in the deep learning community, namely transposed convolution [10] and sub-pixel [11] convolution to learn the mapping function. Unlike interpolations, transposed convolution layer learns weights from the training data to upscale the image. Hence, it can capture more variations in the input images and can be used for improving the quality, independent of the languages. We have created the training dataset in such a way that it covers maximum possible variations in the input image space.

3   Motivation

The issue addressed in this work is an actual industrial problem, where a huge number of documents have been scanned at a low resolution, and unfortunately, the original documents have been destroyed and hence, are no longer available for better scanning. The images are of poor quality, also because they are obtained from very old newspapers. Since the available regional language OCRs have a very poor recognition performance on these images, we would like to improve the quality of such images so that humans find it easy to read. It is very difficult and laborious to manually type and/or correct such a high volume of documents. Also, some of the images are so degraded that even native people find it uneasy to read them. Even if we are able to slightly improve the human readability, it is immensely useful.

Our studies have shown that documents scanned at a resolution of 100 dpi result in an average OCR accuracy of less than 50%; when the same documents are scanned at 200 dpi, the same OCR performs reasonably well and gives CLAs of 80–98%. Hence, it is sufficient for our task, if we can find an optimal model to upscale the document by a factor of 2. And our proposed architecture gives a relative improvement of around 51% in terms of CLA.

4   Dataset Created

The training dataset has been created from binary scanned images of documents in three different scripts (languages), capturing the multitude of variations in the input image space. The LR patches are created in three ways: (i) by taking *alternate pixels* from the HR patches, (ii) degrading the LR patches by *multiplying them element-wise by random masks of zeros and ones*, and (iii) by selecting the patches directly from images scanned at a LR setting of the scanner. The total number of LR-HR patch pairs created for training is around fifty million. The LR patches are of size 16 × 16 and the corresponding ground truth patches are of size 32 × 32. A patch pair is removed, if the LR patch contains only background pixels.

5   Architecture Advanced

The convolution-transposed-convolution-subpixel (CTCS) architecture proposed by us for upscaling a document image by a factor of 2 is illustrated in Fig. 1. The architecture has a convolution block, followed by an upscaling block. The upscaling block is formed by a transposed convolution, a convolution, and a sub-pixel layer. The initial convolution block extracts the relevant features from the input low resolution, binary image. This block uses 48 filters of size 5 × 5, followed by another layer of 16 filters of size 5 × 5. The transposed convolution layer has 16 filters of size 9 × 9 to upscale the feature maps to double their size. The convolution layer between the transposed convolution and sub-pixel layers is to reduce the number of features to the desired size. The size of the extracted feature map is decreased because we have used convolution filters without padding, to remove the artifacts. This convolution layer provides more non-linearity in the system. These controlled feature maps are passed on to the sub-pixel layer for upscaling the features to double their size. The sub-pixel layer takes in a tensor of size 16 × 16 × 4 and gives an output of size 32 × 32 × 1.

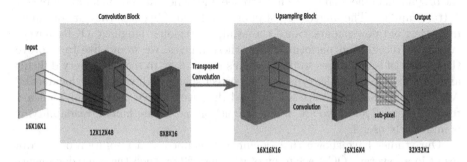

Fig. 1. The convolution-transposed convolution-subpixel (CTCS) architecture for 2X upscaling of binary document images. It is designed to deal with true LR images, scanned at a low resolution and not simulated LR images obtained by downsampling original HR images.

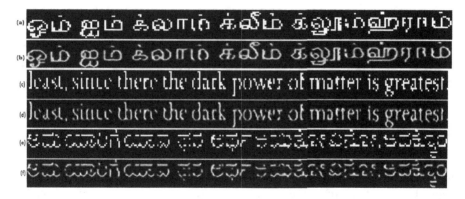

Fig. 2. (a), (c), (e): Tamil, English and Kannada input images originally scanned at 100 dpi. (b), (d), (f): the corresponding 200 dpi output images created by our CTCS architecture, with moderately enhanced human readability. It can be seen that the letter /ம/ that appears in every Tamil word ((a) and (b)) has a clearly improved perceptual quality. In the English images, the letters, "d, o, o, s, g, s" are better enhanced.

Fig. 3. (a), (c), (e): English, Tamil and Kannada input images originally scanned at 200 dpi. (b), (d), (f): the corresponding output images created by our CTCS architecture with increased perceptual quality.

5.1   Transposed Convolution (TC)

To train the model, where the input is in a low resolution space and the output is in a high resolution space, we need a mechanism to first move to the high resolution space. In an earlier work [4], interpolation techniques such as bicubic, bilinear and nearest-neighbor, or their combination have been used to take the LR images to the HR space. Unlike interpolations, transposed convolution [10] learns the weights to upscale the size of the feature map. Thus, the learning based techniques may contribute more than the interpolations in taking the image from a LR to a HR space. The size of the output obtained from the transposed convolution layer with stride $= 1$ and padding $= 0$ is given by $o = i + (f_s - 1)$, where i, o, and f_s are, respectively, the spatial sizes of the input feature map, output feature map and the filter.

Fig. 4. (a), (b): Input binary image and the corresponding output images (gray scale) created by our CTCS architecture. (c): image in (b) after gamma correction [20], displaying significantly improved perceptual quality and enhanced readability.

5.2 Sub-pixel Convolution (SC)

Unlike the TC, the sub-pixel convolution [11] is a technique that only rearranges the feature maps to increase its size or resolution. Hence it is faster than the TC and achieves good results, if properly positioned as a layer in the architecture. So, we avoid the use of multiple TC layers and replace the last layer with SC.

6 Training the CTCS Architecture

Given a low resolution, binary document image our goal is to construct a high resolution image with enhanced human readability and better quality in terms of CLA and WLA. Let the training set be $\{I_l^i, O_h^i\}$, $1 \leq i \leq N$, where N is the total number of patch pairs. The model weights are initialized by the technique in [13]. For each input I_l^i, the model builds a high resolution counterpart image, R_h^i. If there is an error between the produced output and the corresponding ground truth O_h^i, the weights of the model are adjusted to minimize the mean square error loss function.

The CTCS model is first trained with the patch pairs from the Tamil document image. The trained model is then further tuned on English, and then on Kannada. During the process of training, we save the weights at different stages and perform extensive experiments in order to identify the model that works best on any input image, scanned at any resolution. The model is trained for a maximum of 25 epochs. To calculate the gradients required while training, normal back-propagation is used. For adjusting the weights, Adam optimizer [9] is used with a learning rate of 0.0001, β_1 of 0.9 and β_2 of 0.99.

7 Results and Discussion

Figure 2 shows the output images for one sample input image each for three different languages, namely, Tamil, English and Kannada. In each pair of images, the top ones are the inputs scanned at a resolution of 100 dpi and the bottom one are the outputs obtained from our model. The improvement in the perceptual quality of the images is evident from observing the letter ' ம் ' that occurs as the last letter of every word in Fig. 2 (a) and (b). Better enhancement is seen in the English letters, "d, o, g" and the "s" that occurs in the last two words. Figure 3 shows that our CTCS model works independent of the input resolution, where the perceptual quality of the outputs generated by our model is significantly better.

Figure 4 (a) shows a sample, Tamil binary image scanned at 150 dpi. (b) and (c) show the CTCS-reconstructed and gamma corrected images respectively. The strokes of the characters have become smoother, while preserving the structure, thus resulting in better readability.

Table 1. Mean character level accuracy of 150 images from the 3 languages, each containing around 1k characters scanned at different resolutions (100, 200 and 300 dots per inch).

	Scanned at 100 (dpi)	Scanned at 200 (dpi)	Scanned at (300 dpi)
Average CLA	31.2	90.26	95.06

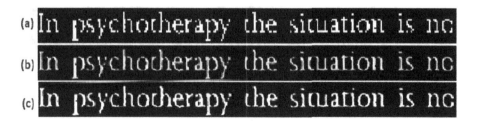

Fig. 5. (a): Input English image originally scanned at 150 dpi. (b): the corresponding output images created by our CTCS architecture with increase in perceived quality and ease of readability. (c): image in (b) after gamma correction [20]

Figures 5 (a) and 6(a) show a sample English and Kannada image each, scanned in binary mode at a resolution of 100 dpi. (b) and (c) show the CTCS-generated and gamma corrected respectively. In both cases, almost all the characters are better enhanced and hence the readability improves significantly.

Figure 7 illustrates how the loss of pixels changes the shapes of characters. The loss of one or two pixels from the character 'm' of the word 'problem' makes it appear as two characters i.e. r and n. Similar analysis can be done for the other characters too. In the characters reconstructed by the CTCS architecture, the strokes are smooth, not pixelated. Also the gaps are filled, preserving the structure and the meaning of the characters.

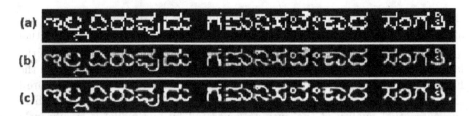

Fig. 6. (a) Input Kannada image originally scanned at 100 dpi. (b) the corresponding output images created by our CTCS architecture. (c) image in (b) after gamma correction [20]

Fig. 7. (a), (b), (c): Input English, bicubic interpolated and the corresponding output images created by our CTCS architecture with increase in perceived quality and ease of readability. In the word 'love' character 'o' and in the word 'problem' character 'm' is broken in both input and the bicubic interpolated output. But the same is reconstructed in the CTCS architecture

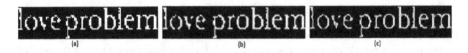

Fig. 8. (a) Input (b) bicubic (c) CTCS (d) input (e) bicubic (f) CTCS (g) Input (h) bicubic (i) CTCS: All images are scanned at 200 dpi: Shows our model is resolution independent.

We performed our initial experiments by scanning the images in binary mode at resolutions of 100, 200 and 300 dpi. Table 1 lists the OCR character level accuracy on 150 document images, each containing over 1000 characters.

Table 1 show that the images scanned in binary mode with less than 200 dpi can be treated as low resolution images. Also, results mentioned in the Table 1 suggest that increasing the resolution help in improving the quality of the input images, that will eventually leads to better readability and OCR character level accuracy. As it can be seen that just by increasing the resolution by factor of 2 the OCR recognition accuracy has significantly improved. But these are at hardware level (means scanner skipping or taking some pixels randomly). Suppose in the situation where the documents are originally scanned at low resolution say at 100 dpi and the original document is destroyed. Can we increase the resolution such document images that the OCR perform well? The first thing that strikes our mind is the interpolation such as bicubic, bilinear or any other super-resolution algorithm can can be used. Traditional interpolation helps in increasing the OCR accuracy slightly and marginally the readability. To tackle this real challenging problem we created the dataset mentioned in Sec. 4 and proposed the architecture shown in Fig. 1. We obtained the mean opinion score on the perceptual quality of the input and output images from 10 subjects each, for each of the three languages. The overall MOS of all the evaluators on all the language documents are listed in Table 2. The HR images obtained by the CTCS architecture have been evaluated at a MOS of 7.5, as compared to 4.5 for the input.

Table 2. Mean opinion score (10 point scale) of the enhancement in the perceptual quality of the document images scanned at 200 dpi

	Input	CTCS
Average MOS	4.5	7.5

Table 3. Mean character level accuracies (CLA) on 15 document images, 5 each from Tamil, Kannada and English, before and after upscaling by our CTCS architecture. Input images are scanned at 100 dpi.

	Input CLA%	CTCS%
Average CLA	33.1	49.99

Table 3 shows the average character level accuracies obtained from the images scanned at 100 dpi, and the outputs created by our CTCS architecture. The results show significant improvements in CLA. The mean improvements in terms of CLA relative to input is 51%.

Our model is designed to be independent of these three languages and the resolution and hence, can be applied on document images of any one of them. Hence, we do not need to change the design of the OCR for the above mentioned performance gain. The results can further be enhanced, if we incorporate (capture) further possible variations in the input training images.

8   Conclusion

We have created a *unique and diverse dataset* that captures maximum possible variations in the input image space from documents of three different languages (key to the success of our model). After performing extensive experiments, We have obtained an effective architecture based on deep learning, that is trained on this diverse dataset. We have shown that the proposed method enhances the perceptual quality of the input document images in terms of the subjective criterion of *human readability* and the objective criterion of OCR recognition, *independent of the languages and resolutions tested.* To our knowledge, this is the first report on enhancement of document images for better readability. Our method scales across the three languages tested and works for multiple input resolutions.

References

1. Shi, Z., Setlur, S., Govindaraju, V.: Image enhancement for degraded binary document images. In: Document Analysis and Recognition (ICDAR). IEEE (2011)
2. El Harraj, A., Raissouni, N.: OCR accuracy improvement on document images through a novel pre-processing approach. arXiv preprint arXiv:1509.03456 (2015)
3. Kumar, V., Bansal, A., Tulsiyan, G.H., Mishra, A., Namboodiri, A., Jawahar, C.V.: Sparse document image coding for restoration. In: 12th IEEE International Conference on Document Analysis and Recognition (ICDAR), pp. 713–717 (2013)
4. Pandey, R.K., Ramakrishnan, A.G.: Language independent single document image super-resolution using CNN for improved recognition. arXiv preprint arXiv:1701.08835 (2017)
5. Pandey, R.K., Ramakrishnan, A.G.: Efficient document-image super-resolution using convolutional neural network. Sadhana **43**(2), 15 (2018)
6. Pandey, R.K., Maiya, S.R., Ramakrishnan, A.G.: A new approach for upscaling document images for improving their quality. In: 14th IEEE India Council International Conference (INDICON). IEEE (2017)
7. Pandey, R.K., Vignesh, K., Ramakrishnan, A.G., Chandrahasa, B.: Binary document image super resolution for improved readability and OCR performance. arXiv preprint arXiv:1812.02475 (2018)
8. LeCun, Y.A., Bottou, L., Orr, G.B., Müller, K.-R.: Efficient BackProp. In: Montavon, G., Orr, G.B., Müller, K.-R. (eds.) Neural Networks: Tricks of the Trade. LNCS, vol. 7700, pp. 9–48. Springer, Heidelberg (2012). https://doi.org/10.1007/978-3-642-35289-8_3
9. Kingma, D.P., Ba, J.: Adam: a method for stochastic optimization. In: Proceedings 3rd International Conference Learning Representations (2014)
10. Xu, L., Ren, J.S., Liu, C., Jia, J.: Deep convolutional neural network for image deconvolution. In: Advances in Neural Information Processing Systems, pp. 1790–1798 (2014)
11. Shi, W., et al.: Real-time single image and video super-resolution using an efficient sub-pixel convolutional neural network. In: Proceedings of the IEEE Conference on Computer Vision and Pattern Recognition (2016)
12. Shivakumar, H.R., Ramakrishnan, A.G.: A tool that converted 200 Tamil books for use by blind students. In: Proceedings of the 12th International Tamil Internet Conference, Kuala Lumpur, Malaysia (2013)
13. He, K., Zhang, X., Ren, S., Sun, J.: Delving deep into rectifiers: surpassing human-level performance on ImageNet classification. In: Proceedings of the IEEE international conference on computer vision, pp. 1026–1034 (2015)
14. Glasner, D., Shai, B., Michal, I.: Super-resolution from a single image. In: 12th IEEE International Conference on Computer Vision (2009)
15. Yang, J., Wright, J., Huang, T.S., Ma, Y.: Image super-resolution via sparse representation. IEEE Trans. Image Process. **19**(11), 2861–2873 (2010)

16. Timofte, R., De Smet, V., Van Gool, L.: Anchored neighborhood regression for fast example-based super-resolution. In: Proceedings of the IEEE International Conference on Computer Vision (2013)
17. Dong, C., Loy, C.C., He, K., Tang, X.: Learning a deep convolutional network for image super-resolution. In: Fleet, D., Pajdla, T., Schiele, B., Tuytelaars, T. (eds.) ECCV 2014. LNCS, vol. 8692, pp. 184–199. Springer, Cham (2014). https://doi.org/10.1007/978-3-319-10593-2_13
18. Ledig, C., et al.: Photo-realistic single image super-resolution using a generative adversarial network. In: IEEE Conference on Computer Vision and Pattern Recognition (CVPR), pp. 105–114 (2017)
19. Lai, W.S., Huang, J.B., Ahuja, N., Yang, M.H.: Deep laplacian pyramid networks for fast and accurate superresolution. In: IEEE Conference on Computer Vision and Pattern Recognition, vol. 2, no. 3, p. 5 (2017)
20. Kumar, D., Ramakrishnan, A.G.: Power-law transformation for enhanced recognition of born-digital word images. In: International Conference on Signal Processing and Communications (SPCOM), pp. 1–5. IEEE (2012)

Pair-Comparing Based Convolutional Neural Network for Blind Image Quality Assessment

Xue Qin, Tao Xiang$^{(\boxtimes)}$, Ying Yang, and Xiaofeng Liao

College of Computer Science, Chongqing University, Chongqing 400044, China
txiang@cqu.edu.cn

Abstract. The introduction of convolutional neural network (CNN) in no-reference image quality assessment (NR-IQA) gains great success in improving its prediction accuracy, and the performance of CNN relies on the magnitude of training samples. However, many widely-used existing image databases cannot provide adequate samples for CNN training. In this paper, we propose a pair-comparing based convolutional neural network (PC-CNN) for blind image quality assessment. By taking reference images into consideration, we generate more training samples of patch pairs by different combinations of distorted images and reference image. We build a new CNN network which has two inputs for patch pairs and two outputs predicting the scores of patches. We conduct extensive experiments to evaluate the performance of our proposed PC-CNN, and the results show that it outperforms many state-of-the-art methods.

Keywords: No-reference image quality assessment ·
Convolutional neural network · Deep learning · Human visual system

1 Introduction

Image quality assessment (IQA) has become an indispensable part in more and more image related applications. Objective IQA is the way to assess the quality of images by computers automatically without the help of humans. A satisfactory objective IQA method is capable of evaluating the performance of image processing techniques like compression, enhancement and so on. Based on the availability of a reference image during quality predicting process, objective IQA can be divided into three categories: full-reference IQA (FR-IQA), reduced-reference IQA (RR-IQA), and no-reference IQA (NR-IQA).

NR-IQA is a more practical but challenging task compared with FR-IQA and RR-IQA. Because NR-IQA can predict the quality of a distorted image without the access to its reference image, it is much more convenient and applicable in practical applications, especially for the scenarios where the reference image is inaccessible. Although the absence of reference image increases the practicability of NR-IQA, it makes NR-IQA a challenging task at the same time because it is usually difficult to measure the distortion level of a distorted image without accessing to its reference image.

© Springer Nature Switzerland AG 2019
H. Lu et al. (Eds.): ISNN 2019, LNCS 11555, pp. 460–468, 2019.
https://doi.org/10.1007/978-3-030-22808-8_45

NR-IQA based on convolutional neural network (CNN) has gained a great attention in recent years for its improvement on NR-IQA performance. CNN is a deep neural network and finds its first application in NR-IQA in [3], where the authors proposed to use CNN to integrate feature extraction and regression into one process and their method achieves better prediction accuracy than the previous ones. Later in [1], Bosse *et al.* considered the visual importance of an image and proposed to learn patch score and patch weight at the same time. Kim *et al.* [4] proposed a NR-IQA method which gives each patch an initial score by using its reference image and distorted image in training phase. In [6], a large amount of reliable training data in the form of quality-discriminable image pairs are obtained automatically and then RankNet is used for learning. In [2], vector regression and object oriented pooling are used for NR-IQA where a vector of belief scores for input image is yielded during vector regression phase.

Although the introduction of CNN improves the accuracy of NR-IQA remarkably, there are still two issues that should be further explored. First, because the performance of CNN relies on the magnitude of training samples, a large number of training data is desired for CNN based NR-IQA. However, many widely-used existing databases contain a small number of images. Second, most NR-IQA methods do not consider reference images, but reference images are crucial for the task of IQA since FR-IQA usually achieves relatively higher prediction accuracy than NR-IQA. For improving the prediction accuracy, it deserves more attention to explore how to utilize reference images in CNN based NR-IQA during training phase.

Motivated by addressing the above issues, we propose a pair-comparing based convolutional neural network (PC-CNN) for blind image quality assessment. In our proposed method, we adequately utilize the information of reference images during training phase for better comparing learning. In order to tackle the problem of lacking of data for training, we generate more training samples using patch pairs under different combination strategies. The proposed PC-CNN works as follows: images are split into non-overlapping and normalized patches during preprocessing phase, and combination strategies are used for generating patch pairs. Then, these pairs are fed into the CNN for acquiring an optimal model. In testing phase, the model receives two same testing patches and its two outputs are used to calculate the score of this patch. The final image score is the average of scores of all its patches.

2 Proposed Method

2.1 Framework

Our proposed PC-CNN method contains three parts as shown in Fig. 1. Part one is the preprocessing work producing locally contrast normalized patches for subsequent procedures. Part two is responsible for generating training patch pairs via different combination strategies. Part three uses CNN for patch-wise regression which takes a patch pair as inputs and outputs two patch scores. Each part is described in detail in subsequent statements.

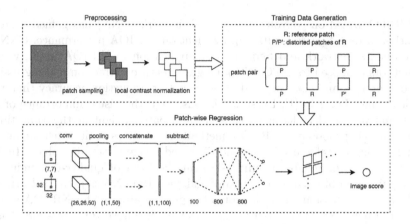

Fig. 1. The framework of our proposed method.

2.2 Preprocessing

The preprocessing work of our method produces sampled patches and applies local contrast normalization to them. First, inspired by the conclusion in [3] that image distortions are often locally homogeneous, we divide images into patches of size 32 × 32 without overlapping in order to expand training data. Then, similar to [3,7], local contrast normalization is performed to these patches to make the network robust to the variation of illumination and contrast. In our implementation, we normalize each patch with the window size of 3.

2.3 Training Data Generation

In order to generate more training data and improve prediction accuracy, we propose to consider reference images in the training phase. We generate patch pairs for CNN training by utilizing the information of distorted patch and its corresponding patch in reference image via different combination strategies. Specifically, we consider the following four combination strategies of patch pairs: two same distorted patches, one distorted patch and its corresponding reference patch, one distorted patch and anther corresponding distorted patch from the same reference image, and two same reference patches. For example, for a distorted patch P and its corresponding reference patch R, we randomly select another different distorted patch P' of the same reference image, and generate four patch pairs: (P, P), (P, R), (P, P') and (R, R). After the generation of training data, all patch pairs are labeled with the score the same as its belonging image.

2.4 Patch-Wise Regression

In order to compare the patch difference between reference image and distorted image in training phase, we design our pair-comparing network. As demonstrated in Fig. 1, our CNN is composed of seven layers: one convolutional, one pooling, one concatenated, one subtract and three fully-connected layers which has two inputs and two outputs. We only choose a seven-layer network because deep-layer network may lead to overfitting problem if the size of training set is limited.

The input of CNN is a patch pair whose size is 32 × 32 after local contrast normalization. There are two parallel ways in the following three layers. The first is a convolutional layer containing 50 kernels of size 7 × 7 with stride of 1 pixel and it produces 50 feature maps with size 26 × 26, following by a pooling layer extracting the max and average values of each feature map to generate two tensors with size 1 × 1 × 50. Then the two tensors are concatenated to a 1 × 1 × 100 size tensor for each way. Next, a subtract layer is responsible for acquiring the difference of the two tensors and flattening them to a vector of size 100. After that are two fully-connected layers with 800 nodes. The last layer outputs two scores. We choose rectified linear units (ReLUs) as the activation function for fully-connected layers in order to prevent none-positive values passing through the network.

In training phase, we optimize the CNN by minimizing objective function where each patch is expected to output the score as similar to its label as possible. For a patch pair (P_i, P_j), let p_i and p_j denote their ground truth scores respectively, $\hat{p}_i$ and $\hat{p}_j$ denote their predicted scores. Then the outputs of CNN can be formulated as

$$(\hat{p}_i, \hat{p}_j) = f(P_i, P_j; \omega) \tag{1}$$

where $f(P_i, P_j; \omega)$ denotes the outputs of learning process with input (P_i, P_j) and network weight ω. We define the objective function as follows by the mean absolute error (MAE), which is a regular evaluation index for regression, between labeled patch score and predicted patch score.

$$Loss = \frac{1}{N_b} \sum_{i,j=1}^{N_b} (|\hat{p}_i - p_i| + |\hat{p}_j - p_j|) \tag{2}$$

where N_b denotes the number of all patch pairs in a mini-batch over an epoch. By minimizing the above objective function, we get the optimal network parameters ω^* of CNN

$$\omega^* = \underset{\omega}{argmin}\ Loss \tag{3}$$

In our implementation, we use stochastic gradient decent (SGD) to train the network for 50 epochs and the nesterov momentum is used in SGD as well, the value is set to 0.9.

In testing phase, two same distorted patches of a test image are fed into the CNN. We average the outputs of the CNN to obtain the patch score, and the final predicted score of test image is the average of all its patches' scores.

3   Experiments and Results

We conduct extensive experiments to analyze the performance of our proposed PC-CNN. We compare our proposed PC-CNN not only with several well-known

FR-IQA methods, but also with many state-of-the-art NR-IQA methods including DIIVINE [9], BLIINDS-II [11], BRISQUE [7], CORNIA [14], etc.

3.1   Databases

We use three databases in our experiments: LIVE [12], TID2013 [10] and CSIQ [5]. The subjective quality of an image is given by mean opinion score (MOS) or difference mean opinion score (DMOS). LIVE and TID2013 are used to perform dependent database evaluation. For cross-database evaluation, we train on LIVE and test on TID2013 and CSIQ respectively; and the MOS in TID2013 and the DMOS in CSIQ are mapped to the same range of DMOS in LIVE in order to make comparable measurement. For each database, 80% randomly selected reference images and their corresponding distorted images are used as the training samples and the rest 20% are used as testing samples.

3.2   Evaluation Measures

Two mainstream measures are used to evaluate the performance of various IQA methods: the linear correlation coefficient (LCC) and the Spearman rank order correlation coefficient (SROCC). LCC measures the linear correlation between subjective and objective scores, and SROCC judges the monotonic property of the relationship between subjective and objective scores. A higher LCC or SROCC value, i.e. a value closer to 1, indicates a better performance of an IQA method. All LCC and SROCC are the average values of experiments based on 10 random splits of samples.

3.3   Combination Strategies

In Sect. 2.3, we introduce the combination strategy in our proposed PC-CNN to generate more training data, actually there are some other combination possibilities. Here we validate that different strategies affect the performance of our method and our chosen strategy is superior to others. We compare the LCC and SROCC on all images on LIVE using six different strategies. To simplify the expression, we referred to patch pairs consisted of two same distorted patches as PP, two same reference patches as RR, one distorted and one reference patches as PR, two different distorted patches of a same reference image as PP'. The six strategies are denoted as (PP, PR, PP', RR), (PP, PR, RR), (PP, PP', RR), (PP, RR), (PP, PR), (PP, PP'). The results are shown in Fig. 2. It can be seen that the more information of reference images in pairs, the better performance our method obtains, and our chosen combination strategy (PP, PR, PP', RR) outperforms the others. We also observe from our experiments that the random selection of two different distorted patches of a same reference image (PP') has negligible impact on experimental results.

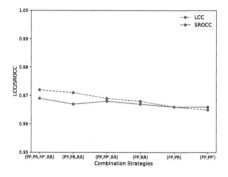

Fig. 2. Performance comparison (LCC and SROCC) of different combination strategies on LIVE.

3.4 Dependent-Database Experiments

We evaluate the performance of our proposed PC-CNN and other existing IQA methods in terms of LCC and SROCC on LIVE and TID2013 respectively. The results on LIVE database are tabulated in Tables 1 and 2. We compare LCC and SROCC of different IQA methods not only based on the whole images in the database, but also based on the images of different distortion types. The best result under each criterion is highlighted in bold. We can see that except for JPEG distortion, our proposed PC-CNN outperforms other IQA methods. The results on TID2013 are listed in Table 3, where we train and test on the whole images in the database. It is clear that our proposed PC-CNN achieves the highest performance in terms of both LCC and SROCC.

Table 1. Performance comparison (LCC) on LIVE.

Method	JP2K	JPEG	WN	GBLUR	FF	ALL
PSNR	0.873	0.876	0.926	0.779	0.870	0.856
SSIM [13]	0.921	0.955	0.982	0.893	0.939	0.906
FSIM [15]	0.910	0.985	0.976	0.978	0.912	0.960
BRISQUE [7]	0.922	0.973	0.985	0.951	0.903	0.942
NIQE [8]	0.937	0.956	0.977	0.952	0.913	0.915
CORNIA [14]	0.951	0.965	0.987	0.968	0.917	0.935
CNN [3]	0.953	0.981	0.984	0.953	0.933	0.953
BIECON [4]	0.965	**0.987**	0.970	0.945	0.931	0.962
dipIQ [6]	0.964	0.980	0.983	0.948	-	-
VROP [2]	0.957	0.984	0.983	0.959	0.951	0.968
PC-CNN	**0.986**	0.982	**0.993**	**0.981**	**0.970**	**0.972**

3.5   Cross-Database Experiments

We also conduct cross-database experiments to evaluate the generalization capability of our proposed method. We train the model using the images in LIVE database and test the prediction accuracy using the images in TID2013 and CSIQ databases respectively. The SROCC of different IQA methods are shown in Table 4. We find that for TID2013, our purposed PC-CNN obtains the second best performance. While for CSIQ, PC-CNN outperforms all other methods. The results illustrate a satisfactory generalization capability of our proposed PC-CNN. Compared with the results on dependent databases, we find that the results here are generally worse. Because TID2013 and CSIQ databases contain distortion types that are not in LIVE, the model has little knowledge of them.

Table 2. Performance comparison (SROCC) on LIVE.

Method	JP2K	JPEG	WN	GBLUR	FF	ALL
PSNR	0.870	0.885	0.942	0.763	0.874	0.866
SSIM [13]	0.939	0.946	0.964	0.907	0.941	0.913
FSIM [15]	0.970	0.981	0.967	0.972	0.949	0.964
BRISQUE [7]	0.914	0.965	0.979	0.951	0.877	0.940
NIQE [8]	0.917	0.938	0.967	0.934	0.859	0.914
CORNIA [14]	0.943	0.955	0.976	0.969	0.906	0.942
CNN [3]	0.952	**0.977**	0.978	0.962	0.908	0.956
BIECON [4]	0.952	0.974	0.980	0.956	0.923	0.961
dipIQ [6]	0.956	0.969	0.975	0.940	-	-
VROP [2]	0.963	0.976	0.984	0.956	0.939	0.967
PC-CNN	**0.975**	0.972	**0.988**	**0.980**	**0.975**	**0.969**

Table 3. Performance comparison (LCC and SROCC) on TID2013.

Method	PSNR	SSIM [13]	FSIM [15]	DIIVINE [9]	BLIINDS-II [11]	BRISQUE [7]	NIQE [8]	CORNIA [14]	PC-CNN
LCC	0.675	0.79	0.877	0.654	0.628	0.651	0.426	0.613	**0.788**
SROCC	0.687	0.742	0.851	0.549	0.536	0.573	0.317	0.549	**0.718**

Table 4. Cross-database performance evaluation (SROCC) training on LIVE and testing on TID2013 and CSIQ.

Method	DIIVINE [9]	BLIINDS-II [11]	BRISQUE [7]	CORNIA [14]	PC-CNN
TID2013	0.355	0.393	0.367	**0.429**	0.415
CSIQ	0.596	0.577	0.557	0.663	**0.665**

4   Conclusion

In this paper, we developed a new blind image quality assessment method by proposing a pair-comparing based convolutional neural network (PC-CNN). The network has two inputs for training and testing. In the training, distorted images and their reference images are used to generate patch pairs under different combination strategies, and these patch pairs are fed into the network for training. In the testing, only a test image is needed to create patch pairs for network inputs. Our proposed method not only can increase the number of training samples based on the existing databases, but also has better prediction accuracy compared with many state-of-the-art image quality assessment (IQA) methods.

Acknowledgments. This work was supported by the National Natural Science Foundation of China (No. 61672118).

References

1. Bosse, S., Maniry, D., Wiegand, T., Samek, W.: A deep neural network for image quality assessment. In: ICIP, pp. 3773–3777 (2016)
2. Gu, J., Meng, G., Redi, J.A., Xiang, S., Pan, C.: Blind image quality assessment via vector regression and object oriented pooling. IEEE Trans. Multimed. **20**(5), 1140–1153 (2018)
3. Kang, L., Ye, P., Li, Y., Doermann, D.: Convolutional neural networks for no-reference image quality assessment. In: CVPR, pp. 1733–1740 (2014)
4. Kim, J., Lee, S.: Fully deep blind image quality predictor. IEEE J. Sel. Top. Signal Process. **11**(1), 206–220 (2017)
5. Larson, E.C., Chandler, D.M.: Most apparent distortion: full-reference image quality assessment and the role of strategy. J. Electron. Imaging **19**(1), 011006 (2010)
6. Ma, K., Liu, W., Liu, T., Wang, Z., Tao, D.: dipIQ: blind image quality assessment by learning-to-rank discriminable image pairs. IEEE Trans. Image Process. **26**(8), 3951–3964 (2017)
7. Mittal, A., Moorthy, A.K., Bovik, A.C.: No-reference image quality assessment in the spatial domain. IEEE Trans. Image Process. **21**(12), 4695–4708 (2012)
8. Mittal, A., Soundararajan, R., Bovik, A.C.: Making a "completely blind" image quality analyzer. IEEE Signal Proc. Let. **20**(3), 209–212 (2013)
9. Moorthy, A.K., Bovik, A.C.: Blind image quality assessment: from natural scene statistics to perceptual quality. IEEE Trans. Image Process. **20**(12), 3350–3364 (2011)
10. Ponomarenko, N., et al.: Color image database TID2013: peculiarities and preliminary results. In: EUVIP, pp. 106–111 (2013)
11. Saad, M.A., Bovik, A.C., Charrier, C.: Blind image quality assessment: a natural scene statistics approach in the DCT domain. IEEE Trans. Image Process. **21**(8), 3339–3352 (2012)
12. Sheikh, H.: Live image quality assessment database release 2. http://live.ece.utexas.edu/research/quality
13. Wang, Z., Bovik, A.C., Sheikh, H.R., Simoncelli, E.P.: Image quality assessment: from error visibility to structural similarity. IEEE Trans. Image Process. **13**(4), 600–612 (2004)

14. Ye, P., Kumar, J., Kang, L., Doermann, D.: Unsupervised feature learning framework for no-reference image quality assessment. In: CVPR, pp. 1098–1105 (2012)
15. Zhang, L., Zhang, L., Mou, X., Zhang, D.: FSIM: a feature similarity index for image quality assessment. IEEE Trans. Image Process. **20**(8), 2378–2386 (2011)

PolSAR Marine Aquaculture Detection Based on Nonlocal Stacked Sparse Autoencoder

Jianchao Fan[1]($\boxtimes$), Xiaoxin Liu[2], Yuanyuan Hu[3], and Min Han[3]

[1] Department of Ocean Remote Sensing, National Marine Environmental Monitoring Center, Dalian 116023, Liaoning, China
jcfan@nmemc.org.cn
[2] Computer Science and Engineering, Washington University in St. Louis, Saint Louis, MO 63130, USA
xiaoxinliu@wustl.edu
[3] Faculty of Electronic Information and Electrical Engineering, Dalian University of Technology, Dalian 116024, Liaoning, China
huyy@mail.dlut.edu.cn, minhan@dlut.edu.cn

Abstract. Marine aquaculture plays an important role in marine economic, which distributes widely around the coast. Using satellite remote sensing monitoring, it can achieve large scale dynamic monitoring. As a classic model of deep learning, stacked sparse autoencoder (SSAE) has the advantages of simple model and self-learning of features. Nonlocal spatial information is utilized to assist SSAE construct NSSAE to improve the precision in this paper. Experimental results demonstrate the superiority of nonlocal SSAE methods on marine target recognition.

Keywords: Polarimetric SAR · Remote sensing images · Nonlocal spatial information · Stacked sparse autoencoder · Classification

1 Introduction

The floating raft aquaculture mainly consists of floating balls located in shallow seas and intertidal zones. Scallops, oysters and other shellfish are cultured under hanging cages [1]. The cultured floating rafts are mainly floating balls on the surface of sea water, which are small in size and sparsely distributed. Therefore, they cannot be completely detected in optical remote sensing images. Synthetic aperture radar (SAR) remote sensing images can detect the intensity of floating ball, which have been successfully applied to their recognition [2,3]. Since the

J. Fan—The work described in the paper was supported by the National Key R&D Program of China (2017YFC1404902, 2016YFC1401007); National Natural Science Foundation of China (41706195, 61773087); National High Resolution Special Research (41-Y30B12-9001-14/16); Key Laboratory of Sea-Area Management Technology Foundation (201701).

© Springer Nature Switzerland AG 2019
H. Lu et al. (Eds.): ISNN 2019, LNCS 11555, pp. 469–476, 2019.
https://doi.org/10.1007/978-3-030-22808-8_46

launch of polarimetric synthetic aperture radar (PolSAR) satellite, plenty of fully polarimetric data have been obtained. Four different polarimetric modes provide richer information from each channel.

As the amount of PolSAR image data increases exponentially, automatic learning and extraction of features become the development trend of image interpretation. In recent years, deep learning has achieved significant success in pattern recognition and machine vision because it automatically learns rich and deep features from raw data [4]. As a classic model of deep learning, stacked sparse autoencoder (SSAE) has the advantages of simple model and self-learning of features, and has been widely used in SAR image classification [5,6]. With the number of network layers increases, the potential relationship of data is automatically discovered, so effective features are extracted, which facilitates accurate identification of subsequent targets.

The spatial resolution of PolSAR images continues to increase, and the actual object area corresponding to each pixel is continuously reduced, which results in an increase in the probability of pixels being affected by accidental factors and increasing the number of noises [7]. However, spatial correlations between pixels in the image increase, which helps provide more abundant spatial information and suppress the influence of speckle noise [8]. Bi et al. [9] used Markov random field and three-dimensional wavelet features to perform PolSAR image classification. Zhang et al. [10] proposed local spatial information to improve SSAE classification capability. However, the SSAE network explores the relationship between features of a single pixel or local neighbour spatial information, while ignoring the whole (global) spatial correlation between pixels. Therefore, it is necessary to introduce nonlocal spatial information into the SSAE to improve the classification accuracy.

The nonlocal Means (NLM) algorithm was originally proposed by Buades et al. [11] for optical image filtering and aims to remove noise effects using redundant information present in the image. For any pixel in the image, there must be a large number of pixels with similar neighborhood structure distributed in the whole image, and these pixels are weighted and summed to obtain rich spatial information, which is convenient for subsequent processing. Zhang et al. [12] presented local and nonlocal spatial information based on Markov random field method for hyperspectral image analysis. Therefore, in this paper, nonlocal spatial information are adopted to SSAE model to improve PolSAR marine aquaculture recognition.

The rest of this paper is organized as follows. Nonlocal spatial information extraction is presented in Sect. 2. The combined nonlocal stacked sparse autoencoder is given in Sect. 3. The experimental results on PolSAR marine aquaculture recognition based on the proposed algorithm are delineated in Sect. 4. Finally, concluding remarks are given in Sect. 5.

2   Nonlocal Spatial Information Extraction

In order to directly describe the significance of nonlocal information extraction, the schematic diagram is shown in Fig. 1, which denotes part of the search window of pixel a, and there are two categories, colored in red and blue respectively.

The center pixel a is located at the boundary (yellow line), and its neighborhood window (black square) contains these two categories. Compared with pixel a, pixel b and c belong to the same category, while pixel d belongs to another category. Since all pixels in the black square make up the neighborhood configuration of pixel a, in its search window, only pixels beside the same boundary, like pixel b, have similar neighborhood configurations and have larger weights ω. Although pixel c and a belong to the same category, their neighborhood configurations are partially similar, so pixel c cannot have a large weight ω_{ac}. Moreover, for pixel d, its neighborhood configuration is also partially similar to pixel a. So, pixel d, even belonging to the different category, still has a similar weight to pixel c. When calculating nonlocal means, pixels beside the boundary are difficult to obtain accurate values and easily misclassified. To solve this problem, the neighborhood configuration of the center pixel is adaptively optimized, by picking pixels considered to belong to the same category as the center pixel in the neighborhood window. It can make full use of all pixels by assigning pixels belonging to the same category large weights and reducing weights of pixels with different categories.

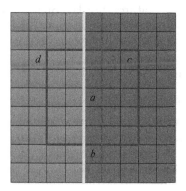

Fig. 1. Part of the search window of pixel a (Color figure online)

For each pixel i in the image, x_i is the value of pixel. Using $R \times R$ size window to search area S_i^R, the nonlocal average value is defined as

$$\bar{x}_i = \sum_{j \in S_i^R} w_{ij} x_j \tag{1}$$

here pixel j belong to S_i^R, and the weight w_{ij} denotes the similarity between pixel i and j, $\sum_{j \in S_i^R} w_{ij} = 1$. Using $r \times r$ size window for pixel i to construct the neighbourhood structure N_i^r, all pixels in N_i^r are expressed as

$$Q_{ij} = \|v_i - v_j\|_{2,\delta}^2 \tag{2}$$

So Q_{ij} is smaller, the similarity between pixel i and j is larger. The corresponding weight w_{ij} becomes larger.

$$w_{ij} = \frac{1}{Z_i} e^{-Q_{ij}/g} \tag{3}$$

here Z_i is a normalized parameter defined as $Z_i = \sum_{j \in S_i^R} e^{-Q_{ij}/g}$. g is the filtering parameter that controls the decay rate of the exponential function.

3   Nonlocal Stacked Sparse Autoencoder

Autoencoder (AE) is an unsupervised feature self-learning algorithm. The network structure commonly consists of three layers, namely input layer, hidden layer and output layer. The calculation process can be regarded as two parts of encoding and decoding. The input data is encoded by the feature representation, and then decoded to obtain the output data. The result obtained by the encoding can be regarded as a different representation of the input data. However, since feature mapping is common nonlinear, deep neural networks can describe more complex function sets in a more concise manner, making it easier to obtain deeper effective features. Stacked Sparse Autoencoder (SSAE) is a multi-layer neural network model formed by superimposing multiple SAEs. The feature representation obtained by the previous SAE is used as the input of the next SAE.

Nonlocal spatial information is adopted in SSAE to perform PolSAR marine aquaculture classification, and the whole schematic is shown in Fig. 2. The extracted nonlocal spatial contents are set as one part of the input data feature, to increase the feature dimension and provide rich and effective information to gain deep features. On the other hand, they are introduced into the first layer structure of the NSSAE network, by approximating each other in the objective function for more robust features.

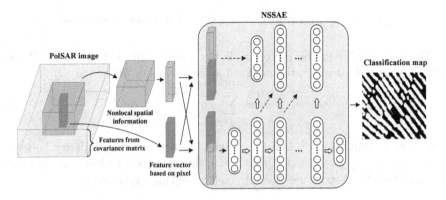

Fig. 2. The model of PolSAR marine aquaculture classification based on NSSAE

4   Experimental Results

In order to verify the marine recognition capability of NSSAE network, the Radarsat-2 PolSAR image of the adjacent sea areas are selected in China. Figure 3 shows the Pauli pseudo-color image. Two regions are selected marked with two boxes. The yellow box indicates research area 1 containing a densely distributed floating carp culture target. The red square indicates research area 2, containing a sparsely distributed scorpion breeding target.

Fig. 3. Two test marine aquaculture areas based on Radarsat-2 PolSAR images (Color figure online)

The Pauli pseudo-color image of the research area 1 and the corresponding ground truth map are shown in Fig. 4, respectively. The size is 300×300. The NSSAE contains 3 hidden layers with 150 nodes in each layer. And the threshold parameter is set to 0.001. In order to make comparison with other art-of-state algorithms, support vector machine (SVM), classical SSAE, nonlocal means filtering with SSAE (NLM-SSAE). Three indexes, the overall accuracy (OA), the average accuracy (Average Accuracy, AA) and the kappa coefficient, are used for qualitative evaluation of each algorithm. So classification results are shown in Fig. 5 with four algorithm, and specifical indexes are tabulated in Table 1. As the same analysis process, the experiment of area 2 are shown in Figs. 6 and 7, tabulated in Table 2, respectively.

From these experimental results, NSSAE has the highest precision, and isolated noises in the classification result images are the least. The foreground

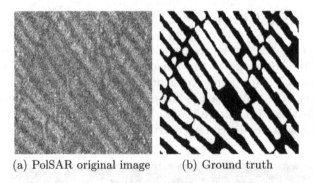

(a) PolSAR original image   (b) Ground truth

Fig. 4. Research Area 1

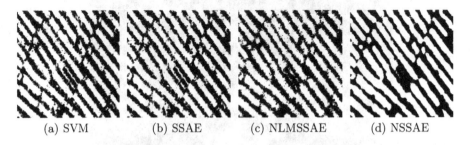

(a) SVM   (b) SSAE   (c) NLMSSAE   (d) NSSAE

Fig. 5. Classification results of Area 1

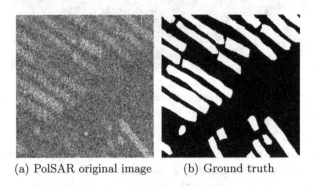

(a) PolSAR original image   (b) Ground truth

Fig. 6. Research Area 2

floating raft targets are completely continuous inside, and the edges are smooth and clear. According to three indexes in Tables 1 and 2, proposed NSSAE has the best results. SVM, SSAE and NLM-SSAE are still sensitive to noises. Introduced nonlocal spatial information can effectively assist SSAE reduce isolated noises, suppress speckle noises and improve the recognition accuracy in the end.

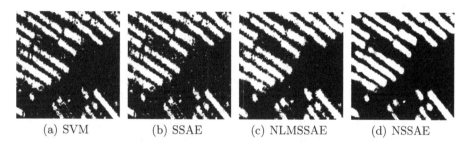

(a) SVM   (b) SSAE   (c) NLMSSAE   (d) NSSAE

Fig. 7. Classification results of Area 2

Table 1. Area 1 results of four algorithm on PolSAR marine aquaculture recognition

Class	SVM	SSAE	NLM-SSAE	NSSAE
Sea water	82.78	88.84	88.48	**95.03**
Floating raft	92.85	86.99	89.52	**93.88**
OA%	87.72	87.80	88.96	**94.50**
AA%	87.54	87.92	89.00	**94.45**
Kappa	0.7533	0.7537	0.7784	**0.8792**

Table 2. Area 2 results of four algorithm on PolSAR marine aquaculture recognition

Class	SVM	SSAE	NLM-SSAE	NSSAE
Sea water	92.85	95.88	93.59	**93.88**
Floating raft	87.25	82.41	87.26	**95.26**
OA%	88.69	90.77	91.59	**94.52**
AA%	90.05	89.15	90.43	**94.57**
Kappa	0.7300	0.8014	0.8063	**0.8798**

5   Conclusions

This proposed NSSAE network for PolSAR classification images can effectively suppress the influence of speckle noises, and automatically obtain deep, abstract and divisible features. The adaptive non-local spatial information is input into the NSSAE network as part of features to greatly reduce isolated noises and improve the recognition accuracy. The experiment of Radarsat-2 PolSAR images in the adjacent sea area of Changhai verify the validity and practicability of the NSSAE network.

References

1. Fan, J., Chu, J., Geng, J., Zhang, F.: Floating raft aquaculture information automatic extraction based on high resolution SAR images. In: 2015 IEEE International Geoscience and Remote Sensing Symposium, Milan, pp. 3898–3901 (2015)
2. Hu, Y., Fan, J., Wang, J.: Target recognition of floating raft aquaculture in SAR image based on statistical region merging. In: 2017 Seventh International Conference on Information Science and Technology, Da Nang, pp. 429–432 (2017)
3. Geng, J., Fan, J., Wang, H.: Weighted fusion-based representation classifiers for marine floating raft detection of SAR images. IEEE Geosci. Remote Sens. Lett. **14**, 444–448 (2017)
4. Shao, Z., Zhang, L., Wang, L.: Stacked sparse autoencoder modeling using the synergy of airborne LiDAR and satellite optical and SAR data to map forest aboveground biomass. IEEE J. Sel. Top. Appl. Earth Obs. Remote Sens. **10**, 5569–5582 (2017)
5. Hu, Y., Fan, J., Wang, J.: Classification of PolSAR images based on adaptive nonlocal stacked sparse autoencoder. IEEE Geosci. Remote Sens. Lett. **15**, 1050–1054 (2018)
6. Tao, C., Pan, H., Li, Y., Zou, Z.: Unsupervised spectral-spatial feature learning with stacked sparse autoencoder for hyperspectral imagery classification. IEEE Geosci. Remote Sens. Lett. **12**, 2438–2442 (2015)
7. Wan, L., Zhang, T., Xiang, Y., You, H.: A robust fuzzy C-means algorithm based on Bayesian nonlocal spatial information for SAR image segmentation. IEEE J. Sel. Top. Appl. Earth Obs. Remote Sens. **11**, 896–906 (2018)
8. Song, W., Li, M., Zhang, P., Wu, Y., Jia, L., An, L.: Unsupervised PolSAR image classification and segmentation using Dirichlet process mixture model and Markov random fields with similarity measure. IEEE J. Sel. Top. Appl. Earth Obs. Remote Sens. **10**, 3556–3568 (2017)
9. Bi, H., Xu, L., Cao, X., Xu, Z.: PolSAR image classification based on three-dimensional wavelet texture features and Markov random field. In: 2017 IEEE International Geoscience and Remote Sensing Symposium, Fort Worth, pp. 3921–3928 (2017)
10. Zhang, L., Ma, W., Zhang, D.: Stacked sparse autoencoder in PolSAR data classification using local spatial information. IEEE Geosci. Remote Sens. Lett. **13**, 1359–1363 (2016)
11. Buades, A., Coll, B., Morel, J.: A non-local algorithm for image denoising. In: 2005 IEEE Computer Society Conference on Computer Vision and Pattern Recognition, San Diego, vol. 2, pp. 60–65 (2005)
12. Zhang, X., Gao, Z., Jiao, L., Zhou, H.: Multifeature hyperspectral image classification with local and nonlocal spatial information via Markov random field in semantic space. IEEE Trans. Geosci. Remote Sens. **56**, 1409–1424 (2018)

Robust Object Tracking Based on Deep Feature and Dual Classifier Trained with Hard Samples

Bo Xu and Chengan Guo[(✉)]

School of Information and Communication Engineering,
Dalian University of Technology, Dalian 116023, Liaoning, China
dutxubo@mail.dlut.edu.cn, cguo@dlut.edu.cn

Abstract. Visual tracking has attracted more and more attention in recent years. In this paper, we proposed a novel tracker that is composed of a feature network, a dual classifier, a target location module, and a sample collecting and pooling module. The dual classifier contains two classifiers, called long-term classifier and short-term classifier, in which the long-term classifier is to maintain the long-term appearance of the target and the short-term classifier is for prompt response to the sudden change of the target. The training samples are divided into positive samples, negative samples, hard positive samples and hard negative samples and are used to train the two classifiers, respectively. Furthermore, in order to overcome the unreliability in locating the target by highest score, a density clustering method is introduced into the target locating process. Experimental results conducted on two benchmark datasets demonstrate the effectiveness of the proposed tracking method.

Keywords: Visual tracking · Ensemble tracking · Hard samples · Density clustering

1 Introduction

Visual tracking, as a basic computer vision task, has a wide range of applications, such as automatic driving, video surveillance, sports analysis, etc. Although remarkable progress has been made in recent years, there are still many challenges in designing a robust tracker that can well handle appearance deformations, occlusions, target-alike object interferences, and many other complex real-world scenarios.

Tracking-by-detection [1] is the current state-of-the-art tracking framework. A rectangular box is given by the first frame in the video sequence to represent the tracked target. In the new frame, candidate samples around the previous target are selected randomly, and a classifier is adopted to distinguish each sample into target object or background. In the online tracking process, most of the existing tracking algorithms update the target model slowly on the assumption that the appearance of the target changes smoothly over time. However, this strategy often fails in tracking the target in which images of adjacent frames vary dramatically. In addition, this kind of algorithm does not perform well when occlusion occurs or there are some interferences

© Springer Nature Switzerland AG 2019
H. Lu et al. (Eds.): ISNN 2019, LNCS 11555, pp. 477–487, 2019.
https://doi.org/10.1007/978-3-030-22808-8_47

caused by other target-alike objects in the image. It is difficult for a single tracker to solve all the problems mentioned above. The ensemble tracking has been proved to be effective in handling such problems [2–4]. Hong et al. [4] employed Integrated Correlation Filter (ICF) for short-term tracking and integrated key point matching-tracking and RANSAC estimation for long-term tracking. In [2], target appearance models are managed based on CNNs in a tree structure to learn more persistent models through smooth updates. Han et al. [3] proposed multiple branches of fully connected layers that are randomly updated to alleviate lack of model diversity. However, these constructed trackers are still unable to significantly improve their robustness in anti-occlusion and against interferences of target-alike objects. For tracking problems, there are few positive samples available for online training, and although a lot of negative samples can be obtained, most of them are redundant, which leads to insufficient diversity for the tracker. And since the importance of the role played by different samples has not been sufficiently taken into account in the training process, the trackers cannot obtain enough information to distinguish the true target from the other target-alike objects that leads to tracking failure in some cases where there have target-alike object interference or severe occlusion. In addition, the tracking algorithm directly selects the candidate sample with the highest classification score as the true target. However, the scores given to the candidate samples by the tracker may not always be reliable enough and the true target may not always get the highest score in a complex tracking environment. This is also one of the important reasons that limit the robustness of the tracker.

To address the above mentioned issues, we propose a modified ensemble tracking method that is more robust in anti-occlusion and against the interference of other target-alike objects. Our tracker contains two classifiers, called long-term classifier and short-term classifier, in which the long-term classifier is to maintain the long-term appearance of the target and the short-term classifier is for prompt response to the sudden change of the target, respectively, and the two classifiers work collaboratively to cope with the occlusion and interference problems. Furthermore, in order to overcome the unreliability of the location by highest score, a density clustering method is introduced into the target locating process to get the cluster center over the candidate samples with top scores. The target is then located based on the clustering center instead of basing on the only sample with the highest score. Experiments conducted in the paper show that using these methods can prevent the new tracker from taking the false detected sample as the true target and hence its robustness is enhanced.

Details on the proposed tracking method will be given in Sect. 2. Section 3 presents the experimental results, and Sect. 4 draws conclusions and points out possible directions of the work.

2 Method

The overall structure of the proposed target tracking method is illustrated in Fig. 1 that consists of a feature network, a dual classifier, a target location module, and a sample collecting and pooling module. In this tracking system, the feature network (i.e., the Feature Net) is for providing general representations for the candidate samples selected randomly in the current frame over the target search area. The feature data of all the candidate samples are then sent to the Dual classifier and classified into target or non-target in evaluative scores (with value between 0 and 1) by the Long-term classifier and the Short-term classifier of the Dual Classifier collaboratively. Based on the scores, a clustering operation is conducted by the target location module to obtain the cluster center of the candidate samples that gain high scores from the Dual Classifier, in which the target is located according to the cluster center, not only according to the sample with the highest score. After the target in the current frame is located, the training samples $\{S^+, S^-, S^{++}, S^{--}\}$ are updated in the frame by the sample collecting and pooling module for training the Dual Classifier to work in next frame. Details on the tracking method will be given in the sequel subsections.

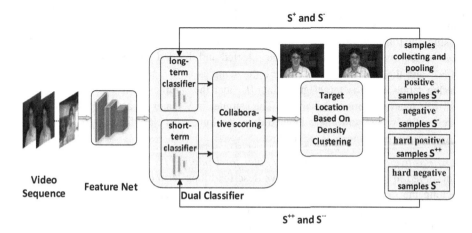

Fig. 1. Block diagram of the proposed tracking method

2.1   Feature Net

It has been widely recognized that features extracted by convolution neural network are more robust than hand-crafted features [11]. Considering the high computational complexity of existing deep networks, such as the VGG16 [18] and ResNet50 [20], we use the first three layers of the lighter VGG-M [19] pre-trained on ImageNet [10] as the feature network. We only employ the network to extract the appearance of the target and do not fine-tune or retrain the network in the whole tracking process to lighten the burden of computation.

2.2   Design of Dual Classifiers

The performance of classifier directly affects the quality of tracking results. In order to make the tracker both able to maintain the long-term appearance of the target and also able to adapt itself to rapid changing scenarios, we design a dual classifier that consists of a long-term classifier and a short-term classifier as shown in Fig. 1, in which the two classifiers have the same network structure of three fully connected layers with two soft-max output nodes.

At first, the two classifiers are initially trained with hand-selected samples in the first frame, then they are employed to track the target in each incoming frame and meanwhile are updated in the tracking process. Effective selection of training samples is very important for the two classifiers to maintain sufficient discriminant ability in a varying tracking environment. In the proposed tracking method, the obtainment of the training samples is accomplished by the sample collecting and pooling module under cooperation of the dual classifier.

Obtainment of Training Samples in Online Tracking

Positive and Negative Samples for Updating the Long-Term Classifier. The positive samples and negative samples are selected based on the IoU overlap ratios as in TCNN [2], Branchout [3] and MDNet [11]. In each current frame, after the target is located by using the dual classifiers trained in previous frame and the density clustering processing, the training samples for updating the long-term classifier are drawn around the target randomly. The samples are then selected and divided into two sets: the positive set S^+ and the negative set S^-, where the samples in S^+ are the ones that their IoU overlap ratios with the target great than 0.7 and the samples in S^- are the ones with their IoU overlap ratios less than 0.3. The samples in S^+ and S^- are labeled with 1 and 0, respectively, and with these labeled data, the long-term classifier is updated in each current frame by using SGD algorithm.

Hard Positive and Hard Negative Samples for Updating the Short-Term Classifier. It is observed in our study that the above described long-term classifier performs well in tracking the target object in the cases where the target appearance changes not significantly over time and there are no serious occlusion or target-alike object interference. However, it is also observed that the long-term classifier often fails to track the target while the shape of the target suddenly changes, or when considerable occlusion and target-alike object interference on the target take place. Since there is only one true target object and a large number of negative samples in an image, the total number of positive samples is much smaller than that of negative samples. In addition, for the positive samples drawn from the same image, they are spatially overlapped, and hence lack diversity. Therefore, the positive and negative samples selected in the above way cannot make the long-term classifier to obtain adequate discriminant ability in a more complicated tracking environment. In order to compensate for this shortcoming, we propose to collect some special samples that are difficult for the long-term classifier to identify them correctly. We call these samples the hard positive/negative samples and shall use them to train a short-term classifier to correctly identify the target in a complicated image environment.

In the tracking process, some non-target samples that obtain relative high scores are confusing for the long-term classifier to identify them from the target, which are regarded as the hard negative samples here. We want to pick this kind of samples out and exploit them to make the tracker more robust to the image variations.

After the t-th frame is tracked, the sample set S_t^+ and S_t^- can be obtained and some extra negative samples $S_t^{-'}$ around the hard negative samples in the previous frame are also extracted, the hard negative samples of the t-th frame are selected as follows:

$$s \rightarrow S_t^{--}, \text{ if } f_1(s) > \tau \tag{1}$$

where s is negative sample from $\{S_t^-, S_t^{-'}\}, f_1$ indicates the score obtained by s from the long-term classifier, S_t^{--} is the set of the hard negative sample in the t-th frame, and τ is the threshold with its value between 0 and 1.

Accordingly, due to occlusion or target-alike object interference, in the tracking process there are some positive samples may obtain relative low scores, and these positive samples are called the hard positive samples in the paper. Since the positive samples extracted around the target in each frame are highly spatially correlated and lack of diversity as mentioned before, it is not conducive to solve the problem of occlusion and target-alike object interference by using this kind of positive samples to train the classifier. In order to overcome this difficulty, we propose to construct hard positive samples as follows:

As shown in Fig. 2, we divided each training sample into 3×3 patches and use the partial patch of hard negative samples to mask the partial patch of positive samples to construct the hard positive samples. In addition, the occluded positions are settled in according to the positional relationship between the positive and negative samples in synthesizing the hard positive samples. For example, as shown in Fig. 2, if the hard negative sample is located at the lower left of the positive sample, the left side of the positive sample may only be occluded by the right side of the hard negative sample, and the lower side may only be occluded by the upper side of the hard negative sample. The synthesized positive samples in the t-th frame are then collected together as the hard positive sample set S_t^{++}.

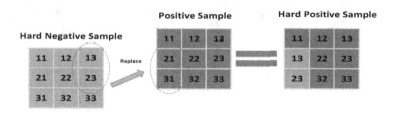

Fig. 2. Construction of hard positive samples

Collaborative Scoring. Having both the long-term classifier and the short-term classifier been updated before each new incoming frame, the two classifiers are used to evaluate the candidate samples collaboratively in the following way:

$$f(x) = \begin{cases} f_1(x), & f_2(x) > \tau_s \\ \alpha * f_1(x) + (1 - \alpha) * f_2(x), & f_2(x) \leq \tau_s \end{cases} \tag{2}$$

where x indicates the candidate sample, f_1 is the score for x assigned by the long-term classifier, f_2 is the score assigned by the short-term classifier, τ_s is the threshold to be suitably selected with $0 < \tau_s < 1$, and α is a parameter with its value between 0 and 1.

If f_2 is higher than τ_s, it means that the current target tracking is stable and the score of long-term classifier is proper. Otherwise, the weighted score of the dual classifier is taken as the final score.

2.3   Target Location Based on Density Clustering

The existing tracking method based on detection framework usually directly uses the candidate sample with the highest score as the tracking target. However, there are always frames in which the scores assigned by the classifiers are unreliable, and tracking failure often occurs in these frames. In this work we have analyzed the distribution of samples with high scores when tracking fails in our experiments. When the surrounding environment of the target is good and the tracking is correct, there is only one cluster center for these top-scored samples, while the surrounding environment is complex or the tracking fails, there exist multiple cluster centers for the top-scored samples as shown in Fig. 3 (a) and (b). In order to effectively handle these problems, we use the density clustering method DBSCAN proposed by Ester, et al. in [8] in the target locating process to get the cluster center over the top scored samples and then locate the target basing on the clustering center instead of basing on the sample with the highest score.

(a)   (b)

Fig. 3. Experimental examples of the distribution of samples with high scores in a tracking process. (a) Distribution of samples with high scores, (b) Red box is groundtruth, green box is the location based on maximum score, and blue box is the location based on density clustering. (Color figure online)

Let $\mathbf{D} = \{p_1, p_2, \ldots, p_k\}$ presents the set of samples in top-k scores that are selected from the N candidate samples $\mathbf{X} = \{x_1, x_2, \ldots, x_N\}$. We can get the cluster center set $\mathbf{C} = \{C_1, C_2, \ldots, C_m\}$ using DBSCAN algorithm. The general clustering process can be divided into two steps. First, we determine all the core objects from the selected samples according to the spatial relationship. Then we search all density-reachable samples from a random selected core object. The set of all samples with reachable densities is considered as a class. The main location operation steps are given in Algorithm 1. For more details on the clustering method, please refer to [8].

Algorithm 1 Location Based On Density Clustering

Input: Candidate samples $X = \{x_1, x_2, ..., x_N\}$ and scores
Output: Estimated target states x_t^*
1: select k samples $D = \{p_1, p_2, ..., p_k\}$ for clustering from X based on scores;
2: get the set of cluster center $C = \{C_1, C_2, ..., C_m\}$ by using DBSCAN method [8];
3: **if** $|C| > 1$ **then**
4:     x_t^* is the clustering center nearest to the target in the previous frame;
5: **else**
6:     x_t^* is the candidate sample with the highest score;
7: **return** x_t^*.

---

3 Experiments

In this section, we first introduce the implementation details, and then report our experimental results in two benchmark datasets: object tracking benchmark (OTB), visual object tracking (VOT) benchmark. Finally, some ablation experiments are conducted to study each part of the proposed tracking system.

3.1 Implementation Details

The proposed tracker is implemented in VS2013 using Caffe toolbox [7] and runs at around 1fps on an NVIDIA GTX 1080ti GPU platform.

We use lightweight VGG-M to extract features, and use full connection of two hidden-layer neural network as classifier. In the online tracking process, the 256 candidates selected with Gauss sampling are evaluated by using the dual classifiers. The location method based on density clustering is used to produce the final location result. Then the long-term sample pool and the short-term sample pool are updated by using the training sample collecting method described in Sect. 2.2. The long-term classifier is updated every 10 frames with positive and negative samples and the learning rate is 0.0001. The short-term classifier is updated with hard positive samples and hard negative samples and the learning rate is 0.0005. The experimental parameter τ_s, τ, α, k are set to 0.5, 0.4, 0.5, 50, respectively.

3.2 Experiments on OTB50 and OTB100

OTB [5] is the main dataset for visual tracking and is widely used at present. OTB100 contains a total of 100 video sequences and OTB50 contains 50 difficult and representative sequences selected from OTB100.

Following the protocol of OTB, we evaluate the performance of different tracking methods with precision plot and success plot in one-pass evaluation (OPE). The proposed method is compared with the state-of-the-art algorithms including MDNet [11],

MCPF [12], CREST [9], PTAV [13], DSST [14], SINT [15], DeepSRDCF [16] and SRDCFdecon [17].

Figure 4 illustrates the precision and success plot on OTB50 and OTB100. The results show that our algorithm is in rank 2 among all the trackers. It is worth noting that MDNet, ranked first, uses video in VOT database for training and is considered to be over-fitting since the videos in these two databases are very similar. Our tracker initializes directly with the weights of VGG-M, and still achieves comparable results. The performance of our algorithm is much better than that of DSST using traditional features, which shows that deep feature can effectively cope with various changes of targets. Our algorithm still have better performance than CREST, DeepSRDCF and SINT, which uses deep features.

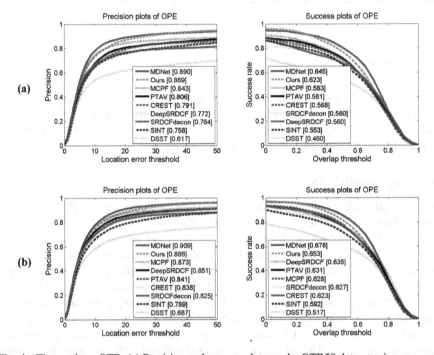

Fig. 4. The result on OTB. (a) Precision and success plots on the OTB50 dataset using one-pass evaluation. (b) Precision and success plots on the OTB100 dataset using one-pass evaluation.

3.3 Experiments on VOT2016

We also use VOT2016 [6] dataset to evaluate the different tracking methods reported in VOT2016 [6]. VOT has two important technical indexes: accuracy and robustness. Following the protocol of VOT2016, Ar and Rr represent accuracy and robustness rank respectively. The expected average overlap (EAO) is the final ranking index.

The proposed tracker is compared with the current advanced algorithms and Table 1 presents the experiment results. The two best results are highlighted in bold and italic, respectively. According to the VOT2016 report [6], EAO over 0.251 is considered as an advanced algorithm. Our EAO is 0.3186, lower than CCOT, but far higher than MDNet [11]. From Table 1 it can be seen that the robustness index of our method ranks first that shows the outperformance of our tracking approach in robustness.

Table 1. Performance comparison for state-of-the-art algorithms on the VOT-2016 dataset. The two best results are highlighted in bold and italic, respectively.

Method	EAO	Ar	Rr
CCOT	**0.3310**	2.28	2.53
TCNN	*0.3253*	*1.82*	2.48
SSAT	0.3232	**1.43**	2.75
MLDF	0.3106	3.48	*2.38*
Staple	0.2952	2.43	4.18
EBT	0.2913	4.68	2.67
SRBT	0.2904	3.50	4.03
DeepSRDCF	0.2763	2.70	3.72
MDNet	0.2583	1.95	3.02
Ours	0.3186	2.18	**2.05**

3.4 Ablation Studies

To verify the effectiveness of each improvement component in our tracking system, we implemented and evaluated several different versions. The ablation study is conducted in 3 implements: use only long-term classifier and target location based on maximum score (Baseline), use dual classifier and target location based on maximum score (Baseline + Dual) and use dual classifier and target location based on density clustering (Baseline + Dual + Cluster). Figure 5 shows the evaluation results on the OTB50. From Fig. 5, one can see that the tracking performance has been improved by using the dual classifier and density clustering method proposed in the paper.

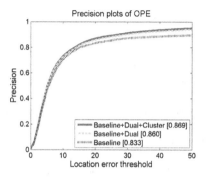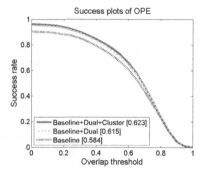

Fig. 5. The ablation analysis on the OTB50 dataset using one-pass evaluation.

4 Conclusion

In this paper, we proposed a new tracking method for robust tracking. The proposed model divides all training samples into positive samples, negative samples, hard positive samples and hard negative samples, and uses dual classifier to cope with challenging frames. We also introduce the density clustering method into the target location process to improve the target positioning reliability and precision. Extensive experiments have been conducted on two commonly-used datasets, and the results validate the proposed tracking method. Further research of the work to be conducted is to speed up the algorithm for achieving real-time tracking performance.

References

1. Kalal, Z., Mikolajczyk, K., Matas, J.: Tracking-learning-detection. IEEE Trans. Pattern Anal. Mach. Intell. **34**(7), 1409–1422 (2012)
2. Nam, H., Baek, M., Han, B.: Modeling and propagating CNNs in a tree structure for visual tracking. arXiv:1608.07242 (2016)
3. Han, B., Sim, J., Adam, H.: Branchout: regularization for online ensemble tracking with convolutional neural networks. In: IEEE Conference on Computer Vision and Pattern Recognition, pp. 3356–3365. IEEE Press, Honolulu (2017)
4. Hong, Z., Chen, Z., Wang, C., et al.: MUlti-store tracker (MUSTer): a cognitive psychology inspired approach to object tracking. In: IEEE Conference on Computer Vision and Pattern Recognition, pp. 749–758. IEEE Press, Boston (2015)
5. Wu, Y., Lim, J., Yang, M.H.: Object tracking benchmark. IEEE Trans. Pattern Anal. Mach. Intell. **37**(9), 1834–1848 (2015)
6. Kristan, M.: The visual object tracking VOT2016 challenge results. In: Hua, G., Jégou, H. (eds.) ECCV 2016. LNCS, vol. 9914, pp. 777–823. Springer, Cham (2016). https://doi.org/10.1007/978-3-319-48881-3_54
7. Jia, Y., Shelhamer, E., et al.: Caffe: convolutional architecture for fast feature embedding. In: ICMM 2014, pp. 675–678. ACM, Orlando (2014)
8. Ester, M., Kriegel, H., Sander, J., Xu, X.: A density-based algorithm for discovering clusters in large spatial databases with noise. In: KDD-1996, pp. 226–231. AAAI Press, Portland (1996)
9. Song, Y., Ma, C., et al.: Crest: convolutional residual learning for visual tracking. In: ICCV 2017, pp. 2555–2564. IEEE Press, Venice (2017)
10. Deng, J., Dong, W., et al.: Imagenet: a large-scale hierarchical image database. In: CVPR 2009, pp. 248–255. IEEE Press, Miami (2009)
11. Nam, H., Han, B.: Learning multi-domain convolutional neural networks for visual tracking. In: CVPR 2016, pp. 4293–4302. IEEE Press, Las Vegas (2016)
12. Zhang, T., Xu, C., Yang, M. H.: Multi-task correlation particle filter for robust object tracking. In: CVPR 2017, pp. 4819–4827. IEEE Press, Honolulu (2017)
13. Fan, H., Ling, H.: Parallel tracking and verifying: a framework for real-time and high accuracy visual tracking. In: ICCV 2017, pp. 5486–5494. IEEE Press, Venice (2017)
14. Danelljan, M., Häger, G., et al.: Discriminative scale space tracking. IEEE Trans. Pattern Anal. Mach. Intell. **39**(8), 1561–1575 (2017)
15. Tao, R., Gavves, E., Smeulders, A.W.M.: Siamese instance search for tracking. In: CVPR 2016, pp. 1420–1429. IEEE Press, Las Vegas (2016)

16. Danelljan, M., Hager, G., et al.: Convolutional features for correlation filter based visual tracking. In: ICCV Workshops, pp. 58–66. IEEE Press, Santiago (2015)
17. Danelljan, M., Hager, G., et al.: Adaptive decontamination of the training set: a unified formulation for discriminative visual tracking. In: CVPR 2016, pp. 1430–1438. IEEE Press, Las Vegas (2016)
18. Simonyan, K., Zisserman, A.: Very deep convolutional networks for large-scale image recognition. arXiv:1409.1556 (2014)
19. Chatfield, K., Simonyan, K., Vedaldi, A., et al.: Return of the devil in the details: delving deep into convolutional nets. arXiv:1405.3531 (2014)
20. He, K., Zhang, X., et al.: Deep residual learning for image recognition. In: CVPR 2016, pp. 770–778. IEEE Press, Las Vegas (2016)

Bio-signal, Biomedical Engineering, and Hardware

Comparison of Convolutional Neural Networks and Search Based Approaches for Extracting Psychological Characteristics from Job Description

Marina Belyanova, Sergey Chernobrovkin, Igor Latkin,
and Roman Samarev$^{(\boxtimes)}$

dotin Inc., Fremont, USA
info@dotin.us

Abstract. The paper discusses two machine learning approaches for analysis of job text descriptions and extracting some psychological characteristics which are required from employees. The Holland Codes (RIASEC model) are used as algorithms target variables. The first approach is based on TF-IDF and WMD. The second approach is based on a convolutional neural network. The results comparison of the proposed algorithms is given.

Keywords: Convolutional neural networks · TF-IDF ·
Word Mover's Distance (WMD) · Psychological characteristics

1 Introduction

The search for talented candidates for employment is a big challenge for any HR department. Nowadays, methods of artificial intelligence and machine learning are being used to automate the work of the HR department. "There's absolutely nothing efficient about sorting through 30,000 resumes by hand. Recruiters are spending months evaluating applicants only to have great promising candidates get lost in the pile." noted the authors of the startup Riminder [1].

Researchers in the field of machine learning are taking focus on solving the most resource-intensive problems of recruitment, such as feature extraction from the plain text of job descriptions and CVs. These features might be used for the candidate to job matching. One of the important tasks is extracting psychological characteristics which are required for the job position. Our work is dedicated to solving this problem.

We have implemented two different approaches to solving them. First is based on TF-IDF document model and WMD metrics. Second is based on convolutional neural networks.

H. Lu et al. (Eds.): ISNN 2019, LNCS 11555, pp. 491–500, 2019.
https://doi.org/10.1007/978-3-030-22808-8_48

2   The Data Source Description

The Holland Codes model [2] is known as the Holland Occupational Themes (or RIASEC) model and widely used in HR. According to [2] types are labeled as Realistic (Doers), Investigative (Thinkers), Artistic (Creators), Social (Helpers), Enterprising (Persuaders), and Conventional (Organizers). Holland codes are used for vocational guidance of students and for staff recruitment.

O*NET is a database which is used in the USA as an official source of job descriptions, required skills and psychological properties of employees. O*NET determines the values of the Holland codes for each profession. It contains 1100 job descriptions, which include tasks that need to be addressed within the profession, as well as the tools, knowledge, and technologies that are necessary for it. There are also values of psychological characteristics: a set of Holland codes preferred for the job. The structure of the used data source is displayed in Fig. 1.

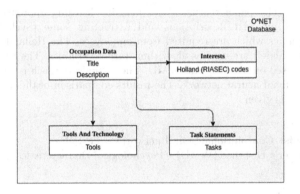

Fig. 1. The structure of the used data source

3   The TF-IDF and WMD Approach

3.1   Common Algorithm Description

Since the O*NET database contains Holland codes for each of the available professions, these professions and related codes can be used to search for suitable professions for a vacancy. After finding the most relevant professions, it is necessary to weigh the psychological attributes related to them in proportion to the relevance of each of the professions and obtain the final result for the vacancy. Thus, the algorithm can be divided into two phases:

1. For the given job information, find the most relevant professions in O*NET.
2. Get a breakdown of Holland codes by professions found.

To search for a given job position among professions, we use TF-IDF vector representation of the position descriptions [3] and next subsequently compare vectors using the cosine distance. In addition to the statistical method based on the frequency of occurrences of words in texts, it is also necessary to take into account text semantics. The most common methods for converting words to a semantic vector are word2vec [4] and GloVe [5]. However, basic word2vec converts only a single word into a vector. We need to convert the entire text for their comparison. The work [6] offers a way to compare texts in which each word is converted into a vector. We use the WMD metric for this task. Because of the computational complexity of the method, we use it only to compare headers of vacancies and professions.

3.2   Algorithm Flow

1. The job title and description are used as an input.
2. Preprocessing of input.
   2.1 Split job title and description into words.
   2.2 Remove punctuation and stop-words.
   2.3 Transform each word to its normal form if needed.
3. For each profession from the O*NET Database:
   3.1 Preprocess title and description texts. This step can be performed once before executing the algorithm.
   3.2 Determine WMD between each O*NET profession and input titles.
   3.3 Determine TF-IDF vectors for O*NET profession and input texts.
   3.4 Calculate the cosine distance between TF-IDF vectors from the previous step.
4. Take top-N O*NET professions with the shortest distance to the input job description (where N is a hyperparameter by default set to 5).
5. For each of N professions determine the Holland codes.
6. Weigh Holland codes proportional to the distances between professions and the input job. The output result of the algorithm is these codes distribution.

3.3   Searching Profession by the Job Description

First of all, all the descriptions from the O*NET for jobs are collected. For this, several CSV-tables are joined by the job code field. Since texts of job descriptions are quite large and reach multiple paragraphs, it is reasonable to use text comparison metrics based on the frequency of word occurrences in texts - TF-IDF.

The dictionary of words is obtained according to the descriptions of all jobs. The text of the job description is represented by a vector where for each word we have TF-IDF value, calculated by the following formula:

$$TF - IDF(t, d, D) = \frac{n_t}{\sum_{k=1}^{N} n_k} \times log \frac{|D|}{\{d_i \in D | t \in d_i\}}, \tag{1}$$

where t – a term in the document, d – the document, D – the documents corpus, n – the amount of words in the document d, N – the amount of words in the corpus D.

Afterwards, these computed vectors are compared using the cosine distance metric:

$$S = \frac{(\mathbf{A} \cdot \mathbf{B})}{||\mathbf{A}|| \cdot ||\mathbf{B}||}, \tag{2}$$

where A – 1st document's vector, B – 2nd document's vector.

Here by documents A and B, we are mentioning vacancy description (query) and O*NET job description accordingly.

After comparing the query with all job description we have a list of pairs distances-description. After sorting that list in ascending order by distance, we have documents ordered by proximity to the query. The first document is the most relevant to the query.

3.4 Searching for the Suitable Profession with Accordance to Semantics

To take semantically similar words into account, we decided to use word2vec [4] approach. Since it is necessary to compare texts but not single words, Word Mover's Distance [6] metric was used.

Previously we have tested WMD based algorithm for solving the problem of clustering of the textual news stream. Area of application and comparison of WMD based with other algorithms are described in the paper [7].

In the current case, each text word is represented by a vector according to the word2vec model, then the normalized bag of words (nBoW) is calculated for each of the documents. After that, the minimum cost of moving all nBoW values of the first document with coordinates represented by word2vec to the second and vice versa is calculated. This task is reduced to the linear programming problem:

$$\min_{T \geq 0} \sum_{i,j=1}^{n} \mathbf{T_{ij}}c(i,j),$$

$$\sum_{j=1}^{n} \mathbf{T}_{ij} = d_i, i \in \{1,\dots,n\}, \tag{3}$$

$$\sum_{i=1}^{n} \mathbf{T}_{ij} = d'_j, j \in \{1,\dots,n\},$$

where d_i – the total flow from the i-th word, $c(i,j)$ – the distance between i-th and j-th words.

As the algorithm's computation complexity is $O(n) = p^3 log(p)$, only distances between the job titles was calculated.

WMD assumes sequential distance calculation, that means maximum value is not defined and may be greater than 1. It is necessary to transform the distance metric to the similarity, so the final result should be in the range 0..1. It can be done after normalizing of these values to the maximum found value.

WMD algorithm was implemented in the Julia language [8].

3.5   Weighing Codes of Suitable Professions

In order to combine the results of the comparison of cosine similarity between the documents in the form of TF-IDF vectors and WMD between word2vec representations of documents, the following steps are performed:

1. Transforming distance to the similarity: $s = \frac{1}{1+d^2}$.
2. If among the five most similar professions (in terms of similarity of TF-IDF vectors and WMD) were the same, weights are increased to get the final similarity between the documents.

4   Approach Based on Neural Networks

The second approach can be a set of neural networks, where each one predicts the value for its own Holland code. In fact, the algorithm described in the previous section can be transferred to the neural network.

O*NET job descriptions are used as a train data set, and codes weights of the profession description are used as target values. Then the network, learning from the texts, does something similar to the previous approach: select the most significant attributes from the given text and find the dependence of these attributes from the target code value.

Convolutional neural networks (CNN) have shown promising results in the field of natural language processing. The idea is that kernels size and outputs are able to detect patterns of words with different length and these patterns are spotted in the text regardless of their position.

For this approach, the CNN architecture represented in Fig. 2 was used. The input data was embedded using word2vec that allows us to encounter semantics, followed by the dropout layer to reduce overfitting, followed by 3 conv1d layers, flattening and a fully-connected layer of 1 out neuron, which predicts the code value.

This software service is implemented in Python using TensorFlow and Keras [9].

5   The Comparison of the Algorithms

Algorithms were evaluated on the dataset from manually labeled Holland codes for vacancies from indeed.com. As a metric, Mean Absolute Error (MAE) for each of the codes is used:

$$MAE = \frac{1}{n} \sum_{i=1}^{n} |p_i - r_i|, \tag{4}$$

where p_i - predicted value; r_i - true value.

The algorithms evaluation results are represented in Table 1.

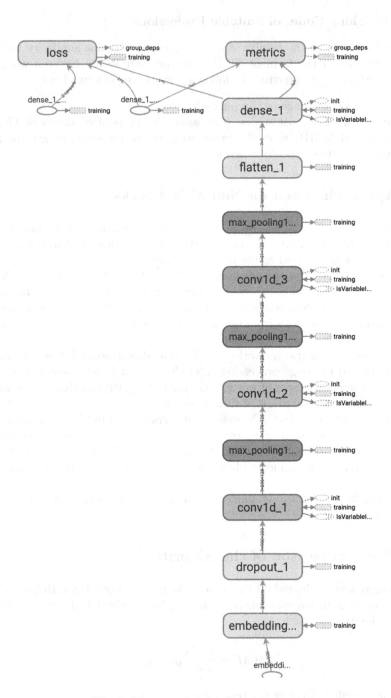

Fig. 2. The neural network architecture

Table 1. The algorithms MAE results

Holland code	MAE TF-IDF + WMD	MAE NN
Realistic	**8.93**	11.08
Investigative	**10.83**	14.50
Artistic	7.86	7.83
Social	10.60	**4.39**
Enterprising	12.94	**9.41**
Conventional	**9.40**	15.75

The considered algorithms can be compared by several characteristics. From the point of view of the speed of work and the subsequent implementation in the existing software system, the approach based on neural networks wins.

The comparison of the algorithms' evaluation was based on manually labeled data.

On the general test sample, the algorithm based on information retrieval has MAE ≈ 4.7, the algorithm based on the neural network has MAE ≈ 5.9, while the neural network shows better results in 37.5% of cases.

It is important to note that the results of the algorithm using the profession search approach can be interpreted and evaluated subjectively. Its results can be used to determine how well the vacancy corresponds to the selected professions. The neural network approach is a "black box", and to get the meaning of selected features is a rather complex task. The interpretability of the results may be an important factor, since the values of psychological features may depend on the subjective assessment of various psychologists.

The example of Holland codes, predicted by both approaches, are represented in the Fig. 3.

6   Discussion

There is nothing surprising in the fact that currently AI and machine learning methods are used mainly to solve the most routine and resource-intensive tasks in the employment process.

The most typical is the task of the candidate to job matching, which is successfully solved by machine learning methods [10]. The authors of paper [10] are building text queries and using the LambdaMART ranking algorithm for accurate result matching. In this approach, the task of identifying psychological characteristics is not considered by the authors.

In most articles devoted to the extraction from the text of the psychological characteristics of the employee, the Big Five personality traits model is used [11,12]. According to [13], the Big Five personality traits, also known as the five-factor model (FFM) and the OCEAN (Openness to experience, Conscientiousness, Extraversion, Agreeableness, and Neuroticism) model, is a taxonomy for personality traits.

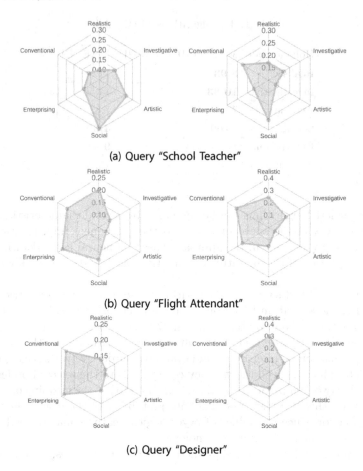

(a) Query "School Teacher"

(b) Query "Flight Attendant"

(c) Query "Designer"

Fig. 3. The examples of predicted Holland codes for two algorithms: at left—TF-IDF + WMD, at right—neural network

In the article [11] the Support Vector Machine (SVM) method is used to match features extracted form micro-blog texts to five FFM factors. The values of accuracy estimation were over 84%.

In the article [12] neural network with convolutional layers is used to match short essay texts to FFM factors. The processing flow in the network comprises four main steps: word vectorization, sentence vectorization, document vectorization and classification from the document vector to the classification result (yes/no). The network comprises seven layers: input (word vectorization), convolution (sentence vectorization), max pooling (sentence vectorization), 1-max pooling (document vectorization), concatenation (document vectorization), linear with Sigmoid activation (classification), and two-neuron softmax output (classification). The values of the best accuracy estimation were about 63%.

Instead of FFM model [13], we use the Holland Codes model [2]. While the FFM model mainly shows the psychological characteristics of an employee's personality, the RIASEC model is more focused on identifying its professional strengths. Therefore, in our work, we use exactly the RIASEC model.

In contrast to articles [11, 12], the problem of code recognition in our work is posed as a regression, not as a classification problem. Therefore, instead of the accuracy metric for classification, we use the MAE metric for regression. The results of our experiments are considered in the corresponding section.

Thus, the task of extracting the psychological characteristics of employees from texts is far from complete. We hope that we were able to take a small step towards solving this problem.

7   Conclusions and Future Work

Now researchers in the field of machine learning focus on solving the most resource-intensive problems of recruitment, such as candidate to job matching. But the important task of extracting the psychological characteristics of employees from textual information has less attention.

The Holland Codes (or RIASEC model) may be used as algorithms target variables for psychological characteristics extraction.

We proposed two algorithms for the extraction of psychological characteristics. The first algorithm is based on TF-IDF and WMD. The second algorithm is based on a convolutional neural network. From the point of view of the performance and the subsequent implementation in the existing software system, the neural networks approach wins.

Since we consider the Julia language as a promising platform for the implementation of our service, the next version of the neural network model is planned to be implemented using the Julia Flux library [14].

As a continuation of the work, it is planned to conduct further experiments and improve the quality of the models, taking into account that the service will be fully implemented in the Julia language.

The task of extracting the psychological characteristics of employees from texts is far from complete. We hope that we were able to take a small step towards solving this problem.

References

1. Riminder uses deep learning to better match people to jobs. https://techcrunch. com/2017/05/15/riminder-uses-deep-learning-to-better-match-people-to-jobs. Accessed 14 Feb 2019
2. Nauta, M.: The development, evolution, and status of Holland's theory of vocational personalities: reflections and future directions for counseling psychology. J. Couns. Psychol. **57**(1), 11–22 (2010). https://doi.org/10.1037/a0018213
3. Ramos, J.: Using TF-IDF to determine word relevance in document queries (2003)

4. Mikolov, T., Chen, K., Corrado, G., Dean, J.: Efficient estimation of word representations in vector space. arXiv preprint arXiv:1301.3781 (2013). http://arxiv.org/pdf/1301.3781v3

5. Pennington, J., Socher, R., Manning, C.: Glove: global vectors for word representation. In: Proceedings of the 2014 Conference on Empirical Methods in Natural Language Processing (EMNLP), pp. 1532–1543, Doha, Qatar (2014). https://doi.org/10.3115/v1/D14-1162

6. Kusner, M.J., Sun, Y., Kolkin, N.I., Weinberger, K.Q.: From word embeddings to document distances. In: Proceedings of the 32nd International Conference on International Conference on Machine Learning, vol. 37, pp. 957–966 (2015)

7. Gapanyuk, Yu.E., et al.: Improvements of the algorithm of clustering of the news stream of text messages. In: Dynamics of Complex Systems – XXI Century, no. 3, pp. 47–53 (2017)

8. Bezanson, J., Karpinski, S., Shah, V.B., Edelman, A.: Julia: a fast dynamic language for technical computing. arXiv preprint arXiv:1209.5145 (2012). http://arxiv.org/pdf/1209.5145v1

9. Chollet, F.: Deep Learning with Python. Manning Publications Co, Shelter Island (2018)

10. Belle, A., Dehling, E., Foster, D.: Improving candidate to job matching with machine learning. https://dshcmorg.files.wordpress.com/2018/08/dshcm_2018_paper_5_job_matching.pdf. Accessed 14 Feb 2019

11. Li, L., Li, A., Hao, B., Guan, Z., Zhu, T.: Predicting active users' personality based on micro-blogging behaviors. PloS One 9, e84997 (2014). https://doi.org/10.1371/journal.pone.0084997

12. Majumder, N., Poria, S., Gelbukh, A., Cambria, E.: Deep learning-based document modeling for personality detection from text. IEEE Intell. Syst. 32(2), 74–79 (2017). https://doi.org/10.1109/MIS.2017.23

13. Rothmann, S., Coetzer, E.P.: The big five personality dimensions and job performance. SA J. Ind. Psychol. 29(1), 68–74 (2003). https://doi.org/10.4102/sajip.v29i1.88

14. The official website of the flux library. https://fluxml.ai/. Accessed 14 Feb 2019

Quasi-Brain-Death EEG Diagnosis Based on Tensor Train Decomposition

Qipeng Chen[1], Longhao Yuan[1,2], Yao Miao[1], Qibin Zhao[2],
Toshihisa Tanaka[4], and Jianting Cao[1,2,3(✉)]

[1] Saitama Institute of Technology, Saitama, Japan
{n850lchq, cao}@sit.ac.jp
[2] RIKEN Center for Advanced Intelligence Project (AIP), Tokyo, Japan
[3] Hangzhou Dianzi University, Hangzhou, China
[4] Tokyo University of Agriculture and Technology, Tokyo, Japan

Abstract. The quasi-brain-death diagnosis based on electroencephalogram (EEG) signal analysis is of great significance for early detection of quasi-brain-death patients which can avoid brain death misjudgment. Tensor is the multi-way array and tensor decomposition is a natural way to analyze high-order data. In this paper, we apply tensor train (TT) decomposition to EEG-based quasi-brain-death diagnosis. By reshaping the EEG data from matrix to higher-order tensor, we use a new algorithm to extract more valuable features from the data. The support vector machine (SVM) classifier is then used to complete the classification task of the extracted features. The experimental result shows that our method is well performed in evaluating the difference between coma patients and brain-death patients.

Keywords: EEG data · Brain-death diagnosis · Tensor · TT decomposition · SVM

1 Introduction

Brain death refers to the complete and irreversible loss of the human brain and brainstem. Brain death is considered to be the end of life, so the diagnosis of brain death is of great significance. Harvard Medical School of the United States has defined a new standard of death and established the first brain death criteria. The present brain death determination procedure is time-consuming and contains some risks [1]. It is dangerous to remove the patient's respirator when performing the "self-breathing" test. The advantage of introducing EEG prediction system is that it can avoid the misjudgment of brain death and reduce the burden on doctors and patients [2, 3]. Therefore, it is meaningful to present a simple and safe method to help doctor diagnose quasi-brain-death patients.

A tensor is a multi-dimensional array, and it can represent high-dimensional data accurately and even simplify the processing and calculation of large amounts of data. In recent years, research on tensor becomes more and more important [4–6]. It has also become one of the most popular research issues because of its natural representation of high-order data. Therefore, it is meaningful to try to apply tensor methods to analyze

H. Lu et al. (Eds.): ISNN 2019, LNCS 11555, pp. 501–511, 2019.
https://doi.org/10.1007/978-3-030-22808-8_49

the EEG. Several studies have applied tensor decompositions in neuroscience. Martí-nez-Montes [7, 8] applies Candecomp/Parafac decomposition (CP) to a time-varying EEG spectrum arranged as a three-order array with modes corresponding to time, frequency and channel. Recently, tensor train (TT) decomposition [9] and tensor ring decomposition [10] has been proposed and have attracted the researchers' attention. They own super linear compression ability and data interpretability.

TT-decomposition is a good method to analyze EEG data because it is naturally a high-order tensor. In this paper, we are the first to diagnose quasi-brain-death patients with TT decomposition. By transforming the EEG data into higher-order tensor and extracting more valuable features by TT decomposition, we obtain better results in evaluating the quasi-brain-death diagnosis.

2 Method of Data Analysis

2.1 Traditional Tensor Decompositions

Tensor is natural multi-dimensional generalization of matrices and have attracted tremendous interest in recent years. Multilinear algebra, tensor analysis, and the theory of tensor approximation plays increasingly important roles in computational mathematics and numerical analysis.

CP decomposition is a commonly useful method at present [11]. The CP decomposition factorizes a tensor into a sum of component rank-one tensors. For example, given a third-order tensor $X \in \mathbb{R}^{I \times J \times K}$, it can be expressed as

$$X \approx \left[\lambda; A^{(1)}, A^{(2)}, \cdots, A^{(N)} \right] = \sum_{r=1}^{R} \lambda_r a_r^{(1)} \circ a_r^{(2)} \circ \cdots \circ a_r^{(N)} \qquad (1)$$

Where R is named tensor rank, which is a positive integer, $\circ$ is the out product operation, $a_r \in \mathbb{R}^I$, $B_r \in \mathbb{R}^J$, and $a_r \in \mathbb{R}^K$, for r = 1, ... , R. CP decomposition is a good method but there is no straightforward algorithm to determine the rank of a given tensor, it is an NP-hard problem and the approximation of the optimal tensor rank is an ill-posed problem, which makes it difficult to apply CP decomposition to mathematical calculation [12].

Tucker decomposition is a form of higher-order principal component analysis (PCA). It decomposes a tensor into a core tensor multiplied by a matrix along each mode. Thus, in the 3rd-order case where $X \in \mathbb{R}^{I \times J \times K}$, we have

$$X \approx \mathcal{G} \times_1 A \times_2 B \times_3 C = \sum_{p=1}^{P} \sum_{q=1}^{Q} \sum_{r=1}^{R} g_{pqr} a_p \circ b_q \circ c_r = [\mathcal{G}; A, B, C] \qquad (2)$$

Where $A \in \mathbb{R}^{I \times P}$, $B \in \mathbb{R}^{J \times Q}$, $C \in \mathbb{R}^{K \times R}$ are the factor matrices (which are usually orthogonal) and can be thought as the principal components for each mode. The tensor $\mathcal{G} \in \mathbb{R}^{I \times Q \times P}$ is called core tensor and its entries can show the level of interaction between the different components. The Tucker model can be generalized to N-th order tensors as

$$X \approx \mathcal{G}_{\times_1} A^{(1)}_{\times_2} A^{(2)} \cdots _{\times_N} A^{(N)} = \left[\mathcal{G}; A^{(1)}, A^{(2)}, \cdots, A^{(N)} \right] \tag{3}$$

The matricized version of (3) is

$$X_{(n)} \approx A^{(n)} G_{(n)} (A^{(N)} \otimes \cdots \otimes A^{(n+1)} \otimes A^{(n-1)} \otimes \cdots \otimes A^{(1)}) \tag{4}$$

The Tucker decomposition has good stability but it is not suitable for the decomposition of large-scale and high-order tensors [13].

2.2 TT Decomposition

In recent years, TT decomposition has been proposed. Compared with traditional CP decomposition and Tucker decomposition, TT decomposition has good calculation convenience. For an element in the order-N tensor $A \in \mathbb{R}^{I_1 \times \cdots \times I_N}$ which is termed as $A(i_1, i_2, \cdots, i_N)$, if it satisfies

$$A(i_1, i_2, \cdots, i_d) = G_1(i_1) G_2(i_2) \cdots G_N(i_N) \tag{5}$$

Where $G_k(i_k)$ is an $r_k \times r_{k-1}$ matrix, $i_k \in \{1, 2, \ldots, I_k\}$, for k = 1, 2, ..., N, and $r_0 = r_d = 1$, then we can define the tensor has tensor train (TT) structure. Furthermore, if 'r' is the minimum number in the TT format and the preceding d − 1 matrices satisfy the left orthogonality

$$\mathcal{L}^T(G_k) \cdot \mathcal{L}(G_k) = I \in \mathbb{R}^{r_k \times r_k}, \forall k = 1, 2, \cdots, N-1. \tag{6}$$

The rank r_k is called compression ranks or TT-ranks, and G_k is defined as the core of the TT-decomposition. In the index form, the decomposition can be written as

$$A(i_1, i_2, \cdots, i_N) = \sum_{\alpha_0, \cdots, \alpha_N} G_1(\alpha_0, i_1, \alpha_1) G_2(\alpha_1, i_2, \alpha_2) G_N(\alpha_{N-1}, i_N, \alpha_N) \tag{7}$$

Since $r_0 = r_d = 1$, this decomposition can also be represented graphically by a linear tensor network, which is presented in Fig. 1

Fig. 1. Tensor train network

This picture looks like a train with carriages and links between them, and that justifies the name tensor train decomposition, or simply TT decomposition. The most effective method to calculate TT decomposition is to apply TT-SVD algorithm [14], which only needs one iteration which includes multiple matrix singular value decomposition (SVD) operation.

2.3 SVM Classifier

Support vector machine (SVM) is one of the most effective supervised learning algorithms. The basic idea of SVM comes from the linear discriminant optimal classification, which means not only separates the two sample planes accurately but also maximizes the classification margin. The partition classification line can be expressed as

$$w^{\mathrm{T}} \cdot x + b = 0, \tag{8}$$

Where w is the normal vector of H, b is the offset of the H which makes the data points on both sides as far as possible from the classification line to make sure $\frac{2}{\|w\|}$ is maximal (Fig. 2).

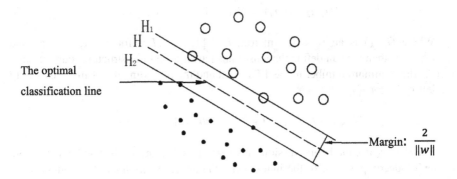

Fig. 2. The optimal classification line of SVM

It can be transformed into a constrained optimization problem as

$$\max_{w,b} \frac{2}{\|w\|}$$
$$s.t. \ y_i(w^t x_i + b) \geq 1, \ i = 1, 2, \cdots, m \tag{9}$$

It can also be expressed as

$$\min_{w,b} \frac{\|w\|^2}{2}$$
$$s.t. \ y_i(w^t x_i + b) \geq 1 \ i = 1, 2, \cdots, m \tag{10}$$

2.4   Collection of EEG Data

The data used in this paper are obtained from the intensive care unit (ICU) in a hospital in Shanghai. The type of recording instrument is NEUROSCAN ESI-32 system. During the data acquisition, seven electrodes which named Fp1, Fp2, F3, F4, F7, F8 were placed on the forehead of comatose patients, and the GND, A1, A2 are reference electrodes on the ear (Fig. 4). The sampling frequency of the EEG signal is 1000 Hz and the impedance of electrodes is less than 8000 Ω (Fig. 3).

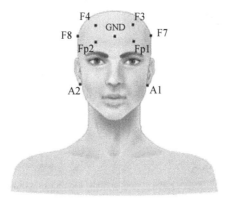

Fig. 3. The location of the electrodes

From June 2004 to March 2006, 91 groups of EEG signals were collected from 33 patients. According to the health status of these patients, different EEG signals were recorded. These patients include 22 coma patients and 11 brain-dead patients according to the diagnosis of the doctors. Next, we apply the proposed method and SVM classifier to determine brain death.

2.5   Computation Procedure and Flow Chart

We propose a novel method which is based on TT decomposition to analyze the quasi-brain-death EEG data. The purpose of our algorithm is to find more features that hide inside the EEG data by changing the order of data. The whole procedure of our method is

Step 1: Input the data to Matlab toolbox named "EEGlab" and obtain data from channel Fp1, Fp2, F3, F4, F7, F8.

Step 2: Set the sampling rate for 1000 points per second and set time for 30 s and then we obtain a matrix of size 6 × 30000.

Step 3: Reshape the 6 × 30000 matrix to a 4th-order tenor of size 6 × 100 × 60 × 5.

Step 4: Apply TT-SVD decomposition algorithm to decompose the tensor.

Step 5: Compute the Frobenius norms of the obtained TT cores and consider them as the extracted features.

Step 6: Compute the singular values of all the unfolding of the TT approximated tensor and also consider them as the features.

Step 7: Choose the obtained Frobenius norms and singular values to constitute a 50×6 matrix in which the last column is formed by '0' and '1' to represent dead patient and coma patient respectively.

Step 8: Apply SVM to classify the extracted features. Choose column 5 as response and column 1–4 as predictors. Set cross-validation fold as 5 folds and plot the result.

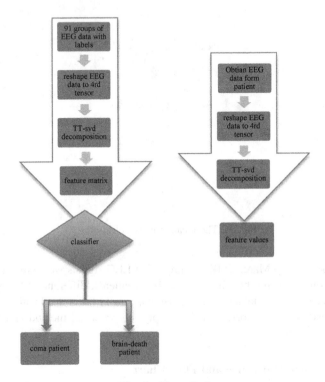

Fig. 4. Flow chart

3 Experiment and Result

We select a small part of EEG data and plot them to analyze the variation of brain wave. Figures 5 and 6 show the EEG signals of two suspected brain-death patients respectively. Comparing the two graphs, we can find that they have different rules of change. Though it seems that the graph of quasi-brain-death and coma states have different forms and coma data is more active, we cannot tell the common difference between them through the observation of the graphs. Nowadays, there are many applications of tensor decomposition to extract features for the classification task. But we are the first to attempt to apply TT decomposition to classify the two different kinds of EEG data. The major work is extracting valuable features and then uses them to classify coma patients and quasi-brain-death patients.

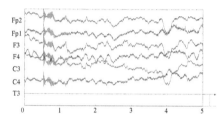

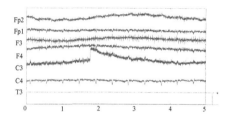

Fig. 5. The EEG of coma patient

Fig. 6. The EEG of brain-death patient

We import 91 groups of EEG data into the Matlab software and tensorize the data into higher-order. Then, the tensor is decomposed by the TT-SVD algorithm [14] and the decomposed tensor in TT-format is the cores $G_1, G_2, \ldots, G_d$. We choose the Frobenius norm of the cores and the singular values of the TT approximated tensor as latent features. Frobenius norm is a very important evaluation index for tensor which is often utilized to define the amount of data. If the Frobenius norm is small, it can be explained that each element is small relatively. Singular values often correspond to the important information implied in the tensor, so we choose the largest singular value of each unfolding as the selected feature. Table 1 enumerates some results of the selected features.

Table 1. Characteristic factors of different patients

Frobenius norm	Singular values 1	Singular values 2	Singular values 3	State
31798	14139	14017	11155	1
31442	14009	13915	10024	1
6212.9	2244.9	2184.7	2383.8	0
5613	2390.3	2407	2409.6	0
⋮	⋮	⋮	⋮	⋮

Column 1 shows the Frobenius norms of the TT cores and column 2–3 is the maximum singular values of the unfoldings of the TT approximated tensor. State means the condition of the patients diagnosed by the doctors. From Table 1 we notice that the Frobenius norm of the dead patient is lower than coma patient obviously. The average Frobenius norm of two different patients was 50980 and 4227. Lower Frobenius norm explicates that there are not too many large values, but many relatively small values inside of the tensor. From the results, we predict that there is less information in the EEG data of the brain-death patient. Similarly, by comparing the singular values, we can also find that there are great differences between them. Significant differences in singular values indicate that the waveforms of two different patients are distinct and there is more energy in the brain waves of coma patients (Fig. 8).

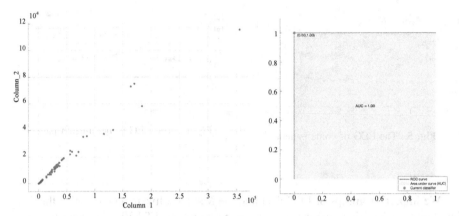

Fig. 7. The scatter plot of two different patients

Fig. 8. ROC curve

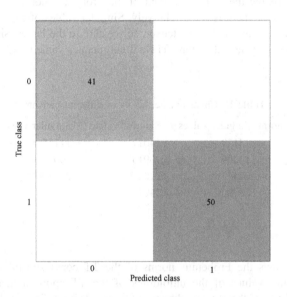

Fig. 9. The result confusion matrix

From Fig. 7 we can find that the brain-death EEG data concentrates on a small area and coma EEG data scatters in the figure. This indicates that the coma data has much more information compared with brain-death EEG data and coma patient has much more brain activity. From the ROC curve, we can find that the curve is near the upper left corner and it shows that the diagnostic accuracy is very high. AUC value is equal to 1 indicates that the diagnosis is effective. Figure 9 indicates that the classification result is satisfactory and the classification result is precise.

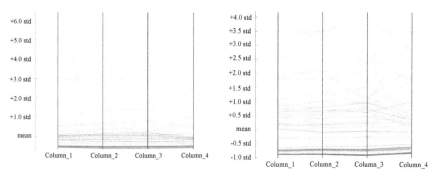

Fig. 10. Parallel Coordinates Plot 1

Fig. 11. Parallel Coordinates Plot 2

From Parallel Coordinates Plot 1 (Fig. 10), it can be concluded that on the same extracted feature attribute, the same color polylines are concentrated, and different colors have certain separation distance. It shows that this attribute is helpful for predicting label categories and each extracted feature is valuable for label classification. We get rid of the patients who have relatively active EEG data and Fig. 11 is obtained. From Fig. 11 we can find a clear gap between blue line and orange line which indicates the effectiveness of SVM classifier.

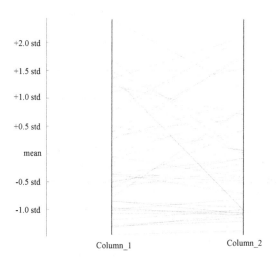

Fig. 12. Parallel Coordinates of matrix-SVD decomposition

At last we use SVD-decomposition to process original matrix and extract two features to diagnose the quasi-brain-death patients. From Fig. 12 we can find that some polylines is misplaced in column 2 and that indicates this attribute is not good as another attribute for predicting label categories. It is concluded that tensor is good method to diagnose quasi-brain-death patients and it has better reliability compared with matrix.

4 Conclusions

We apply a novel method that transforms the EEG data into higher-order, then tensor is first introduced to the diagnosis of quasi-brain-death. We employ TT decomposition to extract more valuable features from the tensor data to diagnose the quasi-brain-death patients. The result shows that EEG data of brain-death patients and coma patients are effectively classified by our method to diagnose quasi-brain-death patients. Most of all, our method can be a good reference to doctors which can reduce the risks and save valuable time for patients.

In this paper, we use tensor method to diagnose the quasi-brain-death patients, and the result is satisfying. In our future work, we aim to use tensor methods to diagnose more brain-related diseases. Also, we are going to apply more effective tensor decomposition model to extract more valuable features from EEG data to improve the performance of our method.

Acknowledgements. This work was supported by JSPS KAKENHI (Grant No. 17K00326, 15H04002, 18K04178), JST CREST (Grant No. JPMJCR1784), and the National Natural Science Foundation of China (Grant No. 61773129).

References

1. Marks, S., Zisfein, J.: Apneic oxygenation in apnea tests for brain death: a controlled trim. Arch. Neurol. **47**(10), 1066–1068 (1990)
2. Cao, J.: Analysis of the Quasi-brain-death EEG data based on a Robust ICA approach. In: Gabrys, B., Howlett, R.J., Jain, L.C. (eds.) KES 2006. LNCS (LNAI), vol. 4253, pp. 1240–1247. Springer, Heidelberg (2006). https://doi.org/10.1007/11893011_157
3. Cao, J., Chen, Z.: Advanced EEG signal processing in brain death diagnosis. In: Mandic, D., Golz, M., Kuh, A., Obradovic, D., Tanaka, T. (eds.) Signal Processing Techniques for Knowledge Extraction and Information Fusion, pp. 275–298. Springer, Berlin (2008). https://doi.org/10.1007/978-0-387-74367-7_15
4. Möcks, J.: Topographic components model for event-related potentials and some biophysical considerations. IEEE Trans. Biomed. Eng. **35**, 482–484 (1988)
5. Muti, D., Bourennane, S.: Multi-dimensional filtering based on a tensor approach. Signal Process. **85**, 2338–2353 (2005)
6. Vasilescu, M., Terzopoulos, D.: Tensortextures: multilinear image-based rendering. ACM Trans. Graph. **23**, 336–342 (2004)

7. Martínez-Montes, E., Valdés-Sosa, P., Miwakeichi, F., Goldman, R., Cohen, M.: Concurrent EEG/fMRI analysis by multi-way partial least squares. NeuroImage **22**, 1023–1034 (2004)
8. Miwakeichi, F., Martínez-Montes, E., Valds-Sosa, P., Nishiyama, N., Mizuhara, H., Yamaguchi, Y.: Decomposing EEG data into space-time-frequency components using parallel factor analysis. NeuroImage **22**, 1035–1045 (2004)
9. Govindaraju, N., Knott, D., Jain, N., et al.: Interactive collision detection between deformable models using chromatic decomposition. ACM Trans. Graph. **24**, 991–999 (2005)
10. Zhao, Q., Zhou, G., Xie, S., et al.: Tensor ring decomposition. arXiv preprint arXiv:1606.05535 (2016)
11. Stegeman, A., Sidiropoulos, N.: On Kruskal's uniqueness condition for the CANDECOMP/PARAFAC decomposition. Linear Algebra Appl. **420**, 540–552 (2007)
12. Hastad, J.: Tensor rank is NP-complete. J. Algorithms **11**, 644–654 (1990)
13. Lathauwer, D., Vandewalle, J.: Dimensionality reduction in higher-order signal processing and rank-(R1, R2,…,RN) reduction in multilinear algebra. Linear Algebra Appl. **391**, 31–55 (2004)
14. Oseledets, I.: Tensor-train decomposition. SIAM J. Sci. Comput. **33**(5), 2295–2317 (2011)

Automatic Seizure Detection Based on a Novel Multi-feature Fusion Method and EMD

Lei Du[1,2], Yuwei Zhang[1,2], Qingfang Meng[1,2(✉)],
Hanyong Zhang[1,2], and Yang Li[1,2]

[1] School of Information Science and Engineering,
University of Jinan, Jinan 250022, China
ise_mengqf@ujn.edu.cn
[2] Shandong Provincial Key Laboratory of Network Based
Intelligent Computing, Jinan 250022, China

Abstract. The analysis of electroencephalogram (EEG) signals is becoming more important because of the time-consuming and large bias of traditional visual detection technology, especially in diagnosis of epilepsy. Based on the nonlinear and non-stationary of the EEGs, empirical mode decomposition (EMD) is applied to decompose the original signals into intrinsic mode functions (IMFs). In this paper, after getting the IMFs, the fluctuation index and variation coefficient are calculated to analyze the amplitude change of IMFs. In order to better reflect the information of EEG signals, Hilbert transform is applied to obtain the instantaneous frequency for each IMF. Then, the novel feature named fluctuation index and variation coefficient of instantaneous frequency for IMFs are calculated. Furthermore, feature based on sample entropy of the first order difference is extracted. Finally, both of the calculated features will put together as a fusion feature into SVM for classification. The proposed method is evaluated using the Boon epileptic dataset and the highest average classification accuracy is 99.59%, showing a powerful method to detect seizure.

Keywords: Epileptic seizure · EMD · Multi-feature fusion · Hilbert transform

1 Introduction

Epilepsy is one of a most common disease and the mortality of epileptic patients is 2–3 times higher than normal people [1]. The electroencephalogram (EEG) is an indispensable auxiliary tool in the detection of epilepsy. Due to the time-consuming and strong subjectivity of the traditional method, exploring the feasible epileptic seizure detection method which can get more information about the EEGs to give a more accurate diagnosis and more appropriate treatment options is necessary.

Recently, to diagnose epilepsy, many automated detection methods have been proposed which can be summarized as the time domain, the frequency domain, time-frequency domain, nonlinear dynamical and methods based on graph theory and so on. Chandaka et al. proposed a method based on cross-correlation to analysis healthy and ictal EEG signals [2]. As for frequency domain analysis, most of them used Fourier spectral analysis to transform time-domain signals into frequency domain [3].

© Springer Nature Switzerland AG 2019
H. Lu et al. (Eds.): ISNN 2019, LNCS 11555, pp. 512–521, 2019.
https://doi.org/10.1007/978-3-030-22808-8_50

However, the aforementioned methods can only respect the characteristics in time domain or frequency domain and they are based on the assumption that the signals are stationary. Hence, to analyze the EEG signals which are typically non-stationary, time-frequency estimation methods such as discrete wavelet transform (DWT) [4], short time Fourier transform (STFT) [5] are used. Furthermore, in [6] a graph theory based on key-point local binary pattern was proposed and in [7] Kumar et al. achieved the automatic detection of epileptic seizure by using multi-level local patterns (MLP).

Additionally, empirical mode decomposition (EMD) is a practical method to analyze non-stationary and nonlinear signals [8]. It has been reported that when analyzing biological signals EMD has a better expression [9, 10]. Compared with the other methods like Fourier transform and wavelet transform, EMD method decomposes the signal based on the time scale characteristics of the data itself, so it is adaptive. By EMD, EEG signal is composed into a collection of intrinsic mode functions (IMFs).

In this paper, a novel multi-feature fusion method in the EMD domain to diagnose epileptic EEG signals is proposed. First of all, the EMD is applied on the raw EEG signal to divide it into several IMFs. Then multi-features are extracted. Fluctuation index and variation coefficient of the IMFs obtained by EMD are calculated to analyze the amplitude change. In order to get the frequency information of the signals for better describing, Hilbert transform is applied to obtain the instantaneous frequency of the IMFs. Then according to this, this paper proposed a novel feature which is to calculate the fluctuation index and variation coefficient of instantaneous frequency. Furthermore, only the linear features may omit some information so an improved feature, sample entropy of first order difference, is calculated. Finally, these features getting from the IMFs are combined together as a fusion feature to increase the classification accuracy. The novelty of this paper is to propose a new method of multi-feature fusion, combining traditional linear feature, Hilbert transform and the nonlinear features. And the proposed method reaches a higher classification performance in diagnosis of seizure and seizure-free EEG signals.

2 Method

Figure 1 shows the flowchart of the proposed automatic seizure detection framework.

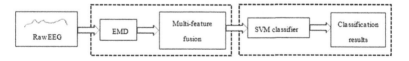

Fig. 1. Flowchart of the automatic seizure detection framework.

2.1 Empirical Mode Decomposition

Generally, EEG signals have obvious non-stationarity. Empirical mode decomposition (EMD) is an adaptive signal time-frequency processing method which is especially suitable for the analysis of non-stationary signals. The purpose of EMD is to decompose a non-stationary signal into multiple single component functions named intrinsic mode functions (IMFs). Each IMF should meet the following two conditions. Firstly, within the entire time range, the number of local extreme and zero crossings must be equal to, or at most diff one. Secondly, there must be a zero mean between the envelope of the local maximum and the envelope of the local minimum. Here is the IMFs extraction process:

(1) Find the local extreme of the raw signal $x(t)$, $t = 1, 2, ..., N$. the maximum and minimum of the signal are respectively interpolated by the cubic spline line, then the upper envelop (e_{max}) and the lower envelop (e_{min}) are formed.
(2) Calculate the mean of the upper envelop and the lower envelop:

$$m_1 = \frac{e_{max} + e_{min}}{2} \tag{1}$$

(3) Set $h_1 = x(t) - m_1$, if h_1 satisfies the two conditions of IMFs, then the first component can be designated as $c_1 = h_1$, otherwise, regard h_1 as a new $x(t)$ and return to (1) until get the first IMF.
(4) Regarding the residue r_1 as:

$$r_1 = x(t) - c_1 \tag{2}$$

(5) r_1 will be taking as a new signal and the sifting process (1)–(4) will go on to get the other IMFs until the residue becomes a monotone function. The original signal $x(t)$ can be represented as:

$$x(t) = \sum_{i=1}^{m} c_i + r \tag{3}$$

where c_i is the ith intrinsic mode function (IMF), r is the final residue.

2.2   Fluctuation Index and Variation Coefficient of Amplitude

Fluctuation index and variation coefficient are two indicators to describe the changes in time series. We calculate the fluctuation index and variation coefficient for the IMF component $c(t)$ to analyze the amplitude change of signals. The fluctuation index (F_a) in each IMF is:

$$F_a = \frac{1}{N} \sum_{t=1}^{N-1} (c(t+1) - c(t)) \tag{4}$$

Additionally, variation coefficient (V_a) in this paper is defined as:

$$V_a = \frac{\delta}{\mu}, \quad \mu = \frac{1}{N} \sum_{t=1}^{N} |c(t)|, \quad \delta = \sqrt{\frac{1}{N} \sum_{t=1}^{N} (|c(t)| - \mu)^2} \tag{5}$$

In Eq. (5), we first get the absolute value of the signal $c(t)$ and then calculate the mean (μ) and standard deviation (δ) of the absolute value signal $|c(t)|$.

2.3 Fluctuation Index and Variation Coefficient of Instantaneous Frequency

Solely extracted features in time domain to analysis the information from amplitude may cause the omission of important frequency information. Considering that each intrinsic mode function contains only one frequency component at each time, Hilbert transform [12], a generally used method is applied here to analyze the instantaneous features of signals. The process of instantaneous frequency calculation is as follows:

(1) The Hilbert transform of $x(t)$ can be obtained by:

$$y(t) = x(t) * \frac{1}{\pi t}$$
$$= \frac{1}{\pi} \int_{-\infty}^{\infty} \frac{x(\tau)}{t - \tau} d\tau \tag{6}$$

(2) Let $z(t)$ be the analytic signal of $x(t)$:

$$z(t) = x(t) + jy(t)$$
$$= a(t)e^{j\theta(t)} \tag{7}$$

where $a(t) = \sqrt{x^2(t) + y^2(t)}$ represents the amplitude of the analytic signal $z(t)$ and $\theta(t) = \arctan(y(t)/x(t))$ is the instantaneous phase representation.

(3) The instantaneous frequency of the signal $x(t)$ can be solved by:

$$\omega(t) = \frac{d\theta(t)}{d(t)} = \frac{y'(t)x(t) - x'(t)y(t)}{x^2(t) + y^2(t)} \tag{8}$$

In this paper, Hilbert transform is applied on the IMFs to get the instantaneous frequency $\omega(t)$. In this work, according to Hilbert transform and abovementioned two features, we proposed a novel feature extraction method which is to calculate the fluctuation index (F_ω) and variation coefficient (V_ω) of instantaneous frequency.

$$F_\omega = \frac{1}{N} \sum_{t=1}^{N-1} (\omega(t+1) - \omega(t)) \tag{9}$$

$$V_\omega = \frac{\delta}{\mu}, \quad \mu = \frac{1}{N} \sum_{t=1}^{N} |\omega(t)|, \quad \delta = \sqrt{\frac{1}{N} \sum_{t=1}^{N} (|\omega(t)| - \mu)^2} \tag{10}$$

2.4   Sample Entropy of First Order Difference

The EEG signals are extremely complex nonlinear and non-stationary signals, so if only the linear features are extracted to analysis the signals, the results will be limited. Hence, sample entropy (SampEn) of first order difference, a nonlinear measure, is used to make the analysis of EEGs more comprehensive and credible [13]. The difference operation reflects a change in the discrete quantity, which can be used for error correction and data stabilization. In this paper, we combined the SampEn algorithm and the first order difference. The sample entropy of first order difference is proposed as:

(1) Compute the first order difference sequence $x'(n)$ of the signal $x(n)$, $x(n) = [x(1), x(2), ..., x(T)]$, and the length of $x'(n)$ is $N = T - 1$.

(2) For the series $x'(n)$ of N points, form a set of vectors with m data points

$$X_m^i = [x'(i), x'(i+1), ...x'(i+m-1)], \{i | 1 \le i \le N - m + 1\} \qquad (11)$$

(3) The distance between two vectors like this is defined as the maximum difference of the corresponding data between them:

$$d[X_m^i, X_m^j] = \max(|x'(i+k) - x'(j+k)|), 0 \le k \le m - 1 \qquad (12)$$

(4) Let B_i be the number of $d[X_m^i, X_m^j]$ within r (a threshold given advanced), and define B^m as:

$$B^m = \frac{1}{N-m} \sum_{i=1}^{N-m} \frac{B_i}{N-m-1} \qquad (13)$$

(5) Let m plus one, repeat the process (2)–(4) to obtain the vectors X_{m+1}^i and A^m, finally, the sample entropy of first order difference is:

$$SampEn(m, r) = \lim_{N \to \infty} [-\ln \frac{A^m}{B^m}] \qquad (14)$$

(6) while N is a finite value, it can be expressed as:

$$SampEn(m, r, N) = -\ln \frac{A^m}{B^m} \qquad (15)$$

In the paper, we applied the sample entropy of first order difference on the IMFs. Usually, the sample entropy of first order difference in normal EEG signals is higher than ictal EEG signals.

2.5   Multi-feature Fusion Method

Given a EEG signal $x(t) = [x(1), x(2), ..., x(T)]$, using empirical mode decomposition to compose the signal into a series intrinsic mode functions, $c_i(t)$ represents the ith intrinsic mode function. Extract features from IMFs. First, according to (4) and (5) calculate the fluctuation index and variation coefficient of amplitude which we denote as $C_a^i = [F_a^i, V_a^i]$. Next, according to (6)–(10), the fluctuation index and variation coefficient of instantaneous frequency can be extracted as $C_\omega^i = [F_\omega^i, V_\omega^i]$. Then extract the SampEn of first order difference C_s^i. Finally, a new fusion feature space $[C_a^i, C_\omega^i, C_s^i]$ which represents the ith IMF is constructed.

2.6   Classification

Support vector machine (SVM) [14], is a well-known supervised algorithm based on the theory of VC dimension. The principle of minimum structural risk that SVM follows makes a good generalization ability, while traditional classifiers which based on the principle of error minimization may cause the risk of over-fitting when the sample size is small. Therefore, the features calculated earlier are fed into the SVM to find a decision boundary. In this paper, we selected a RBF-based SVM to evaluate the effectiveness of the features.

3   Experimental Results

3.1   Dataset

The EEG signals used in this work are the records which are from the epilepsy laboratory of Boon, Germany [11]. There are five subsets in this dataset, which are known as Z, O, N, F and S. Sets Z and O are the EEG signals extracted from five healthy persons, respectively measured with eyes open and closed while the remaining three sets are records from epileptic patients. Sets N and F are the EEGs during the seizure-free intervals respectively taken from hippocampus and the epileptogenic zone, and the set S contains ictal signals from focal zone. Each of the five sets contains 100 single-channel EEG epochs whose sampling frequency is 173.61 Hz and time duration is 23.6 s (each epoch has 4097 points). In our work, all of the five subsets are used and each signal is divided into four parts with 1024 points, hence there are 400 segments in each class.

3.2   EMD and Feature Extraction

For the Boon dataset, the EEG signals from Z, O, N, F and S are all processed by EMD first. For comparison, the original signal and the IMFs are shown in Fig. 2, the samples are respectively from set F and set S. The first five IMFs are chosen by analyzing the spectrum. After EMD, EEG signal features are calculated via the method mentioned above. Thus, there are 5 features in total for each IMF of an EEG signal. Eight cases we considered in this work are shown in Table 1.

Table 1. Different classification cases.

Case	1	2	3	4	5	6	7	8
Group 1	Z	O	N	F	NF	ZNF	ONF	ZONF
Group 2	S	S	S	S	S	S	S	S

N-fold cross-validation is a statistical method used to verify the performance of classifiers. The input dataset is randomly grouped into *N* subsets of the same size, let each of the *N* subset be the test set respectively and the remaining *N* − 1 subsets be the training set. Then *N* models will be built during this procedure, regarding the average accuracy of the *N* models as the final classification accuracy. In this work, both 4 and 10-fold cross-validation are performed to evaluate the performance of the proposed method.

In order to verify the effectiveness of the proposed feature fusion method, several experiments were carried out, Table 2 shows the results in case 4. According to Table 2, it can find that the feature SampEn of first order difference has a better performance than the original SampEn. It can be reported that the classification accuracy used EMD especially the first two IMFs is better than the raw EEG signals, which may due to the more information that the IMFs contain. And the fusion feature has the improved classification performance compared with the single feature.

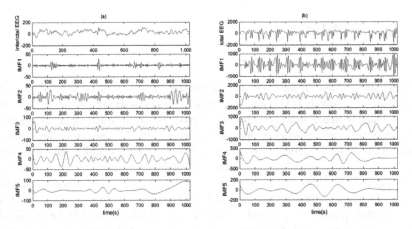

Fig. 2. Different components of IMFs taken from the original EEG signal. (a) an interictal signal from set F, (b) an ictal signal from set S.

Table 2. Classification accuracy of 4-fold CV for different features in case 4.

Component	EEG	IMF1	IMF2	IMF3	IMF4	IMF5
SampEn	76	74.375	74.25	72.375	67.125	66
SampEn of first order difference	92	82.875	93.75	78.75	71.875	68.25
$[F_a, V_a]$	96.5	97.25	95.125	85.25	77.625	67.875
$[F_\omega, V_\omega]$	88.875	95.875	93.25	84.375	70.625	63.5
Fusion feature	97.875	**98.625**	98	92.875	85	78.375

3.3   Classification Results

It is a very challenging problem when it comes to correctly identify ictal EEG signals from healthy and interictal records, therefore, it is very necessary to achieve the automatic epilepsy detection method with good classification effect. In this work, we use the above-mentioned database to evaluate the proposed framework. In order to get a higher performance of the classifier, features from different IMF components are fused to form a new feature vector. From Table 3, results of 4-fold and 10-fold CV based on the fusion feature are shown. It can be observed that the average classification by combining different IMF components is higher than only used the first IMF in almost all the 8 cases, and the average accuracy between the 8 cases can be up to 99.59% (10-fold) and 99.48% (4-fold). The ROC curves of case 4 and case 5 are shown in Fig. 3.

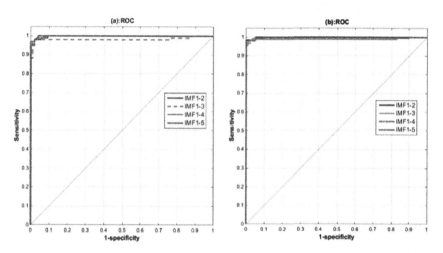

Fig. 3.   (a) ROC curve of case 4, (b) ROC curve of case 5.

Table 3.   Classification accuracy of 4-fold and 10-fold CV for fusion features.

Case	IMF1		IMF1–2		IMF1–3		IMF1–4		IMF1–5	
	4-fold	10-fold	4-fold	10-fold	4-fold	10-fold	4-fold	10-fold	4-fold	10-fold
1	99.875	99.75	**100**	99.875	99.875	99.875	**100**	**100**	**100**	**100**
2	97.125	97	99.75	99.625	**100**	99.875	**100**	**100**	**100**	99.875
3	99.25	99.25	**100**	99.75	**100**	99.75	**100**	99.875	99.75	99.625
4	**98.625**	98.75	98.375	**99**	98.25	98.625	98.25	98.875	98.125	98.5
5	99.25	99.167	**99.333**	**99.25**	99	99.167	99	**99.25**	98.833	98.917
6	99.375	99.438	99.5	**99.5**	**99.625**	99.375	99.563	99.563	99.313	99.375
7	98.188	98.25	99.438	99.438	**99.625**	99.375	99.563	**99.563**	99.375	99.375
8	98.4	98.55	99.35	99.5	99.45	99.45	**99.5**	**99.6**	99.3	99.5
Average	98.76	98.77	99.47	99.49	**99.48**	99.44	**99.48**	**99.59**	99.34	99.40

Table 4 shows the comparison of classification accuracy for the same dataset according to different methods. The results of our proposed method in Table 4 are average accuracy of 10-fold CV. It can be seen from Table 4 that the proposed multi-feature fusion method in EMD domain has a higher accuracy in both the 8 cases compared with the other methods.

Table 4. Comparison from different methods

Authors	Methods	Cases	Accuracy (%)
Zhu et al. [16]	Degree + strength of HVG + K-NN	Z-S	100
		F-S	93.0
		ZONF-S	95.4
Kumar et al. [7]	MLP + NN	NF-S	98.33
Song et al. [13]	MS-SE-FF + ELM	F-S	97.53
Kumar et al. [15]	Fuzzy approximate entropy + SVM	Z-S	100
		O-S	100
		N-S	99.6
		F-S	95.85
		ZNF-S	98.15
		ONF-S	98.22
		ZONF-S	97.38
Proposed method	Multi-feature fusion + SVM	Z-S	100
		O-S	100
		N-S	99.875
		F-S	99
		NF-S	99.25
		ZNF-S	99.5
		ONF-S	99.563
		ZONF-S	99.6

4　Discussion and Conclusions

In this work we aim to perform an efficient automated detection method for diagnosing seizure and non-seizure EEG signals. In this paper, a multi-feature fusion method based on EMD was proposed. First, by EMD the intrinsic mode functions (IMFs) contain more information of the time series. In the proposed work, based on fluctuation index and variation coefficient of amplitude, Hilbert transform is combined with them to obtain a new feature named fluctuation index and variation coefficient of instantaneous frequency. And the improved feature sample entropy of first order difference is used to extract nonlinear dynamic information. Then features mentioned above are fused to obtain more information about the EEG signals and fed into SVM. The average classification accuracy of this proposed method is 99.48% (4-fold CV) and 99.59% (10-fold CV). The result shows that the method has a well classification performance.

The performance of the proposed method has been evaluated by comparing with the other methods. It can be concluded that the proposed method has a higher recognition rate, showing its potential for seizure detection.

Acknowledgment. This work was supported by the National Natural Science Foundation of China (Grant No. 61671220, 61701192, 61640218), the Natural Science Foundation of Shandong Province, China (Grant No. ZR2017QF004).

References

1. Acharya, U.R., Sree, S.V., Swapna, G., Martis, R.J., Suri, J.S.: Automated EEG analysis of epilepsy: a review. Knowl. Based Syst. **45**(3), 147–165 (2013)
2. Chandaka, S., Chatterjee, A., Munshi, S.: Cross-Correlation Aided Support Vector Machine Classifier for Classification of EEG Signals. Pergamon Press, Inc., New York (2009)
3. Polat, K., Güneş, S.: Classification of epileptiform EEG using a hybrid system based on decision tree classifier and fast Fourier transform. Appl. Math. Comput. **187**(2), 1017–1026 (2007)
4. Subasi, A.L.: EEG signal classification using wavelet feature extraction and a mixture of expert model. Experts Syst. Appl. **32**(4), 1084–1093 (2007)
5. Li, Y., Luo, M.L., Li, K.: A multi-wavelet-based time-varying model identification approach for time-frequency analysis of EEG signals. Neurocomputing **193**, 106–114 (2016)
6. Tiwari, A., Pachori, R.B., Kanhangad, V., Panigrahi, B.K.: Automated diagnosis of epilepsy using key-point based local binary pattern of EEG signals. IEEE J. Biomed. Health Inform. **21**, 888–896 (2016)
7. Kumar, T.S., Kanhangad, V., Pachori, R.B.: Classification of seizure and seizure-free EEG signals using multi-level local patterns. In: International Conference on Digital Signal Processing. IEEE (2014)
8. Huang, N., Shen, Z., Long, S.R., Wu, M.C., et al.: The empirical mode decomposition and the Hilbert spectrum for nonlinear non-stationary time series analysis. Proc. Math. Phys. Eng. Sci. **454**, 903–995 (1998)
9. Li, S.F., Zhou, W.D., Cai, D.M., Liu, K., Zhao, J.L.: EEG signal classification based on EMD and SVM. J. Biomed. Eng. **28**(5), 891–894 (2011)
10. Pachori, R.B., Bajaj, V.: Analysis of normal and epileptic seizure EEG signals using empirical mode decomposition. Comput. Methods Programs Biomed. **104**(3), 373–381 (2011)
11. Lehnertz, K.: Epilepsy and nonlinear dynamics. J. Biol. Phys. **34**(3–4), 253–266 (2008)
12. Huang, N., Wu, M.L., Qu, W., Long, S.R., Shen, S.S.P.: Application of Hilbert-Huang transform to non-stationary financial time series analysis. Appl. Stoch. Model. Bus. Ind. **19**(3), 245–268 (2010)
13. Song, J.L., Hu, W., Zhang, R.: Automated Detection of Epileptic EEGs Using a Novel Fusion Feature and Extreme Learning Machine. Elsevier Science Publishers B.V., Amsterdam (2016)
14. Vapnik, V., Cortes, C.: Support-vector network. Mach. Learn. **20**, 273–297 (1995)
15. Kumar, Y., Dewal, M.L., Anand, R.S.: Epileptic seizure detection using DWT based fuzzy approximate entropy and support vector machine. Neurocomputing **133**(8), 271–279 (2014)
16. Zhu, G., Li, Y., Wen, P.: Epileptic seizure detection in EEGs signals using a fast weighted horizontal visibility algorithm. Comput. Methods Programs Biomed. **115**(2), 64–75 (2014)

Phenomenology of Visual One-Shot Learning: Affective and Cognitive Components of Insight in Morphed Gradual Change Hidden Figures

Tetsuo Ishikawa[1], Mayumi Toshima[2], and Ken Mogi[3(✉)]

[1] Medical Sciences Innovation Hub Program, RIKEN, 1-7-22 Suehiro-cho,
Tsurumi-ku, Yokohama, Kanagawa 230-0045, Japan
tetsuoishikaha@gmail.com
[2] Digital Content and Media Sciences Research Division,
National Institute of Informatics, 2-1-2 Hitotsubashi, Chiyoda-ku,
Tokyo 101-8430, Japan
mamitako@gmail.com
[3] Sony Computer Science Laboratories Inc., Takanawa Muse Bldg. 3-14-13,
Shinagawa-ku, Tokyo 141-0022, Japan
kenmogi@qualia-manifesto.com

Abstract. People sometimes gain insight into an innovative solution of problem. In the visual domain, one-shot learning in hidden figures is a prominent instance of such Eureka moments. However, the nature of conscious experience accompanying the visual one-shot learning has not been well studied. Here we show the phenomenology of visual one-shot learning scrutinized through an experiment considering more diverging aspects of subjective feelings. Correlation and exploratory factor analysis were performed on the participants' recognition time, accuracy, and subjective judgments of hidden figure recognition in morphing gradual change paradigm. As a result, two salient factors were found, which were interpreted as "Aha!" experience and task difficulty. Furthermore, the "Aha!" experience consists of affective and cognitive components of insight. The results suggested that insight can be characterized by multidimensional factors in the case of visual one-shot learning as in common with other problem domains and modalities.

Keywords: One-shot learning · Visual object recognition · Hidden figures · "Aha!" experience · Insight · Affective · Cognitive · Phenomenology

1 Introduction

Visual object recognition is one of the routine cognitive processes of our daily life. In the dim light, the visual system tries to collect information out of the scarce context. Similarly, hidden figures, such as a Cow [1] and a Dalmatian [2], do not have visually rich context. Hidden figures seem to have no definite interpretation at first; once the viewer realizes the image hidden in the blurred picture, there is no going back. This special type of abrupt perceptual learning is called *one-shot learning* [3–5]. The subject typically experiences an insightful sensation ("Aha!" experience) at the moment of perceptual learning [6, 7]. When and how this Eureka moment of learning happens is

© Springer Nature Switzerland AG 2019
H. Lu et al. (Eds.): ISNN 2019, LNCS 11555, pp. 522–530, 2019.
https://doi.org/10.1007/978-3-030-22808-8_51

not well known, partly because this type of experience typically happens just once in a special condition.

A new method to analyze this insightful experience with morphing pictures was introduced [4]. This method was contrived with applying black and white two-tone ("Mooney") images [8, 9] gradually changing in gray scale. Behavioral characteristics and quality of subjective experiences were not fully examined in the previous study, because only the reaction time (RT) and the confidence rating were recorded [4]. Some recent research has addressed multi-dimensional features of insight and revealed the phenomenology of "Aha!" experience by using multifaceted scoring method [10–13].

By modifying these new methodologies to be applied to visual one-shot learning, here we report experiments in which the subjects viewed morphing hidden figures on the screen. After each trial, the subjects were asked to provide subjective ratings in six-point scale about ten aspects (Table 1): Suddenness, Confidence, Vividness, Three-dimensional perception, Fun (gratifying), Delight, Surprise, misperception (false alarm, FA), tip-of-the-tongue (TOT) phenomena (a state in which only the name does not come out although it is understood), and willingness to recommend (WTR).

Table 1. Subjective assessment questionnaire

#item	Description
1.	How suddenly did you find the perceived picture? (Suddenness/Sudden)
2.	How much are you convinced about your answer? (Confidence/Sure)
3.	How vividly did you see the answer picture? (Vividness/Vivid)
4.	How sterically did you feel about the picture? (Stereoscopicity/Sterical/3D)
5.	How interesting did you feel about the picture? (Fun)
6.	How pleased were you when you got the right answer? (Delight)
7.	How surprised did you feel at the answer? (Surprise)
8.	How long were you having a wrong figure before you got the right answer? (FA)
9.	How much did you feel the-tip-of-the-tongue state? (TOT)
10.	How much do you prefer to show this picture to your friends? (WTR)

The reasons why these subjective assessments were adopted are as follows. There are four prominent features that typically occur at the time of the "Aha!" experience accompanying insight: (i) noticing the answer suddenly, (ii) understanding quickly and easily when it is the correct answer, (iii) experiencing positive emotions, (iv) being convinced that it is the correct answer [7, 14]. We selected the items that are expected to reflect the prominent features. Surprise and Suddenness are related to (i), 3D-perception and Vividness which may be grounds for judging as answer are in relation to (ii), FA or misrecognition of distractor which inhibits correct recognition, sharing the same experiences with others (WTR), Fun (enjoyment) or Delight of understanding something are related to (iii), while the TOT phenomenon and Confidence of clearly understanding are related to (iv).

We assumed that some of these rating scores might be related to the "Aha!" experiences. There are several other possible options. As the number of questions is increased too much, the cognitive load will be too high or the impression will fade during the answer, so we narrowed the questions down to just ten items. Also, as objective indexes, the RT was measured and the correct answer rate (accuracy) was calculated. For these subjective and objective variables, we investigated the underlying factor structure by exploratory factor analysis (EFA) to verify whether the hypothesis that insight can be best described by multidimensional perspectives [10–13] is also true in the visual one-shot learning of hidden figures.

2 Methods

2.1 Participants

Twenty-four undergraduate/graduate students (12 females and 12 males; mean ± SD age: 24 ± 7 years old) took part in this experiment. All participants had normal or corrected-to-normal vision and all but one were right-handed by self-report. All procedures were performed with the participants' informed written consent and in accordance with the protocols approved by the Ethics Committee of National Institute of Informatics.

2.2 Stimuli

Twenty-four grayscale pictures (300 × 300 pixels) with familiar objects [15] were blurred and binarized to be unrecognizable. The resultant black and white ambiguous images are called Mooney objects [8, 9]. By morphing each Mooney object and its blurred original grayscale counterpart, intermediate images between them with various ambiguity levels were made [4]. By connecting these series of morphing images together, we got hidden movies with 101 frames gradually changing from original blending ratio 0% to 100% in increments of 1% (Fig. 1).

Presentation time of each frame was 500 ms, and thus each movie length was 50.5 s. The stimuli (12.5° × 12.5°) were presented against middle gray background on the 17-in. Tobii T60 display (Tobii Technology, Stockholm, Sweden).

2.3 Procedure

The participants' task was to detect and recognize hidden object in stimulus as soon as possible. After right-clicking the mouse button to indicate recognition or reaching the video end (time out) and answering the object name verbally, ten types of subjective rating (Table 1) in 6-point (options: 0–5) scale were asked in pseudo-random order. We did not give a linguistic description using adjectives and adverbs expressing each option. Instead, the participants were asked to judge the extent a particular feeling is true by the magnitude of the numerical value from 0 to 5. The order of stimulus videos was counterbalanced among the participants. Since the total number of stimuli was small, we analyzed the data from the practice and experiment sessions without distinguishing them.

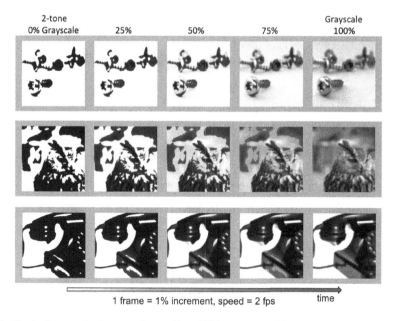

Fig. 1. Typical example frames of morphing hidden movies (The answer: Screw, Owl, and Telephone from top to down). The grayscale images before degradation were adapted from http:// www.freeimages.com/

The condition for judging that the subjects correctly recognized hidden objects was whether or not a synonym of object name was included in the answer. Verbal reports of object name using different words across the participants such as "high heels", "heels of shoes", or "pumps", for example, were all regarded as correct answers.

3   Results

On average (±SD), 91.7 ± 8.2% (22.0 ± 1.97 movies) of all the stimuli were correctly recognized. RT was 36.0 ± 4.3 s in correct trials and 32.3 ± 9.0 s in incorrect trials including no response. In comparison between correct and incorrect/time out trials, all subjective ratings were higher in correct trials than in incorrect trials (all $ps < 1.0\mathrm{e}{-5}$, two-tailed t-tests), except for WTR ($p = 0.34$), FA ($p = 0.95$), and TOT ($p = 0.25$) (Fig. 2).

Because the number of error trials were not sufficient for further analysis, henceforth, only data in correct trials were analyzed. We performed the Pearson correlation analysis between the 10-item subjective ratings and two task performance measures, i.e., RT and accuracy (correct rate, abbreviated as Correct in Figures and Tables) calculated for each stimulus.

There were many combinations showing positive correlations among them, while negative correlations were observed only in four combinations: (i) RT and confidence, (ii) RT and 3D, (iii) RT and accuracy, and (iv) confidence and TOT. Only one variable showed a positive correlation with RT was FA (Table 2).

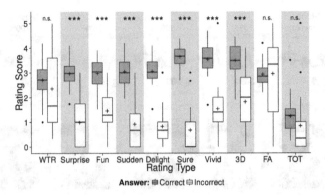

Fig. 2. Comparison between correct and incorrect answers for each subjective assessment. Mean (+), median (thick line in the box). ***p < 1.0e−5, n.s.: not significant

Table 2. Pearson correlations between subjective evaluations and objective performances

	WTR	Surprise	Fun	Sudden	Delight	Sure	Vivid	3D	Correct	FA	RT
WTR											
Surprise	.83♭										
Fun	.88♭	.82♭									
Sudden	.84♭	.84♭	.91♭								
Delight	.81♭	.73♭	.91♭	.88♭							
Sure	.52♮	.44♮	.65♭	.66♭	.82♭						
Vivid	.59♮	.59♮	.74♭	.77♭	.80♭	.84♭					
3D	.49♮	.47♮	.68♭	.68♭	.66♭	.72♭	.84♭				
Correct	.24	.22	.38	.45♮	.39	.49♮	.48♮	.49♮			
FA	.47♮	.74♭	.38	.46♮	.24	−.12	.14	.01	−.18		
RT	.11	.26	−.05	−.03	−.26	−.56♮	−.35	−.42♮	−.48♮	.74♭	
TOT	−.13	−.12	−.19	−.11	−.37	−.53♮	−.12	−.09	−.16	.15	.30

# p < 0.05, ♮ p < 0.01, ♭ p < 0.001

In order to reveal latent structures behind observed variables, EFA was carried out. For the first step, it was necessary to infer the number of factors. According to the Scree plot (Fig. 3), in which the eigenvalues were arranged in order of magnitude, variance of the first factor was accounted for more than twice of the variance of the other factors. The Scree test to estimate factor numbers from location of "elbow" in the graph shape suggested that there existed an underling three factor structure.

There is another rule of thumb, called the Kaiser–Guttman (KG) criterion, proposing that the number of eigenvalues equal to or larger than 1.0 is regarded as the factor number. In light of this KG criterion, the estimated number of factors was 3. Therefore, it was reasonable to conclude that a three-factor model was most likely.

The factor loadings (ML1, ML2, ML3, corresponding to the first, second, and third factors, respectively) and the commonality (h^2) were estimated (Table 3) through maximum likelihood method with the Promax rotation.

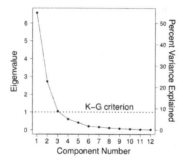

Fig. 3. Scree plot to estimate factor number. Three-factor structure is plausible

Table 3. Factor loadings and commonalities in three-factor structure

	ML1	ML2	ML3	h^2
WTR	0.84	0.33	0.00	0.79
Surprise	0.80	0.56	-0.06	0.92
Fun	0.94	0.18	-0.01	0.89
Sudden	0.97	0.20	0.08	0.93
Delight	0.91	0.01	-0.15	0.92
Sure	0.73	-0.36	-0.27	0.91
Vivid	0.90	-0.21	0.15	0.82
3D	0.80	-0.30	0.18	0.70
Correct	0.47	-0.35	0.04	0.35
FA	0.34	0.90	0.01	0.89
RT	-0.18	0.87	0.04	0.83
TOT	0.01	0.04	0.99	1.00

Next, to investigate whether there was a substructure in the biggest factor, nine variables strongly related to the first factor were selected and further analyzed. Results of the Scree test and the KG criterion (Fig. 4) consistently suggested that there might be two sub-factors (ml1 and ml2) in the original main factor (Table 4).

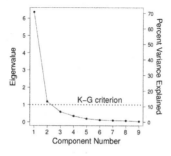

Fig. 4. Scree plot to estimate sub-factor number. Two-factor structure is plausible

Table 4. Factor loadings and commonalities in two sub-factor structure

	ml1	ml2	h^2
WTR	0.99	-0.08	0.87
Surprise	0.96	-0.11	0.79
Fun	0.81	0.21	0.94
Sudden	0.75	0.27	0.90
Delight	0.58	0.45	0.90
Sure	-0.05	0.94	0.83
Vivid	0.10	0.86	0.87
3D	0.03	0.83	0.72
Correct	-0.09	0.60	0.30

4 Discussion

In seven out of ten items in the subjective ratings, rated scores were significantly higher in correct answer trials than in incorrect ones, suggesting that the participants had stronger feelings when they reached the correct answers: they were surprised and convinced to find suddenly a vivid 3D object in the stimulus and pleased to have such a fun experience. There was, however, no difference in the WTR, FA (e.g., mental fixation) and TOT phenomenon ratings between correct and incorrect conditions.

The WTR judgment could be regarded as an indicator of word-of-mouth. In marketing research, it is known that there is a U-shaped relationship between product satisfaction level and frequencies of word-of-mouth generation. When the product is very satisfactory or otherwise not very satisfactory for customers, the word-of-mouth reactions are most likely to be induced [16]. If the correct answer gives a sense of satisfaction and, on the other hand, incorrect/unsolved (or at least not fully achieved disambiguation) case provides a feeling of dissatisfaction, it would be a natural consequence that the urge to spread word-of-mouth or willingness to recommend becomes high in either case.

In the debriefing after experiment, some of the participants wanted to know the answers and the results of the problems and they also wanted to know whether others could solve them. According to the cognitive dissonance theory [17], in the case when the participants cannot solve the problems, they tend to rationalize that it is not a lack of ability of themselves but because these problems are so much difficult that nobody can solve them.

The factors obtained by EFA can be interpreted as follows. Variable group of the major factor (ML1) with the highest factor load (mainly appearing on the right-hand side in Fig. 5) can be interpreted as an axis of "Aha!" experience, because this axis has relevant features characterizing the "Aha!" experience, such as feelings of suddenness, confidence and positive emotions, with other several insight problem domains [7, 14]. On the other hand, variables constituting the second factor (ML2) can be interpreted as an axis of Task difficulty, since this direction has positive correlations with RT and FA, and negative correlations with confidence and accuracy. The first "Aha!" experience axis is a factor related to the state of the subjective experience that occurs at the moment, and the second Task difficulty axis is a factor related to the state until the

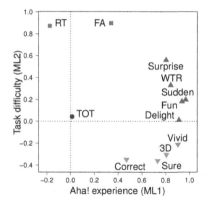

Fig. 5. Factor loadings placed on the 1st–2nd factor plane. Variables having a strong connection to ML1 were depicted by triangles (▲ or ▽). Other variables having a strong influence on ML2 were depicted by squares (■). Those related to ML3 (represented by the axis orthogonal to the plane on the figure) were shown by circles (●). The sub-factor ml1 (▲) and ml2 (▽) had positive and negative correlation with ML2, respectively

answer is known. The third small factor (ML3) consisting of the TOT phenomenon is rather independent from the other factors. Moreover, as a substructure of the first factor, there were two sub-factors. The first sub-factor (ml1) is considered to reflect the more affective aspect of "Aha!" experiences, while the second sub-factor (ml2) reflects more objective judgment on perception and (re)cognition, which is consistent with [11, 13].

In summary, we found two salient factors describing both subjective and objective features of visual one-shot learning in morphed hidden figures, which were interpreted as "Aha!" experience and Task difficulty. Furthermore, the "Aha!" experience consists of two sub components: Affective and Cognitive components of insight. The fact that WTR is a strong indicator of the "Aha!" experience, particularly its Affective components, is, to our best knowledge, a novel finding. The results suggested that insight can be characterized by multidimensional factors in the case of visual one-shot learning, as in common with other problem domains and modalities. In conclusion, we characterized the phenomenology of "Aha!" experience in the visual one-shot learning for further creative journey.

Acknowledgments. We greatly appreciate Viktors Garkavijs and Rika Okamoto for their technical supports and helpful comments. We also very thank Noriko Kando for her kind hospitality and financial support.

References

1. Dallenbach, K.M.: A picture puzzle with a new principle of concealment. Am. J. Psychol. **64**, 431–433 (1951)
2. Gregory, R.L.: The Intelligent Eye. Weidenfeld & Nicolson, London (1970)

3. Giovannelli, F., et al.: Involvement of the parietal cortex in perceptual learning (Eureka effect): an interference approach using rTMS. Neuropsychologia **48**(6), 1807–1812 (2010)
4. Ishikawa, T., Mogi, K.: Visual one-shot learning as an 'anti-camouflage device': a novel morphing paradigm. Cogn. Neurodyn. **5**(3), 231–239 (2011)
5. Dudai, Y., Morris, R.G.: Memorable trends. Neuron **80**(3), 742–750 (2013)
6. Bowden, E.M., Jung-Beeman, M., Fleck, J., Kounios, J.: New approaches to demystifying insight. Trends Cogn. Sci. **9**(7), 322–328 (2005)
7. Topolinski, S., Reber, R.: Gaining insight into the "Aha" experience. Curr. Dir. Psychol. Sci. **19**(6), 402–405 (2010)
8. Mooney, C.M.: Age in the development of closure ability in children. Can. J. Psychol. **11**(4), 219–226 (1957)
9. Mooney, C.M., Ferguson, G.A.: A new closure test. Can. J. Psychol. **5**(3), 129–133 (1951)
10. Danek, A.H., Fraps, T., von Müller, A., Grothe, B., Öllinger, M.: It's a kind of magic—what self-reports can reveal about the phenomenology of insight problem solving. Front. Psychol. **5**, 1408 (2014)
11. Shen, W., Yuan, Y., Liu, C., Luo, J.: In search of the 'Aha!' experience: elucidating the emotionality of insight problem-solving. Br. J. Psychol. **107**(2), 281–298 (2016)
12. Webb, M.E., Little, D.R., Cropper, S.J.: Insight is not in the problem: investigating insight in problem solving across task types. Front. Psychol. **7**, 1424 (2016)
13. Shen, W., et al.: Defining insight: a study examining implicit theories of insight experience. Psychol. Aesthetics Creativity Arts **12**(3), 317–327 (2018)
14. Gick, M.L., Lockhart, R.S.: Cognitive and affective components of insight. In: Sternberg, R. J., Davidson, J.E. (eds.) The Nature of Insight, pp. 197–228. MIT Press, Cambridge (1995)
15. Snodgrass, J.G., Vanderwart, M.: A standardized set of 260 pictures: norms for name agreement, image agreement, familiarity, and visual complexity. J. Exp. Psychol. Hum. Learn. **6**(2), 174–215 (1980)
16. Anderson, E.W.: Customer satisfaction and word of mouth. J. Serv. Res. **1**(1), 5–17 (1998)
17. Festinger, L.: A Theory of Cognitive Dissonance. Stanford University Press, Palo Alto (1957)

Protein Tertiary Structure Prediction Based on Multiscale Recurrence Quantification Analysis and Horizontal Visibility Graph

Hui Jiang[1,2], Anjie Zhang[1,2], Zaiguo Zhang[3], Qingfang Meng[1,2(✉)], and Yang Li[1,2]

[1] School of Information Science and Engineering,
University of Jinan, Jinan 250022, China
ise_menqf@ujn.edn.cn
[2] Shandong Provincial Key Laboratory of Network Based Intelligent
Computing, Jinan 250022, China
[3] CET Shandong Electronics Co., Ltd., Jinan 250101, China

Abstract. As protein types continue to increase, there are more and more methods for predicting protein structure. In this paper, a feature extraction method based on multiscale coarse-grained time series recurrence quantification analysis and horizontal visibility graph is proposed. First, the chaos game representation is used to map the protein secondary structure sequence into two time series. Multiscale coarse granulation time series. Then feature extraction by combining recurrence quantification analysis and horizontal visibility graph. Thereby a 30-D feature vector is obtained. This paper uses support vector machine to predict protein tertiary structure. In this paper, the prediction results of the two low homologous protein datasets were 95.33% and 93%, respectively.

Keywords: Protein structure prediction ·
Multiscale coarse-grained time series · Recurrence quantification analysis ·
Horizontal visibility graph

1 Introduction

Protein tertiary structure prediction is an important research topic in Bioinformatics. Different types of proteins have different patterns of existence and according to their patterns, proteins can be classified as all-α, all-β, α/β and $\alpha + \beta$. The all-α and all-β indicate that they are mainly composed of α-helix and β-strands, respectively [1].

At present, there are more and more methods of protein structure prediction, which mainly focus on two aspects of feature extraction and classification algorithm. There are many methods for feature extraction, such as pseudo-amino acid composition (PseAAC) [2], polypeptide composition [3]. Although these methods can predict protein structure and produce certain effects, when they encounter low homologous proteins, the predictive efficiency becomes very low. The chaos game representation (CGR) method was originally used to predict secondary structure types of proteins [13], but the limitation of this approach is that it does not apply to short protein sequences. In order to solve this problem, recurrence quantification analysis (RQA) was introduced to solve the problems

H. Lu et al. (Eds.): ISNN 2019, LNCS 11555, pp. 531–539, 2019.
https://doi.org/10.1007/978-3-030-22808-8_52

caused by different protein sequence lengths [9]. In recent years, the concept of complex network theory has been applied in various fields. In [12], RQA is combined with complex network to improve the prediction accuracy of protein structure to a new level. Then, a feature extraction method based on horizontal visibility graph was proposed [11], which further proves the feasibility of complex network in the field of protein structure prediction. After feature extraction, different classifiers are used to classify features.

In this paper, we have a structural prediction of two low homologous proteins 25PDB and 1189 based on the protein secondary structure. The secondary structure sequence is mapped into two time series according to CGR. In this paper, multiscale coarse granulation of time series is used to obtain a new time series. And this paper combines RQA and horizontal visibility graph to extract a 30-D feature. The feature vector is input to the SVM for classification. The jackknife test on two low homology datasets showed that the method was effective.

2 Methods

2.1 Secondary Structure Prediction

We predict each amino acid in the protein sequence to be one of the three elements of the secondary structure of the protein: C (coil), E (strand) and H (helix). There are many computational methods to predict the secondary structure of proteins. In this study, PSIPRED [8] was used to predict the secondary structure of proteins.

2.2 Chaos Game Representation

The chaos game representation (CGR) method is a technique that can transform a one-dimensional sequence into two dimensions. It is able to visualize the sequence while preserving the sequence structure. This method has often been used for prediction in recent years. A sequence of protein secondary structures is represented in a unit equilateral triangle. Its three vertices correspond to three letters H, C, and E, respectively. The first point in the figure is at the midpoint of the vertices corresponding to the first letter of the protein sequence in the center of the triangle; the i-th point in the figure is at the midpoint of the vertices corresponding to the $(i - 1)$th point and the i-th point. Thus, a CGR graph is obtained, as shown in Fig. 1.

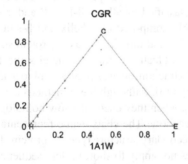

Fig. 1. CGR of 1A1W predicted secondary structure protein1A1W.

We decompose the CGR graph into two time series, that is, each point in the CGR graph is determined by the coordinates of x and y. The two time series shown in Fig. 2 correspond to the points in Fig. 1, which are CGRX and CGRY, respectively.

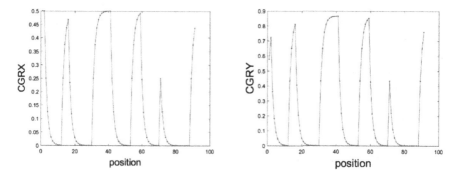

Fig. 2. Two time series related to Fig. 1.

2.3   Multiscale Coarse-Grained Time Series

In this paper, we adopt the method of multiscale coarse-grained time series to further process the converted time series [15]. This method is an order pattern in a time scale range. It considers the spatial proximity of the phase space of time series. Given the time series $\{x_1, x_2, \ldots, x_n\}$, a new time series is obtained according to Eq. (1).

$$z_j^s = \frac{1}{s} \sum_{i=j}^{j+s-1} x_i, 1 \leq j \leq j - s + 1 \tag{1}$$

Where s is time scale, and z_j^s is new time series. The length of the new time series is N − s + 1.

The multiscale coarse-grained time series extend the time scale range of measurement and can capture more abundant information than before. Different time scales have different effects on protein structure prediction. The effects of different time scales will be mentioned later.

2.4   Recurrence Quantification Analysis

Recurrence plot (RP) [10] uses phase space reconstruction technology to map one dimensional time series to high dimensional space. It can represent the dynamic behavior of the original time series. For one time series $\{x_1, x_2, \ldots, x_n\}$ with length N, its time series point in the phase space is $X_i = (x_i, x_{i+\tau}, \ldots, x_{i+(m-1)\tau})$ for i = 1, 2 ... N_m and $N_m = N - (m-1)\tau$, where m and τ denote the embedding dimension and time delay, respectively [14]. Calculates the Euclidean distance between all vector pairs in the reconstructed phase space. Select the appropriate threshold ε so that the recurrence matrix is defined as follows.

$$R = \theta(\varepsilon - \|\overrightarrow{x_i} - \overrightarrow{y_j}\|), i,j = 1,\ldots, N_m \tag{2}$$

where $\theta(.)$ is the Heaviside function and $\|.\|$ is a norm. If $R = 1$, the recurrence point exists, expressed in black dots; if $R = 0$, the recurrence point does not exist, without the black dot representation.

RQA is a nonlinear index quantization method developed on the basis of RP, which is used to analyze the dynamic characteristics of recurrence points. Here, we mainly apply the following 8 recurrence variables [13]:

The first recurrence variable is the recurrence rate (RR), which describes the ratio of recurrence points in the RP to the total number:

$$RR = \frac{1}{N^2} \sum_{i,j=1}^{N} R_{i,j} \tag{3}$$

Where $R_{i,j}$ is the recurrence points, N is the length of time series.

The second recurrence variable is determinism (DET), which describes the proportion of diagonal lines of different lengths:

$$DET = \sum_{l=l_{min}}^{N} lP(l) / \sum_{i,j=1}^{N} R_{i,j} \tag{4}$$

Where l is the diagonal length. $P(l)$ is the probability of a diagonal structure of length l.

The third recurrence variable is the longest diagonal

$$L_{max} = \max(\{l_i\}_{i=1}^{N_l}) \tag{5}$$

The fourth recurrence variable is entropy (ENTR), which describes the Shannon entropy of the diagonal length distribution. The larger the Shannon entropy value is, the stronger the complexity of the time series is. The minimum value of l_{min} is usually 2.

$$ENTR = - \sum_{l=l_{min}}^{N} P(l) \ln P(l) \tag{6}$$

The fifth recurrence variable is the average diagonal length

$$L_{mean} = \sum_{l=l_{min}}^{N} lP(l) / \sum_{l=l_{min}}^{N} P(l) \tag{7}$$

The sixth recurrence variable is the laminarity (LAM), which refers to the proportion of line segments in the form of vertical lines to recursive points

$$LAM = \sum_{v=v_{min}}^{N} vP(v) / \sum_{i,j=1}^{N} R_{i,j} \tag{8}$$

Where the v is the length of vertical lines, and the $P(v)$ is the is probability of a vertical lines of length v.

The seventh variable, trapping time (TT), is the average length of vertical lines. The eighth recurrence variable is maximal length of the vertical lines in RP (V_{max}), which is similar to L_{max}.

2.5   Horizontal Visibility Graph

In recent years, complex networks have been introduced to analyze biological problems, and the method of graph is used to analyze time series [18]. In this paper, we use the horizontal visibility graph to study the protein sequence. The construction method of horizontal visibility graph is shown in Fig. 3. The construction of horizontal visibility graph mainly lies in the establishment of nodes and the connection of edges. Suppose the time series $\{x_1, x_2, \ldots, x_n\}$ with a length of N, and each sample point as a node n_i of graph G. If the node n_i and the node n_j satisfy the Eq. (9), there is an edge connection between the two nodes.

$$n_k < \min(n_i, n_j), \forall k, i < k < j \tag{9}$$

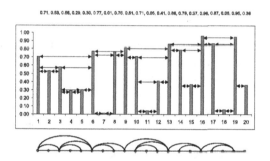

Fig. 3. Sample points of 20 data and its associated horizontal visibility graph

Based on the graph, we extract the following features to represent the protein sequence:

The first features is degree k. The degree of any node i is shown in Eq. (10). This paper uses the maximum value (k_{max}) of degree as our feature.

$$k_i = \sum_j a_{ij} \tag{10}$$

The second and third features are the average shortest path (L) and diameter (D) between the nodes, respectively. The definition is as follows. The diameter defines as the maximum value of all the shortest path in a network. It is a measure of the compactness in a network.

$$L = \frac{1}{N(N-1)} \sum_{i \neq j} d_{ij} \tag{11}$$

The clustering coefficient (C) gives the probability that two neighbors of any node are also neighbors.

$$C = \frac{3N_\Delta}{N_3} \tag{12}$$

The energy (E) and the Laplacian Energy (LE) [11] is defined as follows. Where μ_i denotes the eigenvalue of the adjacency matrix.

$$E = \sum_{i=1}^{n} |\mu_i| \tag{13}$$

$$LE = \sum_{i=1}^{n} \left| \mu_i - \frac{2m}{n} \right| \tag{14}$$

From each time series, 6 features were extracted. They are the maximum value (k_{max}), the average shortest path length (L), the network diameter (D), the clustering coefficient (C), the energy (E), and the Laplace energy (LE).

2.6   Prediction of Protein Structure Class

Each secondary structure sequence of protein is mapped to two time series CGRX and CGRY by CGR, and the average values of $\bar{x}$ and $\bar{y}$ are obtained. The multiscale coarse-grained time series are obtained by coarse-graining. The next step is to construct the recurrence matrix and horizontal graph respectively. The total number of features extracted for each multiscale coarse-grained time series is 14, among which 8 feature vectors come from recurrence quantization analysis and the others come from horizontal graph. Plus the average of $\bar{x}$ and $\bar{y}$, there's $14 \times 2 + 2$ features.

In this paper, we adopt SVM as classifier. SVM is a supervised machine learning algorithm based on statistical learning theory proposed by Vapnik et al. [5]. In our study, we choose Gaussian kernel function. We select the initial parameter range is $c = 2^{-10} \sim 2^{10}, g = 2^{-10} \sim 2^{10}$. Then we use genetic algorithm to optimize the initial parameters, and select the optimal SVM parameters c and g. As protein structural class prediction is a multi-class classification problem, SVM usually adopts One-Versus-One (OVO) or One-Versus-Rest (OVR) strategy to solve multi-class problems. This paper used LIBSVM software developed by Chang and Lin [6].

3   Experimental Results

3.1   Dataset

In this paper, in order to compare with the existing methods, two commonly used low homologous proteins were selected as the benchmark datasets, namely 25PDB and 1189 datasets. The 25PDB dataset consists of 1,673 proteins with a similarity of approximately 25%. This dataset has 443 class all-α proteins, 443 class all-β proteins, 346 class α/β proteins, and 441 class $\alpha + \beta$ proteins. The 1189 dataset consists of 1092 proteins with similarity of less than 40%, with 223 class all-α proteins, 294 class all-β proteins, 334 class α/β proteins, and 241 class α + β proteins [4].

3.2   Evaluation

Among many other methods, the jackknife test is an effective evaluation index. Therefore, the jackknife test is still adopted in this paper. In jackknife test, each protein sequence in the dataset is selected as a test sample, and the predictor is trained by the remaining protein sequences. Then we calculate the overall accuracy (OA) of each data set and select Sensitivity (Sens), Specificity (Spec) and Mathew's Correlation Coefficient (MCC) [7] as our standard performance measures.

$$\text{Sens} = \frac{TP}{TP + FN} \tag{15}$$

$$\text{Spec} = \frac{TN}{FP + TN} \tag{16}$$

$$\text{MCC} = \frac{TP \times TN - FP \times FN}{\sqrt{(TP + FP)(TP + FN)(TN + FP)(TN + FN)}} \tag{17}$$

$$\text{OA} = \frac{TP + TN}{TP + FN + FP + TN} \tag{18}$$

The experimental results are shown in Table 1. In addition to the OA, the results of MCC, Sens and Spec of each structural class are also shown in the table.

Table 1. The predicted quality results of our method on two datasets

Dataset	Class	Sens (%)	Spec (%)	MCC
25PDB	All-α	99.03	98.08	0.9458
	All-β	1	97.62	0.9323
	α/β	81.56	1	0.9835
	α + β	95.25	98.09	0.9456
	OA	**95.33**		
1189	All-α	99.46	95.79	88.75
	All-β	80.73	99.59	85.15
	α/β	1	97.92	96.55
	α + β	97.36	97.79	93.24
	OA	**93.0**		

This paper has multiscale coarse-grained processing of time series, so different time scales have different prediction results. Through experimental comparison, for the 25PDB dataset, we choose the time scale s = 9 to have the best prediction effect, and the overall accuracy rate is 95.33%. For the 1189 dataset, we choose the best time scale s = 6, the overall accuracy rate is 93%. Therefore, the further processing of the time series greatly improves the prediction of protein structure.

3.3   Comparison with Existing Methods

In this part, we compare the methods used in other literatures, and the results are shown in Table 2. According to Table 2, the proposed method in this paper is feasible for datasets 25PDB and 1189, with the overall prediction accuracy reaching 95.33% and 93%, respectively. Compared with RQA [9] and horizontal visibility graph, respectively [11], the results show that the combination of RQA and weighted horizontal visibility graph is better.

Table 2. Performance comparison of different methods

Dataset	Method	Accuracy (%)	Dataset	Method	Accuracy (%)
25PDB	[9]	82.9	1189	[9]	81.3
	[17]	86.6		[17]	85.0
	[16]	85.0		[16]	85.2
	[12]	90.08		[12]	86.02
	[11]	82.85		[11]	79.4
	This paper	**95.33**		This paper	**93.0**

4   Conclusion

This paper predicts the tertiary structure of proteins based on recurrence quantification analysis of multiscale coarse-grained time series and complex networks. In this paper, multiscale coarse granulation is performed on the time series mapped by CGR, which makes the new time series content more abundant. Combines RQA with complex networks. Each resulting eigenvector representing a protein is input into the SVM algorithm to predict the protein structure class. The experimental results show that the proposed method has higher overall prediction accuracy and MCC for the 25PDB and 1189 datasets.

Acknowledgments. This work was supported by the National Natural Science Foundation of China (Grant No. 61671220, 61701192, 61640218), the Natural Science Foundation of Shandong Province, China (Grant No. ZR2017QF004).

References

1. Levitt, M., Chothia, C.: Structural patterns in globular proteins. Nature **261**, 552–558 (1976)
2. Chou, K.C.: Prediction of protein cellular attributes using pseudo-amino acid composition. Proteins: Struct. Funct. Bioinform. **43**, 246–255 (2001)
3. Jin, L., Fang, W., Tang, H.: Prediction of proteins structural classes by a new measure of information discrepancy. Comput. Biol. Chem. **27**, 373–380 (2003)

4. Kurgan, L.A., Homaeian, L.: Prediction of structural classes for protein sequences and domains-impact of prediction algorithms, sequence representation and homology, and test procedures on accuracy. Pattern Recogn. **39**, 2323–2343 (2006)

5. Vapnik, V.: The Nature of Statistical Learning Theory. Springer, New York (1995)

6. Chang, C.C., Lin, C.J.: LIBSVM: a library for support vector machines. ACM Trans. Intell. Syst. Technol. **2**, 27 (2011)

7. Liu, T., Geng, X., Zheng, X., Li, R., Wang, J.: Accurate prediction of protein structural class using auto covariance transformation of PSI-BLAST profiles. Amino Acids **42**(6), 2243–2249 (2012)

8. Jones, D.T.: Protein secondary structure prediction based on position-specific scoring matrices. J. Mol. Biol. **292**, 195–202 (1999)

9. Yang, J.Y., Peng, Z.L., Chen, X.: Prediction of protein structural classes for low-homology sequences based on predicted secondary structure. BMC Bioinform. **11** (2010)

10. Eckmann, J.P., Kamphorst, S.O., Ruelle, D.: Recurrence plots of dynamical systems. Europhys. Lett. **4**, 973–977 (1987)

11. Zhao, Z.Q., Luo, L., Liu, X.Y.: Low-Homology protein structural class prediction from secondary structure based on visibility and horizontal visibility network. Am. J. Biochem. Biotechnol. **14**(1), 67–75 (2018)

12. Olyaee, M.H., Yaghoubi, A., Yaghoobi, M.: Predicting protein structural classes based on complex networks and recurrence analysis. J. Theor. Biol. **404**, 375–382 (2016)

13. Yang, J.Y., Peng, Z.L., Yu, Z.G., Zhang, R.J., Anh, V., Wang, D.: Prediction of protein structural classes by recurrence quantification analysis based on chaos game representation. J. Theor. Biol. **257**, 618–626 (2009)

14. Zbilut, J.P., Webber, C.L.J.: Embeddings and delays as derived from quantification of recurrence plots. Phys. Lett. A **171**, 199–203 (1992)

15. Xu, M., Shang, P., Lin, A.: Multiscale recurrence quantification analysis of order recurrence plots. Physica A **469**, 381–389 (2017)

16. Kong, L., Zhang, L., Lv, J.: Accurate prediction of protein structural classes by incorporating predicted secondary structure information into the general form of Chou's pseudo amino acid composition. J. Theor. Biol. **344**, 12–18 (2014)

17. Ding, H., Deng, E.Z., Yuan, L.F., Liu, L., Lin, H., Chen, W., et al.: iCTX-type: a sequence-based predictor for identifying the types of conotoxins in targeting ion channels. Biomed. Res. Int. (2014)

18. Lacasa, L., Luque, B., Ballesteros, F., Luque, J., Carlos-Nuno, J.: From time series to complex networks: the visibility graph. Proc. Natl. Acad. Sci. U.S.A. **105**, 4972–4975 (2008)

Classification of Schizophrenia Patients and Healthy Controls Using ICA of Complex-Valued fMRI Data and Convolutional Neural Networks

Yue Qiu[1], Qiu-Hua Lin[1(✉)], Li-Dan Kuang[2], Wen-Da Zhao[1],
Xiao-Feng Gong[1], Fengyu Cong[3,4], and Vince D. Calhoun[5,6]

[1] School of Information and Communication Engineering,
Dalian University of Technology, Dalian 116024, China
qhlin@dlut.edu.cn
[2] School of Computer and Communication Engineering,
Changsha University of Science and Technology, Changsha 410114, China
[3] School of Biomedical Engineering, Dalian University of Technology,
Dalian 116024, China
[4] Department of Mathematical Information Technology,
University of Jyvaskyla, Jyväskylä, Finland
[5] The Mind Research Network, Albuquerque, NM 87106, USA
[6] Department of Electrical and Computer Engineering,
University of New Mexico, Albuquerque, NM 87131, USA

Abstract. Deep learning has contributed greatly to functional magnetic resonance imaging (fMRI) analysis, however, spatial maps derived from fMRI data by independent component analysis (ICA), as promising biomarkers, have rarely been directly used to perform individualized diagnosis. As such, this study proposes a novel framework combining ICA and convolutional neural network (CNN) for classifying schizophrenia patients (SZs) and healthy controls (HCs). ICA is first used to obtain components of interest which have been previously implicated in schizophrenia. Functionally informative slices of these components are then selected and labelled. CNN is finally employed to learn hierarchical diagnostic features from the slices and classify SZs and HCs. We use complex-valued fMRI data instead of magnitude fMRI data, in order to obtain more contiguous spatial activations. Spatial maps estimated by ICA with multiple model orders are employed for data argumentation to enhance the training process. Evaluations are performed using 82 resting-state complex-valued fMRI datasets including 42 SZs and 40 HCs. The proposed method shows an average accuracy of 72.65% in the default mode network and 78.34% in the auditory cortex for slice-level classification. When performing subject-level classification based on majority voting, the result shows 91.32% and 98.75% average accuracy, highlighting the potential of the proposed method for diagnosis of schizophrenia and other neurological diseases.

Keywords: Deep learning · fMRI · ICA · Schizophrenia · Model order · Data argumentation

© Springer Nature Switzerland AG 2019
H. Lu et al. (Eds.): ISNN 2019, LNCS 11555, pp. 540–547, 2019.
https://doi.org/10.1007/978-3-030-22808-8_53

1 Introduction

Deep neural networks recently have exhibited great superiority in the investigation of neurological disorders [1–4]. Resting-state functional magnetic resonance imaging (rs-fMRI) is able to provide non-invasive measures for brain function and abnormalities [5]. Several deep learning methods have been applied to rs-fMRI data analysis, including deep auto-encoders (DAE) [6], deep belief networks (DBN) [7] and convolutional neural networks (CNN) [8, 9]. Recently, CNN has exhibited promising applications by virtue of its ability to leverage information from adjacent voxels [8, 9].

There are some problems in existing CNN frameworks for fMRI-based disease diagnosis. For example, Sarraf and Tofighi successfully classified Alzheimer's subjects from normal controls using observed fMRI data, and achieved an accuracy of 96.85% by the LeNet-5 architecture [8]. However, the observed fMRI data contains not only components related to brain activity, but also physiological noise which may contaminate the results. Kam et al. took advantage of brain functional networks to build a deep learning framework for accurate diagnosis of early mild cognitive impairment (eMCI) [9]. This framework combined group-information-guided ICA (GIG-ICA) with 3D CNN. The use of brain functional networks reduced the training cost of the network, nevertheless, the limited fMRI data may cause overfitting since the 3D CNN model contains a large amount of parameters. Moreover, to the best of our knowledge, a CNN framework for classification of schizophrenia patients (SZs) and healthy controls (HCs) using fMRI data has not been reported.

In this study, we propose a new framework combining ICA and CNN to classify SZs from HCs. Complex-valued fMRI data is employed to obtain more contiguous spatial maps using ICA than magnitude fMRI data [10]. We select and label functionally informative slices from components of interest which have been previously implicated in SZs. These slices are fed into 2D CNN for feature learning and classification, which mitigates the interference of physiological noise involved in observed fMRI data. To enhance the training, we leverage the ICA results for multiple model orders to perform data argumentation.

The rest of this paper is organized as follows. Section 2 describes materials and preprocessing. Section 3 presents the proposed method, including ICA, component selection, and CNN feature learning and classification. Section 4 consists of experiments and results. Section 5 is the conclusion.

2 Materials and Preprocessing

The fMRI datasets used in this study have been described in [11]. Resting-state complex-valued fMRI data were collected from 82 subjects, including 40 HCs and 42 SZs. Data preprocessing was performed using the SPM software package[1]. Following motion correction and spatial normalization, the data was slightly sub-sampled to $3 \times 3 \times 3$ mm^3, resulting in $53 \times 63 \times 46$ voxels. Both magnitude and phase images

---

[1] available at http://www.fil.ion.ucl.ac.uk/spm.

were spatially smoothed with an $8 \times 8 \times 8$ mm^3 full width half-maximum Gaussian kernel. A total of 146 scans per subject were entered into the ICA analysis.

3 Proposed Method

Figure 1 shows the framework of the proposed method for classification of SZs and HCs. We first perform ICA, component selection, phase de-ambiguity and denoising to obtain spatial maps of components of interest. Informative slices of these selected components are then fed into CNN to learn features and perform classification.

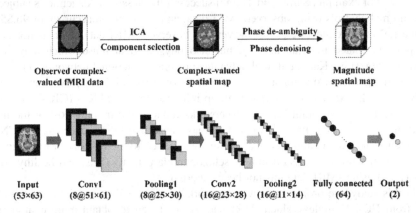

Fig. 1. The proposed framework for schizophrenia classification.

3.1 ICA and Component Selection

Given the observed complex-valued fMRI data, we adopt multiple runs of entropy bound minimization (EBM) [12], an efficient and flexible complex-valued ICA algorithm, for component separation. Due to the relatively small number of subjects, ICA with multiple models orders is performed for data argumentation. Components of interest are then identified for each individual based on their spatial references. Spatial maps of these complex-valued components are denoised using phase de-ambiguity and phase denoising proposed in [10]. The best spatial map estimates for each component across all runs are selected based on combining cross-run averaging and a one-sample t-test [11]. Then the magnitude of complex-valued spatial map estimates is used for sequential feature learning and classification.

3.2 CNN Feature Learning and Classification

For each component, the magnitude of 3D spatial maps is concatenated along the z axes and converted to a stack of 2D slices. We remove slices with no functional information and normalize the range of the remaining slices to [0, 1], respectively. By adding a label of HC or SZ for binary classification, we generate a sample $\{x_i, y_i\}$ for the ith slice, where x_i is a normalized slice, y_i is its corresponding label (0 for HC and 1 for SZ).

The training (60%), validation (20%) and testing (20%) samples are randomly selected by shuffling subjects. In other words, none of training, validation and testing samples is selected from the same subject. We perform five-fold cross validation, in order to enhance the robustness and reproducibility of our proposed method.

We propose to use 2D CNN to learn high-level features from each slice. The network structure used in this paper consists of two convolutional layers with a kernel size of 3×3, two max-pooling layers with a kernel size of 2×2, one fully connected layer with 64 nodes, and one output layer with two nodes, as shown in Fig. 1. The non-linear activation function used in the fully connected layer is rectified linear unit (ReLU). The two nodes in the output layer represent labels for the HC or SZ groups, and binary classification is determined by the softmax function. We select the cross-entropy as the loss function. An L_2-norm regularization is used in each convolutional layer and fully-connected layer to reduce overfitting; the parameter λ for regularization is set to be 0.1. Thus, the loss can be calculated as:

$$L(\mathbf{W}, \mathbf{b}) = \frac{1}{M} \left(\sum_{i=1}^{M} (-y_i \ln H_{\mathbf{w},\mathbf{b}}(x_i) - (1 - y_i) \ln(1 - H_{\mathbf{w},\mathbf{b}}(x_i))) + \frac{\lambda}{2} \|\mathbf{W}\|^2 \right), \quad (1)$$

where $\mathbf{W}$ and $\mathbf{b}$ are parameters for the network; M is the number of samples; $H_{\mathbf{w},\mathbf{b}}(\cdot)$ is the function learned by the network, and $H_{\mathbf{w},\mathbf{b}}(x_i)$ devotes the output of the network given the input x_i. We also add a batch normalization layer followed each convolutional layer and fully-connected layer.

The model is adjusted for 50 epochs with batch size of 64 and optimized by the Adam algorithm [13]. Diagnostic performance is computed by the accuracy, sensitivity, and specificity:

$$ACC = \frac{TP + TN}{TP + TN + FP + FN}, \quad (2)$$

$$SEN = \frac{TP}{TP + FN}, \quad (3)$$

$$SPEC = \frac{TN}{TN + FP}, \quad (4)$$

where TP, TN, FP and FN denote true positive, true negative, false positive and false negative, respectively. We finally select the model showing the best accuracy given in Eq. (2) on the validation set. During testing, the binary output of model is the result of slice-level classification of SZ and HC. For subject-level classification, the final decision is based on majority voting of slices. Specifically, for each subject, the numbers of slices classified as HC and SZ are compared, and the class (HC or SZ) with more slices is regarded as the result for the subject.

4   Experiments and Results

4.1   Slice-Level Classification vs. Subject-Level Classification

We repeated ICA with 13 different model orders (from $N = 20$ to $N = 140$, with an interval of 10) 10 times. Two components of interest, the default mode network (DMN) and auditory cortex (AUD) were selected via spatial references [14, 15], since these two components are high-level cognitive function-related networks that have been previously implicated in schizophrenia. Figure 2 shows the spatial references for two selected components. After phase de-ambiguity, phase denoising, and slice removal, 25 slices of spatial maps were obtained for DMN and AUD at the best run, respectively, as shown in Fig. 3. In summary, a total of $82 \times 13 \times 25 = 26650$ samples with size of 53×63 were generated, including 13000 HCs' and 13650 SZs' samples.

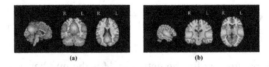

Fig. 2. Spatial references of two selected components for classifying SZs and HCs: (a) default model network (DMN), (b) auditory cortex (AUD).

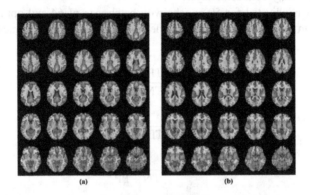

Fig. 3. Slices from the (a) default model network (DMN) and (b) auditory cortex (AUD) for classification of schizophrenia.

Table 1 shows the comparison of mean accuracy, sensitivity, and specificity defined in Eqs. (2)–(4) of five-fold cross validation for the proposed method and relevant methods proposed in [9]. For slice-level classification, the proposed method shows an average accuracy of 72.65% in the DMN and 78.34% in the AUD, both are higher than the best accuracy (71.13%) of 3D CNN obtained from a single network, i.e. SB-CNN in [9]. The proposed method also demonstrates better specificity in the DMN (80.75% vs. 68.75%) and higher sensitivity and specificity in the AUD (SEN: 79.11% vs. 73.47%; SPEC: 77.25% vs. 68.75%). Furthermore, the results of AUD can even compete with the merged networks based on multiple networks, i.e. MB-CNN in [9] (ACC: 78.34% vs. 74.23%; SEN: 79.11% vs. 75.51%; SPEC: 77.25% vs. 72.92%). We analyzed the samples that were classified wrongly and found they were mostly from slices located at the edge of the network. In other words, errors in these samples may be due to less feature information contained in the less spatial activations, which is natural and reasonable. For subject-level classification, accuracy, sensitivity, and specificity were further improved, as shown in Table 1. Compared to DMN, AUD reaches much better performance, i.e., an accuracy of 98.75%, sensitivity of 100%, and specificity of 97.50%.

Table 1. Comparison of mean accuracy, sensitivity, and specificity of five-fold cross validation for the proposed method and relevant methods proposed in [9] (FPN1 and FPN2 are two frontoparietal networks, AN1 and AN2 are two attention networks, and ECN is an executive control network).

Method	Networks	ACC (%)	SEN (%)	SPEC (%)
SB-CNN in [9]	DMN	70.10	71.43	**68.75**
	FPN1	**71.13**	73.47	**68.75**
	FPN2	70.10	**73.47**	66.67
	AN1	68.04	71.43	64.58
	AN2	67.01	69.39	64.58
	ECN	67.01	67.35	66.67
MB-CNN in [9]	All of above	**74.23**	**75.51**	**72.92**
Proposed method (slice-level)	DMN	**72.65**	64.51	**80.75**
	AUD	**78.34**	**79.11**	**77.25**
Proposed method (subject-level)	DMN	**91.32**	**85.00**	**97.50**
	AUD	**98.75**	**100.00**	**97.50**

4.2 Effects of Multiple Model Orders for Data Argumentation

As mentioned above, we performed 13 different model orders (from $N = 20$ to $N = 140$, with an interval of 10). To analyze effects of multiple model orders for data argumentation, we changed the interval of model orders, thus changed the number of samples. Specifically, we had three cases: (1) $l = 1$ ($N = 120$), the number of samples was $82 \times 1 \times 25 = 2050$; (2) $l = 7$ (from $N = 20$ to $N = 140$, with an interval of 20), the number of samples was $82 \times 7 \times 25 = 14350$; (3) $l = 13$ (from $N = 20$ to $N = 140$, with an interval of 10), the number of samples was $82 \times 13 \times 25 = 26650$. Mean results of five-fold cross validation are shown in Fig. 4. It is observed that the accuracy, sensitivity, and specificity of the proposed method increased with an increase in l, both for slice-subject classification and subject-level classification. It is reasonable to predict that the performance of our proposed method could be better as the amount of data increase. On the other hand, when using the largest amount of data, i.e., $l = 13$, the standard deviations of accuracy, sensitivity, and specificity were the smallest, indicating the model was more stable when trained by more model orders.

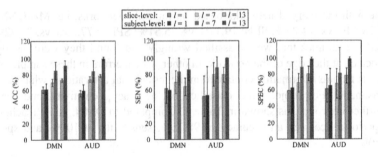

Fig. 4. Results of multiple model orders for data argumentation.

5 Conclusion

In this paper, we proposed a novel ICA-CNN framework for classifying SZs and HCs. To our best knowledge, this is the first spatial network-based 2D CNN framework for disease classification. Compared to the 3D CNN, 2D CNN improved the accuracy of classification while reducing training parameters. In contrast to the most existing CNN methods using the observed fMRI data, our method using spatial maps with sparse but informative activations also significantly reduced training loads. In addition, we confirmed the effectiveness of multiple model orders for data augmentation, which enables us to solve the difficulty of obtaining large quantities of fMRI data. In the future, we will enhance our framework for merging multiple components of interest and apply it to fMRI phase data.

Acknowledgements. This work was supported by National Natural Science Foundation of China under Grants 61871067, 61379012, 61671106, 61331019 and 81471742, NSF grants 1539067, 0840895 and 0715022, NIH grants R01MH104680, R01MH107354, R01EB005846 and 5P20GM103472, the Fundamental Research Funds for the Central Universities (China, DUT14RC(3)037), China Scholarship Council, and the Supercomputing Center of Dalian University of Technology.

References

1. Plis, S.M., et al.: Deep learning for neuroimaging: a validation study. Front. Neurosci. 8(219), 1–11 (2014)
2. Vieira, S., Pinaya, W.H.L., Mechelli, A.: Using deep learning to investigate the neuroimaging correlates of psychiatric and neurological disorders: methods and applications. Neurosci. Biobehav. Rev. 74, 58–75 (2017)
3. Madsen, K.H., Krohne, L.G., Cai, X.L., Wang, Y., Chan, R.C.K.: Perspectives on machine learning for classification of Schizotypy using fMRI data. Schizophr. Bull. 44(2), 480–490 (2018)
4. Kim, J., Calhoun, V.D., Shim, E., Lee, J.H.: Deep neural network with weight sparsity control and pre-training extracts hierarchical features and enhances classification performance: evidence from whole-brain resting-state functional connectivity patterns of schizophrenia. Neuroimage 124, 127–146 (2016)

5. Vemuri, P., Jones, D.T., Jack, C.R.: Resting state functional MRI in Alzheimer's disease. Alzheimer's Res. Ther. **4**(2), 1–9 (2012)

6. Suk, H.I., Wee, C.Y., Lee, S.W., Shen, D.G.: State-space model with deep learning for functional dynamics estimation in resting-state fMRI. Neuroimage **129**, 292–307 (2016)

7. Aghdam, M.A., Sharifi, A., Pedram, M.M.: Combination of rs-fMRI and sMRI data to discriminate autism spectrum disorders in young children using deep belief network. J. Digit. Imaging **31**(6), 895–903 (2018)

8. Sarraf, S., Tofighi, G.: Deep learning-based pipeline to recognize Alzheimer's disease using fMRI data. In: Future Technologies Conference, pp. 816–820. IEEE Press, San Francisco (2016)

9. Kam, T.-E., Zhang, H., Shen, D.: A novel deep learning framework on brain functional networks for early MCI diagnosis. In: Frangi, A.F., Schnabel, J.A., Davatzikos, C., Alberola-López, C., Fichtinger, G. (eds.) MICCAI 2018. LNCS, vol. 11072, pp. 293–301. Springer, Cham (2018). https://doi.org/10.1007/978-3-030-00931-1_34

10. Yu, M.C., Lin, Q.H., Kuang, L.D., Gong, X.F., Cong, F., Calhoun, V.D.: ICA of full complex-valued fMRI data using phase information of spatial maps. J. Neurosci. Methods **249**, 75–91 (2015)

11. Kuang, L.D., Lin, Q.H., Gong, X.F., Cong, F., Sui, J., Calhoun, V.D.: Model order effects on ICA of resting-state complex-valued fMRI data: application to schizophrenia. J. Neurosci. Methods **304**, 24–38 (2018)

12. Li, X.L., Adalı, T.: Complex independent component analysis by entropy bound minimization. IEEE Trans. Circ. Syst. I Regul. Pap. **57**(7), 1417–1430 (2010)

13. Kingma, D., Ba, J.: Adam: a method for stochastic optimization. arXiv:1412.6980 (2014)

14. Smith, S.M., et al.: Correspondence of the brain's functional architecture during activation and rest. Proc. Natl. Acad. Sci. U.S.A. **106**(31), 13040–13045 (2009)

15. Allen, E.A., et al.: A baseline for the multivariate comparison of resting-state networks. Front. Syst. Neurosci. **5**(2), 1–23 (2011)

A Novel Prediction Method of ATP Binding Residues from Protein Primary Sequence

Chuyi Song[1], Guixia Liu[2,3], Jiazhi Song[2,3], and Jingqing Jiang[4(✉)]

[1] College of Mathematics, Inner Mongolia University for Nationalities,
Tongliao 028000, China
songchuyi@sina.com
[2] College of Computer Science and Technology, Jilin University,
Changchun 043000, China
lgxl034@163.com, songjz671@nenu.edu.cn
[3] Key Laboratory of Symbolic Computational and Knowledge Engineering,
Changchun 043000, China
[4] College of Computer Science and Technology,
Inner Mongolia University for Nationalities, Tongliao 028000, China
jiangjingqing@aliyun.com

Abstract. ATP is an important nucleotide that provides energy for biological activities in cells. Correctly identifying the protein-ATP binding site is helpful for protein function annotations and new drug development. With the innovation of machine learning, more and more researchers start to predict the binding sites from protein sequences instead of using biochemical experiment methods. Since the number of non-binding residues is far from the number of binding residues, a popular method to deal with the ATP-binding dataset is to apply the under-sampling to construct training subset which will inevitably lose the negative samples. However, a lot of valuable information for ATP binding properties is hidden in negative samples which should be carefully considered. In this study, the dataset which contains full negative samples are applied in training process. In order to avoid biased in prediction result, the decision tree classification algorithm which shows stable performance in imbalanced data is applied. The prediction performance on five-fold cross validation has demonstrated that our proposed method improves the performance compared with using under-sampled data.

Keywords: ATP-binding site · Protein primary sequence · Decision tree · Binary classification

1 Introduction

Adenosine-5'-triphosphate (ATP) as an important molecule has been found to play a significant role in many biological processes such as energy metabolism, membrane transportation, signaling and replication and transcription of DNA [1]. However, ATP needs to interact with proteins to accomplish its functions. By the time of January, 2019, 6520 structures in the Protein Data Bank have been detected binding with ATP, which occupy about 4.4% in known protein structures [2]. Besides, the ATP binding

© Springer Nature Switzerland AG 2019
H. Lu et al. (Eds.): ISNN 2019, LNCS 11555, pp. 548–555, 2019.
https://doi.org/10.1007/978-3-030-22808-8_54

sites have significant potential to be developed as drug targets for antibacterial or anti-cancer chemotherapy [3]. Therefore, identifying the binding sites of ATP in protein sequences and structures is of great value.

Traditional methods for binding sites detection utilize the biochemical experiments such as X-ray crystallography and nuclear magnetic resonance (NMR). However, these methods come with huge time-consuming and money-consumption which make it difficult for wide application. With the development of machine learning and artificial intelligence, an increasing number of researches start using computational methods to predict the binding sites in proteins and make assists for experimental approaches. In the field of ATP-binding site prediction, Raghava et al. created ATPint [4] to predict the ATP-binding sites in protein sequences based on Position Specific Scoring Matrix (PSSM) and support vector machines (SVM) which makes the pioneering contribution. The results in their dataset have shown that ATPint is capable of accurately predicting the ATP interacting residues based on protein sequential information. Later on, Kurgan et al. developed two prediction methods named ATPsite [5] and NsitePred [6] which applied the SVM and PSSM with the combination of several predicted structural information. However, the number of ATP interacting residues are much less than the non-interacting residues in protein sequences. Therefore, in the prediction methods mentioned above, they made the under-sampling on their datasets which means they randomly keep the same number of non-interacting residues as the binding residues and combine them with the binding residues to construct new training subsets. This process can provide relatively balanced data for SVM and reduce the time consumption during training process. However, a lot of valuable information of ATP binding properties is contained in the non-interacting residues, losing them may cause the training process incomplete and the classifier is not sensitive enough for non-interacting residues.

Based on this background, in our study, a prediction method is developed based on the full training information. The decision tree is selected as classification algorithm which is less sensitive for imbalanced data than SVM. The five-fold cross validation results have shown that the performance of full training data is superior to the performance with under-sampling technique.

2 Dataset

In the present study, the ATP168 which was constructed by Raghava et al. is adopted. The dataset was extracted from SuperSite encyclopedia [7] and then reduced the sequence identity below 40% using the program CD-HIT [8]. After removing the non-ATP-interacting proteins by the software Ligand Protein Contact (LPC) [9], 168 non-redundant ATP protein chains are collected which are available at [10].

In order to show the effectiveness of our proposed method, the five-fold cross validation is utilized where the protein sequences are randomly divided into five parts. Four parts are used for training the model and the remaining one part is used to examine the prediction performance. This process is repeated for five times which makes sure that each protein sequence has been predicted. The final performance is calculated by averaging the performance of all five parts.

3 Feature Extraction

The proposed method utilizes both sequential and predicted structural information as inputs. The sliding window technique is also applied to take consideration of the adjacent residues. The window will be labelled as ATP binding if the central residue is the ATP binding site. In this study, the window size is set to 15 by experimental attempts. The features involved include:

- Position specific scoring matrix

The PSSM profile is generated by running PSI-BLAST [11] against Swiss-Prot [12] database with default parameters. For each residue in the sequence, the PSSM profile provides 20 values to reflect the evolutionary conservation corresponding to 20 types common amino acids. We normalize these values to the range between 0 and 1 by formula (1) where $S_{i,r}$ represents the original value of the ith residue at rth column in PSSM, $S_{i,r}^{(N)}$ is the normalized value.

$$S_{i,r}^{(N)} = \frac{1}{1 + e^{-S_{i,r}}} \tag{1}$$

For the window size of 15, we collect 15 * 20 = 300 dimensional vectors to represent the evolutionary conservation from protein sequences.

- Predicted secondary structure

Previous studies have proved that the secondary structure is relevant to ATP binding sites. Therefore, appropriately adding the secondary structure into the feature vector is valuable to improve the prediction performance. In this study, PSIPRED [13] is applied to predict the protein secondary structure based on protein sequences. The output of PSIPRED including three probabilities which represent the residue belonging to three secondary structure classes: Coil (C), Helix (H) and Strand (E). Thus, for a sliding window of size 15, the dimension of predicted secondary structure is 15 * 3 = 45.

- Amino acid physicochemical properties

Based on the dipoles and volumes of the side chain of amino acids, the 20 types of amino acids are divided into seven groups [14] including class a = {Ala, Gly, Val}, class b = {Ile, Leu, Phe, Pro}, class c = {His, Asn, Gln, Trp}, class d = {Tyr, Met, Thr, Ser}, class e = {Arg, Lys}, class f = {Asp, Glu}, class g = {Cys}. As a result, a seven-dimensional vector using one-hot key encoding technique is applied to reveal the class that the residue belongs to. In the sliding window, the physicochemical properties are 7 * 15 = 105 dimensional features.

After feature extraction, for each residue in protein sequences, 30 dimensional vectors are extracted including 20 for PSSM, 3 for predicted secondary structure and 7 for physicochemical properties.

4 Architecture of Proposed Method

The decision tree is a predictive machine learning model. It predicts the label based on known various feature values from available data and is commonly used for sample classification [15]. A classical decision tree consists of many leaves and branches, where leaves represent classification and branches represent conjunctions of features that lead to the classification. The advantages of decision tree over SVM can be reflected from two aspects: firstly, the decision tree shows better performance than SVM in imbalanced data prediction; secondly, for high-throughput data, the decision tree has faster training and prediction process which efficiently reduces the time consumption of proposed method. In this study, the decision tree is introduced from Scikit-learn package, and the grid-search algorithm is applied to collect the best parameters for the classifier. The architecture of proposed method is shown in Fig. 1.

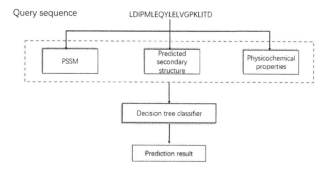

Fig. 1. The architecture of proposed method

5 Results and Discussion

5.1 Evaluation Criterion

In order to access the prediction performance of our proposed method, a series of evaluation indexes are applied in present study including the overall accuracy (ACC), the sensitivity, the specificity and the Matthews Correlation Coefficient (MCC). Among them, the MCC value is a comprehensive evaluation criterion for binary classification and can be regarded as powerful standard to reflect the overall performance. The range of MCC value is between −1 and +1, where −1 represents the prediction results are totally opposite to the true label, 0 means the prediction results are almost random and +1 means the prediction results are identical to the true label. These evaluation indexes can be calculated as follows:

$$ACC = \frac{TP + TN}{TP + TN + FP + FN} \tag{2}$$

$$\text{Sensitivity} = \frac{TP}{TP + FN} \tag{3}$$

$$\text{Specificity} = \frac{TN}{TN + FP} \tag{4}$$

$$MCC = \frac{TP * TN - FP * FN}{\sqrt{(TP + FP)(TP + FN)(TN + FP)(TN + FN)}} \tag{5}$$

Where TP, TN, FP and FN represent the number of true positive sample, true negative sample, false positive sample and false negative sample in prediction results respectively.

5.2   Effectiveness of Applying Full Training Data

Previous works often utilize under-sampling technique to construct training subset which contains same number of positive samples and negative samples for classifiers. However, the negative samples are same valuable and indispensable for identifying ATP binding sites since the negative samples contain key binding properties as the positive samples. Therefore, applying under-sampling has the potential of losing that important information. Taking advantage of decision tree classifier, we can fully use the training data instead of cutting off the majority of negative samples or worry about the time consumption when applying the large scale of training data.

In order to show the improvement of using full training information, we apply the under-sampling technique to our dataset to construct training subset which is similar to previous studies. The new training subset is then sent into the decision tree classifier and five-fold cross validation is utilized to show the prediction performance. The performance is shown in Table 1 and we make comparison with the performance of using full training information.

Table 1. Performance comparison between using under-sampled training data and full training data with five-fold cross validation

Training data involved	ACC	Sensitivity	Specificity	MCC
Under-sampling	0.673	0.677	0.672	0.160
Full training data	0.947	0.240	0.984	0.295

It can be observed that with the increase of training data, two main evaluation indexes ACC and MCC have greatly improved. The ACC climbs from 0.673 with under-sampled data to 0.947 with full training data. The MCC rises from 0.160 of under-sampled data to 0.295 of full training data. The improvement of prediction performance has demonstrated that valuable information of ATP binding is hidden in the negative samples. Our proposed method which takes fully consideration of them is an efficient way to improve the prediction performance of ATP binding sites.

5.3   Comparison with Existing Method

The ATPint [5] is designed for predicting protein-ATP binding sites from protein primary sequence. The ATPint is developed based on PSSM-based features and SVM classifier to accomplish the prediction on the same dataset. To fairly compare the performance of ATPint and our proposed method, the five-fold cross validation is applied on both methods. Table 2 shows the prediction performance of two methods on ATP168 dataset with five-fold cross validation.

Table 2. Performance comparison of ATPint and our proposed method on ATP168 with five-fold cross validation

Method	ACC	Sensitivity	Specificity	MCC
ATPint	0.751	0.744	0.758	0.249
Proposed method	0.947	0.240	0.984	0.295

From Table 2, our proposed method shows better performance than ATPint with 0.046 higher in MCC and 0.196 in overall accuracy. Note that higher MCC value often indicates that the classifier shows better identifying ability for both positive samples and negative samples. Therefore, the improvements have shown that in prediction of protein-ATP binding site, our proposed method which applies full training data with decision tree classifier achieves better performance than ATPint using under-sampled data with SVM classifier. The improvement is mainly from two aspects: firstly, we take fully use of the training data which makes the classifier collects as much information as possible including important properties hidden in negative samples; secondly, the decision tree classifier shows more stable performance under imbalanced data than SVM classifier.

5.4   Case Study

In order to better explain the improvement of prediction performance with full training data, we select ID of 1G3I_A from Protein Data Bank (PDB) to show the improvement of prediction performance in 3D structure. The protein 1G3I_A shares the sequence identity less than 40% with proteins in ATP168 dataset which can be regarded as independent data. The 3D structure of protein is generated by Pymol [16] software. Figure 2(a) and (b) display the prediction performance of classifiers on 1G3I_A based on under-sampled data and full training data respectively.

In Fig. 2, the green color, blue color and cyan color represent the true positive samples (TP), false negative samples (FN) and false positive samples (FP) respectively in prediction result. Others are true negative samples (TN). It can be obviously observed that the number of false positive (FP) samples is greatly reduced in the prediction result of classifier based on full training data. The reduction of FP samples in actual application can greatly improve the accuracy for positive samples and reduce the time-consuming for further experimental identification which is a main shortcoming for previous works. For negative samples in the dataset, we can find that the number of FN

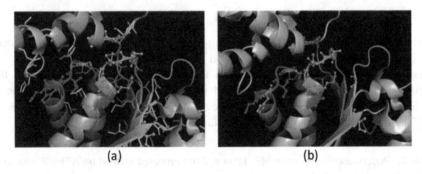

Fig. 2. Prediction results on 1G3I_A of classifiers based on under-sampled data and full training data (Color figure online)

samples doesn't greatly increase which means applying full training data will not lead to biased prediction result for negative samples. The improvements of prediction results can be explained that the classifier obtains better ability to recognize both positive and negative samples because of its larger scale of training data. Among 16 ATP-binding sites in 1G3I_A's sequence, the classifier of full training data correctly predicts 10 of them with 1 false positive sample, whereas the classifier of under-sampled data correctly predicts 9 of them with 87 false positive samples which demonstrate that applying full training data is helpful to improve the performance in ATP-binding site prediction. It's worth mentioning that the difference of prediction performance on 1G3I_A is similar to the difference during five-fold cross validation.

6 Conclusion

In this study, we develop a sequence-based prediction method for protein-ATP binding site. All the negative samples are applied during training process instead of using under-sampling to construct a training subset which was utilized in previous studies. With the combination of decision tree classification algorithm, our proposed method achieves better performance than ATPint with the overall accuracy of 0.947 and MCC value of 0.295. In order to show the improvement on prediction performance of using full training data more vividly, the 3D structure of query protein is displayed. The improvement has demonstrated that utilizing full training data and decision tree classifier can accurately predict the ATP binding sites in protein sequences and provide assistance to biological researchers.

Acknowledgement. This work was supported by The National Natural Science Foundation of China (Project No. 61662057, 61672301) and Higher Educational Scientific Research Projects of Inner Mongolia Autonomous Region (Project No. NJZC17198).

References

1. Andrews, B.J., Hu, J.: TSC_ATP: a two-stage classifier for predicting protein-ATP binding sites from protein sequence. In: Computational Intelligence in Bioinformatics and Computational Biology (CIBCB). IEEE, Niagara Falls (2015)
2. Burley, S.K., Berman, H.M., Christie, C., et al.: RCSB protein data bank: sustaining a living digital data resource that enables breakthroughs in scientific research and biomedical education. Protein Sci. **27**(1), 316–330 (2018)
3. Johnson, Z.L., Chen, J.: ATP binding enables substrate release from multidrug resistance protein1. Cell **172**(1-2), 81 (2018)
4. Chauhan, J.S., Mishra, N.K., Raghava, G.P.: Identification of ATP binding residues of a protein from its primary sequence. BMC Bioinform. **10**(1), 434 (2009)
5. Chen, K., Mizianty, M.J., Kurgan, L.: ATPsite: sequence-based prediction of ATP-binding residues. Proteome Sci. **9**(S1), S4 (2010)
6. Chen, K., Mizianty, M.J., Kurgan, L.: Prediction and analysis of nucleotide-binding residues using sequence and sequence-derived structural descriptors. Bioinformatics **28**(3), 331–341 (2012)
7. Bauer, R.A., Günther, S., Jansen, D.: SuperSite: dictionary of metabolite and drug binding sites in proteins. Nucleic Acids Res. **37**(Database issue), D195–D200 (2009)
8. Li, W., Godzik, A.: CD-HIT: a fast program for clustering and comparing large sets of protein or nucleotide sequences. Bioinformatics **22**(13), 1658 (2006)
9. Sobolev, V., Sorokine, A., Prilusky, J.: Automated analysis of interatomic contacts in proteins. Bioinformatics **15**(4), 327–332 (1999)
10. Chauhan, J.S.: Identification of ATP binding residues of a protein from its primary sequence. http://www.imtech.res.in/raghava/atpint/atpdataset. 16 Oct 2018
11. Altschul, S.F., Madden, T.L., Shaffer, A.: Gapped BLAST and PSI-BLAST: a new generation of protein database search programs. Nucleic Acids Res. **25**, 3389–3402 (1996)
12. Bairoch, A., Apweiler, R.: The SWISS-PROT protein sequence data bank and its supplement TrEMBL. Nucleic Acids Res. **25**(1), 31–36 (1997)
13. Mcguffin, L.J., Bryson, K., Jones, D.T.: The PSIPRED protein structure prediction server. Bioinformatics **16**(4), 404–405 (2000)
14. Shen, J., Zhang, J., Luo, X.: Predicting protein-protein interactions based only on sequences information. Proc. Natl. Acad. Sci. **104**(11), 4337–4341 (2007)
15. Yedida, V.R.K.S., Chan, C.C., Duan, Z.H.: Protein function prediction using decision trees. In: IEEE International Conference on Bioinformatics and Biomedicine Workshops, pp. 193–199. IEEE, Philadelphia (2008)
16. Shringi, R.P.: PyMol software for 3D visualization of aligned molecules. Biomaterials **26**(1), 63–72 (2005)

A Novel Memristor-CMOS Hybrid Full-Adder and Its Application

Hui Yang[1,2], Shukai Duan[1,2(✉)], and Lidan Wang[1,2]

[1] Southwest University, Chongqing 400715, China
duansk@swu.edu.cn
[2] Brain-Inspired Computing and Intelligent Control of Chongqing Key Lab,
Chongqing 400715, China

Abstract. Memristor is a nano-scale component with information storage capability and binary characteristics. The memristive logic circuit composed of the structure is simple in structure and complete in logic function, and can be applied to logic operation and storage. However, the existing memristive logic circuit has a single function, the component size is too large, and the delay step is too much, so that the circuit efficiency is low. This paper proposes a novel memristor-CMOS hybrid full adder. Compared with MAD Gates, IMPLY logic circuit significantly reduces the operation steps, the circuit has no time delay, and optimizes the requirements of circuit components. Based on the proposed circuit, a novel N-bit subtractor is designed, which can be combined with the full-adder to implement composite logic operations.

Keywords: Memristor · Hybrid-CMOS · Full-adder · Subtractor

1 Introduction

In 1971, Chua proposed the fourth basic circuit component in addition to capacitance, inductance, and resistance according to the theory of circuit completeness. It is called a memristor [1], which theoretically defines a memristor and describes its physical properties. However, the physical material of the memristor has not been developed, so it has not caused too much attention from researchers. In 2008, HP Labs' researchers announced the development of the world's first memristor physical device [2, 3], which immediately attracted widespread attention from researchers and industry. Because of its memory characteristics, it can store the calculation results by its own advantages. A large number of scholars have begun to study the potential of such basic circuit components with memory characteristics in the design of modern circuit systems, and the enormous application potential in the fields of information storage and logic computing. Some researchers have proposed the SPICE simulation model of memristor [4, 5]. Shaarawy et al. combined the memristor with the diode and proposed a 1T2M, 2T2M memory structure to replace the original random memory cell composed of 6 diodes [6, 7]. Luo et al. analyzed the performance of the memristor for different series-parallel connections [8]. Duan et al. used memristors to form a random resistive memory for information storage [9–11]. Researchers such as Zhang used memristors to form the basic logic gates, and based on this, built 4M1M memory cells for information

© Springer Nature Switzerland AG 2019
H. Lu et al. (Eds.): ISNN 2019, LNCS 11555, pp. 556–564, 2019.
https://doi.org/10.1007/978-3-030-22808-8_55

storage [12, 13]. Sarwar et al. combined memristors with CMOS devices to form a 3T2M hybrid structure for non-volatile random storage [14]. Kvatinsky et al. designed a full-adder circuit based on memristive logic [15, 16]. Guckert combines resistors with memristors, proposes a memristor-drived gate (MAD Gates) logic circuit, implements a half adder and a full adder [17].

However, most of the existing memristive logic circuits have their own drawbacks. For example, MAD Gates, whose logic circuits contain large-sized components such as switches and resistors, are not conducive to large-scale integration, and due to the structure of the circuit itself, the logic response has a two-step delay, greatly increasing the circuit response time.

Most of the above studies use memristors for information storages or logic operations only, ignoring the memory characteristics of memristors. In this paper, memristors are combined with traditional FETs to propose a new full-adder circuit, which overcomes the problem that the traditional memristor memory circuit or the memristor logic operation circuit has relatively simple functions, and combines logic operation and storage to realize logic operation and storage in the same circuit. The circuit is opposite to the memristor-drived gate logic circuit (MAD Gates). The circuit proposed in this paper is composed only of memristor and field effect transistor, and the circuit size is greatly reduced, which is convenient for large-scale integration. Moreover, the proposed full-adder circuit is applied to the subtraction circuit and a novel subtractor is designed, which can combine subtraction and addition operations to achieve composite logic operations.

2 Memristor-CMOS Hybrid Full-Adder

Since the memristor has a resistive property, a different voltage is applied across it, and the resistance value varies. Different logical values are represented.

The existing full-added circuit logic expression is:

$$V_{Sum} = V_A \, XOR \, V_B \, XOR \, V_{Cin} \tag{1}$$

$$V_{Cout} = (V_A \, AND \, V_B) \, OR \, (V_B \, AND \, V_{Cin}) \, OR \, (V_A \, AND \, V_{Cin}) \tag{2}$$

Where V_A and V_B are two inputs and V_{Cin} is the carry signal of the previous bit. After simplifying the logical expression of the full-add logic, the following relationship can be obtained:

When $V_{Cin} = 0$ ('0' means low voltage V_{Low}):

$$V_{Sum} = V_A \, XOR \, V_B \tag{3}$$

$$V_{Cout} = V_A \, AND \, V_B \tag{4}$$

When $V_{Cin} = 1$ ('1' means high voltage V_{High}):

$$V_{Sum} = V_A \, NXOR \, V_B \tag{5}$$

$$V_{Cout} = V_A \, OR \, V_B \tag{6}$$

According to the simplified logical expression, a memristor-CMOS hybrid full-adder circuit is designed.

2.1   Proposed Full-Adder Circuits

Compared with the half adder, the full adder has an initial carry signal, which is a three-input and two-output circuit. The input is the initial carry signal V_{Cin}, the initial voltage V_A and V_B. The output is the carry signal V_{Cout} and the resulting voltage signal V_{Sum}. According to the logical expressions (3)–(6), a new full-adder circuit is proposed, as shown in Fig. 1.

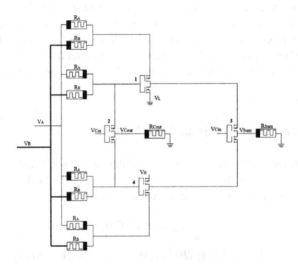

Fig. 1. CMOS-Memristor hybrid full adder

The full-adder circuit proposed in this paper is composed of XOR logical circuit, a *NXOR* logical circuit and two inverters. The *XOR* logical circuit and the *NXOR* logical circuit are composed of four memristors and one inverter. The circuit has four inverters and ten memristors, among which eight memristors realize the gate function, and the other two memristors R_{Sum} and R_{Cout} are respectively connected to the output terminals V_{Sum} and V_{Cout} to realize the output voltage storage function. The initial carry signal V_{Cin} is the input voltage of the inverters No. 2 and No. 3, The output voltage of the

XOR gate circuit is connected to the upper port of the inverter No. 3, and the output of the same *OR* circuit is connected to the lower port of the inverter No. 3. The output of inverter No. 2 is the carry output signal V_{Cout}. The output of inverter No. 3 is the resulting voltage signal V_{Sum}.

According to the truth table of full-adder, when $V_{Cin} = 0$ ('0' means low voltage V_{Low}). Both inverters No. 2 and No. 3 controlled by the carry input signal are turned on at the upper port. At this point, the circuit response has the following four cases:

1. The two input voltage $V_A = 0$, $V_B = 0$, therefore the output voltages of the AND gates and OR gates are low voltage V_{Low}. The upper ports of the inverter No. 1 and No. 4 are turned on. The output of full-adder circuits V_{Sum} and V_{Cout} are low voltage V_{Low}. The output memristors are high resistance, where $R_{Sum} = R_{off}$, $R_{Cout} = R_{off}$.
2. $V_A = 0$, $V_B = 1$, it means input V_A is low voltage V_{Low}; V_B is high voltage V_{High}. Therefore the output voltages of AND gates are low voltage V_{Low}, and the output voltages of OR gates are high voltage V_{High}. The upper port of inverter No. 1 is turned on, and the output is high voltage. The lower port of inverter No. 4 is turned on, and the output is high voltage. The outputs of full-adder circuits can be obtained: $V_{Sum} = V_{High}$, $V_{Cout} = V_{Low}$. The output memristor $R_{Sum} = R_{on}$, $R_{Cout} = R_{off}$.
3. $V_A = 1$, $V_B = 0$, it means input V_A is low voltage V_{High}; V_B is high voltage V_{Low}. It's the same to case 2. The outputs of the circuit is the same to the case 2, $V_{Sum} = V_{High}$, $V_{Cout} = V_{Low}$. The output memristor $R_{Sum} = R_{on}$, $R_{Cout} = R_{off}$.

The two input voltages $V_A = 1$, $V_B = 1$, both are high voltage V_{High}. Therefore the output voltages of the AND gates and OR gates are low voltage V_{High}. The output of inverter No. 1 is low voltage because of its lower port is turned on. The output of inverter No. 4 is high voltage because of its lower port is turned on. We can know the outputs of full-adder circuit: $V_{Sum} = V_{Low}$, $V_{Cout} = V_{High}$. The output memristor $R_{Sum} = R_{off}$, $R_{Cout} = R_{on}$.

2.2 Simulation Results of Proposed Full-Adder Circuits

The SPICE model of the HP memristor is used in this paper. According to the CMOS and memristor hybrid circuit schematic, the SPICE circuit simulation software is used to build the simulation circuit. Let 0.1 V–0.6 V be low voltage and 4 V–5 V be high voltage (Fig. 2).

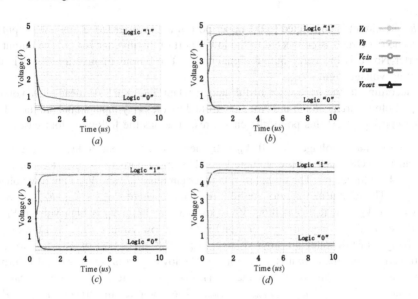

Fig. 2. Simulation results of Memristor-CMOS hybrid full-adder when $V_{Cin} = V_{Low}$

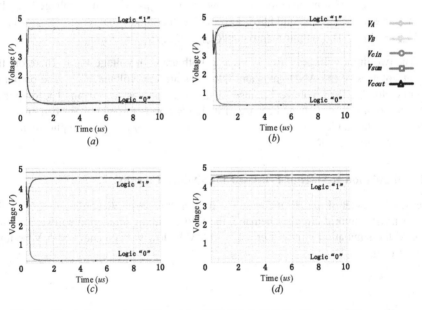

Fig. 3. Simulation results of Memristor-CMOS hybrid full-adder when $V_{Cin} = V_{High}$

2.3   Comparison of Proposed Full-Adder Circuit

By comparing the above Fig. 3, it is found that the IMPLY logic addition circuit has the most operation delay, and the CMOS circuit has the following addition circuit compared with the previous circuit, and has no response step delay, which greatly optimizes the response of the circuit (Table 1).

Table 1. Number of components and delay comparison of the proposed circuits and versus prior work for N-bit adder

	MAD gates logic	MRL logic	IMPLY	Proposed full-adder
Delay	N + 1	2N + 4	2N + 19	0
Number of components	8N memristor 2N + 3 drivers 9N resistor 14N switches	34N MOSFETs	7N + 1 memristors 7N drivers N resistors 8N − 1 switches	10N memristors 2N + 1 drivers 8N MOSFETs

Again, note the area of the proposed memristor-CMOS hybrid circuit is also significantly reduce the number of components. The proposed circuit requires only 10N memristors, 2N + 1 drivers, and 8N MOSFETs as compared with 34N MOSFETs for the traditional CMOS full adder. The proposed circuits dose not require resistors or switches, which are large size components compared the MAD Gates and IMPLY logic.

Energy comparisons are not given because prior work do not report energy consumption for the full adder.

3 Subtractor Based on Full-Adder Circuit

3.1 One-Bit Subtractor

According to the subtractor algorithm in digital circuit, we design a subtractive circuit based on memristor, as shown in the Fig. 4:

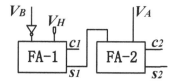

Fig. 4. Proposed one-bit subtractor

As can be seen from the Fig. 4, the proposed subtractor is composed of two full-adder. Where V_A and V_B are the subtracted and subtraction respectively, and V_H is high voltage (representing a logical "1"). In the subtractor, the complement of the subtraction is obtained by the full-adder FA-1, and the complement of the subtraction and the subtracted is added in the full-adder FA-2, whereby the entire circuit can implement the subtraction operation. For the one-bit subtractor, we only consider the case where the subtracted number is greater than the subtraction:

1. $V_A = 1$, $V_B = 1$. At this time, the inverse of V_B can be obtained as logic "0". The inverse of V_B and V_H are the two inputs of full-adder FA-1, and output s1 is the complement of V_B. V_A and the complement of V_B are the inputs of full-adder FA-2. The output s2 of FA-2 is logic "0", which is the operation result of a one-bit subtractor.
2. $V_A = 1$, $V_B = 0$. At this time, the inverse of V_B can be obtained as logic "1". The inverse of V_B and V_H are the two inputs of full-adder FA-1, and output s1 is the complement of V_B. V_A and the complement of V_B are the inputs of full-adder FA-2. The output s2 of FA-2 is logic "1", which is the operation result of a one-bit subtractor.
3. $V_A = 0$, $V_B = 0$. At this time, the inverse of V_B can be obtained as logic "1". The inverse of V_B and V_H are the two inputs of full-adder FA-1, and output s1 is the complement of V_B. V_A and the complement of V_B are the inputs of full-adder FA-2. The output s2 of FA-2 is logic "0", which is the operation result of a one-bit subtractor.

According to the circuit diagram, we performed circuit simulation in SPICE. The simulation results are shown in the Fig. 5:

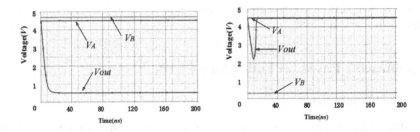

Fig. 5. The simulation of proposed subtractor

3.2 N-Bit Subtractor

Based on the one-bit subtractor, we extended the circuit and designed a N-bit subtractor as shown in Fig. 6.

Where $V_{A1}, V_{A2}, V_{A3}, \ldots \ldots, V_{An}$ are the subtracted number of N-bit subtractor and $V_{B1}, V_{B2}, V_{B3}, \ldots \ldots, V_{Bn}$ are the subtraction. Full adder FA-11, FA-12, FA-13… FA-1n constitutes a serial adder. The obtained result is added to the subtracted number to obtain the result of the subtraction circuit.

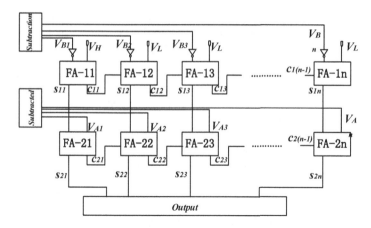

Fig. 6. Proposed N-bit subtractor

4 Conclusions

Firstly, based on the memristor Boolean logic circuit, a memristor-CMOS hybrid full-adder is designed. The feasibility of the proposed addition circuit is verified from the perspective of theory and experimental simulation. Compared with MAD Gates and MRL logic, it has obvious advantages: the circuit uses the non-volatile nature of the memristor, combining computation and storage, requiring fewer components, and the circuit only contains CMOS devices and memristors. The circuit has no time delay and can realize real-time circuit. Secondly, an n-bit subtraction circuit is constructed based on the proposed full-adder, which provides a solution for the combination of addition and subtraction.

References

1. Chua, L.O.: Memristor-the missing circuit element. IEEE Trans. Circ. Theory **18**(5), 507–519 (1971)
2. Strukov, D.B., Snider, G.S., Stewart, D.R., et al.: The missing memristor found. Nature **453** (7191), 80–83 (2008)
3. Williams, R.: How We Found the Missing Memristor. IEEE Press (2008)
4. Biolek, Z., Biolek, D., Biolkova, V.: SPICE model of memristor with nonlinear dopant drift. Radioengineering **18**(2), 210–214 (2008)
5. Vourkas, I., Batsos, A., Sirakouli, G.C.: SPICE modeling of nonlinear memristive behavior. Int. J. Circ. Theory Appl. **43**(5), 553–565 (2015)
6. Emara, A., Ghoneima, M., El-Dessouky, M.: Differential 1T2M memristor memory cell for single/multi-bit RRAM modules. In: Computer Science and Electronic Engineering Conference, pp. 69–72. IEEE, Colchester (2014)
7. Shaarawy, N., Ghoneima, M., Radwan, A.G.: 2T2M memristor-based memory cell for higher stability RRAM modules. In: 2015 IEEE International Symposium on Circuits and Systems, pp. 1418–1421 (2015)

8. Luo, L., Hu, X., Duan, S., Dong, Z., Wang, L.: Multiple memristor series-parallel connections with use in synaptic circuit design. IET Circ. Devices Syst. **11**(2), 123–134 (2017)

9. Jiang, Z., Duan, S., Wang, L., et al.: A threshold adaptive memristor model analysis with application in image storage. In: International Conference on Information Science and Technology, pp. 449–454. IEEE, Jeju Island (2015)

10. Duan, S., Hu, X., Wang, L., et al.: Memristor-based RRAM with applications. Sci. China (Inf. Sci.) **55**(6), 1446–1460 (2012)

11. Hu, X., Duan, S., Wang, L.: Memristive multilevel memory with applications in audio signal storage. In: Deng, H., Miao, D., Lei, J., Wang, F.L. (eds.) AICI 2011. LNCS (LNAI), vol. 7002, pp. 228–235. Springer, Heidelberg (2011). https://doi.org/10.1007/978-3-642-23881-9_29

12. Zhang, Y., Shen, Y., Wang, X., et al.: A novel design for memristor-based logic switch and crossbar circuits. IEEE Trans. Circ. Syst. I Regul. Pap. **62**(5), 1402–1411 (2015)

13. Zhang, Y., Shen, Y., Wang, X., et al.: A novel design for a memristor-based or gate. IEEE Trans. Circ. Syst. II Express Briefs **62**(8), 781–785 (2015)

14. Sarwar, S.S., Saqueb, S.A.N., Quaiyum, F., et al.: Memristor-based nonvolatile random access memory: hybrid architecture for low power compact memory design. IEEE Access **1** (1), 29–34 (2013)

15. Borghetti, J., Snider, G.S., Kuekes, P.J., et al.: Memristive' switches enable 'stateful' logic operations via material implication. Nature **464**(7290), 873 (2011)

16. Kvatinsky, S., Satat, G., Wald, N., et al.: Memristor-based material implication (IMPLY) logic: design principles and methodologies. IEEE Trans. Very Large Scale Integr. Syst. **22** (10), 2054–2066 (2014)

17. Guckert, L., Swartzlander, E.E.: MAD gates—memristor logic design using driver circuitry. IEEE Trans. Circ. Syst. II Express Briefs **64**(2), 171–175 (2017)

Using Coupled Multilinear Rank-$(L, L, 1)$ Block Term Decomposition in Multi-Static-Multi-Pulse MIMO Radar to Localize Targets

Jia-Xing Yang[1(✉)], Xiao-Feng Gong[1], Hui Li[1], You-Gen Xu[2], and Zhi-Wen Liu[2]

[1] School of Information and Communication Engineering,
Dalian University of Technology, Dalian 116023, China
`jiuyunshikong@mail.dlut.edu.cn`
[2] School of Information and Communication Engineering,
Beijing Institute of Technology, Beijing 100081, China

Abstract. We propose a tensorial method for target localization based on multi-static-multi-pulse MIMO radar, which consists of multiple widely separated transmitting and receiving arrays. We show how a set of tensors, which admits a coupled multilinear (ML) rank-$(L, L, 1)$ block term decomposition (BTD), can be constructed from the output signals of different receiving arrays. As such, we compute the coupled ML rank-$(L, L, 1)$ BTD of these tensors to obtain factor matrices. The target locations are then able to be determined from the latent DOA parameters in these factor matrices. The proposed method neither requires prior knowledge, nor assumes orthogonality between probing signals. In addition, by exploiting the coupling, different receiving array outputs are jointly processed, yielding improved performance than uncoupled BTD based methods. Simulation results are provided to illustrate the performance of the proposed method.

Keywords: MIMO radar · Tensor · Block Term Decomposition ·
Multilinear rank-$(L, L, 1)$ · Direction of arrival of arrival

1 Introduction

Recently, multistatic MIMO radar has attracted wide attention [1–5]. In comparison with monostatic MIMO radar [6] and bistatic MIMO radar [7], multistatic MIMO radar consists of multiple widely separated transmitting and receiving arrays, as seen in Fig. 1. As such, by exploiting the rich spatial diversity of the multistatic MIMO radar, the target locations can be determined more accurately.

Traditional methods are mainly matrix-based [1–7], which ignore the potential multilinear structure of the output signal. In the recent several years, there are tensor-based methods for MIMO radar application [8,9]. More precisely, [8]

© Springer Nature Switzerland AG 2019
H. Lu et al. (Eds.): ISNN 2019, LNCS 11555, pp. 565–574, 2019.
https://doi.org/10.1007/978-3-030-22808-8_56

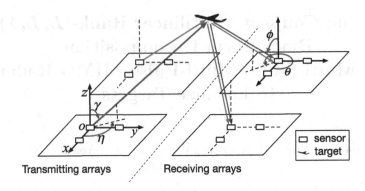

Fig. 1. Diagram of a 2 × 2 MSMP-MIMO radar and the angle definitions

proposed a method based on the canonical polyadic decomposition of the output signal of a bistatic MIMO radar. This method, however, assumes that the probing signals are orthogonal to one another and are known *a priori*. In [9], the multi-static-multi-pulse MIMO (MSMP-MIMO) radar is considered and a multilinear (ML) rank-$(L, M, \cdot)$ block term decomposition (BTD) [10] based method was proposed. Nevertheless, this method does not (at least not explicitly) exploit the coupling among different array signals.

In this paper, we address the problem of target localization in MSMP-MIMO radar from a coupled tensor decomposition perspective. We construct multiple tensors from the observed signals of different receiving arrays, each admitting a ML rank-$(L, L, 1)$ BTD [11]. It turns out that these tensors together admit a coupled ML rank-$(L, L, 1)$ BTD. We perform two coupled ML rank-$(L, L, 1)$ BTD methods to jointly decompose these coupled tensors, yielding factor matrices that carry DOA parameters of each target with regard to different receiving arrays. There parameters can be integrated to localize the targets. In our method we neither require prior knowledge of the probing signals, nor assume orthogonality between them due to the exploitation of coupling.

Notation: we use uppercase boldface, lowercase boldface and uppercase calligraphic letters to denote vectors, matrices and tensors, respectively. We use '$\odot$' and '$\odot_c$' to denote the partition-wise Khatri-Rao product and the column-wise Khatri-Rao product: $A \odot B = [A_1 \otimes B_1, \cdots, A_R \otimes B_R]$ and $A \odot_c B = [a_1 \otimes b_1, \cdots, a_R \otimes b_R]$, respectively, where $\otimes$ is the Kronecker product. We use '$\circledast$' to denote the outer product: $(a \circledast b \circledast c)_{i,j.k} = a_i b_j c_k$. We suppose that A and B consists of R sub-matrices in the partition-wise Khatri-Rao product while R vectors in the column-wise Khatri-Rao product. We use '$\sim$' to denote the estimate of a parameter, e.g., $\tilde{A}$ denotes the estimate of A.

Symbols '$(\cdot)^{\mathsf{T}}$', '$(\cdot)^{\mathsf{H}}$', '$(\cdot)^{*}$', '$(\cdot)^{\dagger}$', and '$\| \cdot \|_{\mathsf{F}}^2$' are used to denote transpose, conjugated transpose, conjugate, Moore-Penrose pseudo-inverse, and Frobenius norm, respectively. We define $\text{vec}(A) = [a_1^{\mathsf{T}}, ..., a_J^{\mathsf{T}}]^{\mathsf{T}}$ to denote the column-wise vectorization of $A = [a_1, ..., a_J]$. MATLAB notation is adopted to represent submatrices of a tensor. For example, we use $\mathcal{T}_{(:,:,k)}$ to represent the kth frontal slice of a third-order tensor $\boldsymbol{T}$. We use $\boldsymbol{T}_1$, $\boldsymbol{T}_2$, and $\boldsymbol{T}_3$ to denote the mode-1, mode-2, and mode-3 matricization of $\boldsymbol{T}$, respectively: $(\boldsymbol{T}_1)_{((j-1)K+k,i)} = (\boldsymbol{T}_2)_{((i-1)K+k,j)} = (\boldsymbol{T}_3)_{((i-1)J+j,k)} = \boldsymbol{T}_{(i,j,k)}$. A decomposition of $\boldsymbol{T} \in \mathbb{C}^{I \times J \times K}$ in a sum of ML rank-$(L,L,1)$ terms is a decomposition of the form: $\boldsymbol{T} = \sum_{r=1}^{R}(\boldsymbol{A}_r \boldsymbol{B}_r^{\mathsf{T}}) \otimes \boldsymbol{c}_r$, $r = 1,...,R$, where $\boldsymbol{A}_r \in \mathbb{C}^{I \times L}$ and $\boldsymbol{B}_r \in \mathbb{C}^{J \times L}$ are rank-L matrices, and $\boldsymbol{c}_r \in \mathbb{C}^{K \times 1}$ is non-zero. Let $\boldsymbol{A} \triangleq [\boldsymbol{A}_1,...,\boldsymbol{A}_R]$, $\boldsymbol{B} \triangleq [\boldsymbol{B}_1,...,\boldsymbol{B}_R]$, and $\boldsymbol{C} \triangleq [\boldsymbol{c}_1,...,\boldsymbol{c}_R]$. The three matrix representations of $\boldsymbol{T}$ can be written as: $\boldsymbol{T}_1 = (\boldsymbol{B} \odot \boldsymbol{C})\boldsymbol{A}^{\mathsf{T}}$, $\boldsymbol{T}_2 = (\boldsymbol{C} \odot \boldsymbol{A})\boldsymbol{B}^{\mathsf{T}}$, and $\boldsymbol{T}_3 = [(\boldsymbol{A}_1 \odot_c \boldsymbol{B}_1)\boldsymbol{1}_L,...,(\boldsymbol{A}_R \odot_c \boldsymbol{B}_R)\boldsymbol{1}_L]\boldsymbol{C}^{\mathsf{T}}$, respectively.

2   Problem Formulation

An illustration of a MSMP-MIMO radar is in Fig. 1. We make several assumptions: (i) the probing signal transmitted from each transmitting array to each target can be regarded as a far-field point source to each target, and so can the signal reflected by each target to each receiving array; (ii) the probing signals are linearly independent, and that the steering vectors are constant during P pulses; (iii) the target reflection coefficients vary from pulse to pulse (swerling II).

For mathematical tractability, we set up the parameters of the system in Table 1. We denote transmitting steering vector from the lth transmitting array to some target as $\boldsymbol{a}^{[l]}(\gamma,\eta)$ and receiving steering vector from some target to the mth receiving array as $\boldsymbol{b}^{[m]}(\theta,\phi)$, which are associated with DOD parameters (γ,η) and DOA parameters (θ,ϕ) (see Fig. 1), respectively, defined as follows:

$$\boldsymbol{a}^{[l]}(\gamma,\eta) = \exp(\frac{2\pi i}{\lambda}[\boldsymbol{k}_1^{[l]\mathsf{T}}\boldsymbol{v}(\gamma,\eta),...,\boldsymbol{k}_{I^{[l]}}^{[l]\mathsf{T}}\boldsymbol{v}(\gamma,\eta)]^{\mathsf{T}}),$$
$$\boldsymbol{b}^{[m]}(\theta,\phi) = \exp(\frac{2\pi i}{\lambda}[\boldsymbol{h}_1^{[m]\mathsf{T}}\boldsymbol{v}(\theta,\phi),...,\boldsymbol{h}_{J^{[M]}}^{[m]\mathsf{T}}\boldsymbol{v}(\theta,\phi)]^{\mathsf{T}}),$$

$$(1)$$

where $\boldsymbol{v}(\gamma,\eta) = [\cos\gamma\sin\eta, \sin\gamma\sin\eta, \cos\eta]$ denotes the Poynting vector associated with the angle parameters (γ,η), and $\boldsymbol{v}(\theta,\phi) = [\cos\theta\sin\phi, \sin\theta\sin\phi, \cos\phi]$ is defined analogously. The vectors $\boldsymbol{k}_i^{[l]} \in \mathbb{R}^3$ and $\boldsymbol{h}_j^{[m]} \in \mathbb{R}^3$ hold the coordinates of the ith sensor in the lth transmitting array and the jth sensor in the mth transmitting array, respectively. The symbol λ denotes the wavelength of the probing signal.

We write the DOD parameter from the lth transmitting array to the rth target and the DOA parameter from the rth target to the mth receiving array as $(\gamma_r^{[l]}, \eta_r^{[l]})$ and $(\theta_r^{[m]}, \phi_r^{[m]})$, respectively. For convenience, we denote $\boldsymbol{a}^{[l]}(\gamma_r^{[l]}, \eta_r^{[l]})$ and $\boldsymbol{b}^{[m]}(\theta_r^{[m]}, \phi_r^{[m]})$ as $\boldsymbol{a}_r^{[l]}$ and $\boldsymbol{b}_r^{[m]}$, respectively. Then the output signal $\boldsymbol{X}_p^{[m]} \in \mathbb{C}^{J^{[m]} \times T}$ of the mth receiving array during the pth pulse can be expressed as:

$$\boldsymbol{X}_p^{[m]} = \sum_{l=1}^{L}\sum_{r=1}^{R} \boldsymbol{b}_r^{[m]}(\varrho_r^{[l,m]})_p \boldsymbol{a}_r^{[l]\mathsf{T}}\boldsymbol{S}^{[l]\mathsf{T}},$$

$$(2)$$

with $l = 1,...,L$, $m = 1,...,M$, $r = 1,...,R$, $p = 1,...,P$, $\boldsymbol{S}^{[l]} \in \mathbb{C}^{T \times I^{[l]}}$. We stack $\boldsymbol{X}_p^{[m]}$ with fixed m and varying p along the third mode into a third-order

Table 1. Parameters for MSMP-MIMO radar

Parameter	Definition
L	Number of transmitting arrays
M	Number of receiving arrays
$I^{[l]}$	Number of antennas in the lth transmitting array
$J^{[m]}$	Number of antennas in the mth receiving array
P	Number of pulses
R	Number of targets
T	Number of time samples in each pulse
$(\varrho_{r,p}^{[l,m]})$	Reflection coefficient of the rth target in the pth pulse, with regard to the lth transmitting array and the mth receiving array
$a_r^{[l]}$	Transmitting steering vector from the lth transmitting array to the rth target
$b_r^{[m]}$	Receiving steering vector from the rth target to the mth receiving array
$S^{[l]}$	Signal matrix that holds in its columns the probing signals from different antennas in the lth transmitting array within one pulse period

tensor $\mathcal{X}'^{[m]}_{(:,:,p)} = X_p^{[m]}$. By denoting $f_p^{[l,m]} \triangleq [(\varrho_{r,1}^{[l,m]}), ..., (\varrho_{r,P}^{[l,m]})]^\mathsf{T} \in \mathbb{C}^P$ and $c_r^{[l]} \triangleq S^{[l]} a_r^{[l]} \in \mathbb{C}^T$, we write $\mathcal{X}'^{[m]}$ as follows:

$$\mathcal{X}'^{[m]} = \sum_{l=1}^{L} \sum_{r=1}^{R} b_r^{[m]} \otimes c_r^{[l]} \otimes f_r^{[l,m]}. \tag{3}$$

We permute the first and third indices of $\mathcal{X}'^{[m]}$ to obtain the data tensor, expressed as:

$$\mathcal{X}^{[m]} = \sum_{r=1}^{R} \left(\sum_{l=1}^{L} f_r^{[l,m]} \otimes c_r^{[l]} \right) \otimes b_r^{[m]}, \tag{4}$$

where $\mathcal{X}^{[m]} \in \mathbb{C}^{P \times T \times J^{[m]}}$. We further denote $F_r^{[m]} \triangleq [f_r^{[1]}, ..., f_r^{[L]}] \in \mathbb{C}^{P \times L}$ and $C_r \triangleq [c_r^{[1]}, ..., c_r^{[L]}] \in \mathbb{C}^{T \times L}$. Then $\mathcal{X}^{[m]}$ can be rewritten as:

$$\mathcal{X}^{[m]} = \sum_{r=1}^{R} (F_r^{[m]} C_r^\mathsf{T}) \otimes b_r^{[m]}. \tag{5}$$

It is shown in (5) that for a fixed m the tensor $\mathcal{X}^{[m]}$ admits a ML rank-$(L, L, 1)$ BTD. For varying m, there is a common factor matrix C among the set of tensors $\{\mathcal{X}^{[m]}, m = 1, ..., M\}$. We say that these tensors together admit a coupled ML rank-$(L, L, 1)$ BTD. We define $F^{[m]} \triangleq [F_1^{[m]}, ..., F_R^{[m]}]$, $C \triangleq [C_r, ..., C_r]$, and $B^{[m]} \triangleq [b_1^{[m]}, ..., b_R^{[m]}]$ as the three factor matrices of $\mathcal{X}^{[m]}$. It is clear that all the tensors are coupled in the second mode by the factor matrix C. Therefore, we can use coupled ML rank-$(L, L, 1)$ BTD algorithms to identify the factor matrices, in which the DOA parameters can be computed. The mode-2 matrix representation of tensor $\mathcal{X}^{[m]}$ can be written as:

$$\boldsymbol{X}^{[m]} = (\boldsymbol{B}^{[m]} \odot \boldsymbol{F}^{[m]})\boldsymbol{C}^{\mathsf{T}}, \tag{6}$$

There exist some trivial indeterminacies for coupled ML rank-$(L, L, 1)$ BTD (5): (i) permutation ambiguity, i.e., one can arbitrarily permute terms if it is done for all coupled tensors consistently: (ii) rotation ambiguity, i.e., one can simultaneously post-multiply $\boldsymbol{F}_r^{[m]}$ by any non-singular matrix $\boldsymbol{\Lambda}_r \in \mathbb{C}^{L \times L}$ and pre-multiply $\boldsymbol{C}_r^{[m]\mathsf{T}}$ by $\boldsymbol{\Lambda}_r^{\dagger}$ in the rth term of every tensor $\mathcal{X}^{[m]}$; (iii) scaling ambiguity, i.e., one can not distinguish $\boldsymbol{F}_r^{[m]}\boldsymbol{C}_r^{\mathsf{T}}$ and $\boldsymbol{b}_r^{[m]}$ from $w_r^{[m]}\boldsymbol{F}_r^{[m]}\boldsymbol{C}_r^{\mathsf{T}}$ and $1/w_r^{[m]}(\boldsymbol{b}_r^{[m]})$ for any non-zero value of $w_r^{[m]}$, and $\boldsymbol{F}_r^{[m]}\boldsymbol{C}_r^{\mathsf{T}}$ and $\boldsymbol{b}_r^{[m]}$ can be arbitrarily scaled if their whole product is not changed. We say that the coupled ML rank-$(L, L, 1)$ BTD is unique if factor matrices $\boldsymbol{F}^{[m]}$, $\boldsymbol{C}$, and $\boldsymbol{B}^{[m]}$ are uniquely identified subject to the above three indeterminacies.

3   Proposed Method

We assume that the shared factor matrix has full-column rank. This assumption holds in the generic sense if the number of samples for one pulse period $T \geq RL$. Roughly speaking, it is required that the pulse period and the sampling rate are sufficiently large. We additionally assume that the uniqueness conditions of ML rank-$(L, L, 1)$ BTD hold for at least one of these coupled tensors. We refer to [11] for details of the uniqueness conditions of a ML rank-$(L, L, 1)$ BTD.

3.1   Identify the Common Factor via Uncoupled BTD

In our method, we first choose a tensor for which ML rank-$(L, L, 1)$ BTD uniqueness conditions hold, and then apply ML rank-$(L, L, 1)$ BTD to this tensor. For example, we can check the generic uniqueness results, as given in [11], to select tensors for which the ML rank-$(L, L, 1)$ BTD uniqueness conditions are expected to hold, and then choose the one for which the decomposition yields the best fit. For convenience, we assume, without loss of generality, that $\mathcal{X}^{[1]}$ is the chosen tensor. By performing ML rank-$(L, L, 1)$ BTD based algorithms to $\mathcal{X}^{[1]}$, we can obtain the shared factor matrix $\boldsymbol{C}$ and factor matrices $\boldsymbol{F}^{[1]}$ and $\boldsymbol{B}^{[1]}$.

We note that there are several options available for the implementation of a ML rank-$(L, L, 1)$ BTD. For instance, we can use the alternating least squares (ALS) based algorithm [12]. This algorithm alternately updates each factor matrix to reduce the LS based cost function. During each step, the updated factor matrix is regarded as unknown while the other factor matrices are fixed. ALS monotonically decreases the cost function along the updating process, but it does not guarantee to converge to a stationary point. We can also implement a quasi-Newton (QN) or nonlinear least squares (NLS) based ML rank-$(L, L, 1)$ BTD via 'll1_minf.m' or 'll1_nls.m' functions in Tensorlab 3.0 [13], respectively. We note that NLS approach is of low per-iteration cost and close to quadratic convergence near a (local) optimum.

3.2 Identify the Remaining Factor Matrices by Making Use of the Common Factor and Coupling Structure

Now that we have obtained the common factor matrix, the remaining factor matrices can be identified by making use of the coupling factor and the ML rank-$(L, L, 1)$ structure of each tensor. For each tensor $\mathcal{X}^{[m]}$, $m = 1, ..., M$, we post-multiply its matrix representation (6) by $(C^{\mathsf{T}})^{\dagger}$, as follows:

$$B^{[m]} \odot F^{[m]} = X^{[m]}(C^{\mathsf{T}})^{\dagger}, \quad m = 2, ..., M, \tag{7}$$

in which $B^{[m]} \odot F^{[m]} = [b_1^{[m]} \otimes F_1^{[m]}, ..., b_R^{[m]} \otimes F_R^{[m]}]$. We write the rth sub-matrix $b_r^{[m]} \otimes F_r^{[m]}$ as:

$$b_r^{[m]} \otimes F_r^{[m]} = \begin{bmatrix} G_{1,r}^{[m]} \\ \vdots \\ G_{J^{[m]},r}^{[m]} \end{bmatrix} = \begin{bmatrix} (b_r^{[m]})_1 F_r^{[m]} \\ \vdots \\ (b_r^{[m]})_{J^{[m]}} F_r^{[m]} \end{bmatrix}, \tag{8}$$

where $G_{j,r}^{[m]} = (b_r^{[m]})_j F_r^{[m]} \in \mathbb{C}^{P \times L}$, $1 \leq j \leq J^{[m]}$, with $(b_r^{[m]})_j$ being the jth element of $b_r^{[m]}$. Hence, we can compute a matrix $P_r^{[m]}$ from $G_{j,r}^{[m]}$ with fixed r, m and varying j:

$$P_r^{[m]} = [\text{vec}(G_{1,r}^{[m]}), ..., \text{vec}(G_{J^{[m]},r}^{[m]})]^{\mathsf{T}} = b_r^{[m]} \otimes \text{vec}(F_r^{[m]}). \tag{9}$$

As shown in (9), $P_r^{[m]}$ is a rank-1 matrix in the exact case. Hence, we can compute $b_r^{[m]}$ and $F_r^{[m]}$ as the vectors that span the column and row space of $P_r^{[m]}$, respectively. In the noisy case, we perform rank-1 approximation on $P_r^{[m]}$ to identify $b_r^{[m]}$ and $F_r^{[m]}$, which usually can be done via singular value decomposition (SVD). By varying m and r, we will obtain all factor matrices $B^{[m]}$ and $F^{[m]}$.

3.3 Localize Targets from the Estimates of Factor Matrices

Up to now, $\{B^{[m]}, m = 1, ..., M\}$, $\{F^{[m]}, m = 1, ..., M\}$, and C have been identified by the coupled ML rank-$(L, L, 1)$ BTD. Next we explain how to localize targets from $\{\tilde{B}^{[m]}, m = 1, ..., M\}$. In (1), we have the prior knowledge of the pattern of each receiving array. One thus can use some angle-searching method to obtain the DOA estimates: $(\tilde{\theta}_r^{[m]}, \tilde{\phi}_r^{[m]})$. For example, $(\tilde{\theta}_r^{[m]}, \tilde{\phi}_r^{[m]})$ can be calculated as those that realize the minimal angle between $b^{[m]}(\theta, \phi)$ and $\tilde{b}_r^{[m]}$:

$$\{\tilde{\theta}_r^{[m]}, \tilde{\phi}_r^{[m]}\} = \underset{(\theta, \phi)}{argmax} \frac{\tilde{b}_r^{[m]\mathsf{H}} b^{[m]}(\theta, \phi)}{\left\|\tilde{b}_r^{[m]}\right\|_{\mathsf{F}}^2 \cdot \left\|b^{[m]}(\theta, \phi)\right\|_{\mathsf{F}}^2}. \tag{10}$$

After the pair $(\tilde{\theta}_r^{[m]}, \tilde{\phi}_r^{[m]})$ is determined, we obtain a line in space that points from the mth receiving array to the rth target. By fixing r and varying m, we can compute M such vectors, and with any two of them we are able to determine the location of the rth target as the intersection point of the two corresponding lines. In the noisy case, there exist errors in the obtained M Poyntine vectors and we can compute the location of each target as the point that is closest to the corresponding M lines in the least squares sense.

4   Experiments

In this section, we provide experiment results to demonstrate the performance of the proposed method. The coupled ML rank-$(L, L, 1)$ BTD is implemented via ALS and NLS, which we denote as C-LL1-ALS and C-LL1-NLS, respectively.

In the experiment, we set $L = 3$, $M = 2$, $R = 3$, $P = 500$, $T = 512$, $I^{[1]} = I^{[2]} = I^{[3]} = 7$, and $J^{[1]} = J^{[2]} = 7$. The carrier frequency is fixed to $1\,\mathrm{GHz}$. Transmitting and receiving arrays are all L-shaped. For each transmitting/receiving array the inter-sensor spacing is set to half the wavelength of the impinging signal. The coordinates of transmitting arrays, receiving arrays and targets are $\{(0, 8, 0), (\pm 8, 8, 0)\}$, $\{(\pm 4, 8, 0)\}$, and $\{(-5, 20, 0), (-5, 0, 0), (10, 0, 0)\}$, respectively. For convenience, without loss of generality, we assume that the transmitting arrays, receiving arrays, and the targets are in the $z = 0$ plane. As such, the elevation angles for the probing signals, $\{\eta_r^{[l]}, r = 1, ..., R, l = 1, ..., L\}$, as well as those for the reflected signals, $\{\phi_r^{[m]}, r = 1, ..., R, m = 1, ..., M\}$, are fixed to $90°$. We draw the real and imaginary part of each entry of the probing signal $\boldsymbol{S}^{[m]}$ and the reflection coefficient matrix $\boldsymbol{F}^{[m]}$ from the normal distribution for $m = 1, ..., M$. Under above settings, we can generate the noise-free data tensors $\{\boldsymbol{\mathcal{X}}^m, m = 1, ..., M\}$ according to (5). We add the white Gaussian noise term $\boldsymbol{\mathcal{N}}^{[m]} \in \mathbb{C}^{P \times T \times J^{[m]}}$ to $\boldsymbol{\mathcal{X}}^{[m]}$ to generate the noisy output signals:

$$\boldsymbol{y}^{[m]} = \sigma_s \boldsymbol{\mathcal{X}}^{[m]} / \left\| \boldsymbol{\mathcal{X}}^{[m]} \right\|_{\mathsf{F}}^2 + \sigma_n \boldsymbol{\mathcal{N}}^{[m]} / \left\| \boldsymbol{\mathcal{N}}^{[m]} \right\|_{\mathsf{F}}^2, \tag{11}$$

where σ_s and σ_n denote the signal and noise level, respectively. The signal-to-noise ratio is defined as SNR$= 20 \log_{10} (\sigma_s / \sigma_n)$.

We include two uncoupled ML rank-$(L, L, 1)$ BTD methods in the comparison, based on ALS and NLS and denoted as LL1-ALS and LL1-NLS, respectively. The ML rank-$(L, L, 1)$ BTD algorithms are applied to each tensor separately to calculate the factor matrices. For the all compared algorithms, the maximal number of iterations is set to 5000 and the tolerance is set to 0.0001.

First we verify the validity of the proposed approach. We fix SNR$=5\,\mathrm{dB}$ and implement 20 Monte-Carlo runs. The estimated target locations obtained via C-LL1-ALS and C-LL1-NLS are plotted in Fig. 2(a) and (b), respectively, showing that both algorithms can localize the three targets.

Then we consider the accuracy of the compared algorithms vs. SNR varying from $-15\,\mathrm{dB}$ to $15\,\mathrm{dB}$. We adopt joint mean square error (JMSE) to measure the accuracy of $\tilde{C}$, defined as follows:

$$\xi \triangleq M^{-1} \sum_{m=1}^{M} \left\| \boldsymbol{C} - \tilde{\boldsymbol{C}}^{[m]} \boldsymbol{\Pi}^\dagger \boldsymbol{\Lambda} \right\|_{\mathsf{F}}^2 / \|\boldsymbol{C}\|_{\mathsf{F}}^2, \tag{12}$$

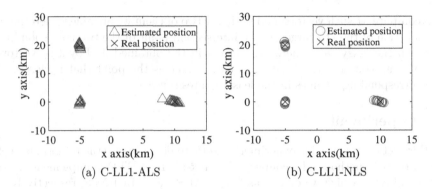

Fig. 2. Position distribution of the three targets estimated via C-LL1-ALS and C-LL1-NLS in 20 runs (SNR = 5 dB)

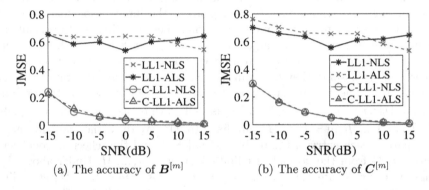

Fig. 3. Comparison of ALS, coupled rank-$(L, L, 1)$ BTD based ALS, NLS, and coupled rank-$(L, L, 1)$ BTD based NLS $(m = 1, ..., M)$

where $\boldsymbol{\varPi}$ is a column block permutation matrix that is common for all the estimates $\boldsymbol{C}^{[m]}$, and $\boldsymbol{\varLambda}$ is a block-diagonal matrix holding $\boldsymbol{\varLambda}_r$ as the rth diagonal submatrix. The matrix $\tilde{\boldsymbol{C}}^{[m]}$ is the estimate of $\tilde{\boldsymbol{C}}$ obtained from the output of the mth receiving array. Obviously, for the coupled decomposition, $\tilde{\boldsymbol{C}}^{[m]}$ remains unchanged with m varying, due to the exploitation of the coupling. However, for the ordinary ML rank-$(L, L, 1)$ BTD, $\tilde{\boldsymbol{C}}^{[m]}$ with different m may be different. In addition, we note by (12) that JMSE does not only measure the accuracy of the estimates, but also indicates if the permutation is aligned for estimates obtained from different datasets. That is to say, a small JMSE only appears when the estimation is accurate and the permutation is aligned among different datasets. Analogously, JMSE also can be applied to evaluate $\{\tilde{\boldsymbol{B}}^{[m]}, m = 1, ..., M\}$.

We perform 100 Monte-Carlo runs for each algorithm to calculate the root mean squared value of JMSE for both the estimates of $\{\boldsymbol{B}^{[m]}, m = 1, ..., M\}$ and $\boldsymbol{C}$. The results are provided in Fig. 3(a) and (b), respectively. It is clearly shown that both C-LL1-ALS and C-LL1-NLS provide correct results while

LL1-ALS and LL1-NLS fail. The above observations clearly show the interests in the proposed methods in a MSMP-MIMO radar application.

5   Conclusions

In this paper, we propose a coupled ML rank-$(L, L, 1)$ BTD based method for target localization in the MSMP-MIMO radar. The output signals from each receiving array can be stacked into a third-order tensor that admits ML rank-$(L, L, 1)$ BTD and these stacked tensors together admit coupled ML rank-$(L, L, 1)$ BTD. Using coupled ML rank-$(L, L, 1)$ BTD based method is able to extract the aligned factor matrices, which contains DOA parameters associated with the position of each target. The method does not require orthogonal probing signals, nor their waveform patterns. Simulation results have shown the validity of the proposed approach, as well as its superiority over the uncoupled ML rank-$(L, L, 1)$ based method.

Acknowledgments. This research is funded by: (1) National natural science foundation of China (Grant No. 61331019, 61379012, and 61601079); (2) Natural science foundation of Liaoning province (Grant No. 20170540169).

References

1. Chernyak, V.: Multisite radar systems composed of MIMO radars. IEEE Aerosp. Electron. Syst. Mag. **29**(12), 28–37 (2014)
2. Li, H., Wang, Z., Liu, J., et al.: Moving target detection in distributed MIMO radar on moving platforms. IEEE J. Sel. Top. Signal Proc. **9**(8), 1524–1535 (2015)
3. Yan, J., Liu, H., Pu, W.: Joint beam selection and power allocation for multiple target tracking in netted colocated MIMO radar system. IEEE Trans. Signal Proc. **64**(24), 6417–6427 (2016)
4. Chen, P., Zheng, L., Wang, X.: Moving target detection using colocated MIMO radar on multiple distributed moving platforms. IEEE Trans. Signal Proc. **65**(17), 4670–4683 (2017)
5. Deligiannis, A., Panoui, A., Lambotharan, S., et al.: Game-theoretic power allocation and the Nash equilibrium analysis for a multistatic MIMO radar network. IEEE Trans. Signal Proc. **65**(24), 6397–6408 (2017)
6. Robey, F.C., Coutts, S., et al.: MIMO radar theory and experimental results. In: Conference Record of the Thirty-Eighth Asilomar Conference on Signals, Systems and Computers, pp. 300–304. IEEE Press, Pacific Grove (2005)
7. Li, J., Stoica, P.: MIMO radar with colocated antennas. IEEE Signal Process Mag. **24**(5), 106–114 (2007)
8. Nion, D., Sidiropoulos, N.D.: A PARAFAC-based technique for detection and localization of multiple targets in a MIMO radar system. In: IEEE International Conference on Acoustics, Speech and Signal Proceedings, pp. 2077–2080. IEEE Press, Taipei (2009)
9. Nion, D., Sidiropoulos, N.D.: Tensor algebra and multidimensional harmonic retrieval in signal processing for MIMO radar. IEEE Trans. Signal Proc. **58**(11), 5693–5705 (2010)

10. De Lathauwer, L.: Decompositions of a higher-order tensor in block terms Part II: definitions and uniqueness. SIAM J. Matrix Anal. Appl. **30**(3), 1033–1066 (2008)
11. Domanov, I., De Lathauwer, L.: Decomposition of a tensor into multilinear rank-$(1, L_r, L_r)$ terms. arXiv preprint arXiv:1808.02423 (2018)
12. De Lathauwer, L., Nion, D.: Decompositions of a higher-order tensor in block terms Part III: alternating least squares algorithms. SIAM J. Matrix Anal. Appl. **30**(3), 1067–1083 (2008)
13. Tensorlab 3.0 - Numerical optimization strategies for large-scale constrained and coupled matrix/tensor factorization. https://www.tensorlab.net

Canonical Polyadic Decomposition with Constant Modulus Constraint: Application to Polarization Sensitive Array Processing

Jin-Wei Yang$^{(\boxtimes)}$, Xiao-Feng Gong, Lu-Ming Wang, and Qiu-Hua Lin

School of Information and Communication Engineering,
Dalian University of Technology, Dalian 116023, China
1837287636@qq.com

Abstract. We consider the joint estimation of direction-of-arrival (DOA) and polarization of constant modulus (CM) signals based on a polarization sensitive array. We propose an algebraic algorithm for canonical polyadic decomposition (CPD) with CM constraint. The proposed algorithm first uses the analytic CM factorization algorithm to calculate the source matrix, and then exploits the CPD structure of the data tensor to compute the remaining factor matrices, from which the DOA and polarization parameters can finally be obtained. Due to the algebraic nature, the proposed algorithm can be used to effectively initialize the optimization based algorithms. We have shown that the proposed algorithm has more relaxed uniqueness conditions from simulation results. We provide simulation results to illustrate the performance of the proposed algorithm.

Keywords: Direction-of-arrival · Polarization · Tensor ·
Canonical polyadic decomposition · Constant modulus

1 Introduction

A polarization sensitive sensor comprises $2-6$ electromagnetic (EM) sensors that provide complete or partial measurements of the EM fields induced by the incident sources [1,2]. A polarization sensitive array consists of multiple polarization sensitive sensors, arranged into a certain spatial configuration, e.g. linear, circular, L-shaped. Compared with the conventional scalar sensor array, the polarization sensitive array is polarization sensitive. As such, during the past several decades, there have been enormous efforts devoted to the development of direction-of-arrival (DOA) and polarization estimation techniques based on the polarization sensitive array [3–7]. These works have revealed the advantages of the polarization sensitive array over scalar sensor array, with regard to parameter estimation accuracy and identifiability.

© Springer Nature Switzerland AG 2019
H. Lu et al. (Eds.): ISNN 2019, LNCS 11555, pp. 575–584, 2019.
https://doi.org/10.1007/978-3-030-22808-8_57

As mentioned above, the polarization sensitive array output signal admits a multi-dimensional (MD) structure in the time-space-polarization domain. However, the early methods, such as multiple signal classification (MUSIC) [3,4] and estimation of signal parameters via rotational invariance techniques (ESPRIT) [5–7], do not fully exploit this MD structure. Tensor based methods for array processing were proposed in the past decades. In [8–11], the canonical polyadic decomposition (CPD) of third order tensor is used for DOA estimation based on a multi-invariance array. Third-order and fourth-order CPDs is widely adopted in a number of polarization sensitive array processing techniques, which are shown to have better performance than the matrix based approaches.

The above methods are mainly based on unconstrained tensor decompositions. In practice, however, prior knowledge of either the source signal or the sensor array is often available. Imposing these priors in tensor decomposition as certain structure or constraint may result in better performance with regard to both accuracy and identifiability.

In this study, we consider DOA and polarization estimation of CM signals based on a polarization sensitive array, using CM constrained CPD. By applying the analytic CM factorization algorithm (ACMA) [14] and imposing the CPD structure, we propose an algebraic CPD algorithm with CM constraint. The proposed algorithm is algebraic. Therefore, It is computationally efficient and stable, and can be used to effectively initialize the optimization based algorithms (e.g., the structured data fusion (SDF) implementation of a CPD with CM constraints). In addition, through simulations we have found that the proposed algorithm can generate correct results even in case where the unconstrained CPD is not unique. This implies that the proposed CPD algorithm with CM constraint has more relaxed uniqueness conditions than the unconstrained CPD.

Notation: Scalars, vectors, matrices and tensors are denoted by italic lowercase, lowercase boldface, uppercase boldface and uppercase calligraphic letters, respectively. The rth column vector and the (i, j)th entry of $\mathbf{X}$ are denoted by $\mathbf{x}_r$ and $x_{i,j}$, respectively. The identity matrix and all-zero vectors are denoted by $\mathbf{I}_M \in \mathbb{R}^{M \times M}$ and $\mathbf{0}_M \in \mathbb{R}^{M \times M}$, respectively. The null space of a matrix $\mathbf{X}$ is denoted as $\ker(\mathbf{X})$. The dimensionality of a vector space $\Im$ is denoted as $\dim(\Im)$. Transpose, conjugate, conjugated transpose, Moore-Penrose pseudo-inverse, Frobenius norm and matrix determinant are denoted as $(\cdot)^T$, $(\cdot)^*$, $(\cdot)^H$, $(\cdot)^\dagger$, $\|\cdot\|_F$, and $|\cdot|$, respectively.

The symbols '$\otimes$', '$\odot$' and '$\circledast$' denote Kronecker product, Khatri-Rao product, and outer product, respectively. The symbol '$\angle$' denotes the phase angle of a complex number, and 'im$(\cdot)$' denotes the imaginary part of its entry. The letter 'i' denotes the imaginary unit.

A polyadic decomposition (PD) of $\mathcal{X}$ expresses $\mathcal{X}$ as the sum of rank-1 terms:

$$\mathcal{X} = \sum_{r=1}^{R} \mathbf{a}_r \circledast \mathbf{b}_r \circledast \mathbf{c}_r = [\![\mathbf{A}, \mathbf{B}, \mathbf{C}]\!]_R, \tag{1}$$

where $\mathbf{A} \stackrel{\Delta}{=} [\mathbf{a}_1, \cdots, \mathbf{a}_R] \in \mathbb{C}^{I \times R}$, $\mathbf{B} \stackrel{\Delta}{=} [\mathbf{b}_1, \cdots, \mathbf{b}_R] \in \mathbb{C}^{J \times R}$, and $\mathbf{C} \stackrel{\Delta}{=}$ $[\mathbf{c}_1, \cdots, \mathbf{c}_R] \in \mathbb{C}^{K \times R}$. We call Eq. (1) a canonical PD (CPD) if R is minimal.

For a matrix $\mathbf{A} \in \mathbb{C}^{I_1 \times I_2}$, $\mathrm{vec}\,(\mathbf{A}) \in \mathbb{C}^{I_1 I_2}$ denotes the vector representation of $\mathbf{A}$: $[\mathrm{vec}\,(\mathbf{A})]_{\tilde{i}} \stackrel{\Delta}{=} a_{i_1, i_2}$, with $\tilde{i} = (i_1 - 1) I_1 + i_2$, while $\mathrm{unvec}\,(\cdot)$ performs the inverse. The matrix representation[1] of $\mathcal{X} \in \mathbb{C}^{\mathbf{I} \times \mathbf{J} \times \mathbf{K}}$ is denoted as $\mathbf{X} \in \mathbb{C}^{IJ \times K}$, and defined by $\mathbf{X}_{((i-1)J+j,k)} = \mathcal{X}_{(\mathbf{i},\mathbf{j},\mathbf{k})}$.

We define $\mathrm{Ten}\,(\mathbf{X}, [I, J, K]) = \mathcal{X}$ as the operation to reshape an $IJ \times K$ matrix $\mathbf{X}$ into a third-order $\mathcal{X}$ of size $I \times J \times K$, such that $x_{i,j,k} = \mathbf{X}_{((i-1)J+j,k)}$. We denote the estimate of a variable $\mathbf{a}$ as $\tilde{\mathbf{a}}$.

2   Data Model and Problem Formulation

2.1   Data Model

We assume that (θ, φ) is the azimuth-elevation 2D DOA of a narrowband planar EM signal, and that (γ, η) is the polarization state of the incident signal, $0 \leq \theta < 2\pi$, $0 \leq \varphi < \pi$, $0 \leq \gamma < \pi/2$, $-\pi \leq \eta < \pi$. The response of a cocentered polarization sensitive sensor can be written as [2]:

$$\mathbf{b}_{\theta,\varphi,\gamma,\eta} \stackrel{\Delta}{=} \mathbf{L} \begin{bmatrix} \mathbf{e}_{\theta,\varphi,\gamma,\eta} \\ \mathbf{h}_{\theta,\varphi,\gamma,\eta} \end{bmatrix}, \tag{2}$$

where

$$\begin{bmatrix} \mathbf{e}_{\theta,\varphi,\gamma,\eta} \\ \mathbf{h}_{\theta,\varphi,\gamma,\eta} \end{bmatrix} = \begin{bmatrix} \mathbf{v}_{(\theta+\pi/2,0)}, & -\mathbf{v}_{(\theta,\varphi-\pi/2)} \\ \mathbf{v}_{(\theta,\varphi-\pi/2)}, & \mathbf{v}_{(\theta+\pi/2,0)} \end{bmatrix} \begin{bmatrix} \cos\gamma \\ \sin\gamma e^{i\eta} \end{bmatrix}, \tag{3}$$

where $\mathbf{e}_{\theta,\varphi,\gamma,\eta} \in \mathbb{C}^3$ and $\mathbf{h}_{\theta,\varphi,\gamma,\eta} \in \mathbb{C}^3$ are vectors holding the electric field components and magnetic field components of the incident EM waves, respectively. We denote the vector $\mathbf{v}_{(\theta,\varphi)} \stackrel{\Delta}{=} [\cos\theta\sin\varphi, \sin\theta\sin\varphi, \cos\varphi]^T$ as the Poynting vector. By definition, $\mathbf{v}_{(\theta,\varphi)}$, $\mathbf{v}_{(\theta+\pi/2,0)}$, $\mathbf{v}_{(\theta,\varphi-\pi/2)}$ constitute a mutually orthogonal triad. The vector $\mathbf{b}_{\theta,\varphi,\gamma,\eta}$ can be denoted as the angular-polarization steering vector.

The binary matrix $\mathbf{L} \in \mathbb{R}^{K \times 6}$ is a selection matrix. It chooses a subset of the complete EM field components according to the type of the employed polarization sensitive sensor. For example, if we use the tripole antenna, then $\mathbf{L} = [\mathbf{I}_3, \mathbf{0}_3]$ and $\mathbf{b}_{\theta,\varphi,\gamma,\eta} = \mathbf{e}_{\theta,\varphi,\gamma,\eta}$.

Now the data model of a polarization sensitive array which comprises N sensors is introduced. We encapsulate the position coordinates of the nth polarization sensitive sensor in a vector $\mathbf{k}_n \in \mathbb{R}^3$, $n = 1, \ldots, N$. The different polarization sensitive sensors can collect the phase delays of the signals. The phase delays can be represented by the so-called spatial steering vector defined as follows:

$$\mathbf{a}_{\theta,\varphi} \stackrel{\Delta}{=} \exp\left(\mathrm{i}2\pi\lambda^{-1}\left[\mathbf{k}_1{}^T\mathbf{v}_{(\theta,\varphi)}, \ldots, \mathbf{k}_N{}^T\mathbf{v}_{(\theta,\varphi)}\right]^T\right), \tag{4}$$

---

[1] Note that there are various types of matrix representation of a third-order tensor, and that the one defined here is indeed the mode-3 matrix representation. In this paper we only consider this type of matrix representation.

where λ is the wavelength of the incident signal. Therefore, we can write the polarization array response to a single incident signal with DOA-polarization parameters $(\theta, \varphi, \gamma, \eta)$ as:

$$\mathcal{X}_{(\theta,\varphi,\gamma,\eta)}(n, k, t) = \mathbf{a}_{(\theta,\varphi)}(n) \cdot \mathbf{b}_{(\theta,\varphi,\gamma,\eta)}(k) \cdot \mathbf{s}(t), \tag{5}$$

where $\mathbf{s} \in \mathbb{C}^T$ is the source vector, it contains the complex envelop collected at T time samples. We can know the response of the polarization sensitive array to a single signal from Eq. (5) is a third-order rank-1 tensor of size $N \times K \times T$: $\mathcal{X}_{(\theta,\varphi,\gamma,\eta)} = \mathbf{a}_{(\theta,\varphi)} \otimes \mathbf{b}_{(\theta,\varphi,\gamma,\eta)} \otimes \mathbf{s}$. We assume that there are R incident signals, then can denote $\mathcal{X}_{\mathbf{r}} = \mathcal{X}_{(\theta_{\mathbf{r}},\varphi_{\mathbf{r}},\gamma_{\mathbf{r}},\eta_{\mathbf{r}})}$, $\mathbf{a}_r = \mathbf{a}_{(\theta_r,\varphi_r)}$, $\mathbf{b}_r = \mathbf{b}_{(\theta_r,\varphi_r,\gamma_r,\eta_r)}$, where $\mathbf{s}_r$ is the rth source vector. We can write the output of the polarization sensitive array as:

$$\mathcal{X} = \sum_r^R \mathcal{X}_r = \sum_r^R \mathbf{a}_r \otimes \mathbf{b}_r \otimes \mathbf{s}_r = [\![\mathbf{A}, \mathbf{B}, \mathbf{S}]\!]_R, \tag{6}$$

where $\mathbf{A} \triangleq [\mathbf{a}_1, \ldots, \mathbf{a}_R]$, $\mathbf{B} \triangleq [\mathbf{b}_1, \ldots, \mathbf{b}_R]$, $\mathbf{S} \triangleq [\mathbf{s}_1, \ldots, \mathbf{s}_R]$ denote the spatial steering matrix, the angular-polarization steering matrix, and the source signal matrix, respectively. For convenience, we have ignored the noise term in the above model. We can see that the polarization sensitive array signal shown in Eq. (6) admits a CPD. Therefore, the matrix representation of Eq. (6) can be written as:

$$\mathbf{X} = (\mathbf{A} \odot \mathbf{B})\, \mathbf{S}^T, \tag{7}$$

where $\mathbf{X}$ denotes matrix representation of tensor $\mathcal{X}$.

2.2   Problem Formulation

We want to estimate the DOA and polarization parameters from the observed tensor $\mathcal{X}$, and we can do it via a CPD based approach. Briefly speaking, we can compute the CPD of $\mathcal{X}$ to obtain the factor matrices $\mathbf{A}$ and $\mathbf{B}$, and then estimate DOA and polarization parameters from $\mathbf{B}$.

In this paper, we consider all source are CM signals, and we incorporate this prior knowledge as a constraint into the CPD decomposition.

We formulate the problem as CPD with CM constraint:

$$\left\{ \tilde{\mathbf{A}}, \tilde{\mathbf{B}}, \tilde{\mathbf{S}} \right\} = \underset{\{\mathbf{A},\mathbf{B},\mathbf{S}\}}{\arg \min} \left\| \mathbf{X} - (\mathbf{A} \odot \mathbf{B})\, \mathbf{S}^T \right\|_F^2,$$
$$s.t.\, |s_{t,r}| = c, r = 1, \cdots, R, t = 1, \cdots, T, \tag{8}$$

where c denotes the modulus of the source signals.

The proposed algorithm mainly makes use of the well-known analytical constant modulus factorization algorithm (ACMA) [14] and the inherent rank-1 structure of the matrix representation of each column of $(A \odot B)$. To facilitate the use of ACMA, we have the additional assumption: $(A \odot B)$ has full column rank.

3   Proposed Algorithm

3.1   Review of ACMA

A brief summary of the ACMA is introduced at first. The readers can find more detains in [14]. The signal model can be written as

$$\mathbf{Y} = \mathbf{M}\mathbf{S}^T, \tag{9}$$

where $\mathbf{X} \in \mathbb{C}^{M \times T}$, $\mathbf{M} \in \mathbb{C}^{M \times R}$, and $\mathbf{S} \in \mathbb{C}^{T \times R}$ denote the observe signal, the mixing matrix, and the source matrix, respectively. We assume that $\mathbf{S}$ has CM constraint. We denote $\mathbf{W} \in \mathbb{C}^{R \times R}$ as the demixing matrix.

The key steps of ACMA are as Table 1:

Table 1. The ACMA (analytic CM factorization algorithm).

1. Estimate the row space of $\mathbf{Y}$: $\{\mathbf{v}_1, \cdots, \mathbf{v}_R\}$
2. Construct matrix $\hat{\mathbf{P}} \in \mathbb{C}^{(T-1) \times R^2}$, estimate $\ker\left(\hat{\mathbf{P}}\right)$: $\{\mathbf{z}_1, \cdots, \mathbf{z}_R\}$
3. Obtain a tensor from $\{\mathbf{z}_1, \cdots, \mathbf{z}_R\}$: $\mathcal{Z} \triangleq \mathrm{Ten}\left([\mathbf{z}_1, \cdots, \mathbf{z}_R], R, R, R\right) = [\![\mathbf{W}, \mathbf{W}^*, \mathbf{F}]\!]_\mathbf{R}$
4. Compute the overdetermined CPD $\mathcal{Z}$ to obtain $\mathbf{W}$
5. Obtain the signal sources matrix: $\mathbf{S}^T = \mathbf{W}\mathbf{Y}$.

3.2   Algebraic CPD with CM Constraint

In this subsection, we propose an algebraic CPD algorithm with CM constraint. The algorithm mainly contains two steps as follow:

(i) Use ACMA to recognize the CM source signals.

We denote $\mathbf{M} \triangleq \mathbf{A} \odot \mathbf{B}$ and rewrite Eq. (7) as $\mathbf{X} = \mathbf{M}\mathbf{S}^T$. Because $(\mathbf{A} \odot \mathbf{B})$ has full column rank, we can recognize the CM source signals $\tilde{\mathbf{S}}$ via ACMA.

(ii) Use rank-1 approximation to compute factor matrices $\tilde{\mathbf{A}}$ and $\tilde{\mathbf{B}}$.

We have estimated the signal sources matrix $\tilde{\mathbf{S}}$, therefore, we can calculate the other factor matrices of the CPD Eq. (6), $\tilde{\mathbf{A}}$ and $\tilde{\mathbf{B}}$, by using the Kronecker-product structure of each column of the matricized columns of [15]:

$$\tilde{\mathbf{A}} \odot \tilde{\mathbf{B}} = \mathbf{X}\left(\tilde{\mathbf{S}}^T\right)^\dagger. \tag{10}$$

In fact, if we ignore the noise term, each column of $\tilde{\mathbf{M}} = \tilde{\mathbf{A}} \odot \tilde{\mathbf{B}}$ is a vectorized rank-1 matrix:

$$\mathrm{unvec}\left(\tilde{\mathbf{m}}_r\right) = \tilde{\mathbf{a}}_r \tilde{\mathbf{b}}_r^T, r = 1, \cdots, R. \tag{11}$$

If we consider the noise term, Eq. (11) holds approximately, and $\tilde{\mathbf{a}}_r$ and $\tilde{\mathbf{b}}_r$ can be computed via the rank-1 approximation of $\mathrm{unvec}\left(\tilde{\mathbf{m}}_r\right)$.

3.3   DOA-Polarization Estimation

Now we give the method to compute DOA and polarization parameters via the estimates $\tilde{\mathbf{A}}$ and $\tilde{\mathbf{B}}$.

For convenience, we use the tripole antenna to receive the source signals. Therefore, the angular-polarization steering vector can be written as $\mathbf{b}_{\theta,\varphi,\gamma,\eta} = \mathbf{e}_{\theta,\varphi,\gamma,\eta}$. According to [16], we can get a result as follow:

$$\mathbf{e}_{\theta,\varphi,\gamma,\eta} \times \mathbf{e}^{*}_{\theta,\varphi,\gamma,\eta} = 2\mathrm{i}\sin\eta\sin\gamma\mathbf{v}_{(\theta,\varphi)}, \tag{12}$$

where the Poynting vector $\mathbf{v}_{(\theta,\varphi)}$ represents the DOA of the incident signal. Then the polarization parameters can be obtained from $\mathbf{b}_{\theta,\varphi,\gamma,\eta}$. For example, we first construct a vector:

$$\rho = \left[\frac{e_x}{e_z}, \frac{e_y}{e_z}\right]^T = \left[\frac{b_x}{b_z}, \frac{b_y}{b_z}\right]^T$$
$$= \left[\begin{array}{c} -\cot\varphi\cos\theta + \cot\gamma\frac{\sin\theta}{\sin\varphi}\cos\eta - \mathrm{i}\cot\gamma\frac{\sin\theta}{\sin\varphi}\sin\eta \\ -\cot\varphi\sin\theta - \cot\gamma\frac{\cos\theta}{\sin\varphi}\cos\eta + \mathrm{i}\cot\gamma\frac{\cos\theta}{\sin\varphi}\sin\eta \end{array}\right]. \tag{13}$$

Then we can obtain (γ, η) from:

$$\begin{cases} \eta = -\angle b_x\sin\theta - b_y\cos\theta \\ \gamma = \mathrm{arccot}\left(\frac{\mathrm{im}(b_y\sin\varphi)}{\cos\theta\sin\eta}\right) \end{cases}. \tag{14}$$

As such, for each column $\tilde{\mathbf{b}}_r$ of matrix $\tilde{\mathbf{B}}$, we can exploit cross-product Eq. (12) to estimate the DOA of each signal and exploit Eqs. (13) and (14) to estimate the polarization parameters of each incident signal.

We note that the triple which consists of three sensors is used in the above derivation. For the polarization sensitive sensors with four to six sensors, the above derivation also applies.

4   Simulation Results

Now we provide simulations to demonstrate the performance of the proposed algebraic CPD algorithm, in comparison with the optimization based algorithm, and unconstrained CPD.

For convenience, the following abbreviations can be used:

- CPD-CM-ALG: the proposed algebraic CPD-CM algorithm.
- CPD-CM-QN (ALG): the quasi-Newton CPD-CM algorithm, initialized with the result of CPD-CM-ALG.
- CPD-CM-QN (RAND): the quasi-Newton CPD-CM algorithm with random initialization.
- CPD: unconstrained canonical polyadic decomposition, initialized with algebraic CPD.

We implement CPD-CM-QN (ALG) and CPD-CM-QN (RAND) with the 'sdf_minf' function, where the CM constraint is incorporated as a regularization term via the use of domain specific language. The tolerance on the relative function value, relative step size and maximum number iterations for 'sdf_minf' are set to TolFun $= 10^{-10}$, TolX $= 10^{-8}$ and MaxIter $= 2000$, respectively. In the implementation of CPD, the tolerance on the relative function value and relative step size in the stopping criteria of 'cpd_nls' are set to TolFun $= 10^{-12}$ and TolX $= 10^{-8}$, respectively.

We assume that R far-field, narrowband signals with identical carrying frequency are impinging upon an L-shaped array of triple antennas. The spacing between adjacent antennas is $d = 6\lambda$. The number of snapshots is set to $T = 1000$. Therefore, the array output signals is a third-order tensor $\mathcal{X}$ of size $3 \times 3 \times 1000$, admitting a CPD of rank R. In simulations, we mainly consider the case $R > 3$. In this case, the first two dimensions of $\mathcal{X}$ are smaller than R, and CPD is usually labelled as "underdetermined". In addition, the noise term is generated as white Gaussian noise. The signal-to-noise ratio (SNR) is defined as follows:

$$SNR \triangleq 20\log_{10}\left(P_s/P_n\right). \tag{15}$$

where P_s and P_n denote the signal and noise levels, respectively. We use the *Root Mean Squared Error (RMSE)* to measure the accuracy of DOA and polarization estimation for each signal, which is defined as follows:

$$RMSE = \sqrt{\sum_{m=1}^{M} \frac{(\tilde{\alpha}_m - \alpha)^2}{M}}, \tag{16}$$

where $\alpha \in \{\theta, \varphi, \gamma, \eta\}$ denotes one of the to-be-estimated parameters, $\tilde{\alpha}_m$ denotes the estimate of α in the mth Monte Carlo experiment, and M denotes the number of Monte Carlo runs, which is set to $M = 200$ in all simulations. For the evaluation of the overall accuracy, we further use the Overall $RMSE$, which is defined as the mean $RMSE$ values of all the signals.

We consider DOA and polarization estimation under CM constraint in the following two settings: (i) a slightly underdetermined case $R = 4$; (ii) a highly underdetermined case $R = 6$;. The corresponding parameter settings are listed in Tables 2 and 3, respectively.

Table 2. Simulation setting of DOA and polarization parameters in the slightly underdetermined case: $N = K = 3$, $R = 4$.

	#1	#2	#3	#4
θ	$\pi/16$	$25\pi/48$	$47\pi/48$	$23\pi/16$
φ	$\pi/16$	$3\pi/16$	$5\pi/16$	$7\pi/16$
γ	$\pi/16$	$3\pi/16$	$5\pi/16$	$7\pi/16$
η	$\pi/16$	$3\pi/16$	$5\pi/16$	$7\pi/16$

We let SNR vary from 0 dB to 40 dB. The Overall RMSE curves versus SNR for the above two settings are plotted in Figs. 1 and 2, respectively.

From Fig. 1, we have observed that the proposed CPD-CM-ALG algorithm and unconstrained CPD all yield accurate DOA and polarization estimates in slightly underdetermined case. We also observe that the optimization based CPD-CM-QN algorithm provides correct results when it is initialized by the results of CPD-CM-ALG, while the optimization based CPD-CM-QN algorithm dose not yield accurate results when it is initialized by random value.

Table 3. Simulation setting of DOA and polarization parameters in the highly under-determined case: $N = K = 3$, $R = 6$.

	#1	#2	#3	#4	#5	#6
θ	$\pi/24$	$3\pi/8$	$17\pi/24$	$25\pi/24$	$11\pi/6$	$41\pi/24$
φ	$\pi/24$	$\pi/8$	$5\pi/24$	$7\pi/24$	$3\pi/8$	$11\pi/24$
γ	$\pi/24$	$\pi/8$	$5\pi/24$	$7\pi/24$	$3\pi/8$	$11\pi/24$
η	$\pi/24$	$\pi/8$	$5\pi/24$	$7\pi/24$	$3\pi/8$	$11\pi/24$

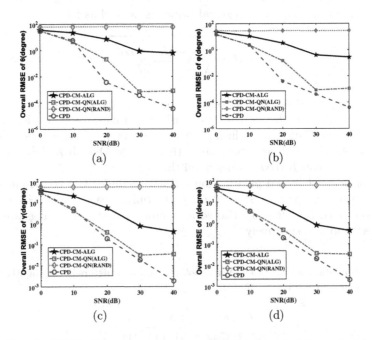

(a)   (b)

(c)   (d)

Fig. 1. Overall RMSE of θ, φ, γ, and η versus SNR in the slightly underdetermined case with CM constraint: $N = K = 3$, $R = 4$, $T = 1000$.

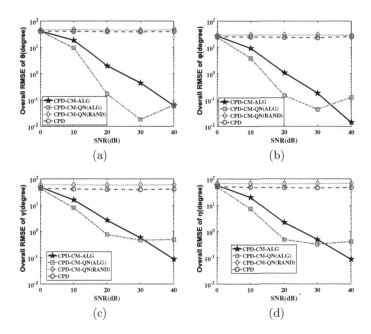

Fig. 2. Overall RMSE of θ, φ, γ, and η versus SNR in the highly underdetermined case with CM constraint: $N = 3$, $K = 3$, $R = 6$, $T = 1000$.

From Fig. 2, we have observed that the proposed CPD-CM-ALG algorithm yields correct results in highly underdetermined case, if SNR is sufficiently high, while unconstrained CPD does not provide correct results. We also observe that the optimization based CPD-CM-QN algorithm provides correct results when it is initialized by the results of CPD-CM-ALG. The optimization based CPD-CM-QN algorithm improves the results of CPD-CM-ALG. However, CPD-CM-QN dose not yield reasonable estimates when it is randomly initialized. The above observation has clearly shown the interests in the proposed CPD-CM-ALG algorithm.

In the above simulations, we observe that the proposed algorithm is faster than the unconstrained CPD.

5 Conclusion

In this study, we have proposed an algebraic algorithm for CPD with CM constraint: CPD-CM-ALG. The proposed algorithm first exploits the analytic CM algorithm to compute the sources matrix, and then uses the CPD structure to calculate the rest factor matrices, form which we can obtain the DOA and polarization parameters. We have shown, through simulations, that the proposed algorithm has more relaxed uniqueness conditions than unconstrained CPD. In comparison with the optimization based quasi-Newton algorithm, which is shown to be unstable and sensitive to initialization, the proposed algorithm is faster and stable, and can be used to effectively initialize the optimization based algorithms.

References

1. Compton Jr., R.T.: The tripole antenna: an adaptive array with full polarization flexbility. IEEE Trans. Antennas Propag. **29**(6), 944–952 (1981)
2. Nehorai, A., Paldi, E.: Vector-sensor array processing for electromagnetic source localization. IEEE Trans. Signal Process. **42**(2), 376–398 (1994)
3. Wong, K.T., Zoltowski, M.D.: Self-initiating MUSIC-based direction finding and polarization estimation in spatio-polarizational beamspace. IEEE Trans. Antennas Propag. **48**(8), 1235–1245 (2000)
4. Wong, K.T., Li, L.S., Zoltowski, M.D.: Root-MUSIC-based direction finding and polarization estimation using diversely polarized possibly collocated antennas. IEEE Antennas Wirel. Propag. Lett. **3**(1), 129–132 (2004)
5. Wong, K.T., Zoltowski, M.D.: Uni-vector-sensor ESPRIT for multisource azimuth, elevation, and polarization estimation. IEEE Trans. Antennas Propag. **45**(10), 1467–1474 (1997)
6. Wong, K.T., Zoltowski, M.D.: Closed-form direction finding and polarization estimation with arbitrarily spaced electromagnetic vector-sensors at unknown locations. IEEE Trans. Antennas Propag. **48**(5), 671–681 (2000)
7. Zoltowski, M.D., Wong, K.T.: ESPRIT-based 2-D direction finding with a sparse uniform array of electromagnetic vector sensors. IEEE Trans. Signal Process. **48**(8), 2195–2204 (2000)
8. Sidiropoulos, N.D., Bro, R., Giannakis, G.B.: Parallel factor analysis in sensor array processing. IEEE Trans. Signal Process. **48**(8), 2377–2388 (2000)
9. Gong, X.F., Liu, Z.W., Xu, Y.G.: Regularised parallel factor analysis for the estimation of direction-of-arrival and polarisation with a single electromagnetic vector-sensor. IET Signal Process. **5**(4), 390–396 (2011)
10. Gong, X.F., Wang, X.L., Lin, Q.H.: Generalized non-orthogonal joint diagonalization with LU decomposition and successive rotation. IEEE Trans. Signal Process. **63**(5), 1322–1334 (2015)
11. Gong, X.F., Lin, Q.H., Cong, F.Y., Lathauwer, L.D.: Double coupled canonical polyadic decomposition for joint blind source separation. IEEE Trans. Signal Process. **66**(13), 3475–3490 (2018)
12. Vervliet, N., Debals, O., Sorber, L., Barel, M.V., Lathauwer, L.D.: Tensorlab 3.0. https://www.tensorlab.net/
13. Lathauwer, L.D., Castaing, J.: Tensor-based techniques for the blind separation of DS-CDMA signals. Signal Process. **87**(2), 322–336 (2007)
14. Van Der Veen, A.J., Paulraj, A.: An analytical constant modulus algorithm. IEEE Trans. Signal Process. **44**(5), 1136–1155 (1996)
15. Sorensen, M., Lathauwer, L.D.: New uniqueness conditions for the canonical polyadic decomposition of third-order tensors. SIAM J. Matrix Anal. Appl. **36**(4), 1381–1403 (2015)
16. Gong, X.F., Jiang, J.C., Li, H., Xu, Y.G., Liu, Z.W.: Spatially spread dipole/loop quint for vector-cross-product-based direction finding and polarisation estimation. IET Signal Process. **12**(5), 636–642 (2018)

Wideband Direction Finding via Spatial Decimation and Coupled Canonical Polyadic Decomposition

Gui-Chen Yu[(⊠)], Xiao-Feng Gong, Jia-Cheng Jiang, Zhi-Wen Liu, and You-Gen Xu

School of Information and Communication Engineering, Dalian University of Technology, Dalian 116023, China
`yuguichen@mail.dlut.edu.cn`

Abstract. We propose an algorithm for direction of arrival (DOA) estimation of wideband array signals based on spatial decimation and coupled canonical polyadic decomposition. First, the wide-band array out is converted to several frequency bins. Second, we choose the bins that have exponential relationship to do spatial decimation. Finally, we use the coupled canonical polyadic decomposition to identify the steering vector and calculate the DOA. The proposed algorithm makes use of the coupled structures between frequency bins and is the deterministic method, which do not depend on statistical properties. We have compared the proposed algorithm with existing wideband DOA estimation algorithms. The numerical results are provided to demonstrate the performance of the proposed algorithm.

Keywords: Array signal processing · Direction of arrival estimation · Spatial decimation · Coupled canonical polyadic decomposition

1 Introduction

Direction of arrival (DOA) estimation is a central problem in array signal processing, in particular for radar, sonar, telecommunications, astronomy and seismology. The array consists of multiple sensors placed at different position in space and receives source signals from different directions [1,2]. Some array processing techniques perform an exhaustive search of the angular range, such as beamforming and multiple signal classification (MUSIC) [3], whereas others rely on algebraic solutions, such as root-MUSIC [4] and ESPRIT [5]. The signal-subspace processing embodied by MUSIC and ESPRIT exploits the algebraic property of the spatial covariance matrix that the eigenvectors corresponding to the largest eigenvalues span the same subspace as the source direction vectors.

For wideband signals, signal-subspaces are different at different frequency bins, and the mentioned narrowband algorithms above need to be adapted to deal with this diversity. The ISM [6] is one of the earliest algorithms for wideband

© Springer Nature Switzerland AG 2019
H. Lu et al. (Eds.): ISNN 2019, LNCS 11555, pp. 585–594, 2019.
https://doi.org/10.1007/978-3-030-22808-8_58

DOA estimation. The wideband signal is decomposed into multiple narrowband components in frequency domain. Then the method does the spectral decomposition in each frequency bin and computes the average of all frequency bins. The ISM only deals with each individual frequency bin and does not make use of the relationship between frequency bins. The CSM [7] makes use of the relationship between frequencies but needs a prior estimate of DOA to form focusing matrices. In this letter, our main contribution is to handle the frequency point data that has a particular frequency relationship as a multidata set signal, which is essentially a method of multi-set data-fusion or joint blind source separation (J-BSS).

J-BSS of multi-set signals has been considered in a number of applications such as multi-subject/multi-model biomedical data fusion and BSS of transformed signals at multiple frequency bins for convolutive mixtures. These approaches usually assume dependence across datasets (inter-set dependence) and independence of latent sources within a dataset (intra-set independence), aiming at BSS of each individual dataset as well as indication of correspondences among decomposed components. A number of J-BSS algorithms have been proposed, e.g. joint and parallel independent component analysis (ICA) [8], independent vector analysis (IVA) [9–11], algebraic methods such as multi-set canonical correlation analysis (M-CCA) [12–14] and generalized joint diagonalization (GJD) [15–18]. Specially, the algorithm for wideband signal based on decimation and coupled canonical polyadic decomposition (DC-CPD) [19] depends on statistical properties between datasets. DC-CPD is not available while the source signals do not meet the assumptions about the statistical properties.

Therefore, our goal is to develop a method for wideband direction finding, that could take advantage of multi-set and does the joint processing without needing prior DOA estimation nor requiring statistical properties. The proposed method could deal with large aperture uniform lines for statistically dependent source signals and only requires a few sampling points.

1.1 Notation

Vectors, matrices and tensors are denoted by lower case boldface, uppercase boldface and uppercase calligraphic letters, respectively. The rth column vector and the (i, j)th entry of $\mathbf{A}$ are denoted by a_r and $a_{i,j}$, respectively. The symbol $\circ$ denotes tensor outer product, defined as follows:

$$(\mathbf{a} \circ \mathbf{b} \circ \mathbf{c})_{i,j,k} \stackrel{\Delta}{=} a_i b_j c_k.$$

Transpose of matrix $\mathbf{A}$ is denoted as $\mathbf{A}^{\mathrm{T}}$. MATLAB notation will be used to denote submatrices of a tensor, e.g., $\mathcal{T}_{(:,:,k)}$ denotes the frontal slice of tensor $\mathcal{T}$ obtained by fixing the third index to k. A polyadic decomposition (PD) of $\mathcal{T}$ expresses $\mathcal{T}$ as a sum of rank-1 terms:

$$\mathcal{T} = [\![\mathbf{A}, \mathbf{B}, \mathbf{C}]\!]_R \stackrel{\Delta}{=} \sum_{r=1}^{R} \mathbf{a}_r \circ \mathbf{b}_r \circ \mathbf{c}_r \in \mathbb{C}^{I \times J \times K}, \tag{1}$$

where $\mathbf{A} \triangleq [\mathbf{a}_1, ..., \mathbf{a}_R] \in \mathbb{C}^{I \times R}$, $\mathbf{B} \triangleq [\mathbf{b}_1, ..., \mathbf{b}_R] \in \mathbb{C}^{J \times R}$, and $\mathbf{C} \triangleq [\mathbf{c}_1, ..., \mathbf{c}_R] \in \mathbb{C}^{K \times R}$. We call (1) a canonical PD (CPD) if R is minimal.

2 Data Model

We assume an uniform linear array (ULA) of N antennas which receive R far-field wide-band signals. Differential travel time from the reference sensor $(n = 1)$ to the nth element of the array is given by $\tau_n = (b_n^T u)/c$, where b_n is the position of the nth antenna, u is the unit steering vector and c is the propagation speed of electromagnetic waves. For an ULA, it becomes:

$$\tau_n(\varphi_r) = \frac{d}{c}(n-1)\sin\varphi_r, \tag{2}$$

where d is the interspacing between two adjacent antennas and φ_r is the direction of arrival (DOA) of the rth signal. In the presence of noise, the model for the nth antenna's wide-band array out is expressed by:

$$\mathbf{x}_n(t) = \sum_{r=1}^{R} s_r(t - \tau_n(\varphi_r)) + \mathbf{n}_n(t), \tag{3}$$

where $s_r(t)$ is the rth source signal and $\mathbf{n}_n(t)$ is noise.

The wide-band array out (3) is transformed into M narrow-band frequency bins by using the Fourier Transform. In each frequency bin, the array signal admins the narrow-band signal model, as follow:

$$\mathbf{X}^{(m)} = \sum_{r=1}^{R} \mathbf{d}_r^{(m)} \mathbf{s}_r^{(m)T}, \tag{4}$$

where $\mathbf{X}^{(m)} \in \mathbb{C}^{N \times T}$ is the array observation in the mth frequency bin, and $\mathbf{s}_r^{(m)} \in \mathbb{C}^T$, $\mathbf{d}_r^{(m)} \in \mathbb{C}^N$ are the narrow-band component of the rth source and its associated array steering vector in the mth frequency bin, respectively. T is the number of temporal samples of the array signal in the mth frequency bin (also for all the other frequency bins). For notational convenience, we omit the noise term. The steering vector $\mathbf{d}_r^{(m)}$ is formulated as:

$$\mathbf{d}_r^{(m)} = \left[1, e^{(-i2\pi d/c)mf_d \sin(\varphi_r)}, \ldots, e^{(-i2\pi d/c)mf_d \sin(\varphi_r)(N-1)}\right]^T, \tag{5}$$

where f_d is the bandwidth of each narrow-band component where $f_d = f_s/M$. As such, mf_d is the actual frequency of the mth narrow-band component.

For convenience, we denote $z_r \triangleq e^{(-i\pi)f_d \sin(\varphi_r)}$ and then $\mathbf{d}_r^{(m)}$ is briefly written as: $\mathbf{d}_r^{(m)} = \left[1, z_r^m, \ldots, z_r^{m(N-1)}\right]^T$. Then (4) can be rewritten as:

$$\mathbf{X}^{(m)} = \begin{bmatrix} 1 & 1 & \cdots & 1 \\ z_1^m & z_2^m & \cdots & z_R^m \\ \vdots & \vdots & \ddots & \vdots \\ z_1^{m(N-1)} & z_2^{m(N-1)} & \cdots & z_R^{m(N-1)} \end{bmatrix} \begin{bmatrix} \mathbf{s}_1^{(m)} \\ \mathbf{s}_2^{(m)} \\ \vdots \\ \mathbf{s}_R^{(m)} \end{bmatrix}^T, \tag{6}$$

where mixing matrix $\mathbf{X}^{(m)}$ has the Vandermonde structure.

3   Proposed Algorithm

In this section, we propose a wideband DOA estimation method based on spatial decimation and coupled canonical polyadic decomposition (C-CPD). We first show how to formulate a C-CPD via spatial decimation of signals at selected frequency bins. Then we introduce algorithms for the computation of C-CPD.

3.1   Spatial Decimation

The traditional method such as MUSIC is a non-unique matrix factorization problem, while unique solutions can be obtained by imposing additional assumptions such as statistical independence. By mapping the matrix data to a tensor and by using tensor decompositions afterwards, uniqueness is ensured under certain conditions. Hankelization has be discussed in [20] which explains the matrix to tensor step and present tensorization as an important concept on itself, illustrated by a number of stochastic and deterministic tensorization techniques.

Consider an exponential signal $f(k) = az^k$ arranged in a Hankel matrix $\mathbf{H}$. The matrix appears to have rank one:

$$\mathbf{H} = \begin{bmatrix} f(0) & f(1) & f(2) & \cdots \\ f(1) & f(2) & f(3) & \cdots \\ f(2) & f(3) & f(4) & \cdots \\ \vdots & \vdots & \vdots & \end{bmatrix} = a \begin{bmatrix} 1 \\ z \\ z^2 \\ \vdots \end{bmatrix} \begin{bmatrix} 1 & z & z^2 & \cdots \end{bmatrix}. \tag{7}$$

These simple exponential functions can be generalized to exponential polynomials, which are functions that can be written as sums and/or products of exponentials, sinusoids and/or polynomials.

Because of the Vandermonde structure of steering vector, we could choose the specific frequency bins and do the decimation in the frequency domain, which different from the method in [20] that is in the time domain. The method we use that is named spatial decimation.

3.2   C-CPD Formulation via Spatial Decimation

We denote the spatial decimation of $\mathbf{X}^{(m)}$ as which is defined as follows:

$$\left(\mathcal{H}_{n,I,J}\left(\mathbf{X}^{(m)}\right)\right)_{:,:,t} = \begin{bmatrix} \left(\mathbf{X}^{(m)}\right)_{1,t} & \cdots & \left(\mathbf{X}^{(m)}\right)_{J,t} \\ \vdots & \ddots & \vdots \\ \left(\mathbf{X}^{(m)}\right)_{1+(I-1)n,t} & \cdots & \left(\mathbf{X}^{(m)}\right)_{J+(I-1)n,t} \end{bmatrix}$$

$$= \sum_{r=1}^{R} \underbrace{\begin{bmatrix} 1 \\ z_r^{mn} \\ \vdots \\ z_r^{mn(I-1)} \end{bmatrix}}_{a_r^{(mn)}} \circ \underbrace{\begin{bmatrix} 1 \\ z_r^{m} \\ \vdots \\ z_r^{m(J-1)} \end{bmatrix}}_{b_r^{(m)}} \cdot s_r^{(m)}(t) \tag{8}$$

where n denotes the factor of spatial decimation, it represents the rate of decimative sampling along the spatial mode. For instance, $(\mathbf{X}^{(m)})_{i,j}$ means (i,j)th element of the matrix $\mathbf{X}^{(m)}$. I,J are the dimensionalities of the first and second modes of the above tensor. (8) is a CPD, thanks to the Vandermonde structure of the mixing matrix in (6):

$$\mathcal{T}^{(mn,m,m)} \triangleq \mathcal{H}_{n,I,J}\left(\mathbf{X}^{(m)}\right) = \left[\!\left[\mathbf{A}^{(mn)}, \mathbf{B}^{(m)}, \mathbf{S}^{(m)}\right]\!\right]_R, \tag{9}$$

where $A^{(mn)} \triangleq \left[\mathbf{a}_1^{(mn)}, \ldots, \mathbf{a}_R^{(mn)}\right] \in \mathbb{C}^{I \times R}$, $\mathbf{B}^{(m)} \triangleq \left[\mathbf{b}_1^{(m)}, \ldots, \mathbf{b}_R^{(m)}\right] \in \mathbb{C}^{J \times R}$, $\mathbf{S}^{(m)} \triangleq \left[\mathbf{s}_1^{(m)}, \ldots, \mathbf{s}_R^{(m)}\right] \in \mathbb{C}^{T \times R}$. In order to acquire enough sample points, the parameters I and J are selected such as:

$$\begin{aligned} n\,(I-1) &< N \\ J &\leq N - (J-1)\,n \end{aligned} \tag{10}$$

By using the requirement (10) is used to check if there are enough spatial samples N for organizing the array observation in the mth frequency bin into a tensor by spatial decimation. However, it does not guarantee the identifiability of the CPD (9). We note that the classical Hankelization operation is a particular case with $n = 1$.

To illustrate the main idea of using C-CPD for wideband ULA processing, we first give an example. We choose N = 50 and 2 frequency bins, where choosing more frequency bins is in the same way.

We regroup the signals in 2 frequency bins as follows:

$$\begin{aligned} \Upsilon^{(1)} &= \left\{\mathbf{X}^{(1)}, \mathbf{X}^{(2)}\right\} \\ \Upsilon^{(2)} &= \left\{\mathbf{X}^{(2)}\right\} \end{aligned}. \tag{11}$$

The set $\Upsilon^{(1)}$ contains array observation matrices $\mathbf{X}^{(l)}, \mathbf{X}^{(2l)}, \ldots$, of which the superscript indices are integer times of l.

For $\Upsilon^{(1)}$, we perform the following:

$$\begin{aligned} \mathcal{T}^{(1,1,1)} &\triangleq \mathcal{H}_{1,I,J}\left(\mathbf{X}^{(1)}\right) = \left[\!\left[\mathbf{A}^{(1)}, \mathbf{B}^{(1)}, \mathbf{S}^{(1)}\right]\!\right]_R \\ \mathcal{T}^{(2,1,1)} &\triangleq \mathcal{H}_{2,I,J}\left(\mathbf{X}^{(1)}\right) = \left[\!\left[\mathbf{A}^{(2)}, \mathbf{B}^{(1)}, \mathbf{S}^{(1)}\right]\!\right]_R \\ \mathcal{T}^{(3,1,1)} &\triangleq \mathcal{H}_{3,I,J}\left(\mathbf{X}^{(1)}\right) = \left[\!\left[\mathbf{A}^{(3)}, \mathbf{B}^{(1)}, \mathbf{S}^{(1)}\right]\!\right]_R \\ \mathcal{T}^{(4,1,1)} &\triangleq \mathcal{H}_{4,I,J}\left(\mathbf{X}^{(1)}\right) = \left[\!\left[\mathbf{A}^{(4)}, \mathbf{B}^{(1)}, \mathbf{S}^{(1)}\right]\!\right]_R \\ \mathcal{T}^{(2,2,2)} &\triangleq \mathcal{H}_{1,I,J}\left(\mathbf{X}^{(2)}\right) = \left[\!\left[\mathbf{A}^{(2)}, \mathbf{B}^{(2)}, \mathbf{S}^{(2)}\right]\!\right]_R \end{aligned} \tag{12}$$

For $\Upsilon^{(2)}$, we perform the following:

$$\begin{aligned} \mathcal{T}^{(2,2,2)} &\triangleq \mathcal{H}_{1,I,J}\left(\mathbf{X}^{(2)}\right) = \left[\!\left[\mathbf{A}^{(2)}, \mathbf{B}^{(2)}, \mathbf{S}^{(2)}\right]\!\right]_R \\ \mathcal{T}^{(4,2,2)} &\triangleq \mathcal{H}_{2,I,J}\left(\mathbf{X}^{(2)}\right) = \left[\!\left[\mathbf{A}^{(4)}, \mathbf{B}^{(2)}, \mathbf{S}^{(2)}\right]\!\right]_R \end{aligned}. \tag{13}$$

In (12) and (13), we let all the tensors have first-mode dimensionality equal to I, and second mode dimensionality equal to J. I and J are selected so that (10) is satisfied all $n = 1, \cdots, 4$. For example we make $I{=}10$, $J{=}10$.

With (12) and (13) we have obtained eight third-order tensors, with triple coupling structures in all the three modes. More precisely, $\mathcal{T}^{(1,1,1)}$, $\mathcal{T}^{(2,1,1)}$, $\mathcal{T}^{(3,1,1)}$ and $\mathcal{T}^{(4,1,1)}$ are coupled in both the second and third modes by $\mathbf{B}^{(1)}$ and $\mathbf{S}^{(1)}$, respectively. $\mathcal{T}^{(2,2,2)}$ and $\mathcal{T}^{(4,2,2)}$ are coupled in both the second and third modes by $\mathbf{B}^{(2)}$ and $\mathbf{S}^{(2)}$, respectively. In addition, $\mathcal{T}^{(2,1,1)}$ is coupled with $\mathcal{T}^{(2,2,2)}$ in the first mode by $\mathbf{A}^{(2)}$, $\mathcal{T}^{(4,2,2)}$ and $\mathcal{T}^{(4,4,4)}$ are coupled in the first mode by $\mathbf{A}^{(4)}$.

3.3   Computation of C-CPD

In this section, we introduce an algebraic C-CPD algorithm that can solve the problem where there is coupling between frequency bins and individual frequency bins themselves. To compare with ESPRIT [21], the algorithm that will be explained next does not need a prior estimate DOA and requires less sampling points, which only need to be greater than or equal to source number. To compare with the CPD algorithm in [22], the method explained that makes better use of the coupled structures in this letter can get more precise results.

If the tensor is overdetermined, we can use the GEVD [23] or GJD [16] to do the tensor decomposition.

If the tensor is underdetermined, by using rank-1 detection mapping [24], we can convert a possibly underdetermined (coupled) CPD into an overdetermined CPD. For more details, the reader is referred to [24]. The rank-1 detection mapping $\Phi^{(R_1)} : (\mathbf{X}, \mathbf{Y}) \in \mathbb{C}^{I \times J} \times \mathbb{C}^{I \times J} \to \Phi^{(R_1)} : (\mathbf{X}, \mathbf{Y}) \in \mathbb{C}^{I \times I \times J \times J}$ is defined as:

$$\left[\Phi^{(R_1)}(\mathbf{X}, \mathbf{Y}) \right]_{i,j,g,h} \triangleq x_{i,g} y_{j,h} + y_{i,g} x_{j,h} - x_{i,h} y_{j,g} - y_{i,h} x_{j,g}. \tag{14}$$

For instance, $\mathcal{T}^{(2,1,1)} \triangleq \left[\mathbf{A}^{(2)}, \mathbf{B}^{(1)}, \mathbf{S}^{(1)} \right]_R \in \mathbb{C}^{I \times J \times T}$ is coupled with $\mathcal{T}^{(2,2,2)} \triangleq \left[\mathbf{A}^{(2)}, \mathbf{B}^{(2)}, \mathbf{S}^{(2)} \right]_R \in \mathbb{C}^{I \times J \times T}$ in the first mode by $\mathbf{A}^{(2)}$, by using the rank-1 detection mapping, we can get an overdetermined CPD

$$\mathcal{M}^{(1,2,\binom{2,1,1}{2,2,2})} \triangleq \left[\!\!\left[\mathbf{S}^{(1)-T}, \mathbf{S}^{(2)-T}, \mathbf{F}^{\binom{2,1,1}{2,2,2}} \right]\!\!\right]_R \in \mathbb{C}^{R \times R \times R}. \tag{15}$$

For this wide-band model, there are two kinds of coupling. In the same mth frequency bin, because of the different factor of spatial decimation n, the coupled tensors are as follows:

$$\begin{aligned} \mathcal{T}^{(mn_1,m,m)} &= \left[\mathbf{A}^{(mn_1)}, \mathbf{B}^{(m)}, \mathbf{S}^{(m)} \right]_R \\ \mathcal{T}^{(mn_2,m,m)} &= \left[\mathbf{A}^{(mn_2)}, \mathbf{B}^{(m)}, \mathbf{S}^{(m)} \right]_R \\ &\;\;\vdots \\ \mathcal{T}^{(mn_k,m,m)} &= \left[\mathbf{A}^{(mn_k)}, \mathbf{B}^{(m)}, \mathbf{S}^{(m)} \right]_R \end{aligned} \tag{16}$$

where tensors have the same second dimension and the third dimension, we can merge them into one tensor:

$$
\mathcal{T}\begin{pmatrix} mn_1, m, m \\ m,m,(\quad \vdots \quad) \\ mn_k, m, m \end{pmatrix} = \left[\!\!\left[\begin{bmatrix} \mathbf{A}^{(mn_1)} \\ \mathbf{A}^{(mn_2)} \\ \vdots \\ \mathbf{A}^{(mn_k)} \end{bmatrix}, \mathbf{B}^{(m)}, \mathbf{S}^{(m)} \right]\!\!\right]_R, \tag{17}
$$

then (17) can do the rank-1 detection mapping with itself. Between the different frequency bins, when the factor of spatial decimation n and frequency bin m satisfy the condition $m_1 n_1 = m_2 n_2$, the coupled tensors are as follows:

$$
\begin{aligned}
\mathcal{T}^{(m_1 n_1, m_1, m_1)} &= \left[\!\!\left[\mathbf{A}^{(m_1 n_1)}, \mathbf{B}^{(m_1)}, \mathbf{S}^{(m_1)} \right]\!\!\right]_R \\
\mathcal{T}^{(m_2 n_2, m_2, m_2)} &= \left[\!\!\left[\mathbf{A}^{(m_2 n_2)}, \mathbf{B}^{(m_2)}, \mathbf{S}^{(m_2)} \right]\!\!\right]_R
\end{aligned}. \tag{18}
$$

By using the rank-1 detection mapping, we can get the overdetermined CPD such as (15). After the transformation of these two forms, we get some overdetermined tensors which also have coupled structures. Then we can use the C-CPD that has be explained in [19] to estimate the DOA.

4   Simulation Results

We provide simulation results to illustrate the performance of the proposed algorithm. We use the uniform linear array that contains 50 omnidirectional sensors with equal inter-element spacing $d = c/2f_0$ is used, where f_0 is the mid-band frequency and c is the velocity of propagation. We consider the 4 far-field wide-band signals. We choose 2 specific frequency bins in the following. The DOA parameters are $(\theta_1, \theta_2, \theta_3, \theta_4) = (10°, 30°, 45°, 60°)$. The white Gaussian noise is added. The signal-to-noise ratio (SNR) is defined as:

$$
\text{SNR} \triangleq 20\log_{10}(P_s/P_n), \tag{19}
$$

where P_s and P_n are the signal and noise levels, respectively. We use the algorithm addressed in Sect. 3 to calculate the DOA and calculate the root mean squared error (RMSE), ε is used to evaluate the algorithm, which is defined as:

$$
\varepsilon = \sqrt{N^{-1} \sum_{n=1}^{N} (\tilde{\alpha}_n - \alpha)^2}, \tag{20}
$$

where N denotes the number of Monte-Carlo runs, which is set to 500 in the simulation, α denotes one of the to-be-estimated parameters and $\tilde{\alpha}_n$ is its estimate in the nth Monte-Carlo run. We fix the number of snapshots T = 200, and let SNR vary from 0 dB to 20 dB. Then we fixed SNR to 10 dB, and let the number of snapshots vary from 50 to 450 with equal interelement spacing 100. The RMSE in both the two above settings are plotted in Figs. 1 and 2.

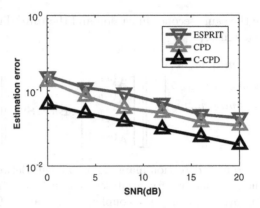

Fig. 1. The RMSE versus SNR. The number of snapshots is fixed to 200

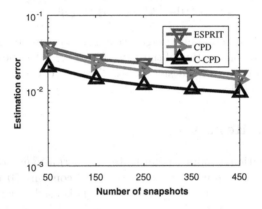

Fig. 2. The RMSE versus the number of snapshots.

The results have shown that, with the increase of SNR for fixed number of snapshots or with the increase of the number of snapshots for fixed SNR, the RMSE is to be less. This verifies the performance of the proposed algorithm.

5   Conclusion

In this paper, we proposed an algorithm for wideband direction finding via spatial decimation and C-CPD. This method makes full use of the complex coupled structures that exist between different frequency bins and in each frequency bin itself. The algorithm does not need a prior DOA estimation and only requires tiny sampling points compared with tradition algorithms and simulation results are given to illustrate the performance of DOA estimator.

Acknowledgments. This research is funded by: (1) National natural science foundation of China (nos. 61331019, 61671106, 61490691, 61601079); (2) Scientific Research Fund of Liaoning Education Department (no. L2014016); (3) Fundamental Research Funds for the Central Universities (nos. DUT16QY07, DUT15RC(3)063).

References

1. Liu, W., Weiss, S.: Wideband Beamforming: Concepts and Techniques. Wiley Series on Wireless Communications and Mobile Computing (2010)
2. Krim, H., Viberg, M.: Two decades of array signal processing research: the parametric approach. IEEE Signal Process. Mag. **13**(4), 67–94 (1996)
3. Schmidt, R.O.: Multiple emitter location and signal parameter estimation. IEEE Trans. Antennas Propag. **34**(3), 276–280 (1986)
4. Barabell, A.J.: Improving the resolution performance of eigenstructure-based direction-finding algorithms. In: IEEE International Conference on Acoustics, Speech, and Signal Processing, ICASSP, vol. 8, pp. 336–339. IEEE Press, Boston (1983)
5. Roy, R., Paulraj, A., Kailath, T.: ESPRIT-estimation of signal parameters via rotational invariance techniques. IEEE Trans. Acoust. Speech Signal Proc. **37**(7), 984–995 (1989)
6. Wax, M., Shan, T.J., Kailath, T.: Spatio-temporal spectral analysis by eigenstructure methods. IEEE Trans. Acoust. Speech Signal Proc. **32**(4), 817–827 (1984)
7. Wang, H., Kaveh, M.: Coherent signal-subspace processing for the detection and estimation of angles of arrival of multiple wide-band sources. IEEE Trans. Acoust. Speech Signal Proc. **33**(4), 823–831 (1985)
8. Sui, J., Adali, T., Yu, Q.B., Calhoun, V.: A review of multivariate methods for multimodal fusion of brain imaging data. J. Neurosci. Meth. **204**, 68–81 (2012)
9. Kim, T., Attias, H.T., Lee, S.Y., Lee, T.W.: Blind source separation exploiting higher-order frequency dependencies. IEEE Trans. Audio Speech Lang. Process. **15**(1), 70–79 (2007)
10. Lee, J.H., Lee, T.W., Jolesz, F., Yoo, S.-S.: Independent vector analysis (IVA): multivariate approach for fMRI group study. NeuroImage **40**(1), 86–109 (2008)
11. Anderson, M., Adali, T., Li, X.L.: Joint blind source separation with multivariate Gaussian model: algorithms and performance analysis. IEEE Trans. Signal Process. **60**(4), 1672–1683 (2012)
12. Li, Y.O., Adali, T., Wang, W., Calhoun, V.: Joint blind source separation by multiset canonical correlation analysis. IEEE Trans. Signal Process. **57**(10), 3918–3929 (2009)
13. Correa, N.M., Eichele, T., Adali, T., Li, Y.O., Calhoun, V.D.: Multi-set canonical correlation analysis for the fusion of concurrent single trial ERP and functional MRI. NeuroImage **50**(4), 1438–1445 (2010)
14. Correa, N.M., Adali, T., Calhoun, V.D.: Canonical correlation analysis for data fusion and group analysis: examining applications of medical imaging data. IEEE Signal Process. Mag. **27**(4), 39–50 (2010)
15. Chatel-Goldman, J., Congedo, M., Phlypo, R.: Joint BSS as a natural analysis framework for EEG-hyperscanning. In: IEEE International Conference on Acoustics, Speech, and Signal Processing, ICASSP. IEEE Press (2013)
16. Li, X.L., Adali, T., Anderson, M.: Joint blind source separation by generalized joint diagonalization of cumulant matrices. Signal Process. **91**(10), 2314–2322 (2011)

17. Congedo, M., Phlypo, R., Chatel-Goldman, J.: Orthogonal and non-orthogonal joint blind source separation in the least-squares sense. In: 2012 Proceedings of the 20th European Signal Processing Conference (EUSIPCO), pp. 27–31. IEEE Press (2012)
18. Gong, X.-F., Wang, X.-L., Lin, Q.-H.: Generalized non-orthogonal joint diagonalization with LU decomposition and successive rotations. IEEE Trans. Signal Process. **63**(5), 1322–1334 (2015)
19. Gong, X.-F., Lin, Q.-H., Cong, F., De Lathauwer, L.: Double coupled canonical polyadic decomposition for joint blind source separation. IEEE Trans. Signal Process. 1 (2018)
20. Papy, J.M., De Lathauwer, L., Huffel, S.V.: Exponential data fitting using multilinear algebra: the decimative case. J. Chemometr. **23**(7–8), 341–351 (2009)
21. Ottersten, B., Kailath, T.: Direction-of-arrival estimation for wide-band signals using the ESPRIT algorithm. IEEE Trans. Acoust. Speech Signal Process. **38**(2), 317–327 (1990)
22. Raimondi, F., Comon, P., Michel, O.: Wideband multilinear array processing through tensor decomposition. In: IEEE International Conference on Acoustics, Speech and Signal Processing, pp. 2951–2955(2016)
23. Parra, L., Sajda, P.: Blind source separation via generalized eigenvalue decomposition. J Mach. Learn. Res. **4**(4), 1261–1269 (2004)
24. De Lathauwer, L.: A link between the canonical decomposition in multilinear algebra and simultaneous matrix diagonalization. SIAM J. Matrix Anal. **28**(3), 642–666 (2006)

Neural Network Methods of HIFU-Therapy Control by Infrared Thermography and Ultrasound Thermometry

Andrey Yukhnev$^{(\boxtimes)}$, Dmitriy Tarkhov, Yakov Gataulin,
Yana Ivanova, and Alexander Berkovich

Peter the Great St. Petersburg Polytechnic University, Saint Petersburg, Russia
a.yukhnev@mail.ru

Abstract. Focused ultrasound-based methods are widely used in various areas of medicine for vascular damage coagulation in limbs and internal organs, venous obliteration, and ablation of breast and thyroid tumors. To plan heat exposure and control its effectiveness, it is necessary to monitor temperature during the treatment procedure. The article proposes two technologies for such monitoring, namely infrared thermography and ultrasound thermometry using neural network methods.

The proposed method of temperature recovery inside the material from surface temperature measurements enhances the potential of infrared thermography and improves the calibration accuracy of ultrasound thermometry.

The proposed method for determining the ultrasound signal shift has not been previously used for ultrasound thermometry. While maintaining acceptable accuracy, it can significantly reduce the computation time, which makes it possible to use it in real time.

Keywords: RBF network · HIFU · Infrared thermography ·
Ultrasound thermometry · Tissue-mimicking phantom

1 Introduction

Wide application of therapeutic devices which use high-intensity focused ultrasound (HIFU) for tissue heating in clinical practice is difficult due to the fact that they require the assessment of heat exposure in real time. Insufficient exposure can lead to recurrence and even accelerated growth of tumors. Excessive exposure may cause damage to surrounding healthy tissues [1].

The magnetic resonance method is generally used to assess heat exposure. However, its equipment is very expensive and large. Over the last decades, other noninvasive methods, such as infrared thermography and ultrasound thermometry (UST), have been actively developed to assess thermal exposure during HIFU-therapy. Both samples of biological tissues and tissue-mimicking materials (TMM) are used for their development and testing [2].

Calibration of ultrasound thermometry near the surface with infrared thermography [3] has errors due to the ultrasound reflection from the air-TMM interface. Signals are generally processed in ultrasound thermometry using cross-correlation methods [4], which requires significant computational resources.

© Springer Nature Switzerland AG 2019
H. Lu et al. (Eds.): ISNN 2019, LNCS 11555, pp. 595–602, 2019.
https://doi.org/10.1007/978-3-030-22808-8_59

The article outlines a method for restoring the temperature field inside a tissue-mimicking material, which was applied for the first time during its HIFU heating by recording the distribution of material surface temperature with an infrared camera, and a method for measuring it using the results of ultrasound scanning in the plane perpendicular to the axis of the focused ultrasound transducer. The original method is used for the first time to process the signal in ultrasound thermometry with HIFU-heating [5]. The results of both methods are compared with the results of direct temperature measurements with thermal sensors. Graphite-filled agar-agar, which simulates the physical properties of biological tissues, was used as a TMM of the phantom.

2 Methods

According to [6], the temperature field in the tissues satisfies the equation

$$\frac{\partial T}{\partial t} = \alpha \Delta T - bT + S(t, \mathbf{x}) \tag{1}$$

Here T is the temperature field, α is the coefficient of thermal diffusivity, b is the coefficient reflecting heat transfer by blood, and $S(t, \mathbf{x})$ is the ultrasonic heat source. The objective is to restore the 3D temperature field using the surface temperature measurements with infrared thermography, which will be discussed later.

Source $S(t, \mathbf{x})$ might be restored from the phantom surface temperature measurements using methods [7–15], but they require quite long computations and are difficult to implement in real time. We use a new method based on neural network decomposition of the source of ultrasonic radiation, which made it possible to reduce signal processing time. In this case, the solution is decomposed by new basis functions corresponding to differential equation under consideration (1). Due to the linearity of Eq. (1), it is logical to form the required temperature field as a result of the total effect of elementary sources, which has the form of an RBF network:

$$S(t, \mathbf{x}) = \sum_{i=1}^{n} c_i S_i(t, \mathbf{x} - \mathbf{x}'_i) \tag{2}$$

In this case, in accordance with [6], it is possible to use S_i as a radial basis function, which in the cylindrical coordinate system has the following form:

$$S_i(t, \mathbf{x}) = Q_i(t) \exp[-r^2/\beta_r(i)] \exp[-z^2/\beta_z(i)] \tag{3}$$

As a result, solution of Eq. (1) takes the following form:

$$T(t, \mathbf{x}) = \sum_{i=1}^{n} c_i T_i(t, \mathbf{x}) \tag{4}$$

where T_i is the solution of Eq. (1) for $S(t, \mathbf{x}) = S_i(t, \mathbf{x})$.

For HIFU heating near the surface of the tissue-mimicking material, we will solve Eq. (1) in half-space. By replacing $T(t, \mathbf{x}) = u(t, \mathbf{x}) \exp[-bt]$, Eq. (1) is reduced to

$$\frac{\partial u}{\partial t} = \alpha \Delta u + P(t, \mathbf{x}) \tag{5}$$

where $P(t, \mathbf{x}) = S(t, \mathbf{x}) \exp[bt]$.

We will solve Eq. (5) in half-space $-\infty \leq x < +\infty, -\infty < y < +\infty, 0 < z < +\infty$ and consider $\mathbf{x}_i' = (0, 0, z_0)$. Due to the fact that the thermal conductivity of air is 25–30 times less than that of the tissue-mimicking material, we take $\frac{\partial u}{\partial z} = 0$ as the boundary condition at $z = 0$. Then, according to [16]

$$u(t, x, y, z) = \int_0^t \int_{-\infty}^{+\infty} \int_{-\infty}^{+\infty} \int_0^{+\infty} P(\tau, \xi, \eta, \zeta) G(t - \tau, x, y, z, \xi, \eta, \zeta) d\tau d\xi d\eta d\zeta \tag{6}$$

where

$$G(t, x, y, z, \xi, \eta, \zeta) = \frac{\exp[-\frac{(z-\zeta)^2}{4\alpha t}] + \exp[-\frac{(z+\zeta)^2}{4\alpha t}]}{8(\pi \alpha t)^{3/2}} \exp[-\frac{(x - \xi)^2 + (y - \eta)^2}{4\alpha t}]$$

For ultrasonic heat source (3) we get $u(t, x, y, z) =$

$$\frac{\beta_r \sqrt{\beta_z}}{2} \int_0^t \frac{\exp[b\tau] Q(\tau)}{\beta_r + 4\alpha(t - \tau) \sqrt{\beta_z + 4\alpha(t - \tau)}} \exp[-\frac{x^2 + y^2}{\beta_r + 4\alpha(t - \tau)}]$$

$$(\exp[\frac{-(z - z_0)^2}{\beta_z + 4\alpha(t - \tau)}] \mathrm{erfc}[\frac{z\beta_z + 4\alpha(t - \tau)z_0}{2\sqrt{\beta_z \alpha(t - \tau)} \sqrt{\beta_z + 4\alpha(t - \tau)}}] \tag{7}$$

$$+ \exp[\frac{-(z + z_0)^2}{\beta_z + 4\alpha(t - \tau)}] \mathrm{erfc}[\frac{-z\beta_z + 4\alpha(t - \tau)z_0}{2\sqrt{\beta_z \alpha(t - \tau)} \sqrt{\beta_z + 4\alpha(t - \tau)}}]) d\tau$$

Using the least squares method, we can reconstruct coefficients $\beta_r, \beta_z, \alpha, b$ and z_0 from infrared thermographic measurements.

The second application of RBF networks is associated with ultrasound thermometry, which is based on the speed of sound change due to increasing temperature. It causes a time shift of the backscatter ultrasound signal from the heated material relative to the reference one from the unheated material. To determine the shift, let us approximate both signals with RBF network $y = \sum_{j=1}^{m} w_j \exp(-a_j(t - t_j)^2)$. The shift of one signal relative to another one will be determined by the difference of the corresponding t_j. In this case, the usual training of the neural network, even with the help of fast algorithms of type [17], turned out to be too resource-intensive to be used in real time. So for the measurement of t_j we used the weighted average method. It operates with each individual half-wave curve, calculating its weighted average by the following formula:

$$t_j = \frac{\sum\limits_{k=N_{j-1}+1}^{N_j} y_k t_k'}{\sum\limits_{k=N_{j-1}+1}^{N_j} y_k} \tag{8}$$

where t_k' is the time instance to which the measurement corresponds, y_k is the signal amplitude at this instance, and index k runs through the points for which the sign is preserved with respect to the average level of the signal. This level remains practically unchanged from wave to wave and is calculated as the average of the half sum of maxima and minima on every wave. The shift itself is calculated as the difference between the weighted average of the corresponding oscillations.

The temperature increase at any depth can be calculated from the calculation of the derivative of the magnitude of the shift in depth $\varepsilon = \frac{d}{dx}(\Delta d)$ using the following formula: $\Delta T = K\varepsilon$, where K is the coefficient of ultrasound signal thermal strain. K - factor is a property of the material and is to be determined.

3   The Results of the Experiment and Calculations

The proposed original methods were tested for effectiveness on an experimental setup with HIFU-heated TMM (Fig. 1). In the setup the focused ultrasound spherical transducer is located at the bottom of a cylindrical sample of the TMM. The heating

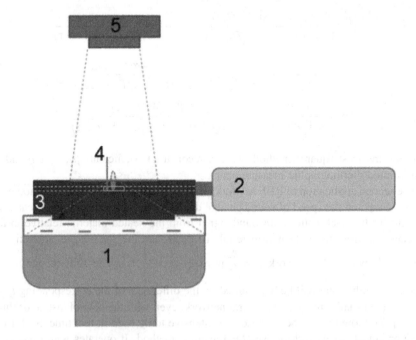

Fig. 1. A setup with the HIFU-heated tissue-mimicking material for testing of the infrared thermography and ultrasonic thermometry techniques: 1 – HIFU transducer, 2 – ultrasound probe, 3 – tissue-mimicking phantom, 4 - temperature sensor, 5 – infrared camera

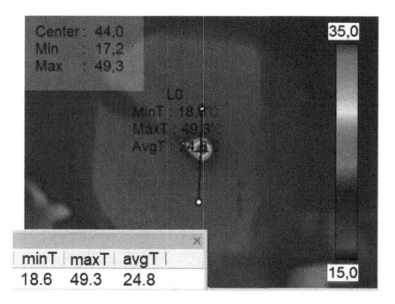

Fig. 2. An infrared image of the HIFU-heated tissue-mimicking material surface

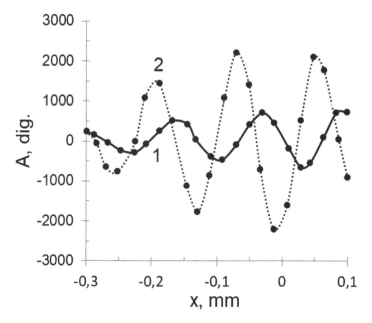

Fig. 3. The shift of the backscatter ultrasound signal from the heated material (1) relative to the backscatter signal from the unheated material (2)

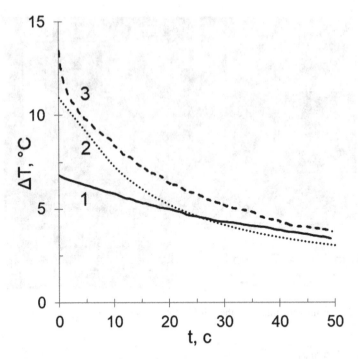

Fig. 4. Temperature decrease in the tissue-mimicking material (at a depth of 4 mm) near the focus of HIFU heating (at a distance of 3 mm): 1 - obtained with ultrasound thermometry, 2 - recovered by infrared thermography, 3 - measurements by a thermal sensor

focus area is located inside the material at a depth of 1 mm from the surface (Fig. 1). An infrared camera is installed above the surface of the TMM and records the infrared video image of the material surface during heating and cooling. It also records temperature distribution along the line passing through the heating center (Fig. 2). The scanning plane of the ultrasound probe is perpendicular to the axis of the HIFU transducer. The ultrasound signal is recorded before HIFU heating is turned on and after it is turned off (Fig. 3). A thermistor sensor, which records the temperature change during heating and cooling, is installed at a distance of 3 mm from the focus and at the same depth as the ultrasound probe scanning plane.

The duration of the HIFU heating was 10 s and the cooling time was 50 s. According to the results of the thermographic visualization of the TMM surface, a 3D temperature field was restored near the material heating region using to the proposed method. According to the results of the backscatter ultrasound signals processing by the proposed ultrasound thermometry program, a 2D temperature field was calculated in the ultrasound probe scanning plane passing through the center of focus.

According to the results of the calculations by two above mentioned methods, the temperature increase time curve was built at a depth of 4 mm and a distance of 3 mm from the center of focus in comparison with the results of temperature sensor measurements (Fig. 4). The difference in the temperature increase restored by thermographic imaging from the one measured with the thermal sensor does not exceed 1 °C. And the difference in the temperature increase calculated using the ultrasound thermography program with the thermal sensor does not exceed 5 °C at the beginning of cooling and 1 °C after 25 s.

4   Conclusion

The verification of the proposed methods of infrared thermography and ultrasound thermometry using an example of the material heating with focused ultrasound showed their effectiveness by improving the accuracy of ultrasound thermometry calibration by infrared thermography and reducing the time of the ultrasound thermometry signal processing by the new program. In the future, the methods are intended to be tested at different depths and in different biological tissues.

Acknowledgements. The study was carried out in the framework of Project **No. RFMEFI57818X0263** supported by the Education and Science Ministry of the Russian Federation for 2018–2019.

References

1. Khokhlova, V.A., Crum, L.A., ter Haar, G., Aubry, J.-F. (eds.): High Intensity Focused Ultrasound Therapy. Springer, Berlin (2017)
2. Zhou, Y.: Noninvasive thermometry in high-intensity focused ultrasound ablation. Ultrasound Q. **33**(4), 253–260 (2017)
3. Hsiao, Y.-S., Deng, C.X.: Calibration and evaluation of ultrasound thermography using infrared imaging. Ultrasound Med. Biol. **42**(2), 503–517 (2016)
4. Ebbini, E.S., Simon, C., Liu, D.: Real-time ultrasound thermography and thermometry. IEEE Signal Process. Mag. **35**, 166–174 (2018)
5. Berkovich, A.E., Smirnov, E.M., Yukhnev, A.D., Gataulin, Y.A., Sinitsyna, D.E., Tarkhov, D.A.: Development of ultrasound thermometry technique using tissue-mimicking phantom. In: IOP Conference Series: Journal of Physics: Conference Series, vol. 1044, p. 12023 (2018)
6. Anand, A., Kaczkowski, P.: Noninvasive measurement of local thermal diffusivity using backscatter ultrasound and focused ultrasound heating. Ultrasound Med. Biol. **34**(9), 1449–1464 (2008)
7. Brenner, S., Scott, R.: The Mathematical Theory of Finite Element Methods. Springer, Heidelberg (2007)
8. Logg, A., Marda, K., Wells, G.: Automated Solution of Differential Equations by the Finite Element Method. The FEniCS Book (2011)
9. Lagaris, I.E., Likas, A., Fotiadis, D.I.: Artificial neural networks for solving ordinary and partial differential equations. IEEE Trans. Neural Netw. **9**(5), 987–1000 (1998)
10. Fasshauer, G.E.: Solving differential equations with radial basis functions: multilevel methods and smoothing. Adv. Comput. Math. **11**, 139–159 (1999)
11. Mai-Duy, N., Tran-Cong, T.: Numerical solution of differential equations using multiquadric radial basis function networks. Neural Netw. **14**, 185–199 (2001)

12. Tarkhov, D., Vasilyev, A.: New neural network technique to the numerical solution of mathematical physics problems. I: Simple Problems Opt. Mem. Neural Netw. (Inf. Opt.) **14**, 59–72 (2005)

13. Tarkhov, D., Vasilyev, A.: New neural network technique to the numerical solution of mathematical physics problems. II: Complicated Nonstandard Prob. Opt. Mem. Neural Netw. (Inf. Opt.) **14**, 97–122 (2005)

14. Kainov, N.U., Tarkhov, D.A., Shemyakina, T.A.: Application of neural network modeling to identification and prediction problems in ecology data analysis for metallurgy and welding industry. Nonlinear Phenom. Complex Syst. **17**(1), 57–63 (2014)

15. Gorbachenko, V.I., Lazovskaya, T.V., Tarkhov, D.A., Vasilyev, A.N., Zhukov, M.V.: Neural network technique in some inverse problems of mathematical physics. In: Cheng, L., Liu, Q., Ronzhin, A. (eds.) ISNN 2016. LNCS, vol. 9719, pp. 310–316. Springer, Cham (2016). https://doi.org/10.1007/978-3-319-40663-3_36

16. Polyanin, A.D.: Handbook of Linear Equations of Mathematical Physics. Physical and Mathematical Literature, Moscow (2001)

17. Tarasenko, F.D., Tarkhov, D.A.: Basis functions comparative analysis in consecutive data smoothing algorithms. In: Cheng, L., Liu, Q., Ronzhin, A. (eds.) ISNN 2016. LNCS, vol. 9719, pp. 482–489. Springer, Cham (2016). https://doi.org/10.1007/978-3-319-40663-3_55

An Improved Memristor-Based Associative Memory Circuit for Full-Function Pavlov Experiment

Mengzhe Zhou[1,2,3], Lidan Wang[1,2,3(✉)], and Shukai Duan[1,2,3]

[1] The College of Electronic and Information Engineering,
Southwest University, Chongqing 400715, China
ldwang@swu.edu.cn
[2] Brain-Inspired Computing and Intelligent Control of Chongqing Key Lab,
Chongqing 400715, China
[3] National and Local Joint Engineering Laboratory of Intelligent Transmission
and Control Technology, Chongqing 400715, China

Abstract. Associative memory networks have been extensively studied to imitate the biological associative learning. The control circuits of most associative memory circuits are more complicated. Using the memristor with forgetting effect as a synapse can significantly reduce the complexity of the circuit. In this paper, an associative memory circuit based on a forgetting memristor is proposed to implement full-function Pavlov associative memory. The learning process and forgetting process in the Pavlov experiment, including forgetting under ringing stimuli, forgetting under food stimuli, and forgetting over time, can be achieved by the proposed circuit. The PSPICE simulation results demonstrate the effectiveness of the proposed circuit.

Keywords: The forgetting memristor · Memristive neural network · Associative memory circuit · Pavlov associative memory

1 Introduction

The memristor was proposed by Chua in 1971 according to the completeness theory of the circuit [1], which is defined as the differential of the magnetic flux to the charge. Then in 1976, the concept of memristor was extended to the memristive system [2], a nonlinear dynamic system. Until 2008, HP Labs made a TiO_2-based memristor [3], providing an experimental and technical basis for the study of memristor. Memristor is a nanoscale passive device, and its resistance can change with the applied voltage, which makes it a natural synapse in artificial neural networks [4]. Its excellent characteristics, such as nonlinearity, low power consumption and good scalability, make it applicable to different fields such as logic operation, neuromorphic system, artificial neural network and so on [5–7]. This also promotes the study of the memristor model. In the past few decades,

© Springer Nature Switzerland AG 2019
H. Lu et al. (Eds.): ISNN 2019, LNCS 11555, pp. 603–610, 2019.
https://doi.org/10.1007/978-3-030-22808-8_60

many memristor models with different characteristics have been proposed, such as VTEAM memristor with threshold characteristic [8], WO_x memristor with forgetting effect [9] and migration diffusion memristor model [10].

Pavlov associative memory is a classic experiment showing biological associative learning, whose circuit implementation has been extensively studied. A circuit based on memristor and microcontroller was first designed to implement Pavlov associative memory [11]. Subsequently, organic transistors were used as synapses to achieve associative memory [12]. Various rules, such as MIF (Maximum Input Feedback) [13], AIF (Average Input Feedback) [14] and WIF (Weighted Input Feedback) [15], were proposed for the implementation of associative memory. Logic gates were used to control the input signal to achieve a more complete associative memory function [16–18]. The emotional associative memory model was also implemented using a memristor with forgetting effect as a synapse [19]. These papers have made important contributions to the development of memristive associative memory networks, but there are also some problems. The function of learning and memory is not perfect, or the peripheral control circuit is too complicated.

Based on this, in this paper, an associative memory circuit based on the forgetting memristor is proposed to realize the Pavlov associative memory. The learning process, the forgetting process under ring stimuli, the forgetting process under food stimuli, and the forgetting process over time are realized. The forgetting performance of the forgetting memristor can reduce the complexity of peripheral circuits. The proposed circuit achieves a balance of circuit complexity and associative memory function.

The organization of the paper is as follows. The second part briefly introduces the forgetting memristor. The third part shows the proposed forgetting memristive neural network for full-function Pavlov associative memory. The fourth part presents the simulation results of the proposed circuit in PSPICE. The conclusion is given in the fifth part.

2   The Forgetting Memristor

A drift-diffusion memristor model with threshold characteristic and forgetting effect was proposed in [10]. Its mathematical expression is as follows:

$$M\left(x\right) = R_{on}x + R_{off}\left(1 - x\right), \quad x = \frac{w}{D} \in (0,1), \tag{1}$$

$$v(t) = M(x)i(t), \tag{2}$$

$$\frac{dx}{dt} = \left(l - \omega\left(x - \varepsilon\right)\right)k\left(i,x\right), \quad l = \frac{\mu_v R_{on}}{D^2}i\left(t\right)stp\left(abs\left(v\right) - vt\right), \tag{3}$$

$$k\left(i,x\right) = \begin{cases} \alpha, & if \ x = 1 \ and \ l - \omega\left(x - \varepsilon\right) < 0 \\ \alpha, & if \ x = 0 \ and \ l - \omega\left(x - \varepsilon\right) > 0 \\ f\left(x\right), & otherwise, \end{cases} \tag{4}$$

$$\frac{d\varepsilon}{dt} = \sigma l f\left(x\right) \tag{5}$$

where M is the memristance of the memristor, R_{on} is the minimum memristance, R_{off} is the maximum memristance, w is the thickness of the doping region, D is the thickness of the memristor, and μ_ν is the drift rate of oxygen holes. In addition to these general parameters, there are two state parameters x and ε, both of which are in the range $[0, 1]$. As shown in Eq. (3), x is determined by the drift term and the diffusion term. The drift term l indicates the effect of ion drift. The threshold characteristic is included and vt is the threshold voltage. When the magnitude of the applied voltage exceeds vt, the drift term has a value, otherwise it is zero. The diffusion term $-\omega (x - \varepsilon)$ indicates the effect of Fick diffusion and Soret diffusion on the state variable x, where ω is the attenuation coefficient. $k (x)$ is a piecewise function that is used to limit the range of x. $f (x)$ is a window function. α and σ are parameters. And the state variables ε is only determined by the drift term.

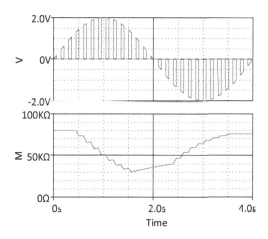

Fig. 1. The characteristic curve of forgetting memristor

Let $R_{on} = 1\,\mathrm{k\Omega}$, $R_{off} = 100\,\mathrm{k\Omega}$, $D = 10\,\mathrm{nm}$, $\mu_\nu = 10^{-14}\,\mathrm{m^2 s^{-1} V^{-1}}$, $M(0) = 75\,\mathrm{k\Omega}$, $\varepsilon(0) = 0.2$, $vt = 0.9\,\mathrm{V}$, $\omega = 0.6$, $\alpha = 0.01$ and $\sigma = 0.09$. When a voltage is applied to the memristor, the corresponding characteristic curve is shown in Fig. 1. It can be seen from the figure that at the beginning, the applied positive voltage is less than the threshold, and the resistance of the memristor does not change. Then when the voltage is greater than the threshold, the memristance decreases. When it is less than the threshold, the resistance increases. Under a negative voltage, when the absolute value of the applied voltage is greater than the threshold, the resistance increases faster. And when it is less than the threshold, the resistance slowly rises. The threshold characteristic and forgetting effect of the forgetting memristor can be clearly seen from Fig. 1.

3　The Proposed Associative Memory Circuit

Pavlov associative memory is a classic experiment that demonstrates the ability of biology to learn and forget. At the beginning, when the dog sees the food, it will secrete saliva. This reaction is unconditional reflex in instinct, and the food is unconditional stimulation. When the dog hears the ring, it does not secrete saliva. After a period of study, that is, the ring and the food appear together for a while, the dog can also secrete saliva when only the ring is heard. This is a conditional reflex, and the ring is a conditional stimulation. Then after the dog gets the conditioned reflex, if only the food stimulus or the ring stimulus is given, or even no stimulus is given, the forgetting phenomenon will occur, and the conditioned reflex will gradually disappear. After a period of time, the dog will not secrete saliva when hearing the ring. This is the process of learning and forgetting in Pavlov associative memory.

According to the above process, only when food and ring stimuli appear at the same time, the ring weight increases, which is the learning process. After learning, the forgetting processes happen, and the ring weight decreases. Based on the circuit in [19], the feedback control term is added to realize the learning and forgetting process in Pavlov associative memory. The circuit diagram is shown in Fig. 2.

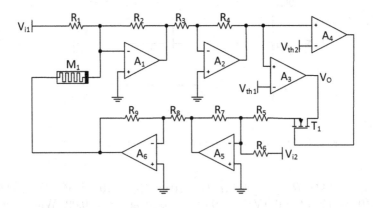

Fig. 2. The proposed associative memory circuit

In Fig. 2, V_{i1} and V_{i2} are the inputs of the network, representing the food stimulus and the ring stimulus, respectively. V_o is the output of the network, which indicates whether saliva is secreted. Since the food is unconditioned stimulus, the food weight does not change throughout the learning and forgetting process. So the resistor R_1 is used as the food synapse. And a conditioned reflex can be formed between the ring neuron and the saliva neuron after learning. The ring weight changes during the process, so the forgetting memristor is used as the ring synapse. Operational amplifiers A_1 and A_2 form a co-directional adder. A_3 and A_4 are comparators. T_1 is an NMOS transistor. A_5 and A_6 form a co-directional adder.

There are three stages before the learning process. When the dog does not receive any stimulation, it will not secrete saliva. This process corresponds to the situation that both V_{i1} and V_{i2} are zero in the circuit. At this time, the voltages on R_1 and M_1 are both zero. The output voltage of A_2 is 0, which is less than V_{th1}, so the output of A_3 is 0. The output of the circuit is 0, indicating that the dog does not secrete saliva. When the dog receives the food stimulus, it will secrete saliva. This process corresponds to V_{i1} having an input and V_{i2} being 0. At this time, the output of A_2 is greater than V_{th1}, and V_o has an output indicating that the dog secretes saliva. The value is less than V_{th2}, the output of A_4 is 0, and T_1 is off. When the dog receives the ring stimuli, it will not secrete saliva. This process corresponds to the circuit in which V_{i1} is 0 and V_{i2} has an input. At this time, the output of A_2 is less than V_{th1}, and V_o is 0, indicating that the dog does not secrete saliva. The value is less than V_{th2}, the output of A_4 is 0, and T_1 is off.

During the learning process, the dog receives both food and ringing stimuli. Both V_{i1} and V_{i2} have inputs in the circuit. The output of A_2 is greater than V_{th1} and V_{th2}, and T_1 is turned on. During this process, the voltage applied on M_1 is greater than the threshold of the forgetting memristor, the resistance of the memristor is reduced, and the synaptic weight between the ring neuron and saliva neuron is increased.

After studying, there are three stages of forgetting. The F1 forgetting process is only applying the ring stimuli, the F2 forgetting process is only applying the food stimuli, and the F3 forgetting process is not applying any stimulation. In all three forgetting processes, the output of A_2 is less than V_{th2}, T_1 is off, the voltage applied on memristor M_1 is less than the threshold, and the resistance of memristor is increased. The synaptic weight between the ring neuron and the saliva neuron is reduced.

4   Simulation Results

The associative memory circuit is simulated in PSPICE. Let $R_1 = 22\,k\Omega$, $R_2 = 20\,k\Omega$, and $R_8 = 100\,k\Omega$. The value of remaining resistances is $10\,k\Omega$. The initial value of the memristor is set to $80\,k\Omega$, the threshold voltage is $0.9\,V$, and the remaining parameters are the same in the second chapter. The results are shown in Fig. 3.

0–3 s is the initial testing phase and shows the responses of the network to individual stimuli. In 0–1 s, both V_{i1} and V_{i2} are 0, the network has no input, and the output is 0, indicating that the dog does not secrete saliva when there is no stimulation. In 1–2 s, V_{i1} is a 0.5 V pulse, and V_o outputs a 0.5 V pulse, indicating that the food stimulation leads to salivation. In 2–3 s, V_{i2} is a 0.5 V pulse. Since there is no learning process before, the ring weight is small, and the network has no output, indicating that the ring stimulation can not cause saliva secretion.

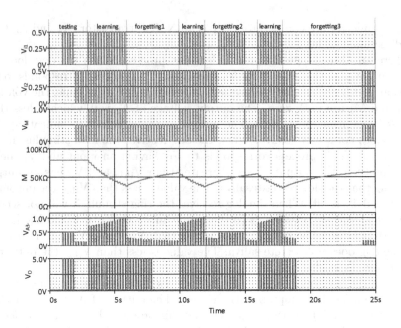

Fig. 3. The results of the associative memory circuit

3–6 s is a demonstration of the learning process. Both V_{i1} and V_{i2} are 0.5 V pulses, and V_o has an output. In this process, the voltage on the memristor is 1 V, which exceeds its threshold. The memristor value is decreased, and the ring weight is increased, so that the network can have an output under V_{i2}. This stage indicates that the simultaneous occurrence of food and ring stimuli gives the dog the ability to salivate when only the ring stimuli are given.

6–10 s is the forgetting process under the ring stimuli. V_{i2} is a 0.5 V pulse, the voltage on the memristor is less than its threshold, and the memristor value is increased with time. At the beginning, V_o has an output, indicating that after the previous learning stage, the ring stimuli can cause the secretion of saliva. After a period of time, V_o no longer has an output, indicating that after a period of forgetting under the ring stimuli, the ring stimuli cannot cause saliva secretion any more.

10–12 s is the process of re-learning. While applying the food and ring stimuli simultaneously, the memristor value is decreased again, and the ring synaptic weight is increased. Compared to the first learning process, the memristor value drops faster. This stage shows that the time required to establish the relationship between the ring neuron and the saliva neuron is shorter in re-learning process than that in the first learning.

12–16 s is the forgetting process under the food stimuli. After the re-learning process, when V_{i2} is a 0.5 V pulse, V_o has an output. Then V_{i1} is a 0.5 V pulse, the voltage applied to the memristor is 0 V, the memristance be increased, and the ring synapse weight is decreased. After a period of time, V_o has no output under V_{i2}. This process suggests that after the forgetting process under the food stimuli, the ring stimuli can not result in the secretion of saliva.

16–18 s is the third learning process. Both V_{i1} and V_{i2} are 0.5 V pulses, and V_o has an output. Similar to the re-learning process, the ring weight is increased, and the network gradually has the ability to associate saliva secretion with the ring stimulus.

18–25 s is the forgetting process over time. At the beginning, V_{i2} is a 0.5 V pulse and V_o has an output. After that, both V_{i1} and V_{i2} are 0, the voltage applied on the memristor is 0 V, the memristor value is increased, and the ring weight is decreased. After a period of forgetting, V_o is 0 when V_{i2} is a 0.5 V pulse. This stage indicates that ring stimuli can no longer cause salivation through the forgetting process over time without any stimulation.

The experimental results show that the proposed circuit can realize the learning and three forgetting processes in the Pavlov associative memory.

5   Conclusion

Based on the memristor with threshold characteristic and forgetting effect, an associative memory circuit was proposed to implement Pavlov associative memory. Using the forgetting memristor as the synapse, the forgetting property of the memristor can be used to realize the weight decay process, which reduces the complexity of the control circuit and realizes a relatively complete associative memory process. The learning process, the forgetting process under ring stimuli, the forgetting process under food stimuli, and forgetting over time can be achieved by the proposed circuit. The PSPICE simulation results demonstrate the effectiveness of the proposed circuit.

Acknowledgments. The work was supported by the National Natural Science Foundation of China under Grant 61571372, 61672436, and 61601376, the Fundamental Science and Advanced Technology Research Foundation of Chongqing cstc2017jcyjBX0050 and cstc2016jcyjA0547, the Fundamental Research Funds for the Central Universities under Grant XDJK2017A005 and XDJK2016A001.

References

1. Chua, L.: Memristor-the missing circuit element. IEEE Trans. Circ. Theory **18**(5), 507–519 (1971)
2. Strukov, D.B., Snider, G.S., Stewart, D.R., et al.: The missing memristor found. Nature **453**(7191), 80–83 (2008)
3. Biolek, Z., Biolek, D., Biolkova, V.: SPICE model of memristor with nonlinear dopant drift. Radioengineering **18**(2), 210–214 (2009)
4. Zhang, Y., Wang, X., Li, Y., et al.: Memristive model for synaptic circuits. IEEE Trans. Circ. Syst. II: Express Briefs **64**(7), 767–771 (2017)
5. Guckert, L., Swartzlander, E.E.: MAD gatesmemristor logic design using driver circuitry. IEEE Trans. Circ. Syst. II: Express Briefs **64**(2), 171–175 (2017)

6. Azghadi, M.R., Linares-Barranco, B., Abbott, D., et al.: A hybrid CMOS-memristor neuromorphic synapse. IEEE Trans. Biomed. Circ. Syst. **11**(2), 434–445 (2017)
7. Wen, S., Xie, X., Yan, Z., et al.: General memristor with applications in multilayer neural networks. Neural Netw. **103**, 142–149 (2018)
8. Kvatinsky, S., Ramadan, M., Friedman, E., Kolodny, A.: VTEAM-A general model for voltage controlled memristors. IEEE Trans. Circ. Syst. II: Express Briefs **62**(8), 786–790 (2015)
9. Chen, L., Li, C., Huang, T., et al.: A synapse memristor model with forgetting effect. Phys. Lett. A **377**(45–48), 3260–3265 (2013)
10. Zhou, E., Fang, L., Liu, R., et al.: An improved memristor model for brain-inspired computing. Chin. Phys. B **26**(11), 118502 (2017)
11. Pershin, Y.V., Di Ventra, M.: Experimental demonstration of associative memory with memristive neural networks. Neural Netw. **23**(7), 881–886 (2010)
12. Bichler, O., Zhao, W., Alibart, F., et al.: Pavlov's dog associative learning demonstrated on synaptic-like organic transistors. Neural Comput. **25**(2), 549–566 (2013)
13. Chen, L., Li, C., Wang, X., et al.: Associate learning and correcting in a memristive neural network. Neural Comput. Appl. **22**(6), 1071–1076 (2013)
14. Wang, L., Li, H., Duan, S., et al.: Pavlov associative memory in a memristive neural network and its circuit implementation. Neurocomputing **171**, 23–29 (2016)
15. Ma, D., Wang, G., Han, C., et al.: A memristive neural network model with associative memory for modeling affections. IEEE Access **6**, 61614–61622 (2018)
16. Liu, X., Zeng, Z., Wen, S.: Implementation of memristive neural network with full-function pavlov associative memory. IEEE Trans. Circ. Syst. I: Regul. Pap. **63**(9), 1454–1463 (2016)
17. Yang, L., Zeng, Z., Wen, S.: A full-function Pavlov associative memory implementation with memristance changing circuit. Neurocomputing **272**, 513–519 (2018)
18. Wang, Z., Wang, X.: A novel memristor-based circuit implementation of full-function Pavlov associative memory accorded with biological feature. IEEE Trans. Circ. Syst. I: Regul. Pap. **65**(7), 2210–2220 (2018)
19. Hu, X., Duan, S., Chen, G., et al.: Modeling affections with memristor-based associative memory neural networks. Neurocomputing **223**, 129–137 (2017)

Author Index

Printed in the United States
By Bookmasters